D0722446

PRINCIPLES OF AQUACULTURE

PRINCIPLES OF AQUACULTURE

Robert R. Stickney
University of Washington, School of Fisheries

JOHN WILEY & SONS, INC.

New York • Chichester • Brisbane • Toronto • Singapore

Copyright © 1994 by John Wiley & Sons, Inc.

All rights reserved. Published simultaneously in Canada.

Library of Congress Cataloging in Publication Data:
Stickney, Robert R.
    Principles of aquaculture / Robert R. Stickney.
        p.   cm.
    Includes bibliographical references and index.
    ISBN 0-471-57856-8 (cloth)
    1. Aquaculture.   I. Title.
SH135.S74   1994
639.3—dc20   93-8870

Printed in the United States of America

10 9 8 7 6 5 4 3

PREFACE

An old Chinese proverb, often repeated, states that if you feed a man a fish, he will have food for a day, but if you teach that man how to raise fish, he will have food for a lifetime. Aquaculture, the art and science of rearing aquatic organisms, has been practiced for millennia. It was initiated by the Chinese between 3,500 and 4,000 years ago. The first known written record describing aquaculture and its benefits was a very short book in Chinese written by Fan Li in 460 B.C. The Japanese reportedly began farming oysters intertidally about 3,000 years ago. The western world may also have been involved in aquaculture during the same period. Pictographs from the tombs of the Pharaohs of Egypt show what appears to have been some form of tilapia culture. Oysters were cultured by the Romans nearly 2,000 years ago. Yet, with its long history (as described by Iverson 1968, Milne 1972, Ling 1974, Pillay 1990, Landau 1992), application of the scientific method to aquacultural production is a recent phenomenon. In the United States, aquaculture goes back at least to the early 19th century (reviewed by Parker 1989), although significant levels of commercial production of such species as trout and channel catfish go back only to about the 1960s.

Government fish production facilities in the United States have been in place since the last half of the 19th century. Research in state and federal hatcheries formed the foundation upon which much of current commercial aquaculture is based. Before many commercial fish farms were in operation, a few governmental laboratories and various academic researchers became interested in aquaculture and the development of a commercial industry. The industries that have arisen in the United States were developed, to a large extent, in response to information produced by researchers, which is counter to how traditional agricultural research began. The sequence of commercial aquaculture development, at least during the early years, more closely parallels the model of a technology in search of a user, such as can sometimes be found in biomedical research and the electronics industry, than research conducted in response to an existing problem.

A decade or two ago, only a handful of colleges and universities in the United States presented courses in aquaculture. That number has grown significantly, and continues to grow as interest in the subject expands. Still, there are few degree programs specifically in aquaculture, and the subject tends to be taught as a part of broader curricula in departments of fisheries, fisheries and wildlife, biology, zoology, or natural resources. Trained professionals with degrees ranging from the B.S. to the Ph.D. have been produced by a number of institutions. Many of those who have entered the profession have engaged in research that has been critical to the rapid expansion of the field.

Overly optimistic projections for the growth of aquaculture in the United States were presented throughout the 1980s and before. While notable growth was seen in some sectors of United States aquaculture, it was insignificant relative to the growth of the commercial industry in other parts of the world. Researchers in this nation have been responsible, in part, for the growth of aquaculture in Asia, Africa, and elsewhere. Programs funded through such entities as the U. S. Agency for International Development, the World Bank, the Asian Development Bank, and the United Nations have made it possible for researchers from the United States to assist in the development of foreign aquaculture expertise and activity. That development has ranged from establishing subsistence farming in rural regions to building the academic and governmental skills required to train and assist commercial producers. As it became apparent that commercial aquaculture had great potential in certain countries, private venture capital appeared and an industry was created.

Part of the motivation for aquaculture research has been to provide mankind with alternative forms of nutritious and plentiful food. Until recently, the concept of growing aquatic organisms to provide food for humanity has been favorably viewed by most people. Aquaculture was seen as a means to maintain an increasing supply of aquatic organisms in the marketplace in the face of a projected maximizing of the commercial fisheries catches worldwide.

With the increasing development of coastal aquaculture, competition for valuable waterfront property, destruction of wetlands and mangrove swamps as a function of pond construction, and utilization of the commons for rearing fish and invertebrates have led to increasing public concern about the consequences of aquaculture development. Resistance to aquaculture, not only in the United States but also in Europe, Asia, and elsewhere, has also developed in response to views that the effluents from culture facilities can have negative impacts on native fauna, that exotic species employed in aquaculture will escape and compete with indigenous species, and that aquacultured organisms transmit diseases to wild populations, to name but a few.

Thus, since about the late 1980s, there has been a shift, at least in some locations, in how commercial aquaculture is viewed. Opinion with respect to certain aspects of public aquaculture (the rearing of fish to augment sport and commercial fisheries) has also begun to change. As a result, the aquaculture research community is becoming increasingly involved in developing ways to rear aquatic organisms without producing negative environmental and social impacts.

A number of very fine books that cover the broad subject of aquaculture or concentrate on one species or species group have appeared over the past decade. Many of those books take a how-to-do-it approach with respect to the species discussed. The approach employed in this book and in its predecessor (Stickney 1979) is to present the underlying principles of aquaculture and to provide examples that underscore the principles and demonstrate the variability that exists among species of culture organisms.

The earlier book (Stickney 1979) emphasized warmwater species. In this volume, the scope is extended to encompass a wider variety of aquatic animals, including such coldwater fishes as trout and salmon. As was true of the previous work, the emphasis is on molluscs, crustaceans, and finfish. While a significant amount of aquaculture around the world is focused on the production of aquatic plants— mostly algae—our emphasis remains on the animals that are being reared.

This book also emphasizes species being reared around the world by commercial aquaculturists. Sport fishes are not ignored, however, and many of the examples presented throughout the text involve species of both sport and commercial interest.

As was true of its predecessor, this book is based largely on course materials developed over a period of more than 15 years of university teaching. The interest and enthusiasm of the many students who have taken those courses and their prob- ing questions have been a source of motivation and pleasure.

The current text updates and expands upon the basic principles that were pre- sented in the earlier work. In addition, more attention is paid in this volume to the social and economic aspects of aquaculture. The science of aquaculture has come a long way since 1979. This book is an attempt to capture some of that development and place it in a general framework that will provide the reader with an appreciation for the scope of the discipline, the state of development that exists, and some of the problems that remain to be solved.

This book is dedicated to the students who helped, albeit perhaps unknown to them, with its development. It is also dedicated to my wife Carolan, without whose continuous love and support for over 30 years, this work would probably not have been developed. Finally, three important people in my life, all deceased, played major roles in my professional development and deserve mention. They were Rob- ert W. (Winston) Menzel, Jonathan Chervinski, and W. A. Isbell. They are gone but certainly not forgotten.

ROBERT R. STICKNEY

Seattle, Washington
September 1993

LITERATURE CITED

Iverson, E. S. 1968. Farming the edge of the sea. Fishing News (Books), Surrey, England. 301 p.

Landau, M. 1992. Introduction to aquaculture. John Wiley & Sons, New York. 440 p.

Ling, S. W. 1974. Keynote address. Proc. World Maricult. Soc. 5: 19–25

Milne, P. H. 1972. Fish and shellfish farming in coastal waters. Fishing News (Books) Ltd., Surrey, England. 208 p.

Parker, N.C. 1989. History, status, and future of aquaculture in the United States. Rev. Aquat. Sci. 1: 97–109.

Pillay, T. V. R. 1990. Aquaculture principles and practices. Fishing News Books, Oxford, England. 575 p.

Stickney, R. R. 1979. Principles of warmwater aquaculture. Wiley-Interscience, New York. 375 p.

CONTENTS

1 Introduction

DEFINITION, SCOPE, AND APPROACH

Aquaculture is the rearing of aquatic organisms under controlled or semicontrolled conditions. More simply, aquaculture is underwater agriculture. While the meaning of the term "aquatic organism" could be broadened, the emphasis in this book is on animals that spend their entire lives in the water and stresses species that are being utilized as human food. Thus, alligators, which are being reared for human food as well as for their hides, are excluded because they are not totally aquatic animals. The same exclusion principle applies to frogs.

In the broad sense, aquaculture includes the rearing of tropical fishes; the production of minnows, koi, and goldfish; the culture of sport fishes for stocking into farm ponds, streams, reservoirs, and even the ocean; the production of animals for augmenting commercial marine fisheries; and the growth of aquatic plants. Plants, such as single-celled algae, are considered to the extent that they may be necessary as a component of some or all of the life history stages of certain aquatic animals. Consideration is also given to the control of nuisance aquatic vegetation, which includes rooted and floating plants as well as filamentous algae.

Sport and commercial fishes are not ignored in this book; in fact, many species are of interest to commercial aquaculturists and recreational as well as commercial fishermen. The emphasis, however, is on species produced for direct human consumption.

Aquaculture species have often been described in relation to their temperature preferences. While many species can survive a broad range of temperatures, aquaculture organisms tend to be classified as having a preference for or demonstrate optimum performance in warm water, cool water, or cold water. While not absolute, warmwater species tend to prefer water at or above 25°C, while coldwater species prefer temperatures below 20°C. Coolwater species have preferences between those two temperatures.

The definition of aquaculture used here encompasses both salt and fresh water. The word "mariculture" has been used to distinguish species that are reared in brackish and marine environments. No complimentary single word for strictly freshwater species has been coined. If a distinction does need to be made, the term "freshwater aquaculture" is commonly used to separate that practice from mariculture.

This chapter includes information on the types and quantities of aquatic animals being commercially reared around the world, addresses such general issues as the contribution of aquaculture to world food production, exotic species issues, permitting, environmental constraints, and some predictions on the future of aquaculture. The chapters that follow examine aquaculture economics, water sources and systems, water quality, feeds and nutrition, reproduction and genetics, disease, and harvesting and processing.

A guiding principle for aquaculturists is that their job is not complete until the product is consumed. Fish or shellfish farmers should not only be concerned about keeping the animals alive and growing well; they also need to ensure that when the product is marketed it represents the best possible quality. While what happens to the organisms once they leave the farm is not often under the control of the aquaculturist, maintaining high quality in the final product is only possible if that product was of excellent quality at the time it left the farm for the processing plant. A consumer who purchases a cultured fish that has off-flavor, for example, may never purchase a similar product again and the farmer will be the big loser. The profitability of many aquaculture ventures is tenuous even under the best conditions. In order to survive economically, it is incumbent upon the producer to provide a product that is judged superior by the consumer.

A FEW WORDS ON WEIGHTS AND MEASURES

With the exception of the United States, the world operates on the metric system of weights and measures. While some industries in the United States have converted to the metric system, farms, including fish farms, are still measured in acres. The scientific community, including scientists involved in aquaculture research, usually report their findings in metric system units, so the transfer of information from the technical literature to the practicing aquaculturist often requires some translation, often through extension fisheries and aquaculture specialists. To make matters even more confusing to all interested parties, authors have sometimes mixed units of measurement in the same book or article. The following paragraph is a contrived example:

> The 0.5 acre pond was stocked with fingerling channel catfish at a density of one fish per square meter. Water temperature was taken at a depth of 1 ft below the water surface each morning at 0630. The temperature averaged 23°C with a range from 45°F to 90°F. Dissolved oxygen (DO) averaged 5 mg/l. Upon termination of the study, the fish were captured with a 10 m-long seine having ¾ inch stretch mesh. Fish were weighed to the nearest gram and measured to the nearest quarter of an inch.

Metric units of weights and measures are employed throughout this book. A conversion table is provided in Appendix A. Equivalent temperatures in Fahrenheit and Celsius are presented in Appendix B.

For most purposes, precise conversions are not necessary. One can, for example, think of a yard as being approximately one meter (1 yd = 3 ft = 0.91 m). Similarly, a one-pound fish (typical market size for many species) weighs about half a

TABLE 1. Representatives of Species within the Phyla Mollusca, Arthropoda, and Chordata That Are Being Cultured or Have Been Considered as Aquaculture Candidates in the United States

Phylum	Common Name	Scientific Name
Mollusca	Bay scallop	*Aequipecten irradians*
	American oyster	*Crassostrea virginica*
	Pacific oyster	*Crassostrea gigas*
	Quahog clam	*Mercenaria mercenaria*
	Blue mussel	*Mytilus edulis*
Arthropoda	American lobster	*Homarus americanus*
	Shrimp	*Penaeus* spp.
	White river crawfish	*Procambarus acutus*
	Red swamp crawfish	*Procambarus clarkii*
Chordata	Channel catfish	*Ictalurus punctatus*
	Striped bass	*Morone saxatilis*
	Rainbow trout	*Oncorhynchus mykiss*
	Coho salmon	*Oncorhynchus kisutch*
	Atlantic salmon	*Salmo salar*
	Tilapia	*Tilapia* spp.

kilogram (2.2 lb = 1 kg). The hectare is perhaps the least common metric unit to most individuals who are beginning to study aquaculture. A hectare is 10,000 m^2, or approximately 2.5 acres (1 ha = 2.47 acres). There are 2,000 lb in a short ton (English measure) and 1,000 kg in a metric ton. While a short ton actually equals 907.2 kg, it is convenient to think of the short ton and a metric ton[1] as equivalents. Also, since the relationships of 2.2 lb/kg and 2.5 acres/ha are similar (2.2:2.5 approximates 1:1), it is convenient to think of 1,000 lb/acre as being equivalent to 1,000 kg/ha. These rough calculations come in handy in casual conversation and for making back-of-the-envelope calculations. For the precision required in technical documents and for properly advising aquaculture producers, more care in converting from one system of measurement to the other is required.

IMPORTANT CULTURE SPECIES

The majority of aquatic animals currently being cultured are representatives of three phyla: Mollusca, Arthropoda, and Chordata. Table 1 presents examples from each phylum that are either being cultured commercially or have at least received some attention by aquaculture researchers in the United States. The list is not exhaustive

---

[1] Some authors use "tonne" to distinguish between a metric ton and an English ton. Since the metric system is used in this book, the word "ton" refers to the metric ton in all instances unless otherwise noted.

Figure 1. An Atlantic salmon at the National Marine Fisheries Service net-pen facility near Manchester, Washington.

by any means, and a number of additional species of culture interest in the United States are mentioned throughout the book. Aquaculturists seem to always be looking for new species to culture, though the bulk of the production typically comes from well-established ones.

Representatives from other phyla could become important to aquaculture in the future. Certain sea urchins (phylum Echinodermata) are examples with potential. They have consumer appeal particularly in southern Europe and Japan.

Carp (family Cyprinidae) are produced in Asia, eastern Europe, Israel, and elsewhere, though there is little interest in carp production in North America. Common carp *(Cyprinus carpio)* are produced primarily in Europe and Israel. In China, various species are reared including grass carp *(Ctenopharyngodon idella)*, mud carp *(Cirrhina molitorella)*, silver carp *(Hypophthalmichthys molitrix)*, and bighead carp *(Aristichthys nobilis)*.

Salmon and trout (family Salmonidae) are produced primarily in North America and Europe, though Chile has become a significant producer nation with respect to salmon in recent years. Atlantic salmon *(Salmo salar)* is raised in North America (Figure 1) and northern Europe, as well as Chile. Pacific species such as chinook salmon *(Oncorhynchus tshawytscha)* and coho salmon *(O. kisutch)* are reared in North America, Chile, Japan, and Russia. There is some chinook salmon production coming from New Zealand, and Japan has a large marine enhancement program underway with chum salmon (Mahnken 1991). China is becoming involved with the reestablishment of salmon in some of the northern rivers. Rainbow trout *(O.*

mykiss, formerly *Salmo gairdneri*) is an important aquaculture species in the United States, Chile, and in northern Europe. Rainbow trout are also produced in New Zealand but only as a sportfish. The culture of that fish for direct human consumption in New Zealand is currently forbidden by law.

United States fish culture is dominated by the channel catfish *(Ictalurus punctatus)*. Other species of importance in the United States are shown in Table 1. Walking catfish *(Clarias* spp.) are reared in Asia, Africa, and Europe.

Additional species of finfish currently under culture include sea bass *(Dicentrarchus labrax)*, sea bream *(Sparus aurata)*, sole *(Solea solea)*, and plaice *(Pleuronectes platessa)* in Europe, along with red sea bream *(Pagrus major)* and yellowtail *(Seriola quinqueradiata)* in Japan. Milkfish *(Chanos chanos)* are being cultured in the Philippines, Thailand, and Indonesia. Some white sturgeon *(Acipenser transmontanus)* culture is being conducted in North America. In addition, there is increasing United States production of tilapia[2] (primarily *Tilapia aurea, T. nilotica,* and red hybrid tilapia), striped bass *(Morone saxatilis)* and striped bass × white bass *(M. chrysops)* hybrids, and red drum *(Sciaenops ocellatus)*.

Norway, Scotland, Canada, and United States researchers have been evaluating the potential of halibut *(Hippoglossus hippoglossus* and *H. stenolepis)* for culture. In Hawaii, research on the culture of the dolphin *(Corhyphaena hippurus)*, also known as mahi-mahi, is presently underway.

Shrimp culture in the Eastern Hemisphere is dominated by the giant tiger shrimp, *Penaeus monodon*, except in China, where *P. orientalis*,[3] followed by *P. chinensis* (Chen 1990) are the primary species under culture. In the Western Hemisphere, the western white shrimp, *P. vannamei*, is currently the most popular species. Other species being produced in significant quantities include *P. stylirostris* and *P. japonicus*. The Malaysian freshwater shrimp, *Macrobrachium rosenbergii*,[4] continues to be cultured but production is not great compared with the marine shrimp species.

Oysters (genera *Crassostrea* and *Ostrea)*, mussels (primarily *Mytilus* spp.), and clams (various genera), abalone (e.g., *Haliotis* spp.), and scallops (e.g., *Pecten* spp. and *Patinopecten yessoensis)* are grown in various places around the world. Their culture sometimes involves merely planting young animals[5] on areas that will support their growth and waiting until the molluscs reach harvest size. More sophisticated culture of molluscs typically involves hanging the animals from rafts or ropes.

---

[2]Trewavas (1973) reclassified many of the tilapia species of culture interest as being members of the genus *Sarotherodon*. Later she recommended reclassification of the same fish into the genus *Oreochromis* (Trewavas 1982). Since the American Fisheries Society has not accepted the proposed changes in generic status for the species of interest, Robins (1991) is followed in this book.

[3]Dean Akiyama, personal communication.

[4]Many individuals refer to freshwater shrimp as "prawns." Technically, any large shrimp is a prawn, so the use of that term only with respect to the freshwater animals should be avoided. In this book, the terms freshwater shrimp and marine shrimp are used throughout, with scientific names provided when necessary.

[5]Oyster spat and the juveniles of oysters and other mollusc species that are outplanted have been referred to as "seed" by many aquaculturists. The view taken here is that "seed" is more appropriately used in conjunction with the plant kingdom and is, thus, avoided throughout the text.

The species mentioned in the above paragraphs do not represent an exhaustive list of aquaculture species by any means. Many sport fishes have been excluded from the list, though their culture for stocking programs in the United States and elsewhere is well developed and significant. Also, the list does not include bait minnows or ornamental fishes and invertebrates.

From the plant kingdom, some of the important organisms being cultured are seaweeds that are representatives of the brown algae (e.g., *Undaria* sp. and *Laminaria* spp.), red algae such as nori (*Porphyra* spp.), and green algae (e.g., *Enteromorpha intestinalis* and *Ulva* spp.). Japan is one of the primary seaweed culturing nations, though various other countries culture seaweeds and have significant harvests from wild populations.

WORLD AQUACULTURE PRODUCTION

Statistics on aquaculture production are compiled each year by a number of nations and by the Food and Agriculture Organization (FAO) of the United Nations. In many instances, landing reports include data from both commercial capture fisheries and aquaculture, and in some cases it is difficult to distinguish between the two. In addition, a considerable amount of fish and shellfish undoubtedly reaches the market without passing through a point where statistical information is collected (e.g., when a fish farmer harvests, processes, and markets fish directly from the farm rather than through a commercial processing plant). Thus, obtaining reliable estimates of aquaculture production is not a simple matter.

What is clear from data available through various sources is that aquaculture production continues to increase. Support for that statement can be found by comparing world aquaculture production figures from 1985 with those from 1989 (Nash 1991). In 1985 total estimated production was 10,587,300 tons. It reached 14,466,306 tons by 1988. Table 2 presents production information for various aquaculture commodities for 1989. It is generally accepted that total global landings from commercial fisheries will plateau at no more than about 100 million metric tons. Aquaculture currently makes up nearly 15% of that amount.

On the whole, production is dominated by Asia, in which about 86% of the world's aquaculture commodities are grown. Table 3 shows the top 10 countries in terms of the economic value of their aquaculture production in 1988. Table 4 shows how the various regions of the world rank in terms of aquaculture production.

China leads the world in aquaculture production. While traditionally a carp-growing nation, China has, in recent years, rapidly become among the world leaders in the production of marine shrimp. There are about 40 commercially valuable species of shrimp found in China, but as previously indicated, *Penaeus chinensis* is the primary species of interest to aquaculturists (Chen 1990). As reported by Rosenberry (1992), estimated shrimp production in China in 1991 was 250,000 tons, which was second only to Thailand with 350,000 tons (the two countries both produced about 250,000 tons in 1990.

Farmed shrimp production was dominated by Latin America during the early

TABLE 2. Aquaculture Production of Selected Seaweeds, Molluscs, Crustaceans, and Finfish During 1989 (FAO 1991a)

Commodity	1989 Production (tons)
Seaweed	
Eucheuma	349,100
Gracilaria	47,300
Kelp	1,519,000
Laver	518,200
Wakame	409,200
Molluscs	
Clams, cockles, arkshells	484,500
Freshwater molluscs	2,100
Marine molluscs, miscellaneous	264,300
Mussels	1,057,800
Oysters	982,500
Pectinids (primarily scallops)	327,700
Crustaceans	
Crabs	3,326
Crayfish	32,908
Freshwater shrimp	28,929
Penaeid shrimp	537,246
Finfish	
Carp	4,008,000
Channel catfish	184,000
Milkfish	334,000
Rainbow trout	245,000
Salmon	276,000
Tilapia	325,000

1980s, with Ecuador being the leading producer. Of the 50,000 tons of shrimp raised in Latin America during 1986, 85% was produced in Ecuador (Weidner 1988). Ecuador's production was 100,000 tons in 1991, and represented 74.9% of total Western Hemisphere production. Rosenberry (1992) reported estimates of 60,000 and 90,000 tons produced in Ecuador in 1990 and 1991, respectively. The United States produced 1,600 tons, or 1.2% of the Western Hemisphere total (Anonymous 1992b).

The crawfish[6] industry in the United States is extremely large, though it is per-

[6]The American Fisheries Society recognizes three species with the common name "crawfish" and a large number (many in the same genera as the crawfish) with the common name "crayfish." This book follows the American Fisheries Society common-name conventions (American Fisheries Society 1989).

TABLE 3. Percentage of Total World Aquaculture Production and Primary Commodities Rearedby the Ten Leading Producer Nations in 1988 (FAO, 1991a)

Country	% of Total	Commodities
China	35.4	Molluscs, shrimp, carp
Japan	20.4	Algae, molluscs, yellowtail
Taiwan	5.4	Molluscs, shrimp, eels
Philippines	3.2	Algae, shrimp, milkfish, tilapia
USSR[a]	2.7	Carp
United States	2.7	Molluscs, catfish, salmon, trout
Norway	2.6	Salmon, trout
Ecuador	2.6	Shrimp
Republic of Korea	2.5	Algae, molluscs
Indonesia	2.5	Algae, shrimp, carp, milkfish, tilapia

[a] Formally disbanded in 1991.

haps not as well recognized as the catfish and trout industries. An estimated 16,100 tons of crawfish were reportedly produced in the United States during 1990 (Anonymous 1991), though de la Bretonne (1988) reported that total production in Louisiana is in excess of 45,000 tons annually with a value of $65 million. About 60% of Louisiana's crawfish are captured from ponds (aquaculture), with the remainder coming from commercial wild harvest in the Atchafalaya River Basin (de la Bretonne 1988). Total crawfish pond area in Louisiana was approaching 55,000 ha by 1988.

By 1986, 20% of the shrimp harvested in Asia came from aquaculture, and Pacific rim nations were producing 80% of the world's farmed shrimp (Niemeier 1988). That percentage was maintained through 1991 (Anonymous 1992b), when world production reached 690,100 tons on an estimated 933,750 ha. In addition to

TABLE 4. World Aquaculture Production by Regional Percentage Contribution (FAO 1991a)

Region	% of Total
Africa	0.4
Asia	77.8
Europe	11.4
North America	3.6
Pacific Island Nations	0.6
South America	3.5
USSR[a]	2.7

[a] Formally disbanded in 1991.

China, Taiwan, Indonesia, the Philippines, India, and increasingly Thailand have become important shrimp-producing nations.

Commercial production of rainbow trout in the United States is centered in Idaho, where an estimated 3,400 tons were harvested in 1988 (Chew and Toba, 1990). Production of rainbow trout in Sweden grew from 4,388 tons in 1987 to 6,634 tons in 1989 (Vattenbruk 1988, 1990). During the period from 1960 to 1989, annual production from the trout industry in Germany increased from 1,000 to 7,000 tons (Steffens 1991). Rainbow trout are also produced in Finland, Chile, Japan, and various other nations.

The greatest production of salmon comes from Norway, with such other countries as England, Ireland, Canada, Iceland, and Chile also producing significant quantities. Estimated production by the major producing nations was 234,000 tons in 1990 (Anonymous 1991). When smaller producer nations are added in, total salmon production from aquaculture in 1990 was 284,000 tons (Anonymous 1992c). That is compared with 33,800 tons in 1984 and 71,800 tons in 1986 (Folsom 1987).

The cultured salmon industry was developed and grew rapidly during the 1980s, with nations such as Canada and Chile becoming important players, though Norway continued to hold a large lead in terms of total production. In 1990 and 1991, the production of wild fish was extremely high, driving prices downward. The combination of heavy debt loads and lower-than-predicted prices carried by aquaculture ventures that were bringing in their first crop or had yet to produce a crop meant that many producers were unable to repay their loans. Foreclosures and restructuring of the industry resulted, with the largest shakeups occurring in Norway and Canada (Anonymous 1992c).

The salmon production industry in Norway recovered, and those remaining in business obtained excellent prices for their product beginning in the summer of 1992. Sales were primarily within the European Common Market, with some sales to Asia. Newly imposed tariffs on fish being imported to the United States effectively eliminated sales of farmed Norwegian salmon into this country, whereas in previous years the United States had been a major purchaser of Norwegian fish.[7]

Many involved in the salmon capture fishery in Alaska, where prices fell as low as $0.26/kg in 1991 for pink salmon *(Oncorhynchus gorbuscha)*, have attacked the U.S. aquaculture industry as being in direct competition with them.[8] In other countries, such as Norway, commercial fishermen also participate in aquaculture. That approach could provide a good economic basis for American salmon aquaculture (Stickney 1988), if prohibitions on aquaculture in Alaska could be overturned and the mind-sets of commercial fishermen could be altered.

Japan, the United States, and Israel are among the most technologically advanced aquaculture nations. Fish farming is important to Israel, and the technology that has been developed in that nation has been applied throughout the world. Carp

---

[7] Ole Torrissen, personal communication.

[8] In reality, U.S. production of farmed salmon is very small relative to the amount of cultured product that is imported to the United States annually.

and tilapia have been the most important species (Hepher and Pruginin 1981), though eels (*Anguilla* sp.), rainbow trout, and *Macrobrachium* have been produced as well. It is interesting to note that a prodigious amount of research and development has occurred in Israel, yet in 1988 there were a total of only 2,818 ha of ponds on 57 farms. Those farms produced 15,113 tons that year, including 382 tons of trout (Sarig 1989).

AQUACULTURE IN THE UNITED STATES

The United States is not, in comparison with many other countries, a major seafood-consuming nation on a per capita basis. For many years, per capita consumption of fish and shellfish hovered at about 4.5 kg/yr. In recent years, there has been an increase in consumption. During the 1980s per capita consumption reached 7 kg/yr in 1987 (Anonymous 1988). With a population of about 250 million, an increase of a kilogram per person annually translates into a considerable amount of additional product that must be harvested and processed, so the increase from 4.5 to 7 kg/yr in recent years represents a major increase.

Aquaculture production in the United States was 402,757 tons in 1988 with a value of $608,858,000. The estimate for 1991 was 543,770 tons with a value of $750,250,000 (Anonymous 1991). The U.S. production level continues to increase but still amounts to only about 3% of the world total. The catfish farming industry, which is the largest in the United States, experienced a 5% growth in 1990 and continued to grow in 1991. The number of growers was placed at 1,664 in 1991 (Anonymous 1991). The majority of the production comes from the Delta region of Mississippi.

A few additional words on the catfish farming industry are in order, as that industry epitomizes the importance of improved production efficiency through technology development and the judicious application of the results of scientific inquiry. In the late 1950s, H. S. Swingle at Auburn University in Auburn, Alabama, developed the methodology for rearing channel catfish to market size in ponds and demonstrated that it was possible to raise catfish at a profit if the producer could receive $1.10/kg (Swingle 1957, 1958). In the three and a half decades that followed, the industry grew and prospered until it became the leading aquaculture industry in the United States (Table 4). Yet, the price of catfish to the producer has increased very little since the 1950s. In 1990, some of the highest prices ever seen were paid by processors (up to $1.72/kg). In 1991, the price fell, reaching as low as $1.17/kg (Anonymous 1992a). Increased production per unit water area, markedly improved feeds, water quality, and disease management, and various other improvements in the way catfish are produced have kept the fish farmers in business in spite of very small increases in the farm gate prices offered for their product over the years. The small difference in price between the 1950s and the present is particularly remarkable when one considers the inflation that took place during the intervening years.

The trout culture industry, which is second to catfish in total dollar value, also continues to grow, with sales reaching $79.6 million in 1990. Other finfish with significant levels of production included hybrid striped bass (680 tons) and tilapia (between 1,820 and 2,270 tons). Production of shrimp was recently estimated at approximately 900 tons, oysters at 5,000 tons, and crawfish at 16,100 tons (Anonymous 1991).

GOVERNMENT ROLE IN U.S. AQUACULTURE

In 1978, the Farm Bill approved by the U.S. Congress acknowledged for the first time the applicability of aquaculture to the goals of the U.S. Department of Agriculture. At about the same time, the U.S. Department of Commerce developed an aquaculture plan (Glude 1977).

During the administration of President Carter, the Joint Subcommittee on Aquaculture (JSA) was formed. It was comprised of various federal agencies that might have roles in the development, regulation, and financing of aquaculture in the United States. Chairmanship of the JSA was to rotate among the three agencies that had the greatest vested interest in the discipline: the U.S. Department of Agriculture (USDA), U.S. Department of Commerce (USDC), and U.S. Department of the Interior (USDI).

The aquaculture pie was to be split in the following manner: the USDC would be involved with all aquaculture in the marine environment, the USDI would have jurisdiction over the aquaculture of species that spawn in fresh water and migrate to the marine environment, and the USDA would become involved with all species which did not fall within the charge of the other two agencies. Clearly, most of the aquaculture being practiced in the United States fell within the mandate of the USDA.

Public Law 96-362, the National Aquaculture Act of 1980, set forth the aquaculture policy of the United States and instructed the Secretaries of Agriculture, Commerce, and Interior to establish a national aquaculture development plan. The aquaculture planning effort languished for a period but was revived, and ultimately a plan was published (Joint Subcommittee on Aquaculture 1983a,b). Revision of the plan was initiated in 1987. In the meantime, both the USDC and the USDI made policy changes that reduced their activities in aquaculture. The USDA became the primary federal agency representing aquaculture. The Department of Agriculture has employed personnel to specifically oversee the aquaculture grants program of the agency and to deal with the overall aquaculture activity within the agency.

Recently, both the USDI and the USDC have increased their visibility in aquaculture. For example, the USDI has created an aquaculture office and named a coordinator for aquaculture. While much of their effort is expended on recreational species and those, such as salmon, that are of interest to both commercial and recreational fishermen, the activities of the USDI and the USDC are clearly within the definition of aquaculture and are significant. The USDC has never totally abandoned commercial aquaculture development as an area of interest. Since its inception in the late 1960s, the Sea Grant Program, which is housed in the USDC, has provided significant levels of grant support in aquaculture research at a number of universities.

In fiscal 1987, the United States Congress passed a revised farm bill that included a section that established four regional aquaculture centers. Those centers, each initially funded at $750,000 annually, were designed to attack regional problems facing the aquaculture industry. The four centers were designated the Northeastern, Southern, Western, and Tropical Pacific Aquaculture Centers. Each center is composed of an administrative office with a director and is governed by a board of directors. Serving the board and director of each center are a technical committee made up of scientists and extension specialists from universities and agencies, and an industry advisory committee of private aquaculturists and suppliers to the aquaculture industry. Working together, the technical and industry advisory committees establish priorities for aquaculture research and extension activities in each of the four regions.

In fiscal 1988, Congress created a fifth center in the north-central region. As a result, all 50 states are represented by the centers. Additional funding came with the creation of the fifth center so that each received $750,000 annually. That level of funding was increased slightly in fiscal 1992. A national research initiative by the USDA, authorized in fiscal 1992, may provide an additional source of grant support for aquaculture research.

In 1993, bills were introduced in Congress to place all commercial aquaculture under the USDA and to create marine aquaculture research centers under the USDC Sea Grant Program. Both bills used the potential of aquaculture expansion in the U.S. as motivations for taking action.

FEEDING THE WORLD THROUGH AQUACULTURE

The expanding population of human beings on earth and maximization of the agricultural production in many regions, plus the fact that politics and economics sometimes prevent nations with food surpluses from finding mechanisms to transfer those surpluses to the needy, have led to very real food shortages in some regions of the world and to predictions of more widespread famine in the future. For example, droughts in parts of Africa have already led to recent extensive starvation.

If we ignore politics and economics, we can still predict that the carrying capacity of the earth for human beings will eventually be reached, unless aggressive population control is implemented—which currently seems unlikely. Until humanity comes to grips with human population expansion, the short-term solution will be to continually attempt to increase the food supply.

One logical environment that would seem to be available for additional exploitation is the world ocean, which covers some 70% of the surface of the globe. This vast expanse of water has often been seen as a limitless source of animal protein for mankind. However, over two decades ago, Ryther (1969) estimated that the rate of exploitation could not be significantly increased because of limits on the capability of primary productivity that would be required to support additional animal life.

Idyll (1970) arrived at a similar conclusion. He believed that some increase in total harvest could be realized if we began to utilize such exotic creatures as krill (small shrimplike animals in the family Euphausiidae) and phytoplankton directly as human food. Some research has been directed toward producing human food from those sources, but palatability problems have made widespread use of them impractical to date. In the meantime, overall harvests from the ocean have increased to some extent, but the upper limit of about 100 million tons per year proposed by Ryther (1969) seems to be upon us. The FAO (1991b) reported that, in 1989, the total marine catch of finfish, crustaceans, molluscs, and other animals, including species raised in aquaculture, was 99,534,600 tons. Inland fisheries and aquaculture accounted for 13,776,700 tons, and there were 4,339,900 tons of seaweeds and other aquatic plants harvested.

If, as many think, we are approaching the limit that can be harvested economically from the world ocean, the opportunity for aquaculture expansion would seem to be significant. That expansion opportunity looks particularly good if, as some predict, the demand for seafood for human consumption will double by the year 2005 (Blackman 1989). There are, as we shall see, some problems associated with expanding aquaculture to fill that void. In any case, if we assume that aquaculture can make a major contribution to meeting the ever increasing demands by people for seafood, what role can the aquatic farmer play in providing inexpensive animal protein for the millions of undernourished people in the world? What kinds of products can aquaculture produce that can be afforded by the world's poor, and will those products be accepted?

Contrary to popular opinion, undernourished and even starving people will not necessarily eat even a highly nutritious food just because it is available. There are many religious, cultural, and psychological factors that influence what individuals and groups of people will accept as edible. Those factors differ from one nation or region of the world to the next. Lack of consumer acceptance was one of the reasons for the failure of fish protein concentrate (FPC) to meet the expectations that had been held out for that product in the 1960s. FPC, a tasteless, odorless, white power made from whole fish, was highly recommended as a protein supplement for malnourished people (Finch 1969). Despite its innocuous appearance and the development of well-organized programs to teach people the value of the product, FPC never won wide acceptance. The death blow came when the use of FPC was banned in the United States, where the process was developed. The ban came because the material was manufactured from whole fish and thus contained entrails, heads, and other body parts not considered suitable for human consumption by the federal government. When people in developing nations learned that the product could not be consumed in the nation that not only developed it but was advocating its widespread use, they refused to accept it and the program died.

While available on the Japanese market for a number of years, relatively new items in the restaurants and food stores of the United States are surimi analog products. Surimi is deboned, minced fish that is prepared, usually at sea, and frozen into blocks. The product is manufactured from what were once considered to be trash fish, like walleye pollock *(Theragra chalcogramma)*. The blocks of frozen surimi are taken to shore-based plants where they are transformed into surimi analog products such as artificial crab legs and shrimp. The pollock, which is worth a few cents per kilogram at capture, is thus transformed into products that sell for a few dollars per kilogram.

The bulk of the fish presently used for surimi analog production in the United States comes from the Bering Sea. Water within the 330 km (200 mile) Exclusive Economic Zone (EEZ) of the United States is currently producing some 2 million tons of walleye pollock and other species for the process. Technology has progressed to the point where it is now possible to turn nearly any fish into surimi, including such unlikely candidates as menhaden (*Brevoortia* spp.), which have historically be processed into fish meal.

Most aquaculture feeds require at least a few percent of animal protein in them to meet the amino acid requirements of the culture species. The primary source of animal protein in aquaculture rations is from fish meal. As increasing quantities of low-value fish are diverted from fish meal to surimi, the potential for a shortage of fish meal becomes increasingly high. Aquaculturists are not the only users of fish meal; in fact, the amount of the world's fish meal supply that goes into aquaculture feeds is relatively insignificant compared with that used in livestock feeds. Approximately 60% of the fish meal supply goes into poultry and 20% into swine feeds, with aquaculture feeds utilizing only about 10% (Barlow 1989). It is likely that livestock will be given priority when fish meal supplies are short. A considerable amount of research by aquaculturists is currently underway in an attempt to find alternative protein sources so the dependency on fish meal can be reduced. After all, it does not make a great deal of economic sense to catch a fish, grind it up, feed it to another fish that is then caught and marketed for human consumption when you can take the first fish and make an entirely acceptable human food therefrom.

Aquaculture has been proposed as a means of supplementing natural marine and freshwater productivity, and has actually increased to the point where it may account for nearly 15% of the plants and animals harvested from water. Under the proper circumstances, densities of animals that can be maintained per unit water volume are considerably higher in aquaculture systems than in those found in nature. Problems do exist, however. As the density of animals per unit water volume increases, there is a concomitant need for increased energy input to the culture system. Energy is provided by aquaculturists primarily in the forms of electricity (to run pumps, aerators, lights, and so forth) and in the food that is provided to the aquaculture organisms. Energy in any form is expensive, and the prospects for less expensive sources of energy in the future are not bright. In general, aquacultured products tend to be relatively expensive, both because of the energy inputs that are often required and because of the technology employed, as well as the need, in many instances, to employ highly skilled labor.

On a global scale, aquaculturists are faced with the problem of finding suitable, nonpolluted water bodies in which to produce their crops. With respect to suitable sites, the typical aquaculturist needs a plentiful supply of high-quality water. In the face of degraded conditions, it will often be the aquaculture species that will first suffer. Pollution and competition with other human activities along coastlines have become problems in many regions. In others, the choice locations for aquaculture have been taken, and what is left can only be termed marginal. The siting of aquaculture operations is discussed in more detail below.

The export of aquacultured products from developed to developing nations is not feasible because the cost of the products in the developing nations would be well beyond the reach of the people who most need the additional dietary protein. Increasingly, aquaculture in developing nations produces fish, shellfish, and even seaweeds that are targeted for sale in the markets of the developed world. While the developing nations gain economically, and additional jobs are created from the production of aquaculture goods for export, in most instances the practice of exporting aquaculture products does little to alleviate local food shortages.

In the vast majority of instances, aquaculture is conducted with a profit motive. Developing countries are often selected for the development of aquaculture enterprises by outside investors interested in export markets because of certain advantages that may exist, such as low land and labor costs. Even small-scale aquaculturists who sell all their products on local markets expect to receive maximum return on their investments in money, time, and labor. They, too, have a profit motive and set prices at whatever the market will bear.

The exception to the pattern described above is subsistence aquaculture, which involves a family, small group of families, or even an entire village wherein people work together to produce an aquaculture crop for their communal use. Subsistence culture is usually conducted in ponds, though cage culture is another option. The technique involves low-density stocking and minimal management. In fact, management may involve no more than fertilization of the pond with manure, which can be done automatically by placing livestock over or adjacent to the pond in the proper numbers. Natural reproduction of the aquatic culture species may be depended on to replace stock that is harvested, though young animals for stocking are often available in even the most underdeveloped nations, either from private or public hatcheries.

Most subsistence aquaculture is conducted in the tropics where a year-round growing season is typically available. Harvest may involve removal of only the quantity of fish required for immediate consumption by the subsistence culturists, or it may involve harvest of the entire crop. In the latter instance there may be too much product for immediate consumption and some of it may be preserved or sold. Since refrigeration and freezing are not available to most subsistence aquaculturists, preservation may involve drying and salting, thereby making the product less valuable than in its fresh condition. From an economic standpoint, placing such aquaculture products into the fresh-food market makes more sense than preserving them.

Various efforts have been put forth by international development agencies to help people in developing countries improve their nutritional status by producing their own aquaculture crops. In many instances, the subsistence approach has not worked very well. Malnourished people, when given the opportunity to enhance their incomes, don't necessarily turn the increased wealth into food. This was demonstrated convincingly in a project with which the author was involved in the Philippines.

The project, funded by the U.S. Agency for International Development (USAID), was designed to provide technical assistance to the government of the Philippines in the development of a fish hatchery. The fish were to be distributed at nominal expense to rice farmers who would add them to their rice paddies, thereby allowing the farmers to produce two different types of crops, one of which would add animal protein to their diets. An additional benefit, and one which was not apparent when the project was developed, was that rice-fish farming has been found to increase rice yields (Sollows and Tongpan 1986).

Many rice farmers made the necessary modifications to their paddies to accommodate fish production. All that was actually involved was the construction of a trench about 1 m deep down the center. Fish introduced into the rice field were able to find refuge in the trench during rice harvest or when it was necessary to drain the paddy for some other reason. The trench also facilitated fish harvesting.

Rice-fish farmers were soon producing fish, which they would often take to town to sell rather than making them available to their families. With the profits from the fish, the farmers would typically purchase color television sets or other luxury items. Their families continued to survive on rice, and continued to be malnourished.

USAID's purpose in developing the project was thwarted, yet the introduction of rice-fish farming was accepted and continues to be practiced. Was the project a success, or did it fail? Certainly, the economic status of the farmers who adopted the technique improved. It can easily be argued that it is their choice as to how and where to spend the additional resources that accrued to them. So, at some level, the project can be viewed as having been successful.

Subsistence aquaculture will continue to have its place in developing countries, but the fact remains that farmers, whether they are in the United States, Rwanda, or Nepal, are basically interested in producing products from sale at the highest possible prices. Most aquaculture products retail at prices that are beyond the means of the world's poor. Thus, the proliferation of shrimp farms in Ecuador, Costa Rica, Panama, Thailand, and most recently, China, has occurred not with the motive of providing food for domestic market but with an eye on sales potential in the developed parts of the world, in particular, the United States, which imports 80% of the shellfish and 65% of the finfish consumed (Blackman 1989). Shrimp prices in developed nations are often much higher than in the developing world, and producers are interested in maximizing profit. That goal is usually achieved by exporting the product. Few business firms are headed by persons who are amenable to reducing or even eliminating profit by selling on local markets in order to feed the needy.

Aquaculture products do not compete in terms of market price with grains, nor even with poultry. From a price standpoint, shrimp, oysters, trout, salmon, and catfish (to name but a few) compete with beef. They are high-quality, even luxury foods. The notion that aquaculture provides a convenient solution to the present and future food problems of the world is thus a myth.

Realistically, aquaculture can be expected to provide an increasing share of the world's aquatic animal protein, but that protein will be expensive. As additional technology is required to meet increasingly tough effluent standards and ensure that incoming water is of proper quality for the animals or plants under culture, the costs of production can be expected to increase instead of decrease. It has only been through technological breakthroughs and adoption of the results of scientific investigation that the price of aquaculture products has been kept as low as current levels. Additional technology development and scientific investigation are going to be needed to meet the challenges of the future if aquaculture is to remain viable.

THE ART AND SCIENCE OF AQUACULTURE

Until about the 1960s, aquaculture could perhaps be best described as an art. While there were some scientific studies of an aquaculture nature being undertaken in government laboratories and a few universities prior to that decade, they were relatively few in number. Most of the progress made in the culture of aquatic organisms came through trial and error, accompanied by careful observation of the animals under culture and from attempts to recreate natural conditions in captive environments.

Much of the knowledge that was gained never made it into print in either the popular press or the technical literature. For example, various state and federal government agencies produced and released the eggs, fry, and juveniles of literally billions of aquatic organisms into the marine, coastal, and fresh waters of the United States well before 1900. The technology to spawn some of the species mentioned in the literature has only been redeveloped within the past few years. For some, no captive breeding is occurring at the present time. A list of species produced by one agency and their numbers for 1897 is presented in Table 5.

Rather than having developed as a distinct scientific discipline, aquaculture, like traditional agriculture, draws upon various sciences and applies the information to aquatic plant and animal production. In addition to involving an array of sciences (some of which are listed in Table 6), various crafts are also integral to the operation of an aquaculture facility. Included in the latter are carpentry, welding, electric wiring, mechanics, and perhaps most importantly, plumbing.

Most aquaculturists with college degrees have been trained primarily in the biological sciences, though some have graduated with degrees in engineering, business, and various other fields. Much of the work that is undertaken by the practicing aquaculturist is not biological in nature. A considerable amount of the technical activity involves chemistry, and, of course, the financial and personnel management aspects are purely business activities. In the final analysis, all of the scientific, engineering, and business acumen cannot lead to the development of a successful aquaculturist if the individual is not also equipped with a good dose of plain old common sense.

It is virtually impossible for any individual to become expert is all of the crafts, to say nothing of the scientific and engineering disciplines, and the business management skills that are involved in an aquaculture venture. An individual, however, can be successful without having developed expertise in each area, depending on the type of culture system that is being operated, the species under culture, and the

TABLE 5. Numbers of Eggs, Fry and Juveniles Produced and Distributed by the United States Commission of Fish and Fisheries during Fiscal 1897 (Brice 1898)[a]

Common Name	Scientific Name	Number
Shad (American shad)	*Alosa sapidissima*	134,545,500
Rock bass	*Ambloplites rupestris*	42,687
Sea bass (black sea bass)	*Centropristis striata*	193,000
Lake herring	*Coregonus artedii*	7,299,000
Whitefish (lake whitefish)	*Coregonus clupeaformis*	95,049,000
Pickerel (northern pike)	*Esox lucius*	1,700
Codfish (Atlantic cod)	*Gadus morhua*	98,258,000
Lobster (American lobster)	*Homarus americanus*	115,606,065
Black (smallmouth) bass	*Micropterus dolomieui*	2,719
Black (largemouth) bass	*Micropterus salmoides*	95,358
Striped bass	*Morone saxatilis*	450,000
Yellow perch	*Perca flavescens*	1,025
Crappie (white crappie)	*Pomoxis annularis*	2,125
Strawberry bass (black crappie)	*Pomoxis nigromaculatus*	3,129
Silver (coho) salmon	*Oncorhynchus kisutch*	298,137
Black spotted trout[b]	*Oncorhynchus mykiss*	42,200
Rainbow trout	*Oncorhynchus mykiss*	768,123
Steelhead (rainbow) trout	*Oncorhynchus mykiss*	499,690
Yellow-fin trout[b]	*Oncorhynchus mykiss*	7,930
Quinnat (chinook) salmon	*Oncorhynchus tshawytscha*	32,104,049
Flatfish (winter flounder)	*Pseudopleuronectes americanus*	64,095,000
Atlantic salmon	*Salmo salar*	2,329,809
Landlocked (Atlantic) salmon	*Salmo salar*	150,566
Loch Leven (brown) trout	*Salmo trutta*	49,709
Von Behr (brown) trout	*Salmo trutta*	23,780
Golden trout	*Salmo aguabonita*	45,000
Swiss lake trout (Arctic char)	*Salvelinus alpinus*	36,082
Brook trout	*Salvelinus fontinalis*	1,359,510
Lake trout	*Salvelinus namaycush*	13,509,149
Mackerel (Atlantic mackerel)	*Scomber scombrus*	652,000
Tautog	*Tautoga onitis*	624,000
TOTAL		568,144,042

[a] Common names in parentheses are the presently accepted names according to Robins (1991). All scientific names have been updated to reflect current taxonomy (Robins 1991).
[b] Species is a variant of other trout (probably either brown or brook) that are no longer recognized as being distinct.

types of technology being utilized within the culture system. Subsistence culturists, for example, may have little or no formal education but can be successful because the culture systems are simple. On the other hand, the operation of certain kinds of aquaculture systems requires the plethora of skills and broad knowledge base outlined above. In instances where all those disciplines come together, there is usually

TABLE 6. Some of the Scientific Disciplines that Impact Aquaculture

Biological Sciences	Chemistry
Botany	Biochemistry
Genetics	Organic chemistry
Ecology	Inorganic chemistry
Endocrinology	Water chemistry
Molecular biology	Hydrology
Nutrition	Limnology
Pathology	Oceanography
Physiology	Physics
Reproductive biology	
Taxonomy	

a cadre of personnel who bring specific kinds of abilities and contribute them to the overall operation. Obviously, the latter approach operates on quite a different technical and economic plane than that on which subsistence culture is built.

Scientists who have been active in aquaculture research and extension over the past three decades have produced a prodigious amount of published material. The amount of information generated has increased almost exponentially with time. Thus, it is incumbent upon practicing aquaculturists to constantly update their skills and increase their knowledge bases. This can be done by reading the pertinent literature, maintaining contact with state agricultural extension service personnel involved with aquaculture, and attending workshops and meetings where new aquaculture information is presented. There are a number of scientific journals devoted to aquaculture in general, as well as to such specialties as fish health and aquaculture engineering in particular. A variety of journals publish the occasional article on aquaculture. Magazines like *World Aquaculture* contain less technical information, in general, than the scientific journals and are good sources interesting and valuable information.

EXOTIC SPECIES IN U.S. AQUACULTURE

Several species that have been popular with aquaculturists in other countries have been introduced into the United States. In recent years, controls on exotic introductions, including the movement of native organisms from one part of the United States where they had historically occurred to another where they were not indigenous, have increased. Regulations on exotic introductions have been developed by both federal and state governmental agencies. A number of aquaculture species are exotic to the United States, and some have even been outlawed years after they had become established and probably cannot be eradicated.

Among the exotic aquaculture animals that have received significant attention by researchers and commercial producers are the freshwater shrimp *(Macrobrachium rosenbergii)*, penaeid shrimp of various species *(e.g., Penaeus stylirostris, P. vannamei,* and *P. monodon)*, grass carp *(Ctenopharyngodon idella)*, and tilapia (species of culture interest include *Tilapia aurea, T. nilotica, T. mossambica,* and red hybrids).

A great deal of controversy exists with respect to further exotic introductions, and while much of it has surrounded finfish species, there has been increasing concern voiced relative to invertebrate introductions, particularly shrimp, where viral diseases, including infectious hypodermal and hematopoietic necrosis (IHHN), have caused severe problems. IHHN was first identified in exotic shrimp in Hawaii by Lightner et al. (1983).

Grass carp and tilapia are perhaps the most controversial exotic fishes currently being commercially cultured in the United States. Both were imported initially as aquatic vegetation control agents. Introduction of the fishes into natural waters has been both intentional and as a result of escapes from culture facilities. Grass carp were once banned in over 30 states, though many states have reevaluated or are currently reassessing their position on the use of the species for aquatic weed control. Some states allow the stocking of only sterile fish.

Grass carp were introduced to the United States in 1963 by government scientists and had spread to 35 states within the next 15 years (Guillory and Gasaway 1978). The fish is native to eastern Asia from the Amur River basin (hence the alternate common name, white Amur) to the West River (Lin 1935). It was first thought that grass carp would not spawn in the United States and that fish that escaped captivity would eventually die without reproducing. Stanley et al. (1978), on the other hand, felt that various rivers in the United States provided conditions suitable for grass carp reproduction. Shortly after that opinion was published, natural spawning was observed in the Mississippi River (Conner et al. 1980). Additional reports of larvae in the lower Mississippi River and the lower Missouri River and their tributaries have appeared in the intervening years (Zimpfer et al. 1987, Brown and Coon 1991), and a commercial fishery for the species has developed (Robinson 1988). Additional reports of natural reproduction have also been received from several rivers in Mexico (Anonymous 1976).

Various techniques have been employed to control grass carp reproduction (Stanley 1979). Presently, stocking triploid fish seems to be the method of choice. Triploid fish contain an extra set of chromosomes ($3n$), and while they may spawn, any fertilized eggs that are produced will not be viable. Triploid grass carp can be produced by physically shocking fertilized eggs with either temperature or hydrostatic pressure change (Allen and Wattendorf 1987, Clugston and Shireman 1987). The stocking of triploid grass carp by permit has been approved in a number of states. Several hatcheries currently produce triploid grass carp and sell them at $4 to $5 for each 200 mm fish (Clugston and Shireman 1987).

Grass carp appear to be strict herbivores, though they are somewhat selective in their food habits and may not remove all the rooted vegetation types that appear in a pond (Colle et al. 1978). Their ability to control undesirable rooted aquatic vege-

Figure 2. Shoals of dead tilapia line the banks of a power plant's discharge canal after the plant was shut down during winter for repair, allowing the water temperature to fall to a lethal level.

tation is well documented, however (Cross 1969, Stott and Robson 1970, Kilgen and Smitherman 1971, Sutton 1977, Sutton et al. 1979). The presence of grass carp in natural waters may have detrimental effects on certain aspects of water quality (Lembi et al. 1978), though from a management standpoint weed control is often the most critical factor.

Tilapia are tropical species native to North Africa and the Middle East. Their history in aquaculture is a long one (Hickling 1963), though it has only been within the past few decades that tilapia have been widely distributed around the tropical world. With modern techniques, various species can be cultured year-round in temperate climates if warm water is made continuously available.

Tilapia have been banned in several states, but there are no general rules in existence concerning culture and transport of the species. Therefore, as is true of other exotic species, producers should contact the appropriate local and state officials before importing or transporting tilapia (Kingsley 1987).

Because they are intolerant of low temperature, tilapia do not pose a threat for long-term establishment except in portions of Florida and south Texas, along with Hawaii and in locations with geothermal water. In addition, tilapia can become established in lakes that receive heated water that has passed through the condensers of power plants (Figure 2). While able to tolerate temperatures approaching 40°C (Allanson and Noble 1964, Gleastine 1974), most tilapia species die when the water falls below about 10°C (Chimits 1957, McBay 1961, Avault and Shell 1968, Gleastine 1974).

Tilapia do not threaten the success of native fish populations in most temperate situations because of the intolerance of the fish to cold water. Interference with the successful nesting of largemouth bass has been reported in a power plant cooling lake, however (Noble et al. 1975), and concerns about the establishment of tilapia have led to prohibitions against the species in some states and even in entire nations (e.g., Belize).

The freshwater shrimp *Macrobrachium rosenbergii* is similar to tilapia with respect to its intolerance to cold water. The species survives well only between 18 and 33°C (Farmanfarmaian and Moore 1978). Aquaculture of freshwater shrimp in the United States has been attempted in South Carolina, Texas, Hawaii, and a few other states, but most of the interest in shrimp has been diverted toward marine shrimp because of the poor postharvest keeping characteristics of *M. rosenbergii* and due to some marketing problems. As of 1987, only about 6% of global shrimp production was attributable to *Macrobrachium*. Much of the production comes from Thailand and Taiwan (New, 1988).

The culture of exotic species in the United States will undoubtedly continue in the future, but special permits and careful control of effluent water to prevent escapement will be required in many states. Persons who are involved in the production and shipment of exotic species, as well as those interested in stocking them, should inquire about local, state, and federal regulations well in advance of transporting or receiving their stock. Transportation of fish from one area where they are legal to another legal area will involve a violation of federal law (the Lacey Act) if the fish pass through any jurisdiction wherein they are banned. Shelton (1986) indicated that the positive aspects surrounding the aquaculture of an exotic species must be accompanied by a protocol that will effect control over that species. Such a protocol should incorporate import permits, the evaluation of potential conflicts with native fauna (and, one would assume, flora), and a period of efficacy testing to ensure that the proper decision was made. There should be adequate safeguards to prevent escape of the exotic while it is being evaluated, including control over reproduction which is one key to maintaining control over potential spreading of what might turn out to be an undesirable species.

An entirely new type of problem that could certainly be considered as an exotic introduction involves the production and distribution of transgenic or genetically engineered fish. It is now possible to transfer genes from one species to another, including such transfers in fish. The resulting fish are genetically distinct from others in the genus and represent organisms new to science. Chen et al. (1988) speculated that the manipulation of growth hormone and growth-factor genes in conjunction with aquaculture species could be revolutionary to aquatic food production.

Objections to the release of genetically engineered organisms have been raised, and concerns are likely to increase as the technology continues to advance. The USDA has developed the National Biological Impact Assessment Program to facilitate safe field testing of transgenic organisms. The USDA takes the view that products developed through biotechnology are not considered to have fundamental differences from products developed through traditional types of research (Mackenzie 1988). Permission to maintain transgenic fish under aquaculture conditions has been granted by the USDA only in instances where it can be demonstrated that the fish and their progeny cannot escape and possibly establish populations in nature. Controls such as those imposed on transgenic fish releases in the United States may not apply in other nations. The issues surrounding use of transgenic fish in aquaculture and public policy development with respect to animals produced through molecular engineering were discussed in an issue of *Fisheries* in 1990 (Kapuscinski and Hallerman 1990, Hallerman and Kapuscinski 1990a,b).

EXTENSIVE AND INTENSIVE AQUACULTURE

One means of distinguishing between aquaculture and the mere hunting and gathering of fish and shellfish is associated with the degree of control that is exerted by humans over the environment in which the organisms live. Freshwater fishery managers attempt to establish stable, complex communities of fishes in natural environments (rivers, streams, and lakes) or artificial impoundments. Managers of inland fisheries exercise regulatory authority, engage in selective stocking, manipulate habitat, and regulate water level, among other techniques, to achieve their aims. Marine fishery managers attempt to sustain commercially harvestable populations largely through regulation. Augmentation of stocks with hatchery fish is a commonly used technique in inland water but has limited application in the marine environment, though it is used in isolated instances (e.g., salmon enhancement) by marine fishery management agencies.

Instead of managing a water system and the various species it contains to obtain an "optimum" or "sustainable" harvest, aquaculturists typically manage for maximum production of one or a small number of species. Attempts are made to eliminate, insofar as is possible, stress on the species being cultured and competition among the organisms of interest.

Natural systems, which are of interest to fishery managers, in most cases have relatively low standing crops and low carrying capacities. As increasing degrees of control over the environment are implemented by the aquaculturist, the level of intensity associated with the culture system is said to increase. Various types of aquatic production systems can be thought of as lying along a continuum of levels of production. Natural systems (e.g., a stream or lake) exist at one end, and recirculating water systems at the other (Figure 3). For some of the types of water systems in between, it can be argued which is more intense than the other since production may be similar, but the level of technology required to develop and operate the systems can vary considerably (Chapter 3 describes the various types of water systems in detail). Even within a given type of culture system (e.g., ponds), there can be a considerable amount of variation in the level of intensity practiced by the culturist. Production within ponds is quite variable, depending on the management strategy that is employed. Some theoretical production values are presented in Table 7. By flowing water through a pond in addition to fertilizing and feeding, and by employing emergency aeration as required, it is possible to reach or even exceed 10,000 kg/ha annual production, as has been achieved by some catfish farmers in Mississippi.[9]

As the intensity of culture increases, initial, as well as operating costs tend to increase, with the exception of cages and net pens, where the operating costs can

[9] It should be noted that production potential varies from one species to another. Walking catfish (*Clarias* spp.) can be reared at extremely high densities compared with many other species. Tilapia are somewhat less tolerant than walking catfish and more tolerant than channel catfish. Salmonids (trout and salmon) will be stressed at low dissolved oxygen or high ammonia levels that are tolerated by the other species mentioned.

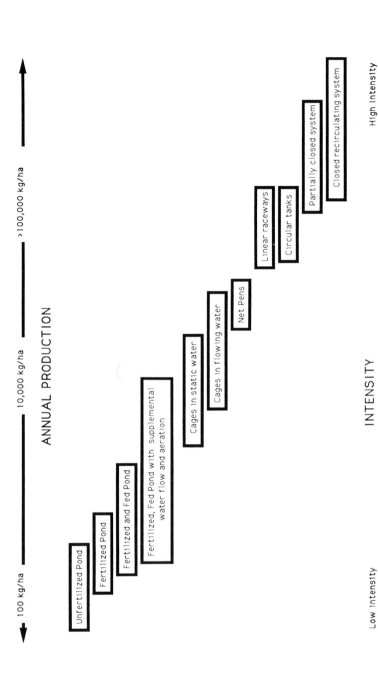

Figure 3. The intensity and production level continuum along which various types of aquaculture systems lie. Depending on the species under culture and various technological modifications in the various systems, their actual locations relative to one another and the range of production maxima for each can only be approximated.

TABLE 7. Fish Production Levels that Can Typically Be Observed as a Result of Various Pond and Lake Management Strategies Using Channel Catfish as an Example. (Values are based on theoretical yields; actual production rates will vary depending on various additional factors)

Management Scheme	Production (kg/ha)
Unmanaged waters	≤100
Stocked with sportfish combinations such as bass, bluegills, and channel catfish, with fertilization	500
Monoculture of catfish with fertilization	1,000
Monoculture of catfish and provision of supplemental feed	1,500
Monoculture of catfish and provision of complete feed	3,000
Monoculture of catfish, provision of complete feed, along with water replacement and supplemental aeration to maintain water quality	≥7,000

be quite low unless frequent cleaning of the enclosures is required. The chances of crop failure increase with increasing intensity because of higher stocking densities, more dependence on technology (particularly on electricity and other forms of energy), and the pressure on water quality exerted by the culture species. By using closed systems, the culturist can actually become independent of local climatic conditions and may have direct control over water quality, though the margin for error is greatly reduced as compared with lower-technology systems.

Another interesting way of looking at intensity, and one that is not widely used by commercial aquaculturists, is to look at the maximum amount of feed that can be added to a particular type of system before water quality problems begin to arise. Rough estimates of maximum feeding rates are as follows:[10]

- Feed added at 5 to 10 mg/l/day   Flowthrough raceways (e.g., trout)
- Feed added at ≈ 15 mg/l/day   Ponds without aeration (e.g., catfish)
- Feed added at ≈ 50 mg/l/day   Ponds with mechanical aeration
- Feed added at 100 to 150 mg/l/day   Ponds to which pure oxygen is added (the upper limit will be reached when low pH and high nutrient levels—e.g., high ammonia—occur)

- Feed at 2,000 mg/l/day or higher   Closed systems with biofiltration (metal balance can eventually become a problem, and factors such as high carbon dioxide and high nitrate may become limiting)

[10] Dallas Weaver, personal communication. Feeding rates are based on liters of water in the system on a daily basis.

To see if any of the figures presented make reasonable sense, we can look at a productive channel catfish pond with no aeration. Mississippi catfish farmers can often produce up to 4,000 kg/ha/yr under such circumstances. At the end of the growing season, when standing crop biomass reaches 4,000 kg/ha/yr, the fish might be fed at 3% of body weight per day:

$$4,000 \text{ kg/ha} \times 0.3\% \text{ daily} = 120 \text{ kg/ha/day} = 120,000,000 \text{ mg/ha/day}$$

A 1 ha pond averaging one m in depth will have a volume of:

$$10,000 \text{ m}^2 \times 1 \text{ m deep} = 10,000\text{m}^3$$
$$10,000 \text{ m}^3 \times 1000 \text{ l/m}^3 = 10,000,000 \text{ l}$$

To determine milligrams of feed added per hectare per day, divide the milligrams of feed offered by the volume of water in the pond:

$$120,000,000 \text{ mg} / 10,000,000 = 12 \text{ mg/l}$$

The value of 12 mg/l of feed offered daily agrees well with the proposed value of up to 15 mg/l/day. An interesting exercise would be to evaluate the other figures in the same way and to extend the figures to cages and net-pen systems.

MONOCULTURE AND POLYCULTURE

The majority of aquaculturists in the United States produce a single species in a given culture chamber, regardless of whether it is a pond, raceway, tank, cage, net pen, or other type of unit. This practice is known as monoculture. Polyculture, the simultaneous rearing of two or more compatible species in the same culture chamber, enjoys popularity in many nations (Bardach et al. 1972) and is perhaps best developed in China, where four or more fish species are commonly grown together in freshwater ponds. The Chinese pond polyculture system depends on stocking species that occupy different feeding niches. Ponds are fertilized (usually with organic fertilizers) and may also be provided with agricultural byproducts. In recent years, the Chinese have also developed increasing interest in using prepared feeds.

Stocking involves the use of mud carp *(Cirrhina molitorella)* to feed on benthic animals, silver carp *(Hypophthalmichthys molitrix)* and bighead carp *(Aristichthys nobilis)* to consume plankton, and grass carp *(Ctenopharyngodon idella)* to consume aquatic plants. Production rates of ponds stocked with those species at the proper densities can be higher than from similarly managed monoculture ponds but do not rival those from more intensive monoculture systems.

In the United States only a limited amount of polyculture has been practiced. Perhaps the most common application of the technique involves the stocking of grass carp at low densities (usually no more than 40 to 80 fish/ha) as aquatic vegetation control agents in ponds wherein some other species is the primary focus of culture. Research has demonstrated that tilapia can be reared with freshwater shrimp *(Macrobrachium rosenbergii)* and that overall pond productivity could be approximately doubled as compared with shrimp monoculture (Rouse and Stickney 1982). Stocking densities were nine shrimp and one tilapia per square meter of pond area. Freshwater shrimp have also been successfully polycultured with other fishes. For example, Sastradiwirja (1986) cultured *M. rosenbergii* with common carp, silver carp, bighead carp, and giant gouramy.

Modifications on traditional polyculture schemes have been developed. One example is the utilization of a series of ponds or raceways. As water quality deteriorates through passage from one raceway to another, species that are increasingly tolerant of degraded water quality can be stocked. At least one commercial aquaculture facility in Idaho employs the technique. Artesian geothermal water enters a series of raceways containing channel catfish. The water flows through a canal where it is aerated to effect partial rejuvenation of quality, then passes through a second series of catfish raceways. After a third such set of raceways, the water flows into the final series, which contains tilapia.

Another approach was used by Stickney and Hesby (1978). They stocked earthen ponds that received the waste from growing-finishing pigs (pigs from weaning to market size) with tilapia. A small amount of recirculated water was continuously added to the tilapia pond, from which it flowed into a second pond stocked with channel catfish. A tertiary pond with an even less tolerant species, such as freshwater shrimp, could also be used in such a system. The catfish and shrimp would receive prepared feeds, while the tilapia would thrive on the natural productivity resulting from the organic fertilization.

Polyculture with plants, such as rice-fish culture, or higher vertebrates such as ducks, has also been developed and is employed in many countries. Rice-fish culture has been described in an earlier section of this chapter. Fish-cum-duck culture is commonly seen in Asia. The ducks serve the same purpose as poultry, cattle, or swine in providing organic fertilization which produces the plankton on which tilapia can thrive. This subject is dealt with in more detail when fertilization is discussed in Chapter 4.

FRESHWATER AQUACULTURE AND MARICULTURE

Both stenohaline[11] and euryhaline[12] species have been successfully reared by aquaculturists. Among the euryhaline species of interest, most are reared throughout the culture period in either fresh or saline water, though some anadromous[13] species, as well as catadromous[14] eels are cultured (Brown 1983) and may be moved from

---

[11] Species which have a narrow range of salt tolerance.

[12] Species which have a broad range of salt tolerance.

[13] Species, such as salmon, which reproduce in freshwater but spend a significant portion of their lives in the marine environment.

[14] Catadromous species are those which reproduce in the marine environment, then migrate to freshwater where they live for much of their lives.

one salinity to another during the course of the culture period. In most cases, however, the entire culture cycle is completed over a fairly narrow range of salinity that mimics that in which the target species can be found in nature. It is possible, on the other hand, to raise euryhaline marine species over a range of salinities. For example, salmon (both Atlantic and Pacific species) can be reared throughout their life cycles in fresh water. The establishment of Pacific salmon populations in the Great Lakes is a clear demonstration of the ability of those fishes to thrive without ever migrating to the sea. Striped mullet *(Mugil cephalus)* are thought of as marine species, but naturally occurring mullet can be found in freshwater springs connected by rivers to the sea. They are a common species in Florida springs, for example, and have been cultured or considered for culture in a number of countries.

In certain instances it may be necessary to provide a range of salinities in response to the stage of the life cycle in which the culture species is found. For example, freshwater shrimp require water of about 1/3 strength seawater salinity during their early development stages, but may then be transferred to fresh water (Bardach et al. 1972).

Various species that inhabit estuaries[15] can be adapted to fresh water, but since they are generally unable to reproduce in low salinity environments, spawning is conducted in seawater. The red drum, *Sciaenops ocellatus,* and flounders of the genus *Paralichthys* have been released by the Texas Parks and Wildlife Department into freshwater reservoirs. Newspaper accounts of returns indicate that growth may sometimes exceed that of the same species in the marine environment. High mortalities have posed a problem with the program, particularly during winter when osmoregulatory stress is exacerbated by cold temperatures. Hard water (water high in calcium and/or magnesium) may mitigate against the stress imposed by low salinity with respect to red drum (Wurts and Stickney 1989).

Certain freshwater species can be introduced into saline waters. Tilapia of several species and certain hybrids tolerate full-strength seawater and even higher salinities (Chervinski 1961, 1966, Chervinski and Yashouv 1971, Chervinski and Zorn 1974). Red hybrid tilapia have been developed in several parts of the world (Galman and Avtalion 1983) that can tolerate high salinities and are being commercially reared in a number of nations and at various salinities. Catfish, including channel catfish, can tolerate in excess of 10 parts per thousand (ppt) salinity (Perry 1967, Perry and Avault 1968, Stickney and Simco 1971) and have been reared in coastal lagoons of relatively low salinity. The salinity tolerance of tilapia and other species of aquaculture interest has been reviewed by Stickney (1986, 1991).

As a general rule, the difficulty of maintaining an aquaculture facility increases with increasing salinity. A saline medium not only introduces new aspects of water quality into the culture system, it also places a tremendous strain on facilities. Salt

[15] The most simple definition of an estuary is the regions where rivers enter the sea; or, the lower, tidal reaches of rivers (reviewed by Stickney 1984). Perhaps the most widely accepted definition is that of Pritchard (1967) who described an estuary as "a semi-enclosed coastal body of water which has free connection with the open sea, and within which sea water is measurably diluted with freshwater from land drainage."

water and the salts in the aerosols that are associated with saltwater facilities are highly corrosive to metal and various other materials; thus, all equipment and structures associated with a mariculture operation must be properly protected and maintained or rapid deterioration will occur.

SITE AND SPECIES SELECTION

An aquaculture facility needs good quality water in sufficient quantity in order to be viable. Depending on the level of technology that is employed, sites that meet those broad qualifications can be found or created almost anywhere. That does not mean that all ventures will show profitability, of course. Requirements for water treatment, or simply the costs of obtaining the necessary water can exceed the profit potential of a venture. The economics of aquaculture are discussed more fully in Chapter 2, and water sources and treatment are discussed in Chapter 3.

One dilemma that sometimes faces a prospective aquaculturist involves whether to select a site first or settle on the species to be cultured before the site is determined. Sometimes the decision is readily made because of circumstances. A person may, for example, have a piece of land that he or she intends to use for an aquaculture venture to the exclusion of any other option. In those cases, the best course of action is to select a culture species that is appropriate for the location.

On the other hand, a person may have training, experience, and the desire to produce a certain species. If that is the case, site selection may become secondary. Problems can, of course, be compounded if a person is obligated to a particular site and species that are incompatible. Avault (1986) developed a checklist to assist persons in selecting a species for culture.

The rearing of tropical species in temperate climates or attempting to produce marine species in locations far removed from seawater would appear on the surface to be examples of inappropriate linkages between site and culture species. However, if the proper technology is applied (e.g., a suitable source of heated water for the tropical species and the use of artificial sea water in a recirculating system for the marine species), it is possible to rear virtually any species in any location. There have actually been attempts to produce oysters in Colorado and penaeid shrimp in Illinois. Tilapia are currently being produced successfully in Idaho using geothermal water. Thus, decisions regarding site selection or species selection are not straight forward.

While there can be many advantages to establishing an aquaculture facility in a developing nation, there are also disadvantages. It is important for the prospective aquaculturist to weigh both the advantages and disadvatages of a given country before a final decision is made. Some of the things that should be taken into consideration are as follows:

1. Labor costs tend to be lower in developing nations, especially with respect to manual labor. However, it is often necessary to provide advanced training for local personnel or import professional technical and managerial staff. Salaries

for immigrant professionals tend to be higher than those paid in the nation of origin because of hardships associated with living in the developing nation.

2. Suitable land is often abundant and inexpensive in developing nations. In addition, many such nations promote the influx of outside investments through tax credits and other forms of dispensation.

3. Many developing countries have large areas of suitable land available that are not currently being subjected to pollution, but that situation could change in the future and there may be few, if any, environmental regulations to protect the aquaculturist.

4. Construction costs, like labor costs, are often much lower in developing nations, at least superficially. The availability of experienced contractors may be a problem, the quality of work can be poor, and delays in construction because of the inavailability of proper construction equipment and supplies can be significant problems.

5. One of the greatest advantages of locating a warmwater aquaculture facility in a tropical developing nation is the year-round availability of water of the proper temperature.[16] This can lead to two or more crops a year as compared with one in temperate climates for such species as tilapia and shrimp and can, therefore, greatly increase the chance for profitability.

6. Many developing nations have histories of political unrest. Not only might the aquaculturists be placed in physical danger in some instances, government instability could lead to significant changes in the initially perceived economic benefits. It is not unheard of for officials to nationalize private businesses following a change in government.

7. In developing nations the avilability of feed, supplies, and equipment may be severely limited, and heavy import duties may be imposed by the host nation. It may also be difficult to get equipment repaired or to find the parts required if the aquaculturist wishes to do the repairs. Maintenance of a large parts inventory may be necessary to help circumvent the problem.

8. Some means of transporting the crop to market at harvest is required. Transport must be available at the appropriate time, and facilities for processing and preserving the crop must be available through the existing infrastructure or developed by the aquaculturist. Distance from markets can impose a significant additional cost for shipping and cold or frozen storage.

9. Energy costs may be considerably higher in developing countries, though in some instances they may actually be lower. The reliability of electricity and the supplies of other sources of energy (e.g., gasoline and diesel fuel) may be suspect, so backup generators and fuel stockpiles may be required.

Many aquaculturists in the United States have established their aquatic farming activities as one part of a larger agrobusiness enterprise. This has been particularly

---

[16]The benefits of abundant water of the proper temperature can be lost if the water is polluted and requires expensive treatment before it is suitable for use in the aquaculture facility.

true in some catfish growing regions of the United States and is true in various other parts of the world. The channel catfish industry began in central Arkansas where farmers were involved with rice farming, which requires levee construction and the use of large volumes of water; thus, the transition to fish farming was accommodated to some extent. Many Mississippi farmers added catfish as an adjunct to existing crops such as cotton.

Other aquaculturists establish facilities that are solely used for the production of aquatic organisms. The trout farms in Idaho and the salmon net-pen facilities in British Columbia and the state of Washington are examples.

Aquaculture ventures may be so-called mom and pop operations with a few hundred kilograms of annual production on less than a hectare of water to sprawling farms with pond or raceway systems that can produce tens to hundreds of thousands of kilograms annually.

Most aquaculture is practiced on land or in the nearshore portions of water bodies. Mariculture in the open ocean was discussed in detail two decades ago by Hanson (1974) and technology that will accommodate the concept has been developed to some extent. Offshore net-pen designs employing spar buoys for stability or suspended from rigid structures that resemble ship hulls have been tested, and there has been some interest in converting offshore oil drilling platforms into sites for aquaculture (e.g., string culture of molluscs from the platform structure or use of the platforms for tank culture).

CONSTRAINTS TO AQUACULTURE DEVELOPMENT

Not everyone views aquaculture as a wise use of resources. Various types of conflicts have arisen over the best use of land and water between aquaculturists and individuals, community groups, and governmental agencies. Commercial harvesters of fish and shellfish have often seen aquaculture as a direct threat to their livelihoods, though it is certainly possible for commercial harvesters to also become aquaculturists and benefit from both activities. This has been proposed for salmon by Stickney (1988) and has been occuring with respect to the Maine mussel industry, where wild harvesting and aquaculture are becoming integrated (Wilson and Fleming 1988, 1989).

One example of the types of conflict that can occur is the opposition that has developed with respect to net-pen culture of salmon in Washington where a battle of sorts has been joined between those who would like to see planned aquaculture development and those who wish to see it curtailed. The subject was addressed in an Environmental Impact Statement (Parametrix, Inc. 1990) and was reviewed by Stickney (1990).

Initial objections to the development of salmon farms in Puget Sound came from upland property owners who objected to the sight of net pens on the basis of esthetics. From a legal standpoint, such property owners were ruled to not have a right to a perpetually unaltered view, so the focus of the objections began to change and expand. Similar objections have been raised in Maine where net-pen salmon culture is a fledgling industry. The issues raised relative to Puget Sound net pens include the following:

1. Feces and waste feeds fall to the sediments creating sterile zones under the net pens and negatively impact local fauna.
2. Nutrients from feces and feed fertilize the water and promote noxious algal blooms.
3. The use of antibiotics in salmon feeds leads to the development of disease-resistant strains of bacteria.
4. Cultured fish transfer diseases to wild fish.
5. Cultured fish that escape will negatively impact wild salmon runs.
6. Net pens are a hazard to navigation.
7. Access to traditional fishing grounds is impaired by the presence of net pens.
8. There is an increased threat of transmission of diseases to humans from the cultured fish.
9. Recreation is impeded by the net pens.
10. Net-pen facilities are sources of excessive noise and odors.

The Environmental Impact Statement addresses each of the above issues, and others. Some are not considered to be problems, while some very real problems, such as the formation of sterile zones beneath net pens, can be avoided by proper siting of the facility so that sufficient flushing occurs.

Use conflicts over sites suitable for aquaculture can cover a broad array of interests, only some of which occur with respect to the Puget Sound salmon net-pen farming industry described in the preceding paragraphs. As discussed by Pollnac (1992), an aquaculture enterprise, depending upon its location, can come into conflict with commercial fishing, subsistence land-use practices (e.g., use of mangroves for building materials or charcoal as opposed to cutting the trees to make room for ponds), agriculture, navigation, reforestation, wildlife enhancement, recreation, and even other aquaculture users. Mariculture issues in the United States have been reviewed by Fridley (1992), the National Research Council (1992), and DeVoe et al. (1992). Aquaculture conflicts also occur in other countries. The subject has recently been examined for Canada (Wildsmith 1992, Dickson 1992), Japan (Murai 1992), and Southeast Asia (Skladany 1992). Conflicts are resolved by policy makers, though the debate often rages in the media. As indicated by Joyce (1992), aquaculturists and their associations can assist policymakers toward rational governmental policies by being aggressive in getting the message across. At the same time, aquaculturists ''must be aware of the genuine concerns that face their industry and be prepared to deal with them in a responsible way.''

Shrimp farmers have recently come under fire for rearing exotic species, as discussed above. Coastal aquaculturists who have pond systems, whether for the rearing of vertebrates or invertebrates, have come into conflict with home, condominium, and shopping center developers; with industries interested in expansion; and with wetlands protection and preservation laws. The demands on land adjacent to the sea coast in the United States are great, and the amount of suitable and available land area for aquaculture, assuming it could be economically purchased for that use, is relatively small.

Fewer objections have been voiced from inland residents who live near aquaculture facilities. One exception has been in the Hagerman Valley of Idaho, where the vast majority of the rainbow trout produced in the United States are reared. The trout industry in the Hagerman Valley is located along an approximately 65 km stretch of the Snake River. Something between 85,000 and 113,000 l/sec of spring water flow through the trout raceways on a continuous basis.[17] The valley has also been colonized in recent years by retirees who have become concerned about degraded water quality as a result of the release of fish wastes and unconsumed feed into the Snake River. Treatment of effluent water is being required to address the issue, and the potential for expansion of the industry would seem to have become limited because of the objections that have been raised.

Catfish, baitfish, and crawfish farmers have thus far escaped from serious objections to their activities. There has been some discussion of enacting laws that would require the treatment of effluents from ponds used for those types of aquaculture, but for the most part the receiving waters are not negatively impacted. In some instances, the water being released is actually of better quality than that in the receiving streams. Unlike the situation in Idaho, where the wastes from the fish are flushed into the receiving stream on a contiuous basis, release of pond water is infrequent and the wastes tend to stay locked up in the pond sediments.

Finally, the development of aquaculture in ponds located on farmland, which is the rule in the catfish, baitfish, and crawfish farming regions of the nation, is viewed as a proper use of private land. On the other hand, the placement of net pens in Puget Sound or a bay in Maine, and the capture of spring water for passage through trout raceways and then release to the Snake River in Idaho are viewed as use of the commons for private gain. While it may be an oversimplification, much of the conflict would appear at least to revolve around common property rights.

The laws of the United States have not been promulgated with aquaculture in mind. In some states, fish and game laws are written in a way that gives the state ownership of the resource. Thus, a fish farmer in one of those states cannot legally sell the product without permission from the state. Those laws were enacted to protect wild fish and wildlife, not to prevent aquaculture. Some states have modified their codes to accommodate the development of aquaculture, but burdensome laws in others persist.

Other constraining laws have hampered aquaculture development. In some areas farmers may be authorized to extract groundwater for only so many days a year (based on the irrigation season for the primary crop of a particular region). Since fish require a constant water supply, many forms of aquaculture would not be viable under that constraint. There are instances where aquaculture has been judged a commercial rather than an agricultural pursuit. Since the water laws give agriculture

---

[17]Michael Falter, personal communication.

first chance at the water, the available supply may be allocated before the aquaculturist gets a chance to obtain the required amount.

There is no uniform permitting system for aquaculture in the United States, though most states require some type of a permit to engage in the culture of aquatic species and may also require a permit to transport the animals. Obtaining a permit may be as easy as sending a few dollars to a particular agency in one state, or as difficult as meeting the requirements of 20 or more agencies, holding public hearings, and spending hundreds of thousands of dollars and a few years in other states. The permits that may be required for aquaculture in the state of Washington are presented in Table 8. In many instances, no procedural guidelines for moving through the permitting system are in place, but that appears to be changing.

Many states now have aquaculture coordinators who serve as contact points for prospective aquaculturists. Each coordinator can explain the permitting system in his or her state and assist the applicant in various ways. In many states, the aquaculture coordinator is located in the state department of agriculture. Some regional information is also becoming available on the permitting situation. For example, the North Central Regional Aquaculture Center has commissioned a report on aquaculture law in that region (Thomas et al. 1992). The region includes Illinois, Indiana, Iowa, Kansas, Michigan, Minnesota, Missouri, Nebraska, North Dakota, Ohio, South Dakota, and Wisconsin.

Depending upon the site on which facilities are to be constructed, federal permits may be also be required. If, for example, the site is located on navigable waters (which are very broadly defined in the United States and can include even very small streams that connect to waters that are truly capable of accommodating a vessel), a permit from the U.S. Army Corps of Engineers may be required under the Rivers and Harbors Act of 1899. The Corps will solicit comment from other agencies, such as the U.S. Fish and Wildlife Service, to determine if the proposed activity will have any negative impact on fish and wildlife resources. Water discharged from aquaculture operations may require a permit from the Environmental Protection Agency under the National Pollution Discharge Elimination Systems (NPDES) permitting requirements. The policy issues, including permitting requirements, associated with United States mariculture activities have been reviewed by the National Research Council (1992).

THE FUTURE OF AQUACULTURE

Aquaculture continues to expand throughout the world, and in some cases, production levels have grown to the point where there has been a softening of the prices paid to producers because of oversupply. In general, the demand for aquaculture products can be expected to grow, though there will be ups and downs because of supply and demand conditions. A glut of shrimp and salmon on the world market during 1991, while certainly only representing a temporary phenomenon, caused a great deal of industry restructuring.

In the United States, interest in developing additional species for culture will continue. The ultimate size of the market for such animals as striped bass, hybrid striped bass, red drum, and sturgeon have as yet to be determined, but production of each of those animals is on the increase.

Though the ultimate ability of aquaculturists to culture transgenic species commercially will undoubtedly be determined in the courts and by regulatory agencies, the potential of genetic engineering to enhance production and improve disease resistance is great (Colwell 1983). Scientists are only beginning to explore the possibilities with respect to aquaculture species.

Expansion of U.S. production of coldwater and even warmwater species in inland waters will be limited because of land and water availability, as well as the current permitting situation. The expansion of salmon production in net pens will also be unlikely unless economically viable offshore technology can be developed and put into place. Water of suitable quantity and quality will be increasingly difficult to find. The delta region of Mississippi, where the catfish industry is centered, was once thought to have an infinite supply of groundwater, but the water table is beginning to show the effects of the large number of wells that have been drilled, and expansion in that region will be limited. Other regions with prodigious supplies of water, soils of good water-holding capacity, and the warm temperatures needed to sustain optimum growth of catfish do not seem to be available.

While much of the commercial aquaculture expansion will continue to occur in other nations, the United States will continue to supply the world with advances in technology. As economical closed water systems are developed, a new round of domestic expansion will be possible. A great deal of work is being conducted on recirculating systems at the present time and significant advances are being achieved.

Recovery of some stocks of fishes from the brink of extinction may be effected through aquaculture. The National Marine Fisheries Service has initiated a recovery program with certain stocks of Columbia River salmon that employs an aquaculture component. Aquaculture may provide the only real hope when populations have fallen so low that only a handful of fish remain.

The quality of effluents from aquaculture facilities has become an issue in recent years, particularly in developed nations, and will continue to demand attention. The application of state-of-the-art water treatment methodology—appropriately adapted from sewage treatment technology in many instances—to aquaculture will increase the cost of doing business. Increased efficiencies of production, including improved performance feeds, better control and prevention of diseases, and improved growth and survival rates will be required to offset some of the additional expense.

Manzi (1989) reviewed the aquaculture industry's needs and came up with a list of problem areas that need to be addressed during the current decade. The types of research that were envisioned as being important was not much different from what had gone before. He identified nutrition, genetics, engineering, disease control, aquaculture ecology (dealing with so-called self-pollution[18] and controlling efflu-

---

[18] Instances have occurred in which the density of animals in a culture system became so high that chronic water quality degradation with resultant losses occurred. Japan, for example, experienced such problems in some bays until limitations on the number of netpens and other types of facilities were imposed.

TABLE 8. Permits that May Be Required for an Aquaculture Venture in the State of Washington. (Source: Cottinghm, K. Undated. Aquaculture permits required. Unpublished mimeo, Washington State Senate Natural Resources Agriculture Committee Staff Attorney, Olympia. 7 p.)

Name of Permit	Issuing Agency	Activities Covered	Process
		State Permits	
Aquatic Land Lease	Dept. of Natural Resources	Any use of publicly owned aquatic lands (tidelands, beds of navigable waters, shorelands)	Complete application, and submit filing fee.
Hydraulic Project Approval	Dept. of Fisheries or Wildlife	Any project that will use, divert, obstruct or change the natural flow or bed of any of the salt or fresh waters of the state	Apply to the responsible agency. Application should contain general project design and plans for the protection of fish life.
Statement of Consistency with the Coastal Zone Management Act	Dept. of Ecology	Activities within the coastal zone	Dept. of Ecology determines if the project is consistent with the the Coastal Zone Mnagement Act.
Water Quality Certification	Dept. of Ecology	In-water construction involving discharge of pollutants and needs a Corps of Engineers permit	Apply for federal permit. Ecology is contacted and can comment before federal permit is issued.
Water Quality Standards Modification	Dept. of Ecology	Discharge of pollutants causing water quality to fall below standards	Apply to Dept. of Ecology.
National Pollutant Discharge Elimination System (NPDES) Permit	Dept. of Ecology	Required of any person who conducts an operation that results in disposal of solid or liquid wastes into waters of the state	Apply to Dept. of Ecology.

Aquacultural Identification Private Sector Products	Dept. of Agriculture	Products cultivated by aquatic farmers that would otherwise be regulated by the Fisheries or Wildlife departments	Follow regulations for labeling or sale documentation.
Registration of Aquatic Farmers	Dept. of Fisheries	Cultivation of private-sector aquatic products	Apply to Dept. of Fisheries.
Fish Disease Control	Dept. of Fisheries	Protection of aquaculture industry and wild fisheries from loss due to aquatic diseases	Apply to Dept. of Fisheries.
Shellfish Certification	Dept. of Social and Health Services	Sale of shellfish	Apply to Dept. of Social and Health Services.

Federal Permits

Dredge and Fill Permit	U.S. Army Corps of Engineers	Placement of structures, excavation, deposit of of dredge or fill material	Apply to Corps. They solicit comments from federal and state agencies.
Navigational Markings	U.S. Coast Guard	Fixed or floating structures in or over water	Apply to Coast Guard.

Local Permits

Shoreline Substantial Development	County or City	Developments in or on the water or on the shoreline having a total cost or fair market value exceeding $2,500	If Dept. of Ecology does not agree with the approval or disapproval, it can appeal the decision to the Shoreline Hearings Board.

ents), and marketing. As a part of the Regional Aquaculture Center activities sponsored by the USDA, the industry in each of the five regions is surveyed periodically to determine research needs. Research projects aimed at addressing regional needs are then developed.

Pending shortages in the world supply of fish meal will also stimulate aquaculture nutrition research. Finding alternative protein sources will be a significant challenge.

Mechanisms for obtaining clearance for the use of drugs and chemicals on aquatic species destined for human consumption need to be streamlined and the arsenal of available compounds needs to be expanded. At the present time, the commercial aquaculturist in the United States has an extremely small arsenal upon which to draw. At the same time, increasing the emphasis on developing vaccines against aquatic animal diseases is of critical importance.

Aquaculture is a highly applied science (and art) that draws heavily on the basic sciences. That dependence is not going to change in the foreseeable future, and will, in fact, increase as the potential application of basic information to aquatic animals becomes apparent. As indicated above, we are only beginning to see the potential of genetic engineering to aquaculture problems.

Thus, the future in terms of expanded production seems positive, though there are some very significant hurdles to be overcome. Aquaculturists have the responsibility of conducting their work in an environmentally sound manner. Research to address public concerns about environmental degradation by aquaculture activities should continue and be redoubled. That is the obligation of the aquaculture community. It is the obligation of policymakers to act judiciously on the scientific findings: halting aquaculture expansion when problems are clearly going to arise, proceeding with caution when the evidence is cloudy, and promoting the organized development of the activity when it can be demonstrated that facilities can be operated in an environmentally responsible manner.

LITERATURE CITED

Allanson, B., and R. G. Noble. 1964. The tolerance of *Tilapia mossambica* Peters to high temperatures. Trans. Am. Fish. Soc., 94: 323–332.

Allen, S. K., Jr., and R. J. Wattendorf. 1987. Triploid grass carp: status and management implications. Fisheries, 12: 20–24.

American Fisheries Society. 1989. Common and scientific names of aquatic invertebrates from the United States and Canada: decapod crustaceans. American Fisheries Society, Bethesda, Maryland. 77 p.

Anonymous. 1976. Status of grass carp. Bull. Sport Fish. Inst., 273: 4–6.

Anonymous. 1988. U.S. catches, eats record amounts of fish and shellfish. Mar. Fish. Rev., 50(3): 51–52.

Anonymous. 1991. Status of world aquaculture 1991. Aquacult. Mag. Buyer's Guide '92. pp. 8ff.

Anonymous. 1992a. Catfish processor report. Aquacult. News, 6(7): 17.

Anonymous. 1992b. Aquaculture accounts for 28% of world shrimp production. Aquacult. Mag., 18(1): 90–91.

Anonymous. 1992c. Salmon surplus—special report. North. Aquacult., Jan./Feb.: 4–5.

Avault, J. W., Jr. 1986. Which species to culture—a check list. Aquaculture Mag. 12(6): 41–44.

Avault, J. W., Jr., and E. W. Shell. 1968. Preliminary studies with the hybrid tilapia *Tilapia nilotica* × *Tilapia mossambica*. FAO Fish. Rep., 44: 237–242.

Bardach, J. E., J. H. Ryther, and W. O. McLarney. 1972. Aquaculture. Wiley-Interscience, New York. 868 p.

Barlow, S. 1989. Fishmeal—world outlook to the year 2000. Fish Farmer, Sept./Oct.: 40–43.

Blackman, N. J. 1989. A view of aquaculture in the United States from the business sector. AAAS Annu. Mtg. Abstr.: 96.

Brice, J. J. 1898. Report of the Commissioner. U.S. Commission of Fish and Fisheries, Washington, D.C. 103 p.

Brown, E. E. 1983. World fish farming: cultivation and economics. AVI, Wesport, CT. 516 p.

Brown, J. D., and T. G. Coon. 1991. Grass carp larvae in the lower Missouri River and its tributaries. N. Am. J. Fish. Manage., 11: 62–66.

Chen, J. 1990. Shrimp culture industry in the People's Republic of China. pp. 70–76, In: K. L. Main and W. Fulks (Eds.). The culture of cold-tolerant shrimp: Proceedings of an Asian-U.S. Workshop on Shrimp Culture, Honolulu, October 2–4, 1989. University of Hawaii Sea Grant Misc. Rep., Honolulu.

Chen, T. T., Z.Y Zhu, D. A. Powers, and R. Dunham. 1988. Fish genetic engineering: a novel approach to aquaculture. J. Shellfish Res., 7: 552.

Chervinski, J. 1961. Study of the growth of *Tilapia galilaea* (Artedi) in various saline conditions. Bamidgeh, 13: 71–74.

Chervinski, J. 1966. Growth of *Tilapia aurea* in brackish water ponds. Bamidgeh, 18: 81–83.

Chervinski, J., and A. Yashouv. 1971. Preliminary experiments on the growth of *Tilapia aurea* Steindachner (Pisces, Cichlidae), in saltwater ponds. Bamidgeh, 23: 125–129.

Chervinski, J., and M. Zorn. 1974. Note on the growth of *Tilapia zillii* (Gervais) in sea water ponds. Aquaculture, 4: 249–255.

Chew, K. K., and D. Toba. 1990. Western region aquaculture industry situation and outlook report. Western Regional Aquaculture Consortium, University of Washington, Seattle. 23 p.

Chimits, P. 1957. The tilapias and their culture. A second review and bibliography. FAO Fish. Bull., 10: 1–24.

Clugston, J. P., and J. V. Shireman. 1987. Triploid grass carp for aquatic plant control. U.S. Department of Interior Fish and Wildlife Disease Leaflet 8. U.S. Fish and Wildlife Service, Washington, D.C. 3 p.

Colle, D. E., J. V. Shireman, and R. W. Rottman. 1978. Food selection by grass carp fingerlings in a vegetated pond. Trans. Am. Fish. Soc., 107: 149–152.

Colwell, R. R. 1983. Biotechnology in the marine sciences. Science, 222: 19–24.

Conner, J. V., R. P. Gallagher, and M. F. Chatry. 1980. Larval evidence for natural reproduction of grass carp *Ctenopharyngodon idella* in the lower Mississippi River. U.S. Fish and Wildlife Service Biological Services Program, FWS/OBS-80/43: 1–19.

Cross, D. G. 1969. Aquatic weed control using grass carp. J. Fish Biol. 1: 27–30.

de la Bretonne, L. W. 1988. Commercial crawfish cultivation practices. J. Shellfish Res., 7: 210.

DeVoe, M. R., R. S. Pomeroy, and A. W. Wypyszinski. 1992. Aquaculture conflicts in the eastern United States. World Aquacult., 23(2): 24–25.

Dickson, F. 1992. Aquaculture conflicts in Pacific Canada. World Aquacult., 23(2): 28–29.

FAO. 1991a. FAO Fisheries Circular 815. Revision 2. Food and Agriculture Organization of the United Nations, Rome. 136 p.

FAO. 1991b. Fishery statistics catches and landings. Food and Agriculture Organization of the United Nations, Rome. Vol. 68. 516 p.

Farmanfarmaian, A., and R. Moore. 1978. Diseasonal thermal aquaculture-1. Effect of temperature and dissolved oxygen on survival and growth of *Macrobrachium rosenbergii*. Proc. World Maricult. Soc., 9: 55–66.

Finch, R. 1969. The U.S. fish protein concentrate program. Comm. Fish. Rev., Jan.: 25–30.

Folsom, W. B. 1987. World salmon aquaculture, 1986–87. National Marine Fisheries Service report NMFS/FIA23/88–6, Washington, D.C. 36 p.

Fridley, R. B. 1992. Mariculture issues in the United States. World Aquacult., 23(2): 20–22.

Galman, O. R., and R. R. Avtalion. 1983. A preliminary investigation of the characteristics of red tilapias from the Philippines and Taiwan. pp. 291–301, In: L. Fishelson and Z. Yaron (compilers), Proceedings International Symposium on Tilapia in Aquaculture, Nazareth, Israel, May 8–13. Tel Aviv University Press, Tel Aviv, Israel.

Gleastine, B. W. 1974. A study of the cichlid *Tilapia aurea* (Steindachner) in a thermally modified Texas reservoir. M.S. Thesis, Texas A&M University, College Station. 258 p.

Glude, J. B. (Ed.). 1977. NOAA aquaculture plan. U.S. Department of Commerce, Washington, D.C. 41 p.

Guillory, V., and R. D. Gasaway. 1978. Zoogeography of the grass carp in the United States. Trans. Am. Fish. Soc., 107: 105–112.

Hallerman, E. M., and A. R. Kapuscinski. 1990a. Transgenic fish and public policy: patenting of transgenic fish. Fisheries, 15(1): 21–24.

Hallerman, E. M., and A. R. Kapuscinski. 1990b. Transgenic fish and public policy: regulatory concerns. Fisheries, 15(1): 12–20.

Hanson, J. A. 1974. Open sea mariculture. Dowden, Hutchinson & Ross, Stroudsburg, Pennsylvania. 410 p.

Hepher, B., and Y. Pruginin. 1981. Commercial fish farming with special reference to fish culture in Israel. Wiley-Interscience, New York. 261 p.

Hickling, C. F. 1963. The cultivation of *Tilapia*. Sci. Am., 208: 143–152.

Idyll, C. P. 1970. The sea against hunger. Crowell, New York. 221 p.

Joint Subcommittee on Aquaculture. 1983a. National aquaculture development plan. Joint Subcommittee on Aquaculture, Washington, D.C. Vol. 1. 67 p.

Joint Subcommittee on Aquaculture. 1983b. National aquaculture development plan. Joint Subcommittee on Aquaculture, Washington, D.C. Vol. 2. 196 p.

Joyce, J. 1992. Aquaculture conflicts and the media—what associations can do. World Aquacult., 23(2): 14–15.

Kapuscinski, A. R., and Hallerman, E. M. 1990. Transgenic fish and public policy: anticipating environmental impacts of transgenic fish. Fisheries, 15(1): 2–11.

Kilgen, R. H., and R. O. Smitherman. 1971. Food habits of the white amur stocked in ponds alone and in combination with other species. Prog. Fish-Cult., 33: 123–127.

Kingsley, J. B. 1987. Legal constraints to tilapia culture in the United States. J. World Aquacult. Soc., 18: 201–203.

Lembi, C. A., B. G. Ritenour, E. M. Iverson, and E. C. Forss. 1978. The effects of vegetation removal by grass carp on water chemistry and phytoplankton in Indiana ponds. Trans. Am. Fish. Soc., 107: 161–171.

Lightner, D. V., R. M. Redman, T. A. Bell, and J. A. Brock. 1983. Detection of IHHN virus in *Penaeus stylirostris* and *P. vannamei* imported into Hawaii. J. World Maricult. Soc., 14: 212–225.

Lin, S. 1935. Life history of Waan Ue (*Ctenopharyngodon idella* Cuv. and Val.). Lingnan Sci. J., 14: 129–135.

Mackenzie, D. R. 1988. National biosafety programs for field testing transgenic organisms. J. Shellfish Res., 7: 558.

Mahnken, C. V. W. 1991. Coho salmon farming in Japan. pp. 131–149, In: Stickney, R. R. (Ed.). Culture of salmonid fishes. CRC Press, Boca Raton, Florida.

Manzi, J. J. 1989. Aquaculture research priorities for the 1990s. World Aquacult., 20(2): 29–32.

McBay, L. G. 1961. The biology of *Tilapia nilotica* Linnaeus. Proc. Southeast. Assoc. Game Fish Comm., 21: 436–444.

Murai, T. 1992. Aquaculture conflicts in Japan. World Aquacult., 23(2): 30–31.

Nash, C. E. 1991. Turn of the millennium aquaculture: navigating troubled waters or riding the crest of the wave? World Aquacult. 22(3): 28–49.

National Research Council. 1992. Marine aquaculture. National Academy Press, Washington, D.C. 290 p.

New, M. B. 1988. Freshwater prawns: status of global aquaculture, 1987. NACA Tech. Man. No. 6. Network of Aquaculture Centers in Asia, Bangkok, Thailand. 58 p.

Niemeier, P. E. 1988. Status of Asian shrimp culture, 1986–1987. National Marine Fisheries Service, Washington, D.C. 37 p.

Noble, R. L., R. D. Germany, and C. R. Hall. 1975. Interactions of blue tilapia and largemouth bass in a power plant cooling reservoir. Proc. Southeast. Assoc. Game Fish Comm. 29: 247–251.

Parametrix, Inc. 1990. Fish culture in floating net-pens. Final Programmatic Environmental Impact Statement prepared for the Washington Department of Fisheries, Olympia. 161 p.

Perry, W. G., Jr. 1967. Distribution and relative abundance of blue catfish, *Ictalurus furcatus,* and channel catfish, *Ictalurus punctatus,* with relation to salinity. Proc. Southeast. Assoc. Game Fish Comm., 21: 436–444.

Perry, W. G., Jr., and J. W. Avault, Jr. 1968. Preliminary experiments on the culture of blue, channel, and white catfish in brackish water ponds. Proc. Southeast. Assoc. Game Fish Comm., 22: 397–406.

Pollnac, R. B. 1992. Multiused conflicts in aquaculture—sociocultural aspects. World Aquacult., 23(2): 16–19.

Pritchard, D. W. 1967. What is an estuary: physical viewpoint. pp. 3–5, In: G. H. Lauff, (Ed.). Estuaries. American Association for the Advancement of Science Publication No. 83. A.A.A.S., Washington, D.C.

Robins, R. 1991. A list of common and scientific names of fishes from the United States and Canada. American Fisheries Society, Bethesda, Maryland. 183 p.

Robinson, J. W. 1988. Missouri's commercial fishery harvest: 1987. Missouri Department of Conservation, Final Report, National Marine Fisheries Service Project 2–I-7–1, Columbia.

Rosenberry, B. 1992. World shrimp farming, Nov./Dec., 17(11): 21.

Rouse, D. B., and R. R. Stickney. 1982. Evaluation of the production potential of *Macrobrachium rosenbergii* in monoculture and in polyculture with *Tilapia aurea*. Proc. World Maricult. Soc., 13: 73–85.

Ryther, J. H. 1969. Photosynthesis and fish production in the sea. Science, 166: 72–76.

Sarig, S. 1989. The fish culture industry in Israel in 1988. Bamidgeh, 41: 50–57.

Sastradiwirja, F. 1986. Production, yield characteristics and economics of polyculture of *Macrobrachium rosenbergii* and various fish species under pond condition. Network of Aquaculture Centers in Asia, Bangkok, Thailand. 66 p.

Shelton, W. L. 1986. Strategies for reducing risks from introductions of aquatic organisms: an aquaculture perspective. Fisheries, 11(2): 18–19.

Skladany, M. 1992. Conflicts in Southeast Asia. World Aquacult., 23(2): 33–35.

Sollows, J. D., and N. Tongpan. 1986. Comparative economics of rice-fish culture and rice monoculture in Ubon Province, Northeast Thailand. pp. 149–152, In: J. L. Maclean, L. B. Dizon, and L. V. Hosillos (Eds.). Proceedings First Asian Fisheries Forum, Manila, Philippines, May 26–31, 1986.

Stanley, J. G. 1979. Control of sex in fishes, with special reference to the grass carp. pp. 201–242, In: J. V. Shireman (Ed.). Proceedings of the Grass Carp Conference. Aquatic Weeds Research Center, University of Florida, Gainesville.

Stanley, J. G., W. W. Miley II, and D. L. Sutton. 1978. Reproductive requirements and likelihood for naturalization of escaped grass carp in the United States. Trans. Am. Fish. Soc., 107: 119–128.

Steffens, W. 1991. On rainbow trout production of the new Federal States (Germany). Fisch. Teichwirt., 42: 42–48.

Stickney, R. R. 1984. Estuarine ecology of the southeastern United States and Gulf of Mexico. Texas A&M Press, College Station. 310 p.

Stickney, R. R. 1986. Tilapia tolerance of saline waters: a review. Prog. Fish-Cult., 48: 161–167.

Stickney, R. R. 1988. Commercial fishing and net-pen salmon aquaculture: turning conceptual antagonism toward a common purpose. Fisheries, 13(4): 9–13.

Stickney, R. R. 1990. Controversies in salmon aquaculture and projections for the future of the aquaculture industry. pp. 455–461, In: Proceedings of the Fourth Pacific Congress on Marine Science and Technology, Tokyo, Japan, July 16–20.

Stickney, R. R. 1991. Effects of salinity on aquaculture production. pp. 105–132, In: D. E. Brune and J. R. Tomasso (Eds.). Aquaculture and water quality. Advances in World Aquaculture, Vol. 3. World Aquaculture Society, Baton Rouge, Louisiana.

Stickney, R. R., and J. H. Hesby. 1978. Tilapia culture in ponds receiving swine waste. pp. 90–101, In: R. O. Smitherman, W. L. Shelton, and J. H. Grover (Eds.). Culture of Exotic Fishes Symposium Proceedings. Fish Culture Section, American Fisheries Society, Bethesda, Maryland.

Stickney, R. R., and B. A. Simco. 1971. Salinity tolerance of catfish hybrids. Trans. Am. Fish. Soc., 100: 790–792.

Stott, B., and T. O. Robson. 1970. Efficiency of grass carp (*Ctenopharyngodon idella* Val.) in controlling submerged water weeds. Nature, 226: 870.

Sutton, D. L. 1977. Grass carp (*Ctenopharyngodon idella* Val.) in North America. Aquat. Bot., 3: 157–164.

Sutton, D. L., V. V. Vandiver, Jr., R. S. Hestand, and W. W. Miley II. 1979. Use of grass carp for control of hydrilla in small ponds. pp. 91–102, In: J. V. Shireman (Ed.). Proceedings of the Grass Carp Conference. Aquatic Weeds Research Center, University of Florida, Gainesville.

Swingle, H. S. 1957. Preliminary results on the commercial production of channel catfish in ponds. Proc. Southeast. Assoc. Game Fish Comm., 10: 160–162.

Swingle, H. S. 1958. Experiments on growing fingerling channel catfish to marketable size in ponds. Proc. Southeast. Assoc. Game Fish Comm., 12: 63–72.

Thomas, S. K., R. M. Sullivan, R. L. Vertrees, and D. W. Floyd. 1992. Aquaculture law in the north central states: a digest of state statutes pertaining to the production and marketing of aquacultural products. Technical Bulletin Series No. 101. North Central Regional Aquaculture Center, East Lansing, Michigan. 76 p.

Trewavas, E. 1973. On the cichlid fishes of the genus *Pelmatochromis* with proposal of a new genus for *P. congicus;* on the relationship between *Pelmatochromis* and *Tilapia* and the recognition of *Sarotherodon* as a distinct genus. Bull. Br. Mus. (Nat. Hist.) Zool., 25: 1–26.

Trewavas, E. 1982. Tilapias: taxonomy and speciation, pp. 3–13, In: R. S. V. Pullin and R. H. Lowe-McConnell (Eds.). The biology and culture of tilapias. International Center for Living Aquatic Resources Management, Manila, Philippines.

Vattenbruk 1988. Aquaculture in Sweden during 1987. Stat. Medd. (Na Naturresur. Miljoe). Stockholm, Sweden. 12 p.

Vattenbruk 1990. Aquaculture in Sweden during 1989. Stat. Medd. (Na Naturresur. Miljoe). Stockholm, Sweden. 15 p.

Weidner, D. 1988. Latin American shrimp culture industry, 1986–90. National Marine Fisheries Service, Washington, D.C. 81 p.

Wildsmith, B. H. 1992. Aquaculture conflicts in Atlantic Canada. Privilege, politics and private rights. World Aquacult., 23(2): 26–27.

Wilson, J., and D. Fleming. 1988. An economic analysis of the Maine mussel industry. J. Shellfish Res., 7: 571.

Wilson, J., and D. Fleming. 1989. Economics of the Maine mussel industry. World Aquacult. 20(4): 49–55.

Wurts, W. A., and R. R. Stickney. 1989. Responses of red drum *(Sciaenops ocellatus)* to calcium and magnesium concentrations in fresh and salt water. Aquaculture, 76: 21–35.

Zimpfer, S. P., C. F. Bryan, and C. H. Pennington. 1987. Factors associated with the dynamics of grass carp larvae in the lower Mississippi River valley. American Fisheries Society Symposium, Bethesda. Maryland, 2: 102–108.

2 Economics

GETTING STARTED

Many people have seen newspaper or magazine articles indicating that there is a lot of money to be made from aquaculture: All you have to do is dig a pond, fill it with water, add some kind of fish, and stand back. The money will quickly start flowing in. Most of us can recognize that such a description is a gross oversimplification of what aquaculture involves. Aquaculture is, afterall, farming, and farmers are known for their hard work and often precarious economic condition. While some have made fortunes from aquaculture, most are able to earn only a living wage, and others have gone into bankruptcy.

Aquaculture is a particularly high-risk type of farming. The aquaculturist is rarely certain as to the exact number of livestock in the system, a situation that would lead terrestrial livestock producers to a state of panic. Since aquatic organisms live under water, they are often not easily observable and are never easy to count. Mortalities may occur that are never discovered. Excessive feed may be provided when the numbers and biomass of organisms in the system are overestimated, which can exacerbate or create water quality problems and is certainly economically wasteful.

To overcome the difficulties of growing what is often a largely or entirely unseen crop, aquaculturists need to attempt to relate to the culture species and to optimize the culture environment. Developing that type of a relationship with animals that live in a completely different environment from human beings is difficult for some people, while others accomplish it with apparent ease. Experience certainly counts. In societies where aquaculture is widely practiced, many people grow up on fish farms. Their ability to predict problems in advance of disaster is legendary and must come from thousands of hours during which they observed and assimilated knowledge about culture systems and the species contained therein.

In the United States, many who become interested in aquaculture develop that interest as adults, with little or no prior exposure to the subject. Before jumping into an aquaculture enterprise, such prospective aquaculturists should go through a checklist to ensure that they have considered as many of the problems and needs associated with the enterprise as possible. One such checklist was developed by Buttner and Flimlin (1991). Their checklist asks a series of questions. As the number of affirmative responses increases, the likelihood that the aquaculture venture will be a success increases, but there is never a guarantee that any facility will be successful. The modified questions of Buttner and Flimlin (1991) are presented in Table 9.

TABLE 9. Questions that a Prospective Aquaculturist Should Ask Before Investing (Adapted from Buttner and Flimlin 1991)

Economic Questions

1. Have you developed a realistic business plan that shows projected cash flows for 3 to 5 years?
2. Do you own or have access to an appropriate site?
3. Have you determined what it will cost to make the necessary improvements on the site to accommodate the aquaculture venture?
4. Do you own or have access to most of the necessary equipment you will need?
5. Can you secure capital for start-up and operation of the facility at a reasonable interest rate?
6. Will the lender accommodate your production and marketing schedule?
7. Is the profit potential higher than that of other possible investments?
8. Will the anticipated profit be adequate compensation for your labor and expenditure of resources?
9. Are you prepared to wait from 6 to 18 months before you obtain any positive cash flow from marketing your product?
10. Do you have adequate cash reserves for unanticipated costs; i.e., have you built in a contingency for equipment failure, crop loss, and so forth?

Personal Questions

1. Are you willing to work long, hard, and irregular hours, often 7 days per week?
2. Are you comfortable with mathematical problem-solving and mechanical trouble-shooting?
3. Are you willing to seek outside help when you need it (and do you know where to go to obtain such help)?
4. Do you have the technical expertise required to manage the operation?
5. Can you afford to hire professionally trained technical personnel?
6. Does your state have an aquacultural association that you can join?
7. Do you receive and read aquaculture literature?
8. Are you willing to attend (or have you been involved with) a course or workshop on aquaculture to become better informed on current practices and new developments?

Marketing Questions

1. Have you examined the existing situation with respect to market size and demand along with level of competition to ascertain whether you can effectively compete with your proposed product?
2. Have you identified both primary and secondary markets for your product?
3. Have you determined the form in which you will market your product (e.g., alive, dressed, value-added)?
4. Can you supply product to your market throughout all or most of the years?
5. Do you have the means to harvest, handle, hold, and transport the product?
6. Are you familiar with legal issues involved with the marketing of your product?
7. Do you have the resources to construct and operate a Health Department-approved processing facility?

Facility Considerations

1. Is the proposed site of your facility in an unrestricted area (zoned for aquaculture, not in a wetland or a location with some other type of protected status)?
2. Is the prospective site located near your processing plant and close to markets?
3. Is the site suitable for aquaculture (free from pesticide contamination, having proper topography and soil type, available electrical and other power sources, and allweather access)?
4. If the answer to number 3 in this section is no, can the site be made suitable for a reasonable cost?
5. Is the site sufficiently large for expansion if that becomes desirable in the future?
6. Have the advantages and disadvantages of leasing as opposed to ownership been investigated?
7. Do you live close enough to the site to ensure security and to get there quickly in case of emergency?
8. Is the system designed and constructed specifically for aquaculture (as opposed to recreation, esthetics, irrigation, etc.)?
9. Is an adequate supply of high-quality water available?
10. Will water quality and quantity remain suitable for year-round production ?
11. Can you control the water flow entering, within, and exiting the system?
12. Can you effectively and efficiently manage the wastes leaving the system?
13. Can you restrict the entry of predators and diseases to your system?

Sociolegal Considerations

1. Is the development of an aquaculture facility at your site an acceptable use with respect to neighbors and others who may use the region?
2. Have you discussed your plans with appropriate state agencies and extension agents?
3. Have you identified the permits required for construction and operation of the facility?
4. Can the required permits be obtained without excessive investment of time, money, and effort?
5. Can you obtain permits for an extended period of time, or do they have to be renewed frequently?

Production Considerations

1. Have you decided on a species for culture, and are you familiar with its biology?
2. Have you explored the different production strategies available and identified one that satisfies your interests and resources?
3. Do you have the resources (financial, technical, and spatial) needed to maintain and spawn adults, spawn and hatch offspring, and rear juveniles?
4. Are dependable sources of fingerling finfish or juvenile shellfish locally available?
5. Can feed and other essential supplies be obtained locally, quickly, and at reasonable costs?
6. Are suitable backup systems available in your culture system?
7. Are disease diagnostic services and dependable technical assistance readily available?
8. Do you have access to a dependable work force?
9. Do you have appropriate predator control, including systems to safeguard against poaching?
10. Do you have adequate dry storage space for feed, chemicals, and equipment?

Since many aquaculturists are biologists and have little or no training in business management, the economic side of the equation is often not given the attention it deserves. Yet, without being on a sound economic footing, no commercial venture can be sustained. It is important for the very survival of a venture to maintain accurate and complete records, utilize a good bookkeeping system, and have at least one individual associated with the business who understands the business side of the equation. An introductory book on aquaculture economics that might be helpful to those who do not have a background in the subject, was recently written by Shang (1990). A guide to preparing feasibility studies for commercial marine shrimp farming was prepared by Aquafood Business Associates (1987).

A number of economic studies have been conducted on various species in recent years. Examples from the tropics include those of Fitzgerald (1988), who evaluated the economic potential of shrimp and fish in Guam, and Mitra et al. (1989), who looked at both social acceptance and economic returns from tilapia culture in West Bengal, India. Economic studies on shrimp farming have also been conducted in South Carolina (Sandifer et al. 1982, Bauer et al. 1983, Liao and Smith 1983, Sandifer and Bauer 1985), Hawaii (Samples and Leung 1985, Wyban et al. 1988), Texas (Huang et al. 1984, Hanson et al. 1986), and with respect to Latin America and the Caribbean (Pretto 1989, Sharfstein 1989). Glude (1983) wrote on the economics and marketing of bivalves in the United States. Ridler and Kabir (1988) and Aiken (1989) examined the economics of salmon culture on the Atlantic coast of Canada. Pacific salmon farming was examined by Lin et al. (1989). The economics of catfish farming have been evaluated in Mississippi (Waldrop and Dillard 1985), Louisiana (Dellenbarger and Vandeveer 1986, Nerrie et al. 1990), and South Carolina (Pomeroy et al. 1987). There has also been an economic study that focused on sturgeon (Shigekawa and Logan 1986).

DEVELOPING A BUSINESS PLAN

In the vast majority of cases, a prospective aquaculturist will need to borrow or raise funds for such things as the purchase of land; the development of facilities; and the purchase of equipment, aquatic livestock, feed, and supplies. The process that prospective aquaculturists should go through in developing their business plans has been discussed by various authors. Two recent useful publications of that type are those of Buttner and Flimlin (1991) and Conte (1992). A summary of the kinds of information needed is presented in Table 9. Bankers and investors will expect to see a complete business prospectus before becoming involved in any type of business enterprise, so the aquaculturist should develop a document that completely describes the proposed project and provides a balance sheet for all anticipated expenses and revenues.

The level of aquaculture development varies considerably from one location to another. Bankers may or may not have familiarity with aquaculture. It is up to the prospective culturist to demonstrate that the business plan is financially sound and that the loan represents a good investment on the part of the bank, that is, one in which there is every probability that the money will be repaid on time.

Some aquaculturists have raised their own venture capital rather than seeking direct loans from banks. Loans may also be available from the Farmers Home Administration (Anonymous 1980), a federal government agency. The government of Puerto Rico, a U.S. territory, also has develped a program to assist not only in financing aquaculture ventures but also to provide technical assistance and market development (Pagan-Font 1981). The financing scene in Canada was described by Dorin (1988).

In any case, the importance of a prospectus cannot be overemphasized. The prospectus should, insofar as possible, anticipate questions that will arise from those who are being approached for financial support. By having considered all the costs of start-up and operation, and by being realistic with respect to projected production and the price that the product will bring, the culturist will establish credibility. Claims of enormous returns on investment can be lures for funding support, but they should be red flags to those who are interested in making sound investments and obtaining a reasonable return therefrom. The development of business plans for aquaculture has been discussed by Bell (1989).

While each aquaculture enterprise is unique, there are types of expenses that are common to nearly all of them. Costs are of two types. One type, called fixed costs (or ownership costs), involve one-time expenses (such as for purchase and modification of the land) and for the purchase of durable equipment (things that have life expectancies measured in terms of years). Fixed expenses do not change as a function of production. They can be depreciated over time on various schedules. While there is the need to renovate ponds and replace equipment, the value of the fixed items will depreciate to zero by the time normal replacement is required.

The other type of expense is known as operating or variable costs. Items that are frequently replaced and that may be required in different amounts from one growing season to the next come under this category. Feed, labor, chemicals, and utility expenses fall into the realm of variable costs. Other variable costs include depreciation and interest.

FIXED COSTS

Fixed-cost items associated with a pond culture system are presented in Table 10. No actual cost figures are presented because they vary from place to place and with time. Actual costs should be determined at the time the prospectus is being developed. That will require contacting a number of suppliers and contractors. Those contacts should be documented in the prospectus so the potential investor can be assured that the figures are accurate. A serious investor may go so far as to verify some or all of the figures independently.

Each of the fixed-cost items listed in Table 10 are discussed below. More complete discussion of many of these items is included, as appropriate, in the chapters that follow.

TABLE 10. Fixed Costs Associated with a Pond Aquaculture Venture (Adapted from Giachelli et al. 1982)

Land acquisition	Disease and weed control
Pond Construction Earth moving Drainage system Gravel Vegetative cover	Other Equipment Boat, motor, and trailer Tractor(s) and accessories Truck(s) Shop equipment Water chemistry testing equipment
Water Supply Wells Pipe Pump	Support building(s)
Feeding System Feeder Bulk storage	

Land

In many cases, the land to be used for an aquaculture facility will have to be purchased. The total cost of investment will be considerably reduced if the land is already owned by the prospective aquaculturist. If a marine net-pen culture facility were being proposed, there would be no cost for land (though there might be a lease fee), nor would there be modifications to the site in terms of earth moving. There would be a significant startup cost for purchase of netpens, and those would be depreciated over time. Land cost might also be less for a raceway or indoor culture system (depending on location) than for a pond system. Again, the cost of constructing facilities would be proportionally higher on a unit area basis for the more intensive types of culture systems.

Pond Construction

Earth Moving. Assuming the land has not been improved, ponds will have to be constructed. The cost will vary depending on the type of equipment used, size of the individual ponds, and the slope of the land. In some cases it may be necessary to move earth significant distances on the site, to bring in additional earth, or to haul excess earth away. Each of those activities adds to the cost of construction.

Drainage System. Properly constructed ponds must be drainable. Various types of drainage structures and the associated plumbing can be utilized, and costs can differ significantly depending on the type of system selected. Drain lines may run entirely underground or ponds may drain into an open canal with the canal option being generally less expensive.

Gravel. Each pond should be accessible by an all-weather road. To ensure that the roadway can be driven on year round, it should be covered with gravel.

Vegetative Cover. Exposed pond levees need to be covered with vegetation to keep them from eroding. Various types of grasses can be planted; suitable types vary with location.

Water Supply

Wells. While other sources of water can be used, many pond culturists rely on wells. The cost of drilling a well and producing water will vary depending on the type of material that is being drilled into, the depth of the well, the diameter of the casing, and whether the water is artesian or has to be pumped to the surface. The number of wells required will depend on total volume requirements and the water-producing capacity of each well.

Pipe. Delivering water from the well(s) to the ponds is usually by way of buried pipelines, though open trenches are an option. The amount of pipe and its diameter will affect the cost. Large diameter pipe requires larger valves, and while increasing the diameter of pipes does not cause a major increase in cost, increasing the size of valves can mean order-of-magnitude price increases in some cases.

Pumps. Unless artesian flow from the well and gravity flow from the wellhead to the ponds can be utilized, water will have to be pumped from the well to the ponds. Pumps vary in cost depending on size. Pump size needed is a function of the amount of head against which the equipment must work and the volume of flow that is required.

Feeding System

Feeder. While it is possible to feed ponds by hand, large operations typically use feed broadcasting equipment. Feed blowers that can be pulled behind a tractor and work off the power takeoff are commonly employed by catfish farmers. Boat-mounted feeding devices can also be used. Automatic or demand feeders are often used in conjunction with a variety of types of water systems.

Bulk Storage. Feed can be purchased in bags or in bulk. Bagged feed is more expensive but may be desired for small operations. Bulk feed is almost essential for large culture operations. Bulk storage bins that can store several tons of feed are required.

Disease and Weed Control

Certain chemicals used in the control of diseases[19] can be applied to the water through a sprayer, as can herbicides. For small ponds, backpack sprayers can be used, but when large amounts of water are going to be treated, a tractor-mounted sprayer is required.

Other Equipment

Boat, Motor, and Trailer. Aquaculture operations with large ponds will find many uses for a boat. In addition to providing a platform from which feed can be distributed, a boat can be used in the application of chemicals and to provide access to various parts of the pond for collecting water quality data. A motor is required unless the ponds are small enough that rowing is an acceptable method of motivation. Depending on the size of the boat, a trailer may be needed to transport it from pond to pond.

Tractor(s) and Accessories. A tractor is one of the most important items of equipment on any fish farm, and can be the most expensive equipment item purchased. The size and number of tractors needed will vary considerably depending on the size of the aquaculture operation and the uses to which the vehicles will be put. Tractors can be used to distribute feed, power aeration equipment, spray chemicals, level and smooth roads, move earth, and mow. An aquaculturist might have two or more tractors that are used for specific tasks and are sized accordingly.

Accessories such as a scraper blade, frontend loader, and various types of mowing attachments will be required. In most cases it makes sense to purchase equipment for which the accessories are interchangeable.

Truck(s). Every aquaculture site requires a pickup truck at the very least. The truck will be used to haul equipment and supplies around the farm, transport items to and from the farm from suppliers, and possibly for the transport of the crop to the processing plant. Some aquaculturists own dedicated hauling trucks, though others depend on custom hauling firms.

Shop Equipment. A well-equipped aquaculture shop will have a full set of tools for mechanics, plumbing, and carpentry. Power tools such as saws and a drill press will find almost continuous use. A cutting torch and welding machine will also find frequent use. The shop should also have an air compressor.

Aeration Equipment. Various types of emergency aeration equipment is available from commercial sources, though some aquaculturists prefer to design and build their own. Such equipment can be operated from the power takeoff of a tractor, or

---

[19] Throughout this book the term "disease" is meant to include all pathogenic organisms that can attack aquatic organisms: viruses, bacteria, fungi, and parasites.

it may be driven by electric, gasoline, or diesel motors. Liquid or bottled oxygen aeration systems have been used with some types of culture systems but are not generally used in conjunction with ponds.

Hauling Tank. Hauling tanks can be permanently mounted on trucks or can be of the type that is placed on the truck only when needed. The tanks may or may not be insulated and can be used with various types of aeration systems. Every pond culture operation should, at the least, have a hauling tank that will fit in the bed of a pickup truck.

Emergency Equipment. Pond systems are usually constructed in rural areas and are often located in places that are subjected to intense thunderstorms during parts of the year. Thus, power outages can occur, often with frustrating frequency. Emergency power should be available in the form of one or more generators powered by diesel or gasoline motors. Emergency power to operate pumps and aeration equipment is sometimes critical for pond culture systems. For intensive systems, emergency power is critical. Loss of power in some systems can result in loss of a crop, sometimes within minutes.

Support Building(s)

Another significant cost can accrue in the form of buildings. At the minimum, a small combination office, shop, chemistry laboratory, and diagnostic facility will be required. Garages for storing vehicles may also be desired. For aquaculturists who plan to spawn broodstock and rear juveniles for stocking, a hatchery building may be required. The types of material employed for construction, size of the buildings, and other factors will influence the costs involved.

OPERATING COSTS

The important operating costs associated with a pond aquaculture facility are listed in Table 11. Each item is briefly discussed below. Once again, many of the items are covered more completely in other chapters.

Depreciation

Equipment items such as pumps, motors, tractors, trucks, boats, and so forth are routinely depreciated over a period of a few years. At the end of the period, typically no more than 7 years, the value has been depreciated to zero. While there is some actual value to many items (such as tractors, trucks, and boats), and there may even be several years of useful service left, it is generally assumed that the equipment will be replaced once it has been depreciated. Whatever value is still in the item may be used as a trade-in, thereby reducing the initial cost of the replacement item.

TABLE 11. Operating Costs Associated with a Typical Catfish Pond Culture Operation (Adapted from Giachelli et al. (1982)

Depreciation	Chemicals
Interest	Fingerlings
Taxes and Insurance	Feed
Repairs and Maintenance	Miscellaneous supplies
Energy	Labor
Chemicals	Management
Fingerlings	Seasonal
	Harvesting and hauling

Ponds and other types of culture chambers can also be depreciated. The depreciation schedule for culture chambers varies depending on type. Ponds, for example, could be placed on a 15 or 20 year schedule, while net pens may be on a 5 year schedule or even less.

Interest

Interest payments will be due annually on both short- and long-term loans. Land may be purchased on the basis of a 15, 30, or even 40 year mortgage. Construction loans are typically shorter-term notes with higher interest rates. In addition, the aquaculturist may have to obtain an operating loan to make purchases and pay staff until the crop comes in. Such loans are often paid off annually, and a new loan is initiated as needed.

Taxes and Insurance

Various types of taxes will be paid during the course of each year by the aquaculturist. They include property taxes on the land, and in most states on motor vehicles and other types of equipment. At the very least, sales taxes will be charged for each item purchased. Employers are also faced with payroll taxes.

The aquaculturist should carry liability insurance, automobile insurance, and insurance against fire and other hazards with respect to the equipment and buildings associated with the aquaculture operation. As an employer, the aquaculturist may provide health and disability insurance as part of the benefits package for the staff, or may share in the cost of such insurance with the employees.

Aquaculture crop insurance is available in many places, but the cost is quite high and most culturists do not carry that type of coverage. Problems that the insurance industry sees relative to aquaculture include the constantly evolving regulatory environment, particularly with respect to effluent standards, and the situation with respect to drug use (Secretan 1988). As we will see in Chapter 8, there are few approved drugs available for use by aquaculturists in many countries, though there are numerous diseases that can devastate a facility.

Repairs and Maintenance

Even if all the equipment is new at the time the operation is initiated, breakdowns will occur beginning in the first year. While it is not possible to accurately predict the costs that might be incurred because of equipment breakdowns, the culturist can predict normal maintenance schedules on mechanical equipment (e.g., oil changes for vehicles) and develop a budget for those activities. A contingency fund for unforeseen repairs should be added to the maintenance budget and might have to be increased annually as equipment ages.

Energy

Various sources of energy are used on the typical aquaculture facility. Electricity, propane, natural gas, gasoline, and diesel fuel may all play a role in the energy costs associated with operating a facility. Initial estimates on energy expenses may be somewhat rough but can be calculated using educated guesses and from past experience, as well as by discussing those costs with other aquaculturists or traditional farmers in the vicinity.[20] Once the farm is operational and a routine has become established, it should be somewhat easier to project energy expenditures in advance.

Chemicals

Herbicides, disease treatment chemicals, fertilizers, and chemicals used in association with obtaining water chemistry information are features of virtually every aquaculture enterprise. There are no hard and fast rules about what the annual cost might be for treating ponds with herbicides or for disease treatment. However, fairly accurate projections of fertilizer and water quality chemical needs can be calculated. Again, experience will provide guidelines once the operation has been ongoing for a few years.

Fingerlings

Depending on the strategy used, the aquaculturist may purchase fingerling fish or juvenile shellfish for stocking into growout ponds. Many culturists maintain broodstock and produce their own animals for stocking. In the latter instance, there should be consideration given to the costs involved in maintaining and spawning the broodstock, and the fixed costs should reflect the need for a hatchery building and associated equipment.

---

[20] In some places, the rate paid for water used for irrigation is less than the rate paid by other consumers. Sometimes aquaculture comes under the irrigation rate, and sometimes it does not. The situation varies from one place to another.

Feed

With some exceptions, such as oysters and clams, aquaculture species receive some type of feed from the culturist. Increasingly, prepared feeds are provided, though there continues to be some use of live foods either harvested from nature or grown specifically for animals being commercially reared. Many species, particularly fishes with small larvae and shrimp, are provided living food early in their life history and are later converted to prepared feed. Feed represents the major variable cost in many aquaculture ventures. Giachelli et al. (1982) and Lee (1991) presented budgets for channel catfish farms of various sizes. The percentage of annual operating cost associated with the purchase of feed ranged from slightly over 40% to nearly 60%, depending on which size of operation was selected.

Miscellaneous Supplies

Dip nets, buckets, brooms, mops, V-belts, nails, PVC[21] pipe, pipe fittings, paint. . . . The list of supply items that are in constant use and seem to constantly require replacement is almost endless. Hand tools never seem to get back to the workbench following use and may be discovered rusted beyond restoration lying in the grass several months after they turn up missing. There never seems to be a screw of the size needed available, so another trip to the hardware store may be required. A considerable supply fund is a virtual necessity.

Labor

In instances where the farm is owner operated, the owner should build in a realistic salary. Depending on the level of production associated with the operation, additional management and clerical staff may be required to oversee the business side of the activity. If the owner serves an investor and is not involved with the day-to-day operation of the facility, it is likely that proceeds will come from stock dividends or distribution of whatever profit is left over after all the bills have been paid at the conclusion of each fiscal year.

A small fish farm may be operated by the owner almost exclusively. Additional labor may be provided by family members as needed, or temporary hired labor may be employed to assist with harvesting, hatchery operations, and so forth. As the size of the operation and the level of technology being employed increase, the need for additional labor also increases. A large fish farm might have staff present 24 hr/day monitoring water quality. A high-intensity culture system that requires the continuous operation of pumps and other water handling and treatment equipment may require full-time maintenance expertise, or at least have someone always on call in case of a failure. Modern shrimp culture companies that engage in brood-

---

[21] Polyvinyl chloride (or PVC) plastic pipe is widely used by aquaculturists throughout much of the world. It is not toxic, can be cut with a hand saw, and its connections can be glued together (although threaded connections can also be used).

stock maturation, spawning, and larval rearing may employ several highly skilled technical people from algae culturists to geneticists. The expense associated with such people is considerably higher than for someone hired to mow levees and broadcast feed.

Harvesting and Hauling

Some aquatic farms have sufficient staff and resources to harvest their own crop. In addition, they may have trucks of appropriate size and capability to haul the fish to the processing plant either alive or on ice. Other farms do their own harvesting but contract shipment of the produce to the processor through a custom hauling contractor. In many instances additional labor may have to be employed during the harvest process, or the operator of the enterprise may contract with people who do custom harvesting. Custom harvesting and hauling firms are only available in regions that have a considerable amount of aquaculture activity.

MARKETING

If the aquaculturist sells directly to the public or to a retail outlet, there will a marketing component that must be considered as a part of the entire economic package. There are associations, such as the Trout Farmers of America and the Catfish Farmers of America, that engage in marketing activities on an industry-wide basis. The marketing efforts of such groups are usually supported by a self-imposed tax to the members that is added to a basic cost of doing business. For example, there may be a fee added to each ton of feed purchased. The revenue can go to support such activities as consumer surveys, advertising, and the development of new markets.

Lack of market information on the catfish industry has been viewed as a critical problem by researchers in the southern United States. A survey of consumer attitudes by Hatch (1988) provided demographic information that showed, not surprisingly, that people in the regions of the United States where catfish are commonly consumed rank farmed fish above wild catfish. The fact that people in New England have the lowest level of familiarity with the product, while also not surprising, provided marketing specialists with useful information for future ad campaigns.

Marketing surveys of aquaculture products, such as the one conducted on catfish by Hatch (1988) and another by Mims and Sullivan (1984), have only begun appearing in the last several years. Other such studies have been conducted with respect to the markets for salmon (Anderson 1988, Anonymous 1990), red drum (Haby 1990) and crawfish (Liao 1984, Roberts and Dellenbarger 1989).

While the United States remains a primary target for such luxury aquaculture goods as shrimp and salmon, Asia and Europe are actually the two largest markets for aquaculture products. The world market situation with respect to aquaculture has been examined by Ratafia and Purinton (1989).

The United States is a large net importer of fishery products, but this does not mean that there is no interest on the part of aquaculturists to become involved in the export market. Some export of crawfish to Europe has occurred, and the notion of exporting catfish has also been presented (Sindelar 1987), though no significant export market for catfish has yet been developed.

Aquaculurists often overlook the marketing aspect of their operation when they begin planning the venture, and few, if any, think about the role that advertising their product might play in increasing sales. Most fish farms do not advertise, and that is also true of other types of farmers. However, there are national and international advertising campaigns ongoing constantly that extol the attributes of one or another agricultural crop. Governments even become involved in promoting the agricultural products produced. The advertising of fish and other aquatic foods as healthy options to more traditional meats is a recent phenomenon.

In 1987, the catfish industry began an advertising campaign that was financed by voluntary farm contributions of $6.00 per ton of feed purchased. Kinnucan and Venkateswaran (1991) summarized the results of the first year of that campaign and reported that between $720,000 and $11,040,000 in additional income was generated to the farmers. That translated into $0.48 to $7.46 of profit from each dollar spent on advertising. Whether the campaign was a success or not cannot be predicted from those returns on investment, but the authors did find that consumer awareness was increased by 15%.

CROP FAILURE

As indicated above, most aquaculturists do not carry crop insurance, primarily because of the high cost associated with such coverage. The general situation relative to aquaculture insurance was reviewed by Secretan (1988), and Angus (1988) discussed the insuring of cultured salmon.

Largely anecdotal information reveals that, on the average, an aquaculturist can expect to experience a crop failure about once every 6 or 7 years. Some aquatic farmers have gone decades without a significant loss, while there are others who operated at a loss virtually every year they remained in business (which typically was not many unless the aquaculture activity was a part of a larger, profitable farming operation).

Crop losses can occur because of disease epizootics,[22] water quality degradation, or natural disasters such as hurricanes and tornadoes. Most disease epizootics and equipment failures do not completely devastate an aquaculture enterprise because they tend to impact only portions of the water system. A bacterial infection may occur in one pond out of 50, for example; or a pond might experience a dissolved-oxygen (DO) depletion that goes unnoticed until the fish are lost, though other ponds may not be affected. A pump failure in association with a recirculating water system can lead to loss of all the animals in the system if there is not an operational

---

[22] When a disease outbreak occurs in humans, it is known as an epidemic. The term ''epizootic'' refers to the same situation in lower animals.

emergency backup system in place and someone available to get it started in the event it is not automatic.

Hurricanes and tsunamis[23] tend to cause more widespread damage than tornadoes, though a fish farm that lies directly in the path of a tornado could be destroyed. During storms, ponds may be inundated allowing the crop to escape, levees can be eroded with the same result, buildings can be leveled, and equipment ruined by water damage. During major storms, people may have to evacuate for their safety, and power losses are routine and sometimes require extended periods of time to repair.

LITERATURE CITED

Aiken, D. 1989. The economics of salmon farming in the Bay of Fundy. World Aquacult., 20(3): 11–19.

Anderson, J. L. 1988. Current and future market for salmon in the United States. p. 6, In: Proceedings Aquaculture International Congress, Vancouver, British Columbia.

Angus, I. D. 1988. Insuring farmed salmonid stocks. pp. 143–148, In: Keller, S. (Ed.). Proceedings Fourth Alaska Aquaculture Conference, Nov. 18–21, 1987. Alaska Sea Grant Rep. No. 88–4, Anchorage.

Anonymous. 1980. FmHA aquaculture loans available. For fish farmers. Mississippi State University Cooperative Extension Service, Mississippi State. pp. 1–2.

Anonymous. 1990. Marketing Norwegian farmed salmon. Bull. Aquacult. Assoc. Can. 90(2): 18–19.

Aquafood Business Associates. 1987. Marine shrimp farming: a guide to feasibility study preparation. Aquafood Business Associates, Charleston, South Carolina. 90 p.

Avault, J. W., Jr. 1986. Which species to culture—a check list. Aquacult. Mag., 12(6): 41–44.

Bauer, L. L., P. A. Sandifer, T. I. J. Smith, and W. E. Jenkins. 1983. Economic feasibility of prawn *Macrobrachium* production in South Carolina, USA. Aquacult. Eng., 2: 181–201.

Bell, B. A. 1989. Making a good business plan (aquaculture economics). N.Z. Fish. Occas. Publ., 4: 87–88.

Buttner, J., and G. Flimlin. 1991. Is aquatic farming for you? Northeastern Regional Aquaculture Center, NRAC Factsheet No. 101. Southeastern Massachusetts University, North Dartmouth. 2 p.

Conte, F. S. 1992. Evaluation of a freshwater site for aquaculture potential. Western Regional Aquaculture Center Publication No. 92–101. University of Washington, Seattle. 33 p.

Dellenbarger, L. E., and L. R. Vandeveer. 1986. Economics of catfish production. La. Agric., 29(3): 4 ff.

---

[23] A tsunami is popularly known as a tidal wave. These extraordinarily large waves, generated by earthquakes under the seafloor rather than by tides, can inundate coastal regions with up to several meters of water.

Dorin, G. 1988. Financing aquaculture: some options available to the fish farmer. Can. Aquacult., 4: 53–55.

Fitzgerald, W. J., Jr. 1988. Comparative economics of four aquaculture species under monoculture and polyculture production in Guam. J. World Aquacult. Soc., 19: 132–142.

Giachelli, J. W., R. E. Coats, Jr., and J. E. Waldrop. 1982. Mississippi farm-raised catfish. Agricultural Economics Research Report No. 134, Mississippi Agricultural & Forestry Experiment State, Bauer et al. 1983, Mississippi State University, Mississippi State. 41 p.

Glude, J. B. 1983. Marketing and economics in relation to U.S. bivalve aquaculture. J. World Maricult. Soc., 14: 576–586.

Haby M. G. 1990. Marketing opportunities for farm-raised red drum. pp. 209–213, In: G. W. Chamberlain, R. J. Miget, and M. G. Haby (Eds.). Red drum aquaculture. Report of the Texas A&M University Sea Grant Program, College Station, Texas.

Hanson, J. S., W. L. Griffin, J. W. Richardson, and C. J. Nixon. 1986. Economic feasibility of shrimp farming in Texas: an investment analysis for semi-intensive pond grow-out. J. World Maricult. Soc., 16: 129–150.

Hatch, L. U. 1988. National survey of U.S. fish consumption. p. 7, In: Proceedings Aquaculture International Congress, Vancouver B.C, Sept. 6–9. Aquaculture International Congress, Vancouver.

Huang, H.-J., W. L. Griffin, and D. V. Aldrich. 1984. A preliminary economic feasibility analysis of a proposed commercial penaeid shrimp culture operation. J. World Maricult. Soc., 14: 95–105.

Kinnucan, H. W., and M. Venkateswaran. 1991. Economic effectiveness of advertising aquacultural products: the case of catfish. J. Appl. Aquacult., 1: 3–31.

Lee, J. S. 1991. Commercial catfish farming. Third edition. Interstate Publishers, Danville, Illinois. 338 p.

Liao, D. S. 1984. Market analysis for crawfish aquaculture in South Carolina. J. World Maricult. Soc., 14: 106–107.

Liao, D. S., and T. I. J. Smith. 1983. Economic analysis of small-scale prawn farming in South Carolina. J. World Maricult. Soc., 14: 441–450.

Lin, B.-H., M. Herrmann, and R. Mittelhammer. 1989. An economic analysis of the Pacific salmon industry: effects of salmon farming. Northwest Environ. J., 5: 160–161.

Mims, S. D., and G. M. Sullivan. 1984. Improving market coordination for development of an aquacultural industry: a case study of the catfish industry in Alabama. J. World Maricult. Soc., 14: 398–411.

Mitra, A., T. K. Nayak, and S. K. Sarkar. 1989. Social acceptance and economic return of tilapia culture in West Bengal. pp. 101–104, In: M. M. Joseph (Ed.). Workshop on Exotic Aquatic Species in India, Mangalore, India, Apr. 25–26, 1988. Special Publication of the Asian Fisheries Society, Indian Branch, No. 1, Bangalore, India.

Nerrie, B. L., L. U. Hatch, C. R. Engle, and R. O. Smitherman. 1990. The economics of intensifying catfish production: a production function analysis. J. World Aquacult. Soc., 21: 216–224.

Pagan-Font, F. A. 1981. Aquaculture: investment opportunities in Puerto Rico. Proc. Gulf Caribb. Fish. Inst., 34: 73–75.

Pomeroy, R. S., D. B. Luke, and T. Schwedler. 1987. The economics of catfish production in South Carolina. Aquacult. Mag., 13(1): 29.

Pretto, R. 1989. Economic and social aspects of penaeid shrimp culture in Latin America. AAAS Ann. Mtg. Abstr.: 97.

Ratafia, M., and Purinton, T. 1989. Emerging aquaculture markets. Aquacult. Mag., 15(4): 32–41.

Ridler, N. B., and M. Kabir. 1988. Economic aspects of salmon aquaculture in Atlantic Canada. Canadian Industry Report in Fisheries and Aquatic Sciences, No. 188. Department of Fisheries and Oceans, Ottawa. 17 p.

Roberts, K. J., and L. Dellenbarger. 1989. Louisiana crawfish product markets and marketing. J. Shellfish Res., 8: 303–307.

Samples, K. C., and P. S. Leung. 1985. The effect of production variability on financial risks of freshwater prawn farming in Hawaii. Can. J. Fish. Aquat. Sci., 42: 307–311.

Sandifer, P. A., and L. L. Bauer. 1985. A preliminary economic analysis for extensive and semi-intensive shrimp culture in South Carolina, U.S.A. p. 173, In: Y. Taki, J. H. Primavera, and J. A. Llobrera (Eds.). Proceedings, First International Conference on Culture of Penaeid Prawns/Shrimps, December 4–7, 1984, Iloilo City, Philippines. Southeast Asian Fisheries Development Center, Iloilo City.

Sandifer, P. A., T. I. J. Smith, and L. L. Bauer. 1982. Economic comparisons of stocking and marketing strategies for aquaculture of prawn *Macrobrachium rosenbergii* (de Man) in South Carolina, U.S.A. Symp. Ser. Mar. Biol. Assoc. India, 6: 12–18.

Secretan, P. A. D. 1988. Trends in aquaculture related to insurance. p. 19, In: Proceedings, Aquaculture International Congress, Vancouver B.C., Sept. 6–9. Aquaculture International Congress, Vancouver.

Shang, Y. C. 1990. Aquaculture economic analysis: an introduction. Advances in World Aquaculture, Vol. 2. World Aquaculture Society, Baton Rouge, Louisiana. 211 p.

Sharfstein, B. 1989. An overview of commercial shrimp culture in Latin America and the Caribbean. AAAS Ann. Mtg. Abstr.: 96.

Shigekawa, K. J., and S. H. Logan. 1986. Economic analysis of commercial hatchery production of sturgeon. Aquaculture, 51: 299–312.

Sindelar, S. 1987. The possibility of catfish exports: a look at Japan. Aquacult. Mag., 13(3): 24–26.

Waldrop, J. E., and J. G. Dillard. 1985. Economics (channel catfish culture). pp. 621–645, In: C. S. Tucker (Ed.). Channel catfish culture. Elsevier, New York.

Wyban, J. A., J. N. Sweeney, and R. A. Kanna. 1988. Shrimp yields and economic potential of intensive round pond system. J. World Aquacult. Soc., 19: 210–217.

3 Water Sources and Systems

GENERAL CONSIDERATIONS

Selection of the proper type of water system is a critical factor for the aquaculturist interested in optimizing production while making the best use of land and water resources. The aquaculturist may have some latitude in selection of a water system, but whatever its design is should provide the best possible conditions for maintenance of the species to be cultured while remaining economical. Various types of water systems have been employed by aquaculturists in attempts to maximize production while avoiding stress to the culture species. It is a basic principle of aquaculture that a healthy population of animals can be maintained only if they remain free from stress.

The water systems considered in this chapter are described in some detail, but no attempt has been made to provide all the engineering details that would be required for actual construction. The purpose here is to familiarize the reader with the basic types of water systems and how they function, as well as to present some of the characteristics of each system that need to be incorporated into the overall design.

While some aquaculturists have the training and ability to design and actually construct all or much of their water system, the majority seek professional assistance in design and construction. There are experienced well-drilling firms throughout the United States that have knowledge of the availability and quality of water in various strata and are familiar with the cost of obtaining water in the required amount. Firms with experience in the actual construction of water systems for aquaculture are less abundant, however. While many earth-moving companies have experience building livestock watering ponds, they may not have constructed aquaculture ponds, which are often quite different in character.[24]

With respect to water movement, aquaculture systems may be either static or flowing. The most widely used static systems use earthen ponds as culture units, though static concrete ponds and cages placed in nonflowing waters are other options. Some of the more common flowing water systems utilize tanks, raceways, cages in streams, or net pens in tidally flushed estuaries.

Static systems should have a reliable source of water available to them, since

---

[24] Most livestock ponds are constructed by forming a dam for the capture of runoff water, whereas most culture ponds do not depend on runoff. The major exception is in Alabama, where much of the catfish production comes from runoff ponds (John Grover, personal communication).

62

seepage and evaporation will result in the need to replace water intermittently. Such systems should also be provided with overflow structures that will allow excess water to exit without permitting the crop to escape. Heavy rainfalls and unusually high tides can inundate ponds, for example.

In flowing culture systems water continuously enters and exits the culture chambers. The turnover rate [25] can vary considerably depending on the type of system as well as the density and type of organisms present.

Flowing systems can be either open or closed. Open systems are those in which the water leaving the culture chambers is released into a receiving water body. [26] Closed systems are also called reuse systems, because the water exiting the culture chambers is recycled back after it receives some type of treatment to improve its quality. An intermediate type of system, of the partial recirculating type, features constant addition of new water along with reuse of some percentage of the water in the system. Often, the amount of new water is a small percentage each day of the total system volume.

Generally, extensive aquaculture is undertaken in static waters and intensive culture is conducted in flowing systems. Pond culture of channel catfish, for example, can produce 4,000 kg/ha/yr, while more than 10^6 kg/ha/yr can be produced in some types of intensive systems. Production figures such as these can be misleading since they are based on the total surface area in the culture chambers and not the total volume of water to which the aquaculture species is exposed during the growing season. In a static pond there may be only one or two volumes of water used to replace the annual losses to evaporation and seepage. In a typical 1 ha pond with an average depth of 1 m to which two volumes are added during the growing season [27] (2 added + 1 original = 3 volumes/yr), the fish would be produced in a total annual water volume of:

$$10,000 \text{ m}^2 \text{ area} \times 1 \text{ m depth} = 10,000 \text{ m}^3 \times 3 \text{ volumes/yr} = 30,000 \text{ m}^3/\text{yr}$$

A raceway system with 1 ha of total surface area and a mean depth of 1 m with a turnover rate of 4 exchanges/hr would expose the fish to a total annual water volume of:

$$10,000 \text{ m}^3 \times 4 \text{ exchanges/hr} \times 8,760 \text{ hr/yr} = 350,400,000 \text{ m}^3/\text{yr.}$$

The amount of water required per kilogram of production at 4,000 kg/ha/yr in ponds would be 7.5 m^3 (90,000/4,000). To produce each kilogram of fish in the raceway system at 10^6 kg/ha/yr (350,400,000/1,000,000) would require 350.4 m^3 of water.

---

[25] Turnover rate is defined as the time required to replace 100% of the volume of the culture chamber. It is usually measured in terms of minutes or hours.

[26] The effluent water can also be put to other uses, such as irrigation, though such applications are not common at the present time.

[27] Water additions are periodically required to replace evaporation and seepage losses.

Thus, the higher production rate requires nearly 50 times more water per kilogram of production ($350.4/7.5 = 46.7$).

Some people believe that the size of the culture chamber can be a major limiting factor to fish growth. In reality, if the water quality is properly maintained, it is at least theoretically possible to virtually fill the chamber with the species under culture. Growth is only retarded because something in the system becomes limiting.[28] Assuming plenty of food is available, the limiting factor that leads to reduction in growth rate is usually a water quality parameter; it is not generally associated with the size of the container in which the animals are being raised. Goldfish provide a good example.

Nearly every child has had a goldfish bowl. It is often placed beside the bed where the child can enjoy watching the fish. In the vast majority of instances, the child either neglects to feed the fish (in which case the fish will not grow because food is limiting), or as commonly, the fish will be fed excessively (with the excess food fouling the water). Water in the bowl is rarely, if ever changed, and the bowl may not even be filled regularly. Fish under such conditions remain small. If a flow of new water is maintained through the bowl and the proper amount of feed is provided, the fish should grow rapidly and may reach adult size in a container that can barely accommodate them.

The more intensive the aquaculture system, the more energy-dependent it tends to be. Up until the early 1970s the cost of heating and moving water was relatively low because of concomitantly low energy costs. That changed dramatically during the energy crisis of 1973. Subsequent energy cost increases have only served to exacerbate the problem. With rising energy costs, interest in improving the efficiency of intensive culture systems has grown, though technological problems remain to be solved before economical production systems become available.

Intensive culture systems have been developed that take advantage of inexpensive or free energy. The rainbow trout industry in Idaho developed in a region that features large volumes of spring water of nearly ideal temperature ($14.8°C$). Thus, there is no cost associated with pumping or temperature modification and the sizes of the various operations reflect the amount of water available at each particular site (Brannon and Klontz 1989). Also available in parts of Idaho and various other locations is geothermal water. In some instances geothermal artesian wells produce water of nearly ideal temperature for the production of warmwater species. Once again, the costs of both pumping and modifying temperature can be avoided. Use of the heated effluent from electrical power generating plants has long been touted for aquaculture, and a significant amount of research has been conducted. That topic is discussed in more detail in conjunction with the section on cage culture later in this chapter.

Intensive culture systems that require water pumping and the heating or cooling of water are not widely used by commercial culturists except during the early life-

---

[28] An alternative possibility is that the animals have reached a point in their life cycle where growth normally slows or ceases. In the case of most commercially produced aquaculture products, harvest and marketing occur while the animals are still capable of rapid growth.

history phases of the animals, when the amount of water being used and biomass present in the systems are both small. Recirculating systems have commonly been used by researchers working with juveniles as well as larvae. Even for replicated experiments, the total amount of water used is insignificant relative to that required for a commercial operation, and while often on a limited budget, the researcher is usually not under the same constraints as the commercial producer with respect to the demonstration of economic feasibility.

Water sources and the various types of water systems that have been developed for aquaculture are discussed in the remainder of this chapter. Since many of the features of one type of water system appear in other types, each is discussed in detail when first introduced and only mentioned briefly thereafter to indicate where that feature may have applicability in other types of water systems. The general designs presented here are not meant to represent specific plans but are provided to assist the reader in visualizing the concepts that are discussed. A brief tour of aquaculture facilities in virtually any region of the world will quickly demonstrate that no two systems are exactly alike but that there are many common design characteristics that can be found in any two systems of the same basic type. Even such systems as net pens that can be purchased from commercial suppliers tend to be laid out in different configurations to conform to the particular culture site.

WATER SOURCES AND PRETREATMENT

Water used by aquaculturists can come from a number of sources including municipal supplies, wells, natural and artificial water bodies, and surface runoff. Costs involved with the use of different water sources can vary significantly depending on the pretreatment that might be required. The quality and quantity of water available not only influences the size and type of water system that might be employed; it also may be an important factor used in determining the appropriate species for culture.

Municipal Water

A few commercial aquaculture ventures and various research facilities have been developed around municipal water supplies. In the United States and various other countries, municipal water is chlorinated to make it safe for human use. As discussed by Wheaton (1977), chlorination is achieved by adding chlorine gas (Cl_2), sodium or calcium hypochlorite [$Na(OCl)_2$ or $Ca(OCl)_2$], or chloramines (e.g., NH_2Cl). Hypochlorite ion reacts with water to form hypochlorous acid (Wheaton 1977):

$$OCl^- + H^+ \rightarrow HOCl$$

Whereas chlorine gas addition was the most common method of chlorination for many years, chloramine addition has now become widespread in municipal water systems (Blasiola 1984).

As any home aquarist knows, fish and other aquatic species are easily killed by chlorine. Adding aquarium fish to tap water without first dechlorinating it leads frequently to total mortality since municipal water may contain a few parts per million (ppm) of chlorine or chloramine, whereas LC_{50} acute toxicity levels[29] for total residual chlorine are typically well below 1 ppm (Kelley 1974, cited in Wheaton 1977, Ctiokeleuchai and Duangswasdi 1981, Hall et al. 1981, Wilde et al. 1983).[30] The same is true for shrimp (Ctiokeleuchai and Duangswasdi 1981). This last study, which tested both fish and shrimp, demonstrated that toxicity varies as a function of ambient water temperature. Resistance to chlorine in Pacific oysters (*Crassostrea gigas*) has been found to increase with increasing salinity and temperature (Chien and Chou 1989). The growth of cultured algae such as *Porphyra* is inhibited by chlorine at very low concentrations (Maruyama et al. 1988). The toxicity of hypochlorous acid and hypochlorite ions varies as a function of pH (Mattice et al. 1981).

Traditionally, water used for the home aquarium is allowed to stand for 24 hr or so, during which the chlorine will dissipate into the atmosphere. Aerating the water hastens chlorine removal. Chloramines cannot be removed by allowing water to stand or with the aid of aeration, nor is it convenient to precondition water to be used in aquaculture systems by allowing it to stand for several hours, particularly in intensive culture systems. Activated charcoal and sodium thiosulfate are commonly used in chlorine and chloramine removal. Charcoal filters and automatic metering systems for sodium thiosulfate are both commercially available. Charcoal filters require backwashing at intervals that vary depending on the size of the filter and the flow rate, and the charcoal needs to be reactivated or replaced periodically. Sodium thiosulfate is metered into the inflowing water as a liquid. If water flow is constant, no adjustment of the flow rate is required, though the tank that holds the thiosulfate solution must be refilled periodically.

The cost of dechlorination would have to be considered as a part of the balance sheet put together when planning a facility in which municipal water will be used. In addition, and generally much more expensive, is the cost associated with purchasing the water. Aquaculturists may be able to obtain a special rate but can still expect to pay dearly for water purchased from a municipality. Commercial ventures with flowthrough water systems may not be economically viable if they depend on a municipal water source.

Well Water

Aquaculture facilities are often sited in locations that have good supplies of groundwater, although the activities of competing users have limited the availability of

---

[29] Acute toxicity studies are typically conducted over periods of 24, 72, or 96 hr. The animals are exposed to various concentrations of the test chemical, and the concentration that produces 50% mortality over the test period is reported as the toxicity level (TL_{50}) or lethal concentration (LC_{50}) for 50% of the population.

[30] Most studies that have evaluated the toxicity of chlorine on aquatic animals were not conducted with aquaculture species, though some such species have been studied. The results demonstrate that residual chlorine in various forms is universally toxic to fish and other aquatic species.

groundwater in some regions where it was once abundant. For example, in eastern Washington, wheat farmers and others have expressed interest in the concept of growing fish in geothermal well water that would be used for irrigation after it passes from the culture chambers. The Washington Department of Natural Resources, which controls the number of wells and amount of water that can be pumped annually, has seen drops in some water tables and curtailed the drilling of new wells. In addition, the number of days per year that wells in some locations can be used is limited to the normal irrigating season, and the availability would not be extended to accommodate fish farming operations. A similar situation exists in parts of the Rio Grande Valley of Texas, where the existing groundwater supply has been fully allocated to terrestrial farmers, leaving none for aquaculture.

In regions of plentiful groundwater, various qualities may be available in vastly different quantities from several depths. Well logs from the immediate vicinity of the proposed aquaculture facility will provide information on the potential flow rates, water quality, and depths from which suitable water can be obtained. That type of information can be obtained from local well drillers or the local water board.

The needs with respect to volumes required per minute should be assessed before the culturist contacts a well-drilling firm. While estimates differ, most pond aquaculturists prefer having enough water available to fill the culture ponds within a few days. As a rule of thumb, it is desirable to have at least 150 l/min available for each hectare of water under culture.

A flowthrough system (tanks or raceways) should have sufficient volumes available to exchange the water in the system at least several times a day. The exchange rate required will be dependent to a large degree on the density and biomass of culture animals in the system and their tolerance to degraded water quality. Turnover rates of two or more times an hour may be required under some circumstances. Such systems require a constant supply of water 24 hr/day, 7 days a week. Pumping the volumes of water required for a large production facility can be quite expensive. The economics improve significantly if artesian water is available.

Closed water systems still require water for replacement of that lost due to evaporation, leakage, and in association with cleaning the various components of the system. An emergency supply of water should also be available in the event of water quality problems caused by equipment failure or attributable to some other cause.

Once the volume requirement for water is known, the well driller, working in conjunction with the aquaculturist, can decide which water stratum to drill into, determine the number of wells that will be needed, and size the diameters of the wells and the necessary pumps accordingly. A pond culturist may, for example, want at least two wells: one to produce the high volumes needed when ponds are being filled, and the second to be used to replace evaporation and seepage losses. The second well could be somewhat smaller and require a smaller pump; this would reduce overall operating costs for the facility, though it would require a larger initial investment as compared with providing only a single well. For large farms the volumes of water required may necessitate the drilling of two or more wells.

Most wells (Figure 4) are cased with pipe to prevent the inflow of water from strata above the one that has been selected for use. Screens are placed in the intake

Figure 4. A truck-mounted drilling rig.

zone to keep sand and other particulate matter out of the water that entering the intake. Pumps may be placed at the bottom of the well or aboveground. A variety of pump types and delivery systems have been discussed by Wheaton (1977). Costs of pumping increase significantly with increasing well diameter and depth. The type of material through which the well must be drilled also influences cost. It is more expensive to drill through rock than sand, for instance.

Once the well has been drilled, it should be test pumped (Figure 5) to determine the volume of flow that can be sustained. The process involves placing a large pump on the well and running it for several hours or even days to determine the amount

Figure 5. Test pumping a new well before determining the size of the pump to be installed.

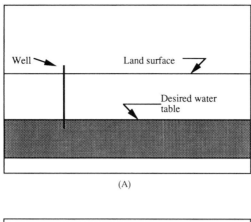

(A)

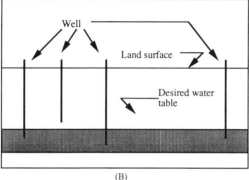

(B)

Figure 6. Hypothetical water table *(A)* with one well and *(B)* with several additional wells that have led to drawdown and the drying up of the original well.

the drawdown [31] that occurs at various rates of removal. The permanent pump can then be properly sized so a constant flow can be sustained.

The long-term sustainability of the water flow assumes that there are no seasonal changes in the availability from the stratum selected and that wells that are subsequently drilled in the vicinity do not cause a general drawdown of the water table, as can happen when water is pumped out of an aquifer more rapidly than it is replaced (Figure 6).

The drawdown of a water table can lead to subsidence of the land surface due to compression of the underlying materials which is caused by void spaces created when the water was removed. In coastal areas, the void spaces created by drawdowns of fresh water may be filled with salt water. Such saltwater intrusion can

---

[31] Water seeps through fractures in rock or through pore spaces in sand, gravel, and so on, to enter the well through the screen. At some pumping rate, more water will be pumped out than flows into the well; therefore the water level within the well will be reduced. This is called drawdown.

destroy the value of a well for domestic use, industrial applications, irrigation, and aquaculture.

Shallow wells, while generally less expensive than deep ones, can become contaminated with pollutants contained in recharge water that seeps down to the water table from the surface. Recharge water can carry biocides, fertilizer, and various other chemicals into the water table.

The temperature of the water pumped from increasingly deep water tables often tends to increase, and the difference between temperature in various water tables can be dramatic even within relatively modest ranges of depth. That is particularly true in geothermal areas where the earth's magma is able to come relatively close to the surface.

Saline wells can be drilled in some regions, and not only near the coast. The Pecos River of western Texas, for example, is brackish. The salinity of the groundwater is sufficiently high in some areas to grow marine organisms. The region, known as the Permian Basin, is a relict seabed, and the soil contains large amounts of salt. There are also extensive regions in other portions of Texas as well as in New Mexico and other states where wells yield salt water.

Saltwater wells are most commonly associated with coastal regions. It is often possible to obtain salt water from wells drilled adjacent to marine waters or by drilling into submerged sediments. Recharge of shallow saltwater wells may be from the overlying or adjacent marine or estuarine water, which is typically very pure after having been filtered through the surrounding sediments.

As it comes from the ground, well water may not provide the proper quality for the maintenance of aquaculture organisms. It is often depleted of oxygen and may have high levels of hydrogen sulfide, iron, or carbon dioxide.

Boyd (1990) indicated that any measurable level of hydrogen sulfide, H_2S, in culture water is probably detrimental to fish. Hydrogen sulfide can be oxidized with potassium permanganate, $KMnO_4$, or the pH can be raised by adding calcium carbonate, $CaCO_3$, in the form of limestone or oyster shell. The toxicity of H_2S increases as pH is reduced, so the more acid the medium, the more deadly the hydrogen sulfide.

Carbon dioxide levels of up to 10 mg/l appear to be acceptable to most fishes if the dissolved oxygen (DO) level is high (Boyd 1990). As indicated, well water may be depleted in oxygen. Also, well water may contain high levels of dissolved nitrogen gas, N_2. If the total volume of the gases present (adding oxygen, nitrogen, and carbon dioxide will give the total) exceeds 100% saturation (which depends on temperature and altitutude, primarily) by more than a few percent, gas bubble disease problems can result. The problem, similar to the bends experienced by divers, occurs when nitrogen comes out of solution in the tissues of the fish or invertebrates exposed to the supersaturated water.

Well water may have excessively high or low alkalinity or hardness and may even be high in ammonia. Aeration can be used to alleviate the problems of low DO and high hydrogen sulfide and carbon dioxide. Iron can be removed by a combination of aeration and filtration. Iron in wells depleted in DO is in the reduced state (Fe^{2+}). When exposed to oxygen, the iron is oxidized to ferric hydroxide, $Fe(OH_3)$, which precipitates. According to Boyd (1990), when carbonate is present, the reaction can occur as follows:

$$4Fe(HCO_3)_2 + 2 H_2O + O_2 \rightleftharpoons Fe(OH_3) \downarrow + 8CO_2$$

Low levels of iron may not be detrimental to aquaculture organisms, but the material will discolor pipes and equipment. At sufficiently high levels, it can lead to stress or be lethal to aquatic animals. Removal can be accomplished by sand filtration in the same manner that silt, clay, algae, and other particulate matter can be eliminated from incoming water.

There are chemical tests available for each of the chemicals mentioned along with hundreds of others. Such testing can be very expensive. For a quick and simple determination of the suitability of a water sample to support the aquaculture species of interest, merely introduce a few of the animals into a sample of the water and observe them for 24 hr or longer. Aeration should be provided to maintain DO.

While survival is, of course, critical, it is also important that the animals grow well. Unless there is some sublethal stress caused by the water, normal growth should occur. Such a stress might be improper salinity that leads to osmoregulatory imbalance. The ability of the water to support maturation and spawning may also be important. For example, a euryhaline species such as Atlantic salmon or red drum might survive and grow well in either salt or fresh water, but the former requires fresh water for proper egg development and larval survival, while the latter requires water with the proper range of salinity for the same activities.

Mariculture facilities should always have a source of significant amounts of fresh water. Fresh water should be used to wash down equipment exposed to the salt water. That can result in adding many years to the life expectancy of some items, such as vehicles. As importantly, evaporative losses from the culture system should be replaced with fresh water. When the water evaporates, it leaves the salts contained behind, thus increasing the salinity of the remaining water. By adding more salt water to replace evaporative loss, the culturist dilutes the salinity to some extent, but since more salt is being added, the net result is an increase from the initial salinity level. The more times salt water is added to replace evaporative losses, the higher the salinity becomes. Eventually, the salinity will rise to the extent that the water becomes stressful or even lethal to the culture animals.

Excessive fresh water resulting from direct heavy rainfall or surface freshwater runoff can significantly dilute the salinity in a mariculture facility to the point that osmoregulatory stress and even death can occur. Keeping rainfall out of ponds is not a simple matter and is not generally even a consideration, though proper construction can assure that runoff water will not enter ponds. Ponds or other types of facilities that have received excessive inflows of fresh water should be flushed with salt water until the salinity is returned to the proper range.

Surface Water

Surface water is used by aquaculturists around the world. Naturally occurring pond and lake water, stream and reservoir water, precipitation runoff, and the water in estuaries, bays, and the open ocean are all utilized for the rearing of aquatic species. Springs or locations where underground rivers erupt are often excellent sources of water for aquaculture. The trout farming industry in Idaho was established on that type of water supply. Coldwater streams, rivers, lakes, and marine waters are typically clear and chemically suitable for aquaculture. Surface waters with temperatures sufficiently high to support at least several months to a year of growth in warmwater species tend to be turbid but may otherwise be of excellent quality.

In order to take best advantage of rivers and streams that might be used by aquaculturists, the intake should be placed upstream of the outflow from the culture system. If the opposite technique is used, water that has been somewhat degraded within the culture system may be recycled back to the culture chambers before it can regain its quality through natural processes. Analogously, the intake to the culture system of one aquaculturist should not be placed immediately downstream from the outflow of a neighboring system. Sufficient downstream distance should be provided to allow the water quality to recover. What that distance should be depends on the quality of the water entering the stream from the culture facility, total stream flow, and a number of other conditions. Each situation should be individually evaluated to ensure that the water is of the proper quality at the point selected for inflow to the culture system.

A major factor in selecting a surface water source involves assurance that there will be sufficient water available throughout the year (Milne 1976). Further, the water needs to be available every year. Some surface water sources can become depleted seasonally or during drought years. Even if sufficient water is available for use by a particular aquaculturist, rationing may be imposed to ensure that all users have an opportunity for at least some water. Restrictions can mean that the aquaculturist has insufficient supplies to meet facilities needs.

Before an aquaculturist in the United States begins removing water from a surface source, the appropriate permits should be obtained. Laws governing the use of surface waters for aquaculture and the return of the water to a receiving water body vary from state to state. The federal government can become involved if the activity occurs in navigable waters. Having a permitting structure in place is to the distinct advantage of the aquaculturist, even though the time and expense involved in obtaining such permits may seem onerous. If such uses were unrestricted, anyone could move in upstream from an existing user and begin withdrawing water. If the amount of withdrawal were sufficient to curtail the activities of downstream users, there would be no recourse in the lack of regulations. A downstream user with a permit has legal recourse against an unauthorized user who moves in upstream or an upstream permit holder who exceeds the amount of water that has been allocated under that permit.

Lakes and reservoirs can be employed for cage culture systems, or the water in them can be flowed into ponds or through raceways. In some instances, water released from culture ponds or raceways has been recycled back to the lake or reservoir for subsequent reuse. In situations where continuous removal and replacement of water that is run through raceways is practiced, the water entering the lake or reservoir should be introduced as far from the outlet to the culture units as possible to allow time for the water to improve in quality. In such systems the volume of water flowed into the culture chambers on a continuous basis should be insignificant relative to the total volume of water in the overall system, unless the culturist is prepared to resort to some form of primary or secondary treatment as described in the section on recirculating water systems below.

When surface water is employed for aquaculture in agricultural regions, the water supply should be checked for the presence of contamination from biocides, fertilizers, and animals wastes.[32] Pesticides and herbicides pose the most immediate threats to aquaculture, though high levels of nutrients from organic and inorganic fertilizers can lead to problems if they promote the growth of undesirable aquatic plants. Surface waters may also be contaminated with domestic sewage in some regions, particularly in nations with poor sanitation.

FOREIGN SPECIES, PREDATORS, AND FOULING ORGANISMS

A significant problem associated with the use of surface water sources involves the introduction into culture chambers of unwanted organisms, including viruses, bacteria, parasites, algae and macrophytes,[33] predaceous insects, and fish that might compete with or prey upon the target culture species. Related problems are from predation by larger organisms that invade via water, land, or air, and from organisms that clog pipes and netting.

Aquatic Invaders

Springs and underground rivers are not a problem with respect to the introduction of undesirable organisms in instances where the culturist captures the water immediately or very soon after it emerges from the ground. Other surface water sources can pose problems, however. It is difficult to eliminate viruses, bacteria, and other pathogens without elaborate and expensive pretreatment facilities that employ sterilization with ultraviolet light, ozone, or chlorination-dechlorination. While surface water sources may carry a pathogen not otherwise present in the culture animals, all aquatic species harbor pathogenic organisms (which may cause no problem until stress occurs), and new ones may be introduced whenever new stock is brought to a facility from another location. Disease problems associated with the introduction of surface water have been reported, but aquatic vegetation, competitors, and predators are usually of greater concern.

In freshwater culture systems, centrarchids such as green sunfish are commonly present as invading species. Small carp, buffalo,[34] gizzard shad, and a variety of

---

[32] As discussed in Chapter 4, livestock manure is sometimes used as a source of fertilizer in aquaculture systems.

[33] Macrophytes are defined here as macroscopic plants and seaweeds (which are algae). The term ''algae'' as used here refers to phytoplankton, filamentous algae, and periphyton (benthic algae).

[34] The term ''buffalo'' refers to a type of fish. Bigmouth buffalo is one species that has received some aquaculture attention.

Figure 7. A simple wooden box with a nylon or metal screen bottom can be utilized to remove small fish and other organisms from incoming water supplies.

other species can also survive passage through pumps and the associated plumbing to enter ponds and other types of culture chambers. Such species can become problems in ponds where they compete for food and oxygen. Since it is not often possible to selectively remove such species, prefiltration of the incoming water is often employed. This can be done by passing the water over a fine-meshed screen (Figure 7). Such structures should be cleaned frequently because they tend to quickly become clogged not only with undesirable fish, but with insects, molluscs, vegetation, and various types of debris.

It is extremely difficult to keep competitors and predators out of cages. Net pens in marine environments often attract small fish that enter through the mesh of the pens and grow sufficiently that they cannot escape. They may compete directly with salmon for feed, though they can also serve as a supplemental food source for the culture species. Otters, which may be considered as primarily aquatic, have been known to climb over the tops of net pens or tear the netting, after which they may prey directly on the culture species or allow them to escape. Large predator nets of material that cannot be breached and that extends well above the waterline have been effectively used around marine net pens. Such measures add to the expense of culture and should be considered during budget preparation. Marine mammalian predators are typically protected by laws from shooting or other lethal measures. Stiff fines and jail sentences can be imposed for violations.

Water snakes and turtles seem to be ubiquitous around culture ponds. Such animals can wreck havoc among the culture species and have been known to bite people who are involved in feeding or harvesting operations. Some species, such as snapping turtles and cottonmouths (water moccasins), are particularly dangerous. In most cases it is permissible to shoot or trap and dispose of such pests.

Terrestrial Mammals, Birds, and Poachers

Raccoons and other terrestrial mammals have been known to remove the tops of cages. Even domestic cats have been found attempting to fish in culture tanks. Wading birds can do significant damage. Herons have been known to pierce large broodfish with their bills. While there is no way a heron can consume a several-kilogram brood catfish, the fish is usually killed as a result of the attack. Various other species of birds prey on fish in various types of culture systems. Most of the bird species that cause problems in the United States are protected under law from control by lethal means. Noise cannons are generally ineffective after a few hours or days, though the stringing of wires over culture chambers has worked well in some cases, and bird netting will work, though it can be expensive and is a bit of a nuisance to work around. A great deal of interest exists in controlling predation in aquaculture, particularly bird predation, but the problem persists and has, in some cases, had significant economic consequences for certain components of the industry, for example, the catfish farming sector in the southern United States.

Two-legged human predators, or poachers, can also pose a significant problem. Some farmers have had to resort to keeping guards on watch 24 hr/day to dissuade poachers from making off with the crop. It is often not practical to install high fences around a large culture facility. Perimeter lighting is also often not a practical or economical solution to the problem. Net-pen and cage sites are particularly difficult to defend against poachers unless there is a human presence on site at all times. Poachers are generally not considered to be ruthless criminals, except by the aquaculturist whose stock is being poached, and the threat of some relatively insignificant punishment by the courts does not offer much in the way of deterrence.

Fouling Organisms

A variety of organisms grow attached to substrates. Some are desirable and even encouraged; an example is the bacteria that colonize biofilters as discussed later in this chapter. Others, such as mussels and oysters, may be the primary species under culture. Those that become a nuisance or actually lead to degradation of conditions within the culture system are known collectively as fouling organisms. Those plants and animals contribute to a situation known as biofouling. Mats of algae, barnacles, bryozoans, sponges, sea anemones, tunicates, and various other types of plants and animals are examples of fouling organisms.

Biofouling is not restricted to the marine environment. Bryozoans and freshwater sponges have been found on cages in freshwater culture situations. Bryozoans growing in pipelines at the Aquaculture Research Center[35] of Texas A&M Univer-

---

[35] References to the Aquaculture Research Center reflect personal observations and experiences of the author.

sity were a significant problem during the summer months. Water in affected pipelines was pumped from the Brazos River into a holding reservoir before being distributed to ponds and laboratories. The bryozoans depleted the oxygen in the water, clogged pipes, valves, and flow regulators, and led to the production of organic debris and hydrogen sulfide when they died and decomposed. A chlorine solution was flushed through affected pipes periodically to keep bryozoan growth under control.

Growth of fouling organisms in freshwater pipelines is not a generally a problem, particularly when well or municipal water sources are employed. In many marine culture systems, water is taken directly from a bay or estuary for use in culture chambers. Pipelines conducting the water to the culture system are continuously being colonized by fouling organisms. Flow can be severely restricted and water quality degraded as a result. Many mariculture facilities utilize dual plumbing systems, each having a one-way or antibackflow valve at the intake end to keep water from draining from the pipeline when the pump is not running. Fresh water is typically pumped into the line that is not in use and allowed to stand. The fresh water kills the marine organisms that have attached to the pipe. Before the switchover from one pipeline to another is made, the dead fouling organisms and fresh water are flushed from the idle pipeline. The time between switchovers can be varied depending on the rate of fouling organism colonization, or a routine schedule may be established. Typical switchover times are once or twice a month.

Net pens and cages (discussed in detail in a later section of this chapter) can experience severe fouling problems. When fouling occurs, water flow through the culture chambers is restricted and water quality problems can result. The problem tends to be much more severe in the marine environment than in fresh water, and some regions have more serious problems than others. Net pens in Puget Sound, Washington, used for both research and commercial production do not quickly foul to the point that flow through the netting is greatly reduced.[36] In contrast, cages used in research conducted in Galveston Bay, near Baytown, Texas in the 1970s became severely fouled with barnacles and various other animals within a few weeks.[37] Fish being reared in those cages had to be moved every 2 or 3 wk into a new cage. The original cage was allowed to dry, and the fouling organisms were scraped off so the cage could be reused. The activity was labor-intensive and time-consuming.

Commercial scale net pens are sufficiently large that the frequent transfer of fish from one pen to another is not practical. A common solution in net pens that are subjected to frequent fouling is to insert a second layer of netting before the fouled netting is removed. The fouled net is cleaned and cycled back into use as the replacement net becomes fouled.

In addition to clogging the openings in cage or net-pen mesh, fouling organisms can significantly increase the weight of the culture chamber and can lead to sinking of the enclosure or failure of the mesh that will allow the culture animals to escape.

---

[36] Personal observations of the author.
[37] Nick C. Parker, personal communication.

Galvanized welded wire has been shown superior to such materials as nylon and polypropylene in resisting the settling of fouling organisms (Milne 1976); however, galvanized wire will eventually corrode. Wire is available that is coated with plastic and thus will not corrode as long as the plastic remains intact, but plastic is readily colonized by fouling organisms. Materials that are impregnated with copper, or alloys high in copper, have been used to reduce fouling. Copper alloy wire is available for cages, but caution must be exercised to avoid copper toxicity to the culture organisms. Also, the use of copper as an antifouling substance in paints and other marine applications has been increasingly restricted because of potential negative impacts on the overall environment.

One innovation in cage design that has promise in areas where biofouling is severe involves rotation in place. Cages have been designed to be turned to expose a different side to the air at frequent intervals. Fouling organisms on the exposed side desiccate and die. A typical rotatable cage might be turned about every 3 days to expose a new side to the air.

SUSPENDED SEDIMENTS

A frequent and often significant problem associated with the use of surface water in aquaculture, particularly in warmwater areas, is the presence of suspended sediments (clay, silt, or fine sand). In sufficient concentration, sediments will deposit in ponds and raceways to the point that the amount of water volume available can be reduced considerably. For example, at a governmental laboratory in Trishuli, Nepal, water is pumped from a river into a holding reservoir before being flowed into a raceway system. Even after being allowed to stand in the reservoir the water contains enough fine sand in suspension that canals leading to the raceways and the upper ends of the raceways themselves quickly fill with the material.[38] Periodic removal is required to keep the system functioning.

While the Nepal situation is an exception, problems with suspended solids are not uncommon. Direct effects on culture species can be burial of benthic organisms such as oysters and mussels, and damage to the gills of various aquatic organisms. For visual feeders, the ability to find and ingest feed may be impaired. Some species (e.g., carp and channel catfish), can naturally be found in turbid water and perform well in water containing high suspended sediment loads.

MECHANICAL FILTRATION

Settling ponds or tanks are usually effective in removing large suspended materials such as sand (except, apparently, in the Nepal situation described above). If the settling chamber is large enough and residence time sufficiently long, silt and clay will also eventually settle out of the water column. For rapid removal of fine sedi-

---

[38] Personal observations of the author.

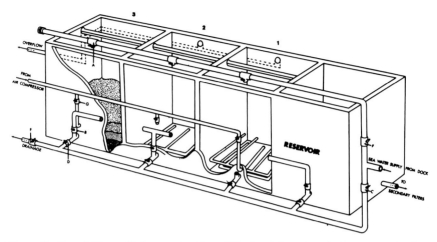

Figure 8. Gravel filter utilizing gravity flow. Filtration takes place in three separate chambers, allowing the culturist to backwash one or two without cutting off the water supply to the culture animals. Water enters through valves above the filter bays *(A)*, and excess water is allowed to overflow at points *1, 2,* and *3.* Valves *(B* through *F)* are used in backwashing (described by White et al. 1973). Original drawing by Daniel Perlmutter.

ments, mechanical filters are often employed. Mechanical filtration is also effective in removing most fouling organisms.

Sand and gravel filters have been used extensively; they are available from commercial sources or can be easily constructed to meet the needs of the particular culture system. Diatomaceous earth[39] filters are also commercially available and perform well, but they tend to become clogged more quickly than sand and gravel filters because of the very small particles that comprise the medium.

As the loading rate of suspended material increases, the rate at which filters become clogged also increases. Water should flow evenly through a mechanical filter, but as clogging of the medium occurs, the water will be forced into confined channels. The efficiency of the filters becomes greatly reduced when channeling occurs. Filters should be backwashed as necessary to remove the accumulated material and prevent channeling. The frequency of required backwashing will depend on the sediment load in the incoming water and the size of the filter. Typically, filtration systems are designed to require backwashing not more than once or twice per day.

Water can be filtered by allowing it to percolate through the filter medium by gravity (Figure 8) or by forcing it through the filter bed under pressure (Figures 9 and 10). Both techniques are effective, though the gravity method may require pumping the water twice: once to get it into the filter and a second time to carry it from the filter to the culture chambers. Pressurized systems typically require only

---

[39]Diatomaceous earth is comprised entirely of the tests (exoskeletons) of diatoms. The material is mined from ancient marine deposits that are found onshore.

Figure 9. A pressurized sand and gravel filter constructed from a steel shipping container. The tank is half-filled with pea gravel and sand. Water flows in the top under pressure, exits at the bottom (large valve), and flows into a wet laboratory located in the building (background). For backwashing, the small valves on the vertical pipes at right are opened, as is a small valve in the horizontal pipe between the vertical ones. Water enters the filter through the bottom by passing down the vertical pipe at the far right. The backwash water exits the top and flows to a drain through the vertical pipe to the left.

Figure 10. Commercially available sand filters designed for use in swimming pools can be readily adapted for use in aquaculture systems. Having two or more filters allows the culturist to backwash one while leaving others in operation.

one pump since the water is taken from its source and pushed through the filter and directly into the culture chambers. With a properly designed system, the same pump can be used both when filtering and backwashing. One or more valves can be used to route the water into the desired mode of operation. In some applications, pressure gauges located at the point of inflow to the filter and at a location downstream of the filter provide an indication of when a filter requires backwashing. A pressure drop between the upstream and downstream gauges signals that filtration efficiency is being compromised. The gauges come in handy particularly in situations where water turbidity changes periodically and a standard backwash schedule cannot be depended upon. Many filtration systems are operated at pressures between approximately 2.1 and 3.5 kg/cm^2. For filters that are not under pressure, the effluent volume will decline when inflow volume is constant and when the medium begins to clog.

Very fine particles (e.g., colloids) may pass through sand filters and later flocculate and settle in culture chambers. In ponds that is not a problem, but colloids can lead to turbid water and undesirable sediment accumulations in tanks and raceways, and is a particular problem in static systems or those having very slow exchange rates such as are often found in hatchery situations. Cartridge filters of various types and sizes are available through plumbing outlets and scientific or engineering supply houses. Such filters come in various pore sizes and can remove particles as small as 1 μm. Such filters impede water flow significantly and must be operated under relatively high pressure if volume is to be maintained. Cartridge filters are typically employed when relatively small volumes of water are required. The cartridges should be changed when they become clogged.

Fluidized bed filter technology has been introduced into aquaculture since about 1980. Fluidized beds are created by maintaining sufficient turbulence within a chamber to keep the medium (sand, ion exchange resins, etc.) in suspension so that the normally solid material behaves like a liquid. Fluidized beds can remove solids or adsorb dissolved nutrients when ion exchange resins are used as the medium. They can also serve as biofilters (Miller and Libey 1986).

RECIRCULATING WATER SYSTEMS

Recirculating or closed water systems continue to be used primarily for experimental work and for the rearing of larval organisms in commercial and research facilities. They are also used in public aquaria. The typical home aquarium is one type of a closed system. There have been many attempts to develop commercial closed systems, and a degree of success has been achieved. The popular press and aquaculture trade magazines have published many articles about systems that, their developers claim, work effectively and profitably. Such systems often do work. One can produce fish that demonstrate good growth rates and feed conversion ratios[40] and that are acceptable to consumers. However, at the time of this writing, there did not

---

[40] Food conversion ratio (FCR) is defined as dry weight of food offered per wet weight per unit time.

appear to be a profitable, entirely closed commercial system producing food-size aquatic animals in the United States.

Given increasing land prices, water shortages that are occurring and are projected to occur in the future, and the increasing governmental regulation on the effluents from aquaculture facilities, there are many who feel that closed water systems will be needed if aquaculture is to continue growing. Many innovations have occurred as a result of research and development, and ultimately, economically viable systems will be developed. In the meantime, claims of economic success should be looked upon with some skepticism until all the data are made available for close scrutiny.

The standard closed water system is typically housed in a building to help maintain a constant environment, though outdoor systems have been designed and operated. When closed systems are placed outdoors, operators tend to experience problems with the growth of unwanted algae throughout the system. Covers over the various components of the system will reduce or eliminate the problem by drastically reducing the level of incident light. Indoors, light levels are usually insufficient to support significant amounts of algae growth, though plants may be purposefully grown in hydroponic chambers for tertiary treatment (described below).

Basic Components

Some of the components of recirculating water systems are unique to closed and semiclosed systems, while others have broad application in aquaculture. Recirculating systems are generally comprised of three basic types of components: one or more settling chambers, a biological filter (or biofilter), and the culture chambers (Figure 11). The appearance of the various components and auxillary types of equipment that have been added to recirculating water systems vary considerably, but most systems operate in basically the same manner. As discussed by Lucchetti and Gray (1988), the functions of such systems include ammonia removal, disease and temperature control, aeration, and particulate removal.

Water leaving the culture chambers usually, but not always, is allowed to enter a primary settling chamber where solids are removed. The water is then passed into a biofilter, where bacteria detoxify ammonia and nitrite. The biofilter is typically followed by a secondary settling chamber where bacteria that sloughs off from the biological filter is removed. After secondary settling, the treated water is returned to the culture chambers.[41] Well-designed systems pump the water only once and utilize gravity as much as possible. If water is pumped more than once, it is necessary to balance the pumping rates precisely to keep portions of the system from either being pumped dry or caused to overflow.

Biofiltration is a form of secondary waste treatment, where the settling of solids represents primary treatment. Tertiary treatment, which removes nutrients like nitrates and may reduce the levels of trace elements and dissolved organics, has been added to some recirculating systems. Supplemental aeration is generally provided

---

[41] Mechanical filters have been used in the place of both primary and secondary settling chambers.

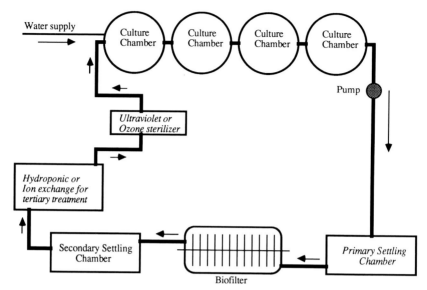

Figure 11. Schematic representation of a recirculating water system showing the three basic components: settling chambers, biofilter, and culture chambers. A pump is required to move water. Tertiary treatment and sterilization are sometimes added to such systems, along with other auxillary components.

in closed systems. Sterilization with ozone or ultraviolet radiation, foam removal, and other auxillary components have also been used. Backup systems for all motors and pumps are critical.

The various components may be quite large or rather small depending on the purpose for which the system has been designed. A closed system designed to produce 0.5 kg trout or catfish, for example, would in all likelihood have larger components than one designed for the rearing of shrimp larvae. Components may be physically separated or at least some may be combined (e.g., a settling chamber and biofilter could be placed within a single unit).[42] A single biofilter and associated settling tanks may service a number of culture chambers, or each culture chamber may be served by its individual biofilter and settling chambers. Various configurations have been designed for both freshwater and marine use. Some designs have employed materials as readily available as plastic pails and garbage cans, while others have been fabricated from concrete, stainless steel, or fiberglass. The use to which a system is to be put, the space into which the system is to be placed, and the resources available play roles in the materials chosen.

Materials used for construction not only of the culture chambers but also the

---

[42] An example is the home aquarium with an undergravel filter. The filter provides space at the bottom of the aquarium in which waste can accumulate, while the gravel resting on top of the filter provide a substrate for beneficial bacteria.

other components of closed systems and other types of water systems should be nontoxic to the culture animals. The same applies to paint used on any of the components. Metal should be avoided as much as possible because of potential toxicity. In saltwater systems, corrosion of metal can be a significant problem. Pumps with plastic impellers are recommended, though if operated continuously, metal impellers will hold up in salt water.

Plumbing in modern aquaculture systems is usually comprised of plastic pipe, usually PVC. PVC pipe comes in various sizes and strengths. Pipe designed for drains should not be used in pressurized systems. Pipe labeled Schedule 40 or Schedule 80 can be placed under pressure and can carry hot water.[43] PVC made for drains should not be used outdoors in climates where freezing temperatures occur unless it is buried. Drain PVC will shatter when water standing in it freezes.

Culture Chambers. Closed, semiclosed, and open culture systems typically employ relatively small culture chambers compared with ponds. Typical culture chambers are circular tanks (also called circular raceways), linear raceways, and silos.[44] The material used for construction of culture chambers can vary widely, depending on such factors as the preference of the culturist, the availability of materials, the cost, and in some cases the species to be cultured. The most commonly used materials are concrete, wood, fiberglass, various types of molded plastic, and sheet metal (usually aluminum or stainless steel, and usually used only in freshwater systems because of the corrosion problems associated with salt water).[45] Glass aquaria have been popular with some researchers and represent yet another type of construction material that can be employed.

Culture-chamber dimensions are quite variable, though most circular and rectangular tanks used in conjunction with closed systems are less than 10 m in diameter and seldom exceed 1 m in water depth (although the sides may be somewhat higher than 1 m). Commercial tanks are generally larger than those used in research.

Circular tanks have an advantage over rectangular culture chambers. When water inflow and drainage are properly designed, circular tanks are characterized by uniform water flow throughout. Raceways tend to have areas of static water in the corners where waste tends to accumulate. Circular tanks usually have center drains and water is typically introduced tangentially to the surface causing it to move clockwise or counterclockwise in the tank. The motion concentrates waste feed and fecal material around the drain, where it can be automatically removed if the proper type of drain is used. Some self-cleaning will occur in linear raceways that are

---

[43] Schedule 80 is significantly stronger and more expensive than Schedule 40 pipe. For most applications, Schedule 40 is acceptable.

[44] Silos are very tall circular raceways. There has been some research on them and limited use in commercial culture. For example, Buss et al. (1970) discussed the use of silos for trout production. The basic principles discussed with respect to circular tanks apply to silos.

[45] Stainless steel is relatively resistant to corrosion by salt water, but eventually it will fail. Galvanized metal is not recommended because of the potential for zinc and cadmium toxicity.

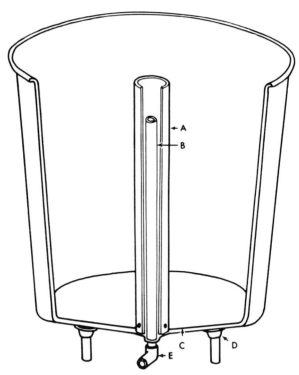

Figure 12. Cutaway of a circular raceway indicating the configuration of the venturi drain: *A*, outer standpipe with holes or slots at the bottom to allow waste to collect inside; *B*, inner standpipe which controls the water level and can be removed to quickly flush away the accumulated waste; *C*, tank bottom sloped toward the drain; *D*, legs sufficiently long to provide room for drainline; *E*, elbow connecting center drain to drainline.

equipped with venturi drains, and venturi drains are very efficient in removing particulate matter from circular tanks. A venturi drain system is shown in Figure 12. Waste material tends to collect between the two standpipes. Some of the material will flow out through the drain, and the remainder can be removed periodically by pulling the inner standpipe and allowing sufficient water to exit to remove the accumulated waste. The self-cleaning action is facilitated, particularly as the diameter of the tank increases, if the bottom of the tank slopes toward the drain (Figure 12). Venturi drains can also be placed on the outside of the tank.

Circular tanks usually drain either into pipelines or into a channel embedded in the floor of the building containing the system. Pipes are usually used when gravity is depended on to carry water to the next component in a recirculating system. Either pipes or floor drain channels can lead to a sump in which a pump moves the water to the next component.

In linear raceways, water typically enters at one end and is discharged through a

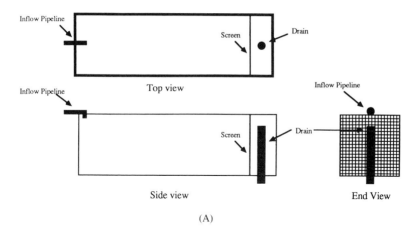

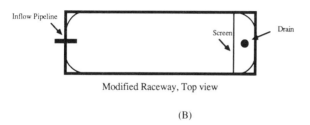

Figure 13. Raceways: *(A)* Top, side, and end view of a typical small linear raceway. *(B)* Top view of a similar chamber with the ends rounded to eliminate dead areas.

standpipe at the other (Figure 13). A venturi drain system like the type shown in Figure 12 can be used in conjunction with a linear raceway and will help in waste removal. Rounded corners can be used in conjunction with linear raceways to avoid dead spaces, but solids will tend to settle throughout much of the long axis of such chambers unless the flow rate is quite high. Screens are often placed in front of the standpipe in linear raceways to keep fish from being lost down standpipes that are not protected with an outside venturi pipe (Figure 13).

In tanks and raceways that have standpipes associated with the drains, water level is maintained in the event of a loss of inflow. Some designs employ bottom or siphon drains with balanced inflow. Depending on the design, such culture chambers may lose all their water in the event of a pump or power failure.

The exchange rate and manner in which water is introduced into raceways influences current velocity. Maintenance of some current not only is important in help-

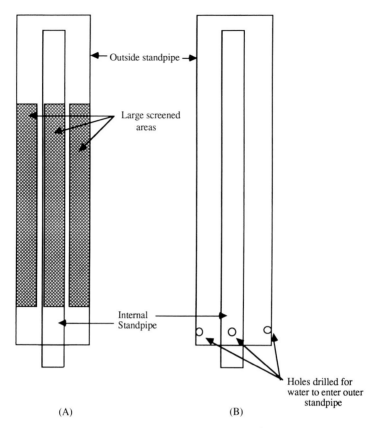

Outside standpipe

Large screened
areas

Internal
Standpipe

Holes drilled for
water to enter outer
standpipe

(A)   (B)

Figure 14. *(A)* Conceptual views of a circular raceway drain with large screened areas in the external standpipe through which exiting water can flow; this configuration might be used in larval fish or shrimp culture tanks. *(B)* The more typical configuration of a self-cleaning drain where water enters the outer standpipe through a few holes near the bottom.

ing flush waste products toward the drain; it may also benefit the culture species by accommodating normal behavior in the case of species that tend to orient themselves with respect to a current, such as trout. There are some who feel that the exercise associated with constant swimming into a current is beneficial in terms of final product quality and may stimulate more active feeding and growth, though scientific evidence is scanty. Too much current can exhaust the culture species, or in the case of larval and early juveniles may push them into screens and lead to heavy mortality. Employing a screen in front of the standpipe in a linear raceway (Figure 13) will prevent the loss of animals down the drain. In circular tanks, large areas associated with the outer standpipe can be screened (Figure 14) which will avoid concentrating the effluent into small areas through which accelerated flow will occur. A constant flow rate of 1 l/min passing through a total pore space of a few square centimeters in the typical venturi drain (Figure 14) will create much more current at the points of exit than when many times that much surface area is available (Figure 14a). The size of the screen mesh in linear raceways and in association with outside standpipes used in circular tanks should be sufficiently fine to prevent escapement of the animals but as large as possible to reduce clogging and the restriction of flow.

Culture tanks and raceways that employ venturi drains should be small enough to allow those drains to function properly. While the diameter of the standpipes can be increased as the size of the culture chamber is increased (often necessary to accommodate the increased volume of flow required to maintain larger raceways), in practice, outside standpipes larger than about 30 cm and inside standpipes larger than 10 cm in diameter are not normally used. Circular tanks larger than about 20 m in diameter may be impractical. Most are no larger than 10 m in diameter.

Raceways with screens placed in front of the standpipe area to prevent escapement can be quite large and can be fitted with two or more standpipes at the drain end if necessary to accommodate the flow. Indoor raceways used in hatcheries and research facilities tend to be relatively small. They are usually no more than 3 to 5 m wide and 25 to 50 m long. Many are less than 1 m wide and only 2 to 3 m long.

It should be possible to drain and refill culture tanks rapidly during harvesting and stocking. Those processes are facilitated when oversized inflow and drain lines are used. During normal operation, inflow lines can be fitted with flow regulation devices to control the amount of water entering. When the raceways are being filled, the flow regulators can be removed or bypassed to allow much higher flow rates. Similarly, while a large internal standpipe and drain lines may not carry anything like their design capacity during normal operation of the raceway system, when culture chambers are being drained, the job can be accomplished quite rapidly by having oversized plumbing. Also, it may be possible to drain several raceways simultaneously without exceeded the capacity of the drains to handle the increased water flow.

For large outdoor raceways, flow is geneally controlled with valves alone. In the laboratory and in the hatchery, it may be necessary or useful to maintain constant, limited volume and equal flows in a number of raceways. For that purpose, homemade flow regulators can be manufactured; commercially produced ones are also readily available through plumbing and irrigation supply houses. A flow regulator can be made by drilling a hole of the desired size through a PVC cap that is then glued onto the end of an inflow pipe. The flow rate through this type of flow regulator will vary if water pressure fluctuates.

Commercial flow regulators are designed to provide constant flow as long as reasonable water pressure is maintained. They adjust to changes in water pressure above some specified minimum. Such changes occur when the amount of water being used is changed, which occurs very commonly when raceways are being filled or cleaned, or when water is diverted to other uses such as filling hauling tanks or washing down the floor.

Some of the most effective commercial flow regulators are made from stainless steel and plastic. The metal ones can be screwed onto the end of a threaded PVC pipe, while the plastic ones are disc-shaped and can be placed into a fitting that will screw onto a faucet (also available in PVC). When the stainless-steel flow regulators

Figure 15. A circular tank of about 1.3 m diameter showing water entering at 3.8 l/min through a stainless-steel flow regulator. Note the turbulent aeration that results from the thin stream of water from the flow regulator. The tank is also equipped with a venturi drain.

are used, it is helpful to place a valve in the water line upstream from each flow regulator so the various raceways can be independently turned on and off. Either type of flow regulator can be easily removed when higher volumes of water are needed to fill the raceways.

Flow regulators come in a variety of sizes. Flow regulators that will deliver less than 1 l/min can be used with small raceways (e.g., 20 l aquaria), while larger units may require flow regulators that deliver several liters per minute. The flow rate through such regulators is controlled by the size of the opening through which the water passes.

Water exiting the types of flow regulators described above flows out in a thin stream under pressure. When that stream is allowed to run directly into the raceway, the incoming water is actually injected into the water within the culture chamber causing turbulence and consequent aeration (Figure 15). Saturated DO can often be maintained in the presence of high fish biomass when such flow control devices are employed. If the culturist wishes to reduce the turbulence, as might be necessary when larvae or sensitive juveniles are being cultured, a length of pipe or plastic tubing can be placed downstream of the flow regulator to dampen the flow.

Each recirculating water system may employ one or more culture chambers in conjunction with each of the other main components of the system. The effluent from several raceways may be collected in a common drain system and flowed into the settling chambers and biofilter before reentering the tanks, or each culture chamber may have its own set of other components (which will add significantly to the expense of construction and operation). In systems wherein a number of culture chambers employ a single set of settling chambers and a commercial biofilter, the latter components need to be properly sized to accommodate the total volumes of water and waste that the system will be required to handle. Advantages to building large settling chambers and biofilters include economical use of space, especially indoors; limitation on the number of backup components required; and reduction in the need to duplicate plumbing. In general, the cost of constructing a few large units is less than that incurred in conjunction with building several smaller ones with the same combined capacity.

When the effluent from several culture chambers is pooled for treatment, an equipment malfunction can result in complete mortality of the animals in the system. Total power failures can lead to serious consequences in any closed water system; however, the failure of a single pump or aerator would be less crucial to the producer if each unit had its own mechanical devices than if several culture chambers shared the same pump and aerator. Duplicating primary and backup systems is expensive, but so is the loss of one or more tanks of fish.

A disease outbreak in a recirculating water system with more than one culture chamber can rapidly spread. Further, if the proper precautions are not taken (as discussed in Chapter 8), diseases can be spread by the culturist from one closed system to another. Disease treatment is often difficult in closed systems because exposing the biofilter to chemicals can kill the beneficial bacteria that reside therein. Isolating the culture chambers for treatment is not generally feasible unless there is a sufficient volume of suitable quality water to put the culture chambers in a flowthrough mode during treatment.

Settling Chambers. In most closed aquaculture systems, effluent water from the culture chambers passes through one settling chamber before entering the biofilter and another settling chamber after leaving the biofilter. Mechanical filtration through sand or other types of filters has also been frequently used to remove particulate matter from culture tank and biofilter effluents in closed systems. Removal of particulates using a hydrocyclone device prior to biofiltration has also been found effective (Scott and Allard 1983, 1984).

A settling chamber is merely a tank in which the exchange rate is sufficiently slow that particulates will settle. In many such chambers, the water is flowed in and out at the top so the material that has already settled does not become resuspended (Figure 16). A valve should be located at the bottom of the settling chamber to provide a means by which settled material can be removed periodically. Sloping the bottom of the chamber to the drain will make solids removal more complete and require less water removal from the system. In some designs, the primary settling chamber is ignored or is incorporated as a part of the biofilter.

Solids removal with a primary settling chamber is important because it cuts down on the loading of the biofilter. The material that is removed from the settling chamber is nutrient-rich and can be used as fertilizer. The material is composed largely of feces (and pseudofeces when certain invertebrates are under culture), unconsumed feed, and bacterial floc.

Water leaving a biofilter will contain bacterial floc that sloughs off the biofilter medium. While bacteria will certainly grow on the walls of pipes and on the walls of culture chambers and settling basins, the bulk of the bacteria in the system is

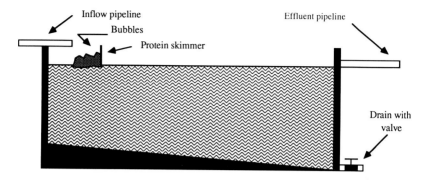

SECONDARY SETTLING CHAMBER

Figure 16. Cross-sectional side view of a settling chamber showing inflow and outflow, drain, and a simple protein skimmer.

contained within the biofilter, and most of the sloughing occurs in that part of the system. Removal of sloughed material in a secondary settling chamber helps keep the water in the culture chambers from becoming turbid.

Proteins dissolved in the water from various sources will form masses of bubbles when the water is sprayed from one of the units in the system to another. Protein skimmers have been designed to remove the bubbles, thereby also reducing loading on the system. A skimmer may simply be a vertical arm that confines the bubbles and causes them to spill over the sides of the settling chamber (Figure 16), or it may be more elaborate, such as a moving belt with paddles on it that pass over the surface of the settling chamber and push the bubbles over the side or into a collection area leading to a drain.

Biofilter Media and Function. Mechanical filters have been used on both experimental and commercial recirculating water systems in place of biofilters, but clogging and resultant channeling and poor filtration efficiency have produced serious water quality problems. Sand filters tend to clog rapidly when subjected to heavy loading. Gravel filters are less subject to clogging, but channeling is not uncommon if they are not frequently backwashed. When a mechanical filter becomes clogged, organic material, including bacteria and other organisms that have colonized the medium, will die and begin to decay. The filter can quickly become anaerobic. Not only does it lose the biological filtration activity that it may have once had; it will also begin to emit hydrogen sulfide, ammonia, and other toxic substances as well as release water that is either very low in or devoid of oxygen. Thus, the filter will begin doing just the opposite of what it was designed to do.

Many functioning biofilters have been designed using media that will not easily clog. Various types of plastics have been used, for example. For small-scale systems, pieces of plastic pipe cut into lengths of a few centimeters have worked well. Closed systems have employed Teflon rings, fiberglass, Styrofoam beads, and all sorts of plastics as media. Anything that bacteria will colonize can serve as a suitable medium for a biofilter. Commercial biofilter media are available as well. Those media are typically honeycombed sheets or cylinders of plastic with holes or slots cut in the sides and with protrusions of various kinds. In all cases, the void space is very large relative to the surface area of the material. Surface area for bacterial growth is enormous in sand or gravel filters compared with that of biofilters containing plastics, but the efficiency of the former tends to be very poor because of the clogging and channeling problem as previously described.

In recent years, advancements have been made in the utilization of more tightly packed media in biofilters. Ion exchange media (Horsch 1984, Hoergensen 1985) and the application of fluidized bed technology with activated charcoal (Paller and Lewis 1988) have been successfuly employed.

The major function of a biofilter is the nitrification of ammonia. Ammonia is excreted into the water by way of the gills of fishes and other organisms, and tends to be one of the most important natural toxic chemicals produced in water systems of all types. The forms of ammonia and their relative toxicities are discussed in more detail in Chapter 4. For now, it is only important to recognize that the nitrification of ammonia occurs in two steps, each undertaken by a different genus of bacteria:

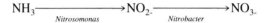

$$NH_3 \xrightarrow[\textit{Nitrosomonas}]{} NO_{2-} \xrightarrow[\textit{Nitrobacter}]{} NO_{3-}$$

The two genera of bacteria *(Nitrosomonas* and *Nitrobacter)* are aerobic and will only live and perform their function when there is oxygen in the water. When deprived of oxygen, even for a brief period, the bacteria will succumb and the biofilter will begin producing ammonia and nitrite. Aerobic conditions can be maintained by bubbling air, compressed oxygen, or liquid oxygen into the biofilter chamber. Splashing water into the biofilter or exposing the bacteria to oxygen periodically (see the section on rotating disc biofilters below) are also effective at maintaining aerobic conditions.

When a new recirculating system is put into use or when a system is restarted after having been harvested, drained, and cleaned, time must be allowed for colonization of the biofilter with the appropriate types of bacteria. At start-up of a commercial system, the biomass of aquatic animals carried in the system may be sufficiently small that water quality deterioration will not occur before the bacteria become active. However, in most instances it is best to ensure that the system is operating effectively before the aquatic organisms are added. This can be done by seeding the system with ammonia (e.g., in the form of ammonium salts), putting in a source of organic material that will deteriorate and form ammonia (e.g., prepared feed), stocking fish or other animals at low biomass, or some combination of those treatments.

The desirable bacteria are cosmopolitan and will soon colonize the biofilter without being inoculated. Commercial solutions containing bacteria are available. Reviews as to the effectiveness of those commercial preparations have been mixed.

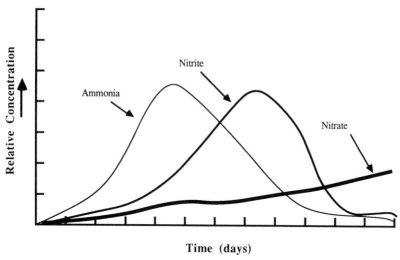

Time (days)

Figure 17. Typical pattern of ammonia, nitrite, and nitrate concentrations in a biofilter that is being colonized and becoming operational.

Typically, *Nitrosomonas* will become active from several hours to several days before *Nitrobacter*; thus, spikes in nitrite production can be expected before the biofilter becomes fully functional (Figure 17). The time required for colonization by biofilter bacteria will vary as a function not only of loading, but also temperature and salinity. More rapid colonization generally occurs in warm as compared with cold water. Marine culture systems require a much longer colonization period than freshwater systems (Nijhof and Bovendeur 1990). Typical colonization rates are from a few to several days in fresh water to in excess of a month in salt water. More rapid colonization has occurred in salt water, but the colonization rate is generally slower than in fresh water.

The sizing of biofilters relates to anticipated ammonia production and the efficiency of the system in converting the ammonia to nitrate (Lucchetti and Gray 1988). At neutral pH (7.0) the activity of the biofilter is a function of temperature and the available surface area (Speece 1973, Hess 1981). Speece (1973) produced a series of graphs that can help aquaculturists determine the amount of biofilter volume needed over a temperature range of 5 to 16°C when the food conversion ratio (FCR)[46] of the aquaculture animals and the surface area of the biofilter medium are known. According to Klontz (1981), flow rates through biofilters should range from 41.5 to 83 l/min.

Biofilters can be installed either indoors or out, but the internal portion should be protected from exposure to sunlight and bright artificial lights to prevent the growth of undesirable algae. Algal growth can lead to clogging of the biological

---

[46] FCR and food conversion efficiency (FCE) are discussed in Chapter 6.

filter and, if blue-green algae become established, there is the potential of off-flavors or even direct toxicity from certain metabolites produced by those organisms. The biofilter should also be protected from large temperature fluctuations, so outdoor filters should be insulated.

Four basic biofilter designs have received the attention of aquaculturists. They are trickling, submerged, updraft, and rotating biodisc filters (Figure 18). In addition, fluidized beds can act as biofilters. All of these designs are briefly discussed below.

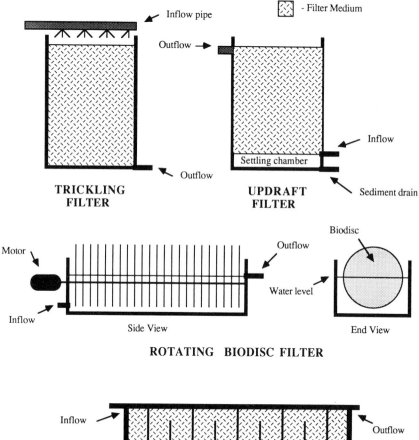

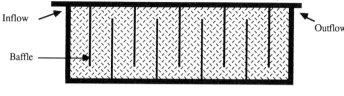

Figure 18. Diagrammatic representation of the four basic types of biofilters used in aquaculture.

Trickling Biofilters. Water enters trickling filters from the top and is allowed to pass by gravity through the filter at a rate that does not allow the medium to become submerged, although all internal portions of the filter are continuously wetted. Municipal sewage treatment plants often employ trickling filters with rock media. Those units are much larger than the ones used for aquaculture because the volume of waste from even small cities is dramatically higher than from even a large aquaculture facility. Whereas large trickling filters are equipped with rotating arms that spread incoming water evenly over the filter medium, most trickling filters that have been developed by aquaculturists have stationary water distribution systems because of their relatively small size (Figure 18).

Submerged Biofilters. The design of submerged biofilters is often similar to that of settling chambers, except that the submerged filter contains a medium on which bacteria become established. Water enters one end of the filter, flows through the medium, and exits from the opposite end in most designs (Figure 18).

Submerged biofilters can be operated by gravity flow or, with the incorporation of a watertight cover, water can be pushed through them under pressure. If pressure is used, inflow and outflow pipes can be at any desired height without danger of losing water because of overflow. In a gravity system, the inflow and outflow lines would logically be placed as shown in Figure 18.

Since the filter medium is constantly underwater in a submerged filter, it is necessary to ensure that sufficient aeration is provided to keep the filter aerobic. If the filter becomes anaerobic, ammonia will begin to be produced instead of eliminated.

Updraft Biofilters. An updraft filter is essentially the same as a submerged biofilter in that the medium is continuously submerged. The difference is that instead of the water moving horizontally through the filter, it enters from the bottom and exits near the top. A sedimentation chamber (primary settling chamber) can be incorporated into an updraft filter, thereby allowing the settling of solids below the influent pipe elevation (Figure 18). A drain valve at the base of the settling chamber allows solids to be removed as necessary. As is the case with submerged filters, those of the updraft variety need to be well oxygenated to maintain aerobic conditions. Alternatively, as described by Paller and Lewis (1982), an updraft biofilter can be "dewatered" periodically to place the bacteria in contact with atmospheric oxygen.

Rotating Biodisc Filters. A rotating biodisc filter or rotating biological contactor (RBC) utilizes a concept somewhat different from that of the filter types previously described. In this case the medium is moved through the water rather than the water being moved past the medium. Rotating biodisc media are composed of numerous circular plates placed on an axle and set in a trough with half of each disc submerged and half exposed to the atmosphere (Figure 18). Fiberglass is commonly used as disc material, though styrofoam and various types of plastic sheets have also been employed. The discs are rotated slowly (only a few revolutions per minute), usually powered by an electric motor and appropriate reduction gears. Bacteria colonize the

Figure 19. A commercial rotating biodisc filter that was used in conjunction with experiments conducted at the U.S. Fish and Wildlife Service laboratory in Marion, Alabama. The biofilter treated water from an underground silo in which various species of fish were cultured.

plates as in other types of biofilters. Alternating exposure to the metabolite-laden water in the trough and to the atmosphere provide the bacteria with a continuous supply of nutrients and oxygen. Early work with experimental rotating biodisc filters in conjunction with aquaculture systems was conducted by Lewis and Buynak (1976). Since then, various systems have been established (both research and commercial scale) employing commercially available rotating biodisc filter units (Figure 19).

Fluidized Beds. Fluidized beds can act as biofilters as previously indicated. Both aerobic and anaerobic fluidized bed biofilters have been used in aquaculture. Anaerobic fluidized beds can remove nitrate (van Rign and Rivera 1990), which, while normally not a problem, can reach undesirable levels in some instances. When nitrate levels reach several hundred parts per million in closed systems (such as has happened in large marine aquarium systems), remedial action may be required.

Given sufficient energy input, many solids can be made to behave as liquids. Fluidized beds have commonly employed sand or ion exchange resins, but they may also employ activated plastic beads, charcoal, limestone, or crushed oyster shell. Depending on the type of medium used, fluidized beds may act as biofilters, mechanical filters, or as a means of adding chemicals to the water.

Biofilter Size. The size of the biofilter required for a particular culture system depends on many factors. Reliable formulas for calculating the relative size required as a function of the number, kind, and biomass of animals in the culture chambers, total water volume in the system, and flow rate remain to be thoroughly developed. Early work on the problem was conducted by Parker and Simco (1973) and Davis

(1977). Speece (1973) and Liao and Mayo (1974) developed design criteria for water treatment in association with salmonid hatcheries. Hess (1981) provided formulas for use in the design of biofilters. Information on the design and performance of biofilters in a mariculture system was presented by Poxton et al. (1981).

Early in the growing season the size requirements for biofilters may be small, since the biomass in the system is low and the amount of waste to be treated is not great. As the culture animals increase in size, the efficiency of the biofilter may increase because proper loading has been reached and then decline as the quantity of waste exceeds the capacity of the biofilter. Additional biofiltration capacity is required when efficiency begins to drop off because of overloading. Routine water quality monitoring is required so the aquaculturist can determine when and if to alter the biofiltration capacity of the system. If a new biofilter is placed on line to augment an existing system, sufficient time for colonization of the medium in the new filter should be provided.

Maintenance of pH

The accumulation of dissolved chemicals in recirculating water systems leads to depression of the pH unless the system is buffered. As the water becomes more acid, stress is placed on the culture organisms, and if the pH becomes too low, death will eventually occur. Microorganisms colonizing the biofilter may also be adversely affected by low pH. Organic acids and carbon dioxide are the primary causes of increased acidity. Chapter 4 treats the chemistry associated with the buffering capacity of water and its control.

Wetzel (1975) noted that ammonia is strongly sorbed to particulate matter at high pH—another compelling reason for preventing the water from becoming acidic. Since bacteria are also associated with surfaces, the nitrification process may be enhanced when adsorbed ammonia and microorganisms are placed in close proximity on a waste particle or on the biofilter medium.

For most freshwater aquaculture systems, the pH should be in the vicinity of 7.0 (with an acceptable range of 6.5 to 8.5), while saltwater systems should be maintained at a pH in excess of 8.0 (reflecting the differences in normal pH between natural fresh and saline waters). To accomplish pH control, calcium carbonate $(CaCO_3)$ is often used as a buffering agent. This material may be in the form of crushed limestone, or more commonly, whole or crushed oyster shell. As hydrogen ions are produced in the system, calcium carbonate slowly dissolves. The carbonate ions remove hydrogen ions from solution to form bicarbonate:

$$H^+ + CO_3^{2-} \rightleftharpoons HCO_3^-$$

Limestone and oyster shell are relatively inexpensive and require little or no attention once incorporated into the water system. Bacterial mats and other types of particulate matter can build up on the surface of the buffering material and may interfere to some extent with dissolution of the calcium carbonate. Cleaning may be necessary at appropriate intervals to ensure that the buffering capacity of the system is maintained.

The amount of buffer material present in the biofilter is not particularly critical. Most culturists utilize a few kilograms per cubic meter of filter capacity. The buffering agent can be located nearly anywhere, but is commonly placed either in conjunction with the biofilter influent or effluent.

Moving Water

As the number mechanical devices in a recirculating system is increased, so are the chances of a failure. One high-quality continuous-duty water pump is often all that is required to move the water within such a system. The pump can be placed between any two components, except when an updraft filter is used, in which case it is desirable to place the pump on the influent side of the biofilter. In all types of systems, units downstream from the pump can obtain water through gravity flow assuming they are placed at appropriate elevations relative to one another. When more than one pump is utilized, the design should ensure that water flow is balanced among the components in the system and that there is no potential for a component to be pumped dry.

Pumps can be expected to run intermittently or continuously for up to several years if they are properly maintained. Submersible pumps that run while submerged in water are handy for some applications, and pumps with impellers that are not corroded by salt water are available for use in mariculture facilities. Pumps of various types were discussed by Wheaton (1977). Metal impellers will function well in salt water if the pump is allowed to run continuously, but trace metals may be released into the water. When present even at low concentrations, some such metals can be highly toxic, particularly to larvae and when used in closed systems where their concentration will continuously increase until the system is drained and refilled with new water.

The amount of water to be moved through a water system is related to the size of the system and the optimum flow rate for the culture organisms and the biofilter. Optimum flow rates for a given species may change depending on the life stage of the organism. For some species, few data are available and flows may have to be determined by trial and error. Turnover times in culture chambers may exceed 24 hr for larvae and may be several times an hour for juveniles.

Water can be moved with airlifts, such as those used in conjunction with submerged filters in home aquaria. Simple airlifts can be constructed from PVC pipe mounted horizontally in the water column (Figure 20). These devices can be used to move water within a culture chamber, as well as to move water between culture chambers. In addition to moving water, airlifts provide aeration. They can operate at relatively low pressures and are quite economical in certain situations. In at least one laboratory scale system, airlifts were used as the means of moving water for recirculation (Murray et al. 1981).

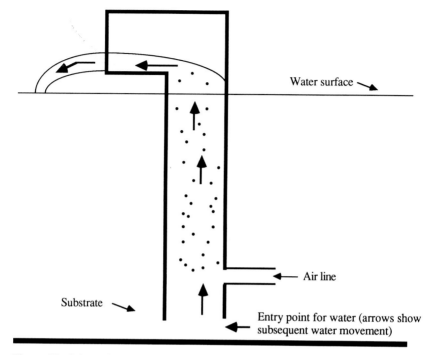

Figure 20. Schematic drawing of a simple airlift that will both move and aerate water.

Tertiary Treatment

There are three phases involved in complete water treatment. The first, which involves the settling of solids and their removal from the waste stream, is known as primary treatment. Secondary treatment involves transformation of toxic chemicals into less toxic forms and is accomplished by biofilters such as those described above. The removal of dissolved chemicals, such as nitrates and phosphates, from water is known as tertiary treatment. Tertiary treatment is rare in sewage treatment plants, which are typically designed for secondary treatment at best, though some tertiary treatment has been employed in aquaculture systems.

Nitrates, phosphates, and micronutrients accumulate in closed water systems even if a highly efficient biofilter is in operation. Although both nitrates and phosphates are required nutrients for plant growth and are not directly toxic to aquaculture animals when present at normal environmental levels, nitrate is known to be toxic at extremely high concentrations (Colt and Armstrong 1981). As reviewed by Colt and Armstrong (1981), the 96 hr LC50 value ranges between 1,000 and 3,000 mg/l for many aquatic animals. More recent information (Muir et al. 1991) indicates that nitrate may be acutely toxic to the larvae of *Penaeus monodon* at concentrations of as low as 1 mg/l. Much higher concentrations can be tolerated by juvenile shrimp such as *P. chinensis* (Chen et al. 1990), but the LC_{50} level was below 100 mg/l for time periods ranging from 24 to 192 hr. Thus, while nitrate has not generally been a problem for fish culturists, it may pose severe limitations with respect to the culture of invertebrates under certain circumstances, such as in recirculating water systems.

Nitrate can be removed from aquaculture systems through denitrification to elemental nitrogen (N_2) as discussed by Meade (1974) and Balderston and Sieburth (1976). Effective removal of both nitrates and phosphates can be achieved by incorporating an additional culture chamber in the system for the production of plants. The plants, either terrestrial or aquatic, might best be located immediately following either the biofilter or secondary settling chamber.

Aquatic plant candidates for tertiary treatment include water hyacinths *(Eichhornia crassipes)* and Chinese water chestnuts *(Eleocharis dulcis)*. Both grow rapidly, and each is efficient in removing dissolved nutrients from the water. Water chestnuts have economic value as human food. Water hyacinths, on the other hand, are generally considered a nuisance when they invade natural waters, but they do hold some potential as livestock feed, as do a large number of other aquatic plant species (Boyd 1968a, b). Since aquatic plants have very high water content, they can only be economically used as livestock feed when they do not have to be transported from the site of harvest. Drying the plants before shipment is not generally considered to be a practical or economical alternative.

An often more attractive type of tertiary treatment with plants involves hydroponics with terrestrial plants. The plants are grown either in water or in an inert medium that is continuously or intermittently wetted with nutrient-rich water. On an experimental level, various types of vegetables (including tomatoes, lettuce, and cucumbers) have been grown in association with the recirculated water of aquaculture systems in which several types of fish have been produced. Greenhouses have been used in conjunction with many such systems to provide the appropriate environmental conditions for the plants. While not a complete list by any means, information on hydroponic vegetable production can be found in the work of Naegel (1977), Lewis et al. (1978, 1981), Rakocy and Allison (1979), Pierce (1980), Sutton and Lewis (1982), Bender (1984), Wren (1984), McMurtry et al. (1986), Zweig (1986), Rakocy (1989a, b), and Rakocy et al. (1989). In all cases where plants are utilized, sufficient natural or artificial light must be provided to maintain active photosynthesis.

Maintaining the proper nutrient concentration for plants raised in conjunction with a recirculating water system is not a simple matter. The amounts of nutrients being produced and taken out of the system are not in equilibrium since both the aquaculture animals and the plants are continuously growing and will be harvested from time to time (but often not at the same time). In many instances it may actually be necessary to supplement the system with nutrients to help maintain plant growth. Macronutrients like phosphates and nitrates and various micronutrients such as trace metals may require supplementation.

Backup Systems and Auxiliary Apparatus

In theory, a closed system can be operated with only the components outlined in the above sections. In reality, redundancy needs to be built in so disaster can be avoided if, for example, there is a pump failure. In addition, various additional features can be added to help the aquaculturist avoid disease problems and increase the efficiency of the system.

Backup Systems. Each mechanical component associated with an intensive aquaculture system of whatever type should have a backup so that the integrity of the system can be maintained in the event of an equipment failure. Aeration is an auxiliary component featured in many water systems, but some aquaculturists depend on flow regulators to provide aeration. A backup pump should definitely be available, and most systems have a backup aeration system such as an air blower (Figure 21) or air compressor (Figure 22). Backup systems need not be identical to the primary system. An air compressor might, for example, take over for a failed blower; bottled gas, liquid oxygen, or agitators could also serve in backup roles. Many backup systems require electricity, so both primary and secondary systems can be lost during a power failure. A gasoline or diesel generator (Figure 23) or such a motor linked to an auxiliary pump should be available to provide power or water pumping capability during electrical outages.

Modern technology has dramatically improved the effectiveness of backup systems. State-of-the-art generators remain continuously warmed up and may even startup periodically for a few minutes to run self-diagnostics and ensure that they are operating properly. Autodialers can be used that will dial one or even several telephone numbers to alert aquaculturists of a power failure during nights, holidays, and weekends when the culture system may be unmanned. The more automated a backup system is, the more expensive it will be, but the cost will be more than offset if the backup components prevent a crop loss.

Auxiliary Apparatus. Backup devices are essential for the operation of intensive aquaculture facilities. In addition there are various types of auxiliary components that can be considered optional, though desirable. Auxiliary apparatus may be used in pretreatment, on-line operation, or posttreatment of the water used in a recirculating system.

Aeration. Various aeration devices are mentioned above, and two are depicted in Figures 21 and 22. Air compressors provide high pressure air, which means that airlifts can be effectively operated when a considerable amount of head is present. Since air-compressor motors cycle frequently (depending on how rapidly the air is bled from the pressure tank), the starters and motors have been known to fail with unacceptable frequency. Air blowers, on the other hand, seem capable of operating continuously and reliably for extended periods of time. They produce large volumes of low pressure air at economical prices. Because the air is delivered under low

Figure 21. Electrically operated air blowers produce large volumes of air at low pressure. They operate continuously.

pressure, such systems cannot operate against significant head pressures, though they are effective in tanks of standard depth.

Liquid oxygen has become popular with many aquaculturists. It is readily available even in most areas, including those that one would generally consider to be quite remote. Large volumes of oxygen can be stored in a fairly small amount of space when liquified. Until the 1980s, bottled oxygen and compressed air were often used as sources of emergency aeration. Liquid oxygen can serve that purpose and in some operations is used routinely as a means of maintaining oxygen at or near saturation in aquaculture systems.[47]

Pretreatment of Water. Some means of pretreating incoming water may be required, particularly when the water is from either a municipal or surface source. Chlorine will have to be removed from municipal water unless only small quantities are being used to replace splashout and evaporation. Filtration may be used in conjunction with surface water to eliminate unwanted organisms and turbidity.

---

[47] Flow through, as well as recirculating systems, may employ liquid oxygen. The same substance has also been used for emergency pond aeration and on live-hauling trucks.

Alternative Ammonia Removal. Clinoptilolite, a natural zeolite with an affinity for ammonia, has been used in aquaculture systems (reviewed by Lucchetti and Gray 1988). Clinoptilolite has been shown to effectively remove ammonia from water at flow rates of up to 330 l/min/m² (Smith et al. 1981). As indicated by Slone et al. (1981), by removing ammonia, clinoptilolite indirectly controls nitrite levels as well. The material can be regenerated by backwashing with a sodium chloride solution when the absorption sites become filled (Lucchetti and Gray 1988). There are ion exchange resins that will remove ammonia from water efficiently, but they are much more expensive than clinoptilolite.

Air stripping, which involves agitation and aeration, can also be used to remove ammonia from water if the pH is greater than 10.0 (O'Farrell et al. 1972, Reeves 1972). Removal of ammonia in either of these ways can reduce the volume of biofilter medium required, or eliminate the need for the biofilter altogether in small systems.

Figure 22. Electrically operated air compressors provide air at high pressure. The motor runs intermittently to pressurize the storage tank.

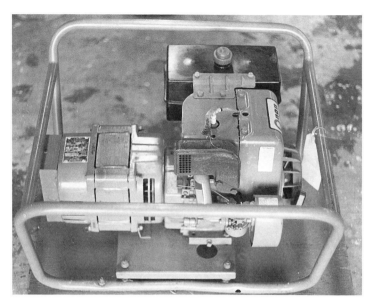

Figure 23. A portable gasoline generator can serve as a backup for electric pumps and motors associated with a small system. Generator capacity should be sized up accordingly as the backup power demands increase in large water systems.

Ultraviolet Treatment. Disease control is difficult in closed water systems as previously mentioned. One of the methods that has been used to keep the level of circulating pathogens below detectable levels in such systems is continuous ultraviolet (UV) irradiation of the water. The effectiveness of UV light is dependent on the size and species of the pathogenic organism and water clarity (Dupree 1981). According to Vlasenko (1969), the dosages required to kill fish pathogens range from 3,620 μW/sec/cm^2 for the bacterium *Aeromonas salmonicida* to 1.7×10^6 μW/sec/cm^2 for the protozoan parasite *Ichthyophthirius multifiliis*.[48] Gratzek et al. (1983) were able to control the spread of Ich from one aquarium to another in a closed water system by exposing the water to 91,900 μW/sec/cm^2 of UV light. Commercially available UV systems typically provide at least 30,000 μW/sec/cm^2 at the flow rates and turbidities common in fish culture systems (Lucchetti and Gray 1988).

Fluorescent bulbs are used in both commercial and homemade UV sterilization systems. The bulbs are kept from contact with the water by housing them in quartz

---

[48] This organism and the disease it causes are commonly called Ich.

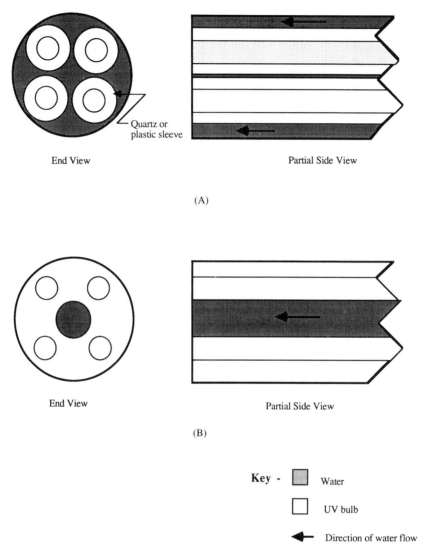

Figure 24. Two common configurations of UV light systems: *(A)* A system in which the UV lights are separated from the water by quartz or plastic sheaths, and *(B)* a system in which the water is contained in a transparent pipe that is surrounded by UV lights.

tubes around which the water flows (Figure 24A), or they may be placed around a transparent quartz or plastic pipe through which the treated water passes (Figure 24B). The UV lights should be in as close proximity as possible to the water. In many cases the water is exposed to the light in a thin film or passes in front of the light in confined channels that are a few millimeters to a few centimeters thick.

Some systems have been designed that allow thin films of water to pass under UV lights suspended from above.

The output of UV bulbs declines with time. According to Lucchetti and Gray (1988), UV output decreases by 10% during the first 100 hr of operation and drops to 70% of the initial level during the first 6 months of use. The bulbs should be replaced every several months or sooner. Particulate matter tends to sediment out on the quartz and plastic tubes that separate the UV bulbs from the water, thereby further diminishing the efficiency of the sterilization system. Prefiltering the water through a sand filter before exposing it to the UV light will help reduce the problem considerably, but it may still be necessary to clean the system periodically and even to replace the protective quartz or plastic on occasion.

Ozonation. Ozone generators have been used in aquaculture systems to both oxidize organic matter and kill pathogenic organisms. Other beneficial effects of ozonation are reduction in biological oxygen demand, ammonia, and nitrite (Colberg and Lingg 1978, Sutterlin et al. 1984). Ozone (O_3) itself is extremely toxic. While little information is available on aquaculture species, the fathead minnow (*Pimephales promelas*) is killed at ozone concentrations of 0.2 to 0.3 mg/l (Arthur and Mount 1975). Sublethal pathology has been observed in rainbow trout exposed for 3 months to 0.05 μg/l O_3 (Wedemeyer et al. 1979). Oyster larvae can be killed by only trace amounts of ozone (DeManche et al. 1975).

At relatively low concentrations, ozone will lead to improved water quality, as was demonstrated with respect to Atlantic salmon (*Salmo salar*) by Sutterlin et al. (1984). If used for pathogen control, higher concentrations may be required, and sufficient time must be allowed for the gas to convert back to molecular oxygen (O_2) before the water is exposed to either the culture chamber or the biofilter after being ozonated. Alternatively, active deozonation may be applied. The half-life of ozone is about 15 min (Layton 1972), so it can take a considerable amount of time for safe levels to occur. Exposure of pathogens to 0.56 to 1.0 mg/l O_3 for 1 to 5 min is effective (Dupree 1981). Aeration assists greatly with the conversion of ozone to molecular oxygen and can reduce the time required for keeping ozonated water away from the culture animals and biofilter. Activated charcoal can also be used to strip ozone from water.

Ozone seems to be much more effective in controlling pathogens in sea water than in fresh water.[49] While the chemistry is apparently not entirely understood, it is believed that ozone reacts with bromine in seawater to produce lethal bromides. The reaction apparently occurs very rapidly, and the efficacy relative to pathogen control is high.

Some systems have been designed in which a portion of the total water flow is ozonated during each pass. When the treated side stream is returned to the main flow, the ozone is diluted to a sublethal concentration. With continuous exposure

[49] John Cussigh, personal communication.

of portions of the flow to ozone, both water quality improvement and pathogen control can be achieved.

Careful monitoring of ozone is necessary, and adjustments to the amount of ozone being added may be required at intervals. While sensors to monitor ozone in water have been developed, they have not been perfected as of this writing, and most monitoring involves sensing ozone in the atmosphere above the deozonating component of the system. Equipment for that purpose is quite effective. Automatic alarms can be set for a desired range of atmospheric ozone in conjunction with the deozonation chamber. If the level moves outside of the desired range, the system will alert the culturist by sounding an alarm. Such systems can also be connected to autodialers that will telephone the culturist to indicate that a problem exists during periods when the facility is not being physically monitored.

Adjusting Water Temperature. Temperature control in closed water systems can be achieved to a degree by placing such systems in a well-insulated building that can provide proper ventilation during hot weather and protection from the cold during winter. Supplemental heating or cooling of the water may be required depending on the species being cultured and on ambient water temperature. Adjusting the temperature of the water is very expensive. Chillers and various types of water-heating devices are available, but the energy costs associated with most of them are prohibitively high. Solar heating has been advocated by some, though when needed the most, they may not be very efficient due to extensive periods of cloud cover. The initial investment for solar systems of the capacity required for an aquaculture system may also be too high to be practical.

Heat can be removed from water with heat pumps as long as the water temperature is above the freezing point. Depending on the needs of the culturist, the water that is cooled may be used in the culture system or discarded. The heat removed from the water by a heat pump can be dissipated into the air or used to warm water that will be used in the culture system. Thus, by employing heat pumps, the culturist can produce quantities of both chilled and heated water and use one or both in conjunction with the aquaculture operation. The application of heat pumps to aquaculture has been described by Fuss (1983) and Stickney and Person (1985).

SEMICLOSED WATER SYSTEMS

A system in which the water is replaced at regular intervals but is at least partly recirculated is referred to as a semiclosed water system. A semiclosed water system may be derived from a closed system merely by initiating some level of water replacement on a continuous or intermittent basis. Early in the growing season it may be quite practical to run such a system in the completely closed mode, but as the animals within the system grow and the loading rate on the biofilter increases, the carrying capacity of the system may be reached. At that time, the replacement of some percentage of the water daily can maintain a suitable culture environment while still conserving water to a large extent. The percentage of replacement water required will probably increase as the animals continue to grow. Careful monitoring of water quality and experience will indicate when the amount of new water being added should be increased.

New water should be added directly to the culture chambers. Pretreatment may be required to remove chlorine (from municipal water); to add oxygen (to well water); to remove undesirable or toxic substances like iron, hydrogen sulfide, and carbon dioxide (from well water); or to remove turbidity and undesirable organisms (from surface water). It will be necessary to drain away water from the system at the same rate that new water is being added. The effluent could come after either the settling chamber or the biofilter. With the implementation of increasingly strict water quality standards on all types of effluents, the best approach would be to release the water following secondary settling, since it is at that point that it will have received maximum treatment (unless tertiary treatment is being employed, in which case the effluent should be released after that final treatment step).

FLOWTHROUGH AND RACEWAY SYSTEMS

Single-pass, or flowthrough, water systems allow the aquaculturist to dispense with one or more of the primary components required in closed and semiclosed systems. Culture chambers are certainly required and settling chambers or mechanical filters are sometimes used, but biofiltration is not necessary. A pump may or may not be required, depending on the water source and the amount of head available to provide gravity flow. The water may require treatment with ozone, UV light, or aeration prior to use, but for many systems the only components are the culture chambers. Those are typically in the form of aquaria, circular tanks, or linear raceways. As previously indicated, circular tanks are normally no larger than about 10 m in diameter. Linear raceways have been constructed in a wide variety of sizes, ranging from 1 m or even less to over 100 m long (Figures 25 to 27). The typical length-width-depth ratios in linear raceways are 30:3:1 (Piper et al. 1982)

Optimum flow rates differ depending on the species and life stage being cultured. Salmonids generally require more rapid water exchange than catfish or tilapia, for example, since trout and salmon are less tolerant of degraded water quality. Larvae and fry may require very low flow rates to keep them from becoming impinged on screens or dashed against the walls of raceways.

As the water moves through a linear raceway, DO will be reduced and ammonia levels will increase because of the metabolic activities of the fish or invertebrates in the system. There are often distinct differences in water quality when samples taken from the upper and lower portions of a raceway are compared. Flow rates should be increased or stocking density decreased when water quality changes in the lower reaches of a raceway indicate that conditions are approaching stressful levels.

Linear raceways may be set up in series or in parallel. When water is passed from one raceway to another in a series (Figure 28, raceways designated A), its quality should be improved between culture units. This can be done by allowing the water to splash over rocks or waterfalls, passing it through a stretch of unpopulated

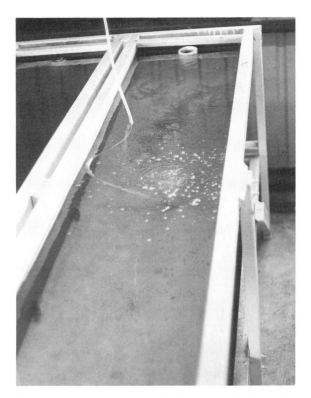

Figure 25. An aluminum raceway approximately 3 m long and 0.5 m wide used for holding fry and fingerlings for experimental use or until they reach stocking size.

channel,[50] or providing supplemental aeration. Ammonia will not be greatly reduced, but DO can be dramatically improved. As the water passes through the series, it will become increasingly degraded. The raceway series should not be so long that fish in the bottom unit are stressed. It may be necessary to reduce stocking density in the lower raceways. Also, it is possible to stock a species more tolerant of degraded water quality in the lower units. Catfish might be stocked in the first few raceways and tilapia in the lower ones, for example.

Parallel raceways (Figure 28, raceways designated B) are more commonly used than are those operated in series. Most trout raceway systems are of the parallel type. In parallel raceways, where the effluent from each unit is discharged from the aquaculture facility, it may still be necessary to provide some form of treatment to the exiting water in order to meet environmental quality standards for receiving waters. Often, settling of solids is all that is required, though secondary treatment may be needed in some instances.

---

[50] Rocks or some other type of material can be placed in the channel to produce riffles and increase aeration.

Figure 26. Fry and fingerling holding raceways in a commercial trout hatchery in Idaho.

Figure 27. Outdoor raceways for trout growout at a commercial farm in Idaho.

Large linear raceways are normally constructed of concrete. Smaller units, such as those found inside laboratory buildings, may be constructed from wood, aluminum, or other materials. Circular raceways are usually constructed from fiberglass, though various types of plastic, along with concrete and metal have also been used. An exception to concrete in large outdoor raceways can be found in conjunction with the Idaho trout industry, where some earthen raceways are used, particularly by small producers. The earthen raceways are basically rectangular ponds through which water is slowly exchanged. While the exchange rate in a concrete raceway

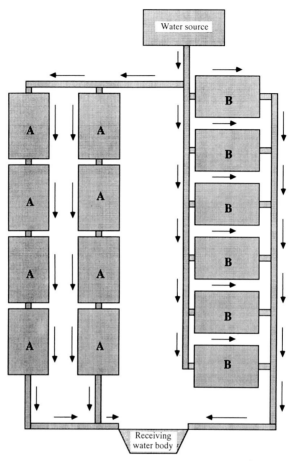

Figure 28. Two rows of raceways run in series are shown on the left (with each raceway designated *A)* and a group of raceways through which water flows in parallel (designated *B*).

may be on the order of one exchange or more per hour, the exchange of the water volume in a pond may take a day or more.

The carrying capacity of a raceway will depend on the rate of water flow, volume of the raceway (which along with flow rate will determine turnover time), temperature, DO content, pH, size of the aquaculture animals, and the species being cultured (Piper et al. 1982). The animals in the system as well as location of the facility will have an influence on some of these parameters. For example, fish with high metabolic rates, such as chinook salmon or dolphin, will have a greater impact on DO on a per-unit-body-weight-of-fish basis than will a fish such as halibut that has a much lower rate of metabolism. The level of DO saturation is dependent on salinity and altitude as well as temperature. For instance, water at sea level and 30°C in a marine system will hold less oxygen than fresh water under the same conditions.

Piper et al. (1982) presented methods of calculating the carrying capacity of trout raceways and discussed the above factors in more detail.

Flowing water through raceways can be extremely expensive, particularly if the water has to be pumped or treated in any way before use. Having an unlimited supply of artesian water of the proper temperature is a luxury shared by few aquaculturists. The trout industry of the Hagerman Valley of Idaho is based on such a water supply, and there are some culturists who have found geothermal springs for use in the production of various types of warmwater fishes.

Diversion of thermally altered water being discharged by electrical generating plants and various types of industry to aquaculture facilities has occurred, often on an experimental basis, but the practice is not widespread, nor are the proper conditions generally available to support an aquaculture activity. The use of power plant effluents for aquaculture is discussed further with respect to cage culture in the following section of this chapter. With respect to their use in conjunction with raceway systems, power plants are not usually designed to accommodate an aquaculture add-on. Even if a raceway system can be designed along with the power plant or added on at some time, there are potential problems. The chemicals used to clean biofouling from the plant's condensers could be toxic to the aquaculture animals, and a shutdown of the power plant could result in the loss of water heating at a critical time. The latter problem can be a particular problem for tropical species being reared outside of their normal range during winter, since even brief exposure to a significant temperature drop can be fatal.

Flowthrough aquaria deserve mention as a specialized type of raceway that has been commonly employed by aquaculture researchers. Holes can be drilled in the sides or bottoms of aquaria to accommodate drains, but it is also possible to establish siphon drains as depicted in Figure 29. Water and air are supplied through pipelines lying above the aquaria, and the effluent water flows through a simple metal or plastic drain gutter of the types used on homes. Flow control regulators can be placed in each of the individual faucets that supply the aquaria. The outside standpipe on the siphon drain maintains water levels in the aquaria in the event of a water failure.

CAGES AND NET PENS

Cages and net pens are basically floating structures that are placed in open water within a pond, lake, reservoir, river, estuary, or the open ocean. They typically have a rigid framework, at least at the water surface, and have a bag suspended beneath the water surface in which the culture species are retained. The upper, rigid framework is provided with floatation devices to keep the upper part of the cage or net pen at or above the water surface. Styrofoam and various other synthetic flotation materials have been used, as have sealed oil drums and bamboo.[51] The bags used in conjunction with cages and net pens are typically constructed of plastic

---

[51] Bamboo is plentiful and inexpensive or free for the taking in many tropical nations.

Figure 29. Aquaria with external siphon drains. The smaller, inside pipe is made in the form of an inverted U. The leg of the inverted U that is inside each aquarium reaches near the bottom, thereby removing settled waste along with water. The leg of the inverted U outside the aquarium can be shorter or the same length of the inside leg. Water being siphoned from the aquarium overflows from the outside pipe, which has a cap glued on its bottom. If the water flow stops, the siphon will also stop and the water level will be maintained at the same height as the outside standpipe. The siphon will resume when the water flow is reactivated.

mesh or nylon netting. Cages sometimes have rigid frames on all sides, while net pens only have rigid frames at the top. In general, cages are smaller than net pens and typically are used in freshwater environments, while net pens are used in salt water.

Cages

Cages of various shapes and sizes have been utilized by aquaculturists. Most have been square, rectangular, or circular at the top and bottom and of various depths. In most cases, cage volumes range from less than one to a few cubic meters. Small cages are often used by researchers (Figure 30), with cages in excess of 1 m^3 volume being used by commercial culturists. Most cages, and particularly those of large size, should not be lifted from the water when stocked with fish, as rupturing of the cage material can occur.

Using cages in water bodies that cannot be drained, are difficult to seine, or that are not capable of containing the aquaculture species can turn an otherwise inappropriate location into a profitable aquaculture site. While leasing public waters is not possible in most states, there are instances, such as in Arkansas, where commercial cage culture leases have been granted in lakes that are otherwise used for recreational fishing (Figure 31). Cage culture is practiced in the public waters of many other nations.

Strings of cages are often tied together and anchored at each end. Each cage

Figure 30. Cages of 1 m³ volume used for experimental catfish culture.

Figure 31. Cages associated with a leased portion of a lake in Arkansas. For harvesting, the cages are towed by boat to the dock and the fish are removed with dip nets.

should have a top constructed either of the same material as that used for containment of the fish or of some rigid material (plywood or aluminum, for example). An opening in each rigid top, fitted with vertical panels on each side that extend above and below the water surface, will provide an opening for feeding and serve as a feed containment ring for floating feed (Figure 30). Alternatively, if mesh material is used for cage tops, a wrapping of material with sufficiently fine mesh to keep feed from escaping can be placed around the top several centimeters of each cage to prevent the escape of floating feed pellets.

Depending on the materials selected, cage costs can vary significantly. Life expectancy will also vary depending on the materials used in construction and the type of environment in which they will be used. Typically, cages have useful lives of about 5 yr.

One of the commonly proposed locations for cages is in association with fossil-fuel and nuclear power plants where, at least in theory, the growing season for the aquaculture animals can be extended by placing cages in the heated water effluent during periods of the year when ambient water temperatures become suboptimal for growth.

Electricity is generated in both fossil-fuel and nuclear power plants by boiling water and producing steam that turns the generators. The steam is then run through condensers where it is cooled and converted back to liquid water. The water is then returned to the boilers for reuse. Water is circulated around the condensers to cool the steam. As a result, the water outside the condensors is warmed by as much as several degrees. In fossil-fuel plants, the water outside the condensers may come directly from a reservoir, river, or estuary. In nuclear power plants, the boiler water and water used to condense the steam are both housed in a containment vessel to prevent the escape of radioactivity. A third water jacket with water from an outside environment is used to cool the containment vessel.

Many power plants use reservoirs as sources of cooling water. Whether constructed in conjunction with a reservoir, stream, or estuary, power plants typically have intake canals from which they draw their cooling water, and discharge canals into which the heated condenser water is discharged (Figure 32). Cages could be placed in the intake canal or in adjacent ambient temperature water during the summer and moved into the discharge canal when warmer water is needed to optimize growth. The cages could be moved up and down the discharge canal as needed when ambient temperature changes. Proper placement of cages in conjunction with heated water effluents can extend the growing season significantly.

Problems have plagued many who have attempted to develop cage culture operations in conjunction with electrical power generating stations. A major concern is that one of more generators may have to be periodically shut down for service. The temperature in the discharge canal can fluctuate fairly drastically as a result, which could lead to mortality in sensitive species. Second, it is sometimes necessary for power plant operators to flush chlorine or other toxic chemicals through their condensers to eliminate fouling organisms. The levels of residual chlorine that enter the discharge canal have sometimes been lethal to caged fishes.

Finally, gas bubble disease has been a serious problem to some cage culturists

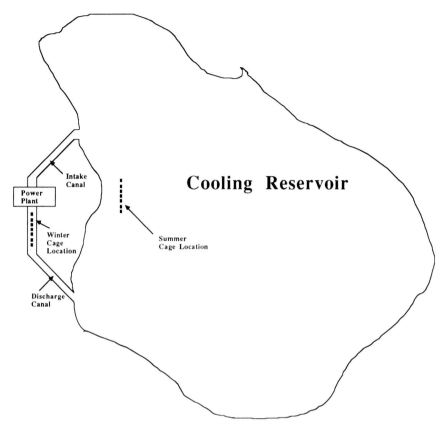

Figure 32. Diagram of a hypothetical power plant showing a lake used as a cooling water supply, along with the intake and discharge canals and possible cage locations.

operating in power plant discharge canals. As the cooling water passes through the condensers, it is placed under pressure and is also heated. Gases that are dissolved in the water can become supersaturated. Of particular concern is nitrogen. Fish exposed to the supersaturated water can develop gas bubble disease as previously discussed with respect to supersatured gas levels in well water. Again, the nitrogen comes out of solution in the tissues of the fish. Bubbles are often seen in the fin rays and also form behind the eyes causing so-called pop-eye (exophthalmia). Exophthalmia is also a sign of certain pathogenic and nutritional diseases (see Chapters 6 and 8), but when it occurs in conjunction with a power plant cage culture operation, it is usually a sign of gas supersaturation.

When ambient water temperature is at its annual low (winter), it has its highest capacity for holding dissolved gases. Thus, when it is rapidly heated under pressure, the supersaturation problem is greatest at just the time when the warm water is most needed by the aquaculturist. To avoid the problem, the aquaculturist may be forced to tow cages into colder water, thereby losing the benefits of being located in association with a power plant. Gas saturometers can be used to measure the levels of dissolved gases in water. By using continuous monitoring and installing an alarm system, the culturist can be made aware of increasing gas levels and take the appropriate action.

An alternative to moving cages away from supersaturated water is to sink them. Increased pressure associated with greater depth lowers the level of gas saturation (Chamberlain and Strawn 1977). If the cages can be lowered to a depth of even a few meters, gas bubble disease may become avoidable. The major disadvantage is that it is difficult to feed fish in submerged cages unless some type of feeding tube can be fabricated or the cage can be brought to the surface for a few minutes when feed is provided. If the period of supersaturation is only a few days in duration, the best approach might be to stop feeding until the cages can be returned to the water surface. Deep cages provide an alternative to submergence. If the cage provides several meters of available depth, the culture animals can seek a level below which gas bubble disease becomes a problem.

Treatment of pathogenic diseases in caged fishes is often difficult. If the diseased animals continue to eat, medicated feed can be provided. Once the fish are off feed, dip or bath treatments may be required. In theory it should be simple to place a plastic exclosure around a cage into which the treatment chemical can be added and contained. In practice the technique is quite difficult. Some cage culturists prefer to mix chemicals in a tank on shore and remove the fish from the cages for treatment. Handling stress can exacerbate the disease problem, however.

Harvesting is usually conducted by moving cages close to shore or up to a floating dock and dipping out the fish with a long-handled dip net. Again, most cages are not sufficiently strong to allow them to be lifted from the water while full of fish.

Poachers find cages an easy target, so there may be security problems associated with a cage culture operation. Power plants offer an advantage in that the public is restricted from entry and security is provided by the utility company.[52]

Net pens

The technology under which salmon could be reared in floating net pens was developed in Washington State during 1969 (Novotny 1975). Perhaps the first salmon net pen constructed in Puget Sound was a structure of approximately 58×14 m that contained a 3 m deep net (Hunter and Farr 1970). Development of the industry in the United States was very slow. By 1992, there were only about a dozen commercial salmon net-pen facilities in Puget Sound, with a small amount of development in Maine. In the meantime, the industry bloomed in Norway and during the late 1980s began to flourish on both the east and west coasts of Canada. Net-pen salmon culture has also been developed in Scotland, Chile, Japan, and New

---

[52]Even what would appear to be highly secure facilities can have poaching problems. According to anecdotal reports, one governmental aquaculture research laboratory in Egypt that is located adjacent to a women's prison had a significant problem with poachers. The poachers were the prison guards!

Figure 33. Salmon farming operation employing commercially manufactured net pens.

Zealand. Rainbow trout have also been cultured in net pens, for example, in Great Britain and Scandinavia (Novotny 1969, 1972).

Commercial net pens typically have galvanized metal platforms that provide access to people involved in tending the fish. Pens of various sizes have been constructed, with most being several meters on a side and about 10 m deep (Figure 33). Some commercial designs provide platforms that will accommodate four net pens, and the units can be bolted together to meet the site requirements. The nets typically extend 1 m or more above the waterline all around the pen to keep fish from escaping.

Bird netting is often placed over the top of each net pen to prevent predation by fish-eating birds. In addition, large mesh block nets that are sufficiently strong to prevent rending by such mammals as sea lions and river otters may be required. Such mammals have been known to cause considerable damage and may lead to the entire escapement of a net-pen salmon population if they tear the primary net that holds the fish.

As with cages, net pens require flotation and are anchored in place to keep them from drifting around. Sites are usually in protected estuaries and obtained by leasing suitable areas from the government. Offshore net-pen designs have been developed, and additional ones are under development. When perfected, and if they can be produced economically, offshore net-pen culture may provide greater potential in the future. At present, problems associated with maintaining offshore facilities during storms have made such systems very expensive and often unreliable.

The number of net pens that can be used in conjunction with a particular site depends on the existing circulation pattern. If a site is well flushed, waste feed and fecal material will be rapidly diluted and removed from the immediate vicinity of the penned fish. In locations that are not well flushed, the wastes can accumulate on the bottom leading to anaerobic sediments. Those accumulations can also lead to significant water quality problems in and around the net pens. The accumulation of wastes in net pens has led to a phenomenon known as self-pollution, which was a serious concern in Japan until limitations were placed on the number of net pens that could be utilized in each water body where culture is undertaken.

Net pens, being significantly larger in surface area and much deeper than cages, are not typically equipped with fine-mesh perimeter netting at the water surface to prevent loss of feed. Both floating and sinking feeds can be effectively used. Because of the depth of net pens, at least with respect to salmon, the feed is virtually all consumed before it can sink sufficiently to exit through the bottom of the enclosure. In the sea bream and yellowtail rearing area of Japan, moist feed is manufactured daily from scrap fish that is mixed with a relatively small percentage of plant meal and then passed through a meat grinder. The resulting material is loaded onto boats and carried directly out to the net pens where it is immediately distributed. Net pens have also been used for the production of sea bass *(Dicentrarchus labrax)* in Italy.[53]

While the most common species cultured in net pens are Atlantic and Pacific salmon, other species have been cultured in such systems. Most notable, perhaps, is the culture of yellowtail *(Seriola quinqueradiata)* and sea bream *(Pagrus major)* in Japan. Net pens of large size (often hundreds of meters on a side) that extend from the water surface to the bottom, are used for the culture of milkfish *(Chanos chanos)* in Laguna de Bay, an embayment off Manila Bay in the Philippines.

Algae blooms have caused problems for net-pen aquaculturists in various parts of the world. Some people have attempted to relate the presence of toxic algae blooms to increased nutrient loading resulting from the penned fish, but the algae blooms may, in fact, be normal phenomena that were not previously detected because they have not caused fish kills in native fish populations that are able to swim away from the algae. The toxic algae tend to concentrate at or near the water surface. Fish in net pens cannot escape because they cannot swim to a depth that allows them to avoid the algae. In most cases, the fish are killed because their gills are damaged by the algae.

Because most of the presently operating net pens have been placed in public waters, they have drawn criticism from various groups, including environmentalists, recreationalists, and commercial fishermen. Strict permitting of net-pen sites in many countries has limited expansion of the practice. Some of the constraints to net-pen aquaculture, particularly with respect to Washington State, were discussed by Stickney (1990).

POLE, RAFT, STRING, AND TRAY CULTURE SYSTEMS

In many instances, molluscs such as oysters, mussels, scallops, and clams are cultured on the bottom in an appropriate location. Natural reproduction is augmented with hatchery-produced young molluscs that are allowed to grow in nature until reaching harvest size. In some instances cages have been placed over clam beds to keep out predators, and oysters have been grown on trays. More widespread, at least in some countries, are pole, raft, and string culture systems. Tray culture is less common.

In each of the types of systems discussed in this section, the molluscs are placed in the water column rather than being allowed to rest on the substrate. Because of

[53] John E. Halver, personal communication.

Figure 34. The algae nori is cultured in Japan on nets that float up and down with the tide by being attached by rings to vertical poles driven into the sediments. Here, workers are cleaning fouling organisms from a net on which nori growth is just beginning.

that, it is necessary for the aquaculturist to be able to have control over the culture environment in terms of exclusive use. Boat traffic through an area where pole, string, or tray culture is being practiced could cause significant damage if the culture system were struck.

Pole Culture

Pole culture has been used for the culture of oysters in the Philippines and for mussel culture in the Philippines and France (Bardach et al. 1972). The French grow mussels on poles in a system known as bouchet culture. Ropes are placed in locations where natural settling of blue mussels *(Mytilus edulis)* occurs. Once the ropes are covered with young mussels, they are taken to the culturing location and wrapped around 4-m-long oak poles of 15 to 20 cm diameter. The poles are then driven about 2 m into the sediment in the intertidal zone. Plastic placed around the base of the poles prevents benthic predators from climbing up to prey on the mussels.

Poles are also used in conjunction with algae culture in Japan. For example, in Kuroshima Bay tens of thousands of poles are driven into the sediments each year to anchor nylons nets in place on which the seaweed nori *(Porphyra* spp.) is grown (Figure 34). Spores from reproducing nori plants are collected on nets in indoor tanks. When the nets are covered with the spores, they are placed into the environment. The nets are able to float at the surface and move up and down with the tide,

Figure 35. Raft in Spain for suspending lines on which molluscs are cultured. (Photo courtesy of Kenneth Tenore)

because they are held in place by rings around the bamboo poles. When the nori is about 15 cm long, it is cut, but the nets are allowed to remain in place until the plants grow sufficiently for a second cutting. Nets placed in cold storage during the early growing season replace those in the bay after the second cutting. Thus, four crops can be obtained during each yearly cycle. The culture of nori and other algae has been reviewed by Stickney (1988).

Raft Culture

Molluscs such as oysters and mussels can be cultured on ropes suspended from floating rafts (Figure 35). Raft culture is perhaps best developed in Spain where coastal bays produce large quantities of oysters and mussels.

The early stages of even those mollusc species that eventually attach to substrates are planktonic during the early stages of their life histories. When the zooplanktonic stages of the culture species[54] settle from the water column, they attach to hard substrates in large numbers. Aquaculturists place what is called cultch material in hatchery tanks containing the zooplanktonic stage of the animals or into a natural spawning area. Once the cultch material, often oyster or scallop shells, has been colonized, it is strung on ropes that are suspended below rafts. It is often necessary to physically remove some of the animals growing on the cultch material because of excessive densities. This is usually done after the animals have been allowed to grow for a period of time.

The ropes may be several meters long, but they should be placed only in the

---

[54] When in the zooplanktonic stage, oysters are known as spat.

Figure 36. A bag of young scallops that is suspended from a horizontal string or longline is brought to the surface of a bay in Japan for inspection.

photic zone[55] where the growing molluscs can obtain the phytoplankton on which they feed. The ropes should not come in contact with the substrate beneath the rafts because predators such as oyster drills and starfish will be able to attack the molluscs by crawling up the ropes.

String Culture

String or longline culture is similar to raft culture except that the cultch material is attached to ropes that lie horizontally in the water suspended from floats. The strings may be tens to hundreds of meters in length. Oysters or mussels can be reared attached to cultch material that is tied to the primary rope, or such culture animals as scallops may be contained in netting or baskets suspended from the string. In Japan, young scallops are reared in baskets suspended from strings (Figure 36) until they are sufficiently large to be hung individually. A hole is drilled in one of the so-called wings of each scallop, and a monofilament line is strung through the hole

[55]The photic zone is defined at the water column to the depth of measurable light penetration.

Figure 37. In Japan, individual scallops may be hung from strings for final growout.

(Figure 37). The individual scallops are then hung on longlines until they reach market size. It is also possible to rear scallops to market size in mesh bags.

Tray Culture

Individual (cultchless) oysters have sometimes been reared in trays that are supported on legs to keep them off the bottom. The spat are allowed to settle on sheets of plastic or other pliable material. By bending the material on which the oysters have settled, the animals will become detached. They can then be raised individually since they will not be able to subsequently attach to another substrate.

While oysters reared on cultch material often grow in clumps and have irregular shapes and sizes, cultchless oysters grown in trays are more uniform and generally bring a premium price in the half-shell trade. As is true of rafts and strings, trays of oysters should be located in unpolluted areas where environmental conditions are conducive to rapid growth and that support a rich phytoplankton population so the oysters will be well nourished.

PONDS

The majority of aquaculture throughout the world is conducted in ponds. The site selected for a pond should have soil that does not have toxic properties and has good water-holding capability. The site should also have a dependable supply of high-quality water. If the choice is between the proper type of soil and a suitable water supply, most culturists elect to take advantage of good water and find ways of dealing with the soil problem.

Site Survey and Development of Pond Layout

Prior to construction, a complete survey of the site should be made to determine where ponds will be placed, the type of construction to be used, and what type of soil is present. Soil borings to at least the depth of projected pond bottoms should be taken at representative locations around the proposed construction site, and the cores obtained should be analyzed to determining the water-holding characteristics of the soil and to look for toxins. Such testing can be expensive. The number of cores required depends to some extent on the amount of money allocated for that task, the presence or absence of available data on that location and adjacent properties, the past history of land utilization on the proposed aquaculture site and adjacent properties, and the general uniformity of soils in the region.

A topographic map should be drawn from which pond elevations can be determined. The slope of the land will help designers determine where to place drain lines and whether to construct aboveground, in-ground, or partly in-ground ponds, or some combination of those configurations (Figure 38). Once the ponds have been laid out and the locations and elevations of levees determined, it will be possible to calculate the amount of soil that will have to be hauled in or hauled away from various portions of the site.

Toxic Soils

As aquaculture has expanded in the tropical world, mangrove communities have often been turned into pond culture facilities. In addition to removing forests that protect coastal regions from erosion and provide nursery grounds for a wide variety of marine organisms, the use of mangrove areas for aquaculture brings with it the problem of toxic soils. Mangrove forest soils typically contain high levels of iron pyrite that leach sulfuric acid and toxic levels of heavy metals into aquaculture pond water when used for levee construction (Simpson and Pedini 1985). Lining ponds with plastic or clay blankets, as described below to prevent leakage, is possible but often expensive. Yunker and Scura (1985) described a technique by which the burrow mound material of a mud lobster *(Thalassina anomala)* was used to cover and seal the banks of ponds constructed in soils containing high iron pyrite levels in Malaysia.

As the importance of mangrove swamps to natural ecosystems becomes better understood by governmental bodies in the developing world, the conversion of such

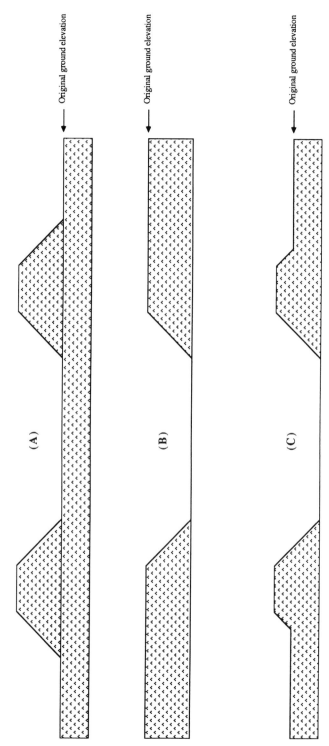

Figure 38. Pond levees can be constructed above (*A*), below (*B*), or partly above and partly below (*C*) original ground elevation.

areas into aquaculture ponds may be reduced, but with the estimated 4.8 million hectares of such mangrove areas in Southeast Asia alone (Yunker and Scura 1985), the practice will not soon be terminated. The poor productivity that has plagued many aquaculturists who have constructed ponds in mangrove areas has not ended the practice, but prospective culturists should be aware of the problem and be prepared to deal with it.

Sealing Ponds against Leakage

The best soils for ponds are those that have a high percentage of clay. Most experts recommend that the soil contain a minimum of 25% clay. Even in regions that feature soils of the proper quality, however, lenses of sand or other porous materials may be present. If one of these is struck during pond construction, leakage may occur in the affected area. If a site is selected because it has a good supply of suitable water but sandy soil, ponds can still be constructed and sealed against leakage. Clay can be hauled in to create a blanket of nonporous material over small areas where sand lenses are uncovered.

When clay is used to line ponds, it should be spread evenly over the sides and bottom, then packed with a sheepsfoot roller or similar device before the pond is filled. Bentonite, an expandable clay mineral used by the petroleum production industry to lubricate drill bits, has been widely used to assist in sealing ponds against leakage. The material is mixed with the soil of the pond levees and bottom, and then the mixture is compacted. When the pond is filled, the clay mineral will expand to several times its original volume and essentially bind the sediment particles together.

Bentonite is available from various commercial sources and is not prohibitively expensive if the problem is not severe and the material does not have to be hauled long distances. For ponds that do not have a severe problem with leakage, bentonite is often effective at the rate of 20,000 kg/ha. Serious leakage problems may require application rates of 125,000 kg/ha or more, which can become prohibitively expensive.

In cases where seepage rates are high and pond sizes are relatively small, or where money is no object, liners may be used to retain water. Various types of plastic sheeting will seal ponds against leakage. Some that are popular include chlorinated polyethylene (CPE), polyvinyl chloride (PVC), high-density polyethylene (HDPE), and chlorosulfonated polyethylene (CSPE). More information about each of these materials can be found in a paper by Cadwallader and Springer (1992). Liner materials come in various thicknesses and strengths. Liner material that does not puncture or tear easily should be used, since holes will destroy the effectiveness of the liner. The material should be capable of being welded at the seams to retain watertight integrity. Plastics used for pond liners should be resistant to ultraviolet light, or they will begin to disintegrate within an unacceptably short period of time.

Clay blankets may be placed over plastic liners, or the plastic may be exposed directly to the water. Some materials tend to float up creating uneven pond bottoms.

Clay blankets or some other means of keeping those materials in smooth contact with the sediments may be required.

Plastic liners vary considerably in cost and in life expectancy. If it is necessary to use a pond liner, the economics of paying a higher initial price for one that can be expected to last for several years should be weighed against the lower cost of a liner that will need to be replaced at more frequent intervals.

Concrete has also been used for pond construction, but the cost is generally prohibitive. Brick walls, covered with mortar, have also been used in pond construction in some countries. They are common in China, for example.

Newly constructed ponds may eventually seal themselves against leakage, to a large extent within a few years after being placed in production. The presence of increasing levels of organic material and the reworking of sediments by benthic organisms and some bottom-dwelling aquaculture species tend to seal ponds over time. For this type of sealing to remain effective, the pond bottoms should not be disturbed, but increasing organic loading has a depressive effect on production. Many aquaculturists drain their ponds and allow them to dry once a year or at least once every few years. They may also disk [56] the bottom to allow the oxidation of organic matter. Production may increase, but so will the rate of seepage.

Proper construction is an important consideration in the prevention of seepage problems. The section ''Pond Configuration and Levee Design'' (later in this chapter), contains a discussion on such factors as the removal of organic material from the site prior to construction and the function of core trenches.

Saltwater ponds are often constructed in sandy soils because of the absence of other sediment types in coastal regions. If such ponds have bottom elevations below that of the original ground level (configuration B or C in Figure 38), hydrostatic pressure from the adjacent saltwater body may be sufficient to maintain pond water level. In many instances saltwater ponds will partially or totally refill after being pumped out because of the movement of water through the sediments under hydrostatic pressure.

Pond Size

There is no actual best size for an aquaculture pond, though extremely large and very tiny ponds should usually be avoided. The amount of earthwork per unit area of pond increases as the size of the individual pond decreases (Figure 39), so construction economics favor larger ponds. Not only is the linear length of levees increased for containment of the same amount of water as pond size decreases; the amount of area occupied by levees also increases. Thus, more raw land must be available for the development of small ponds than for larger ponds having the same total water surface area. In addition to increased costs of levee construction, there is the cost of additional plumbing, as each pond will require its own inflow line and outflow structure.

---

[56] Disking a pond bottom involves the same procedure that a farmer uses when disking a field preparatory to planting.

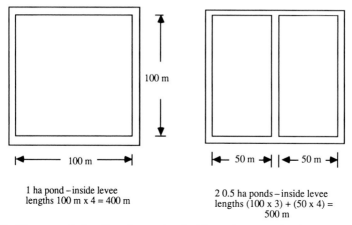

1 ha pond – inside levee
lengths 100 m x 4 = 400 m

2 0.5 ha ponds – inside levee
lengths (100 x 3) + (50 x 4) =
500 m

Figure 39. The amount of earthwork associated with pond construction increases as the size of individual ponds decreases.

Ponds used for research are generally smaller than production ponds. Small ponds are also often used for holding broodfish and for rearing larvae and fingerlings of various species. Most such ponds range in size from about 0.05 to 0.25 ha. Production ponds may be as small as 0.25 ha, but ponds of 1.0 to 4.0 ha are more common. Some channel catfish farmers once constructed ponds as large as 10 ha, but because of difficulties with regard to management, many of the ponds have been subdivided into two or four ponds each.

Feeding can be a problem in a large pond, particularly when the species under culture is relatively sedentary and the feed must be evenly distributed to ensure that each animal has the same chance of finding sustenance.[57] Also of importance is the fact that a water quality or disease problem that leads to high mortality in a single pond will be much more devastating if it is a large pond than a small one, given the same density of animals per unit area.

It should be possible to fill a culture pond of any size from the available water supply within a few days. In addition, it should be possible to drain each pond completely and relatively quickly (e.g., usually less than 72 hr). Constraints on filling ponds relate not only to the volume of water that can be obtained per unit time from the source but also on the capacity of the inflow plumbing to carry that water to the ponds. While the water supply and its associated plumbing might be able to carry sufficient water to fill one pond within a reasonable period of time, many days may be required for the same supply to fill several ponds. Similarly, the drain lines associated with a pond facility might be able to handle sufficient volume to drain a single pond rapidly, but would become overloaded if several ponds were drained simultaneously. Pond draining should be accomplished by gravity wherever

[57] Shrimp ponds should be evenly fed, for example, in order to ensure that all animals have an opportunity to find feed.

possible. Significant additional expense will be acquired if it is necessary to pump water out of ponds. Proper facility planning will help the aquaculturist ensure that pond bottoms are at an elevation that will allow them to drain by gravity.

We have already seen in Chapter 1 that pond production values of several hundred to several thousand kilograms per hectare are typical. What is not shown in such figures is the volume of water contained in each hectare of pond surface. A logical assumption is that if a culturist can produce 4,000 kg/ha in a pond that is 2 m deep, it should be possible to produce 8,000 kg/ha in a 4-m-deep pond.

The expression, "My pond isn't very large, but it is very deep . . . ," is commonly heard. The reality of the situation is that nearly all aquaculture ponds are quite shallow, averaging no more than a meter or two in depth. Catfish ponds are typically 0.9 to 1.8 m deep in the southern United States, for example, though fish culture ponds in the North may be twice as deep. The difference is largely dictated by whether the ponds can be expected to freeze over for long periods during the winter. Winterkill due to oxygen depletion can occur under heavy ice cover; thus, greater depth is important in the North as a means of increasing the volume of available water and oxygen during winter. Shallow ponds can even freeze to the bottom in northern climates—another reason to provide greater depth.

Thermal stratification typically occurs in deep ponds (Figure 40), while very shallow ponds are continually mixed by the wind and do not stratify. Ponds as shallow as 2 m may demonstrate some thermal stratification, though complete mixing will occur in the presence of strong winds. When stratification occurs, a warm layer of water (the epilimnion) is retained at the pond surface, with an underlying thermocline[58] in which there is a rapid decrease in temperature with depth. If the pond is sufficiently deep (several meters), a hypolimnion may form, in which the water is quite cool (well below the optimum temperature required by warmwater fishes and invertebrates), and oxygen depletion can occur since the hypolimnion does not have contact with the surface and may lie below the photosynthetic zone.[59] It is quite possible for the hypolimnion to become anoxic, so while a deep pond may have a large volume of water in it compared to a shallow pond of the same surface area, a good part of that additional volume is not inhabitable by the culture species.

Pond Configuration and Levee Design

Most ponds are rectangular, though irregular and even round ponds have been used effectively. Rectangular ponds tend to be preferred, with the long axis placed to take advantage of wind mixing in benevolent climates and perpendicular to the prevailing wind in places where erosion from high winds is a problem.

---

[58] In many instances the temperature will fall 1°C for each meter of depth within the thermocline. Weak thermoclines often form during the summer even in shallow ponds.

[59] Oxygen enters water through diffusion from the atmosphere and as a byproduct of photosynthesis in the photic zone by aquatic macrophytes and algae. The hypolimnion may be cut off from both sources of oxygen. Utilization of the oxygen in the hypolimnion through respiration by various organisms can lead to an oxygen depletion.

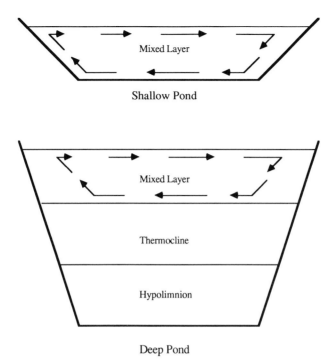

Figure 40. Thermal stratification can occur in deep ponds during summer, leading to oxygen depletions in the deeper waters. Shallow ponds remain at nearly constant temperature from top to bottom as a result of wind mixing.

Before a pond is constructed, the site should be carefully surveyed and the elevations of all pipes, levees, and pond bottoms determined. The amount of earth that will have to be moved will be a function of the type of construction (Figure 38) and the slope of the land. It is very expensive to move earth long distances. Properly designed pond facilities minimize the need to haul earth in or move it off site. Earth can be moved short distances (a few meters, generally) with bulldozers. When longer-distance hauling is required, frontend loaders and trucks may be needed.

Once the facility has been designed, all surface vegetation should be removed. Topsoil is often piled up on the site to be used to topdress the levees following construction. The topsoil will help support the growth of vegetation on pond banks. Organic debris should always be removed from the site. If a pond levee is constructed over a log, for example, eventual deterioration of that log will open a void space, potentially leading to weakening and possible failure of the levee.

If a levee is constructed on top of undisturbed earth, there will be a difference in the level of consolidation or compaction of the earth in the levee and underlying soil (Figure 41A). Water within the pond will tend to seep along the discontinuity and can eventually find its way through the base of a levee, causing leakage and possibly washout of the levee. Seepage at the base of levees can be prevented by construction of a core trench (Figure 41B). Core trenches are usually dug to a depth of about 1 m with a backhoe. Once the trench is dug, it is backfilled with the same material that will be used for levee construction and the material is then compacted. Water that seeps along the base of the completed levee will encounter the area where the core trench is located and will tend to follow the discontinuity downward. The hydraulics of water flow are such that the seepage will terminate at the core trench.

Core trenches may be dug under each individual levee on an aquaculture site, or a perimeter core trench may be relied upon to contain seeping water on site.

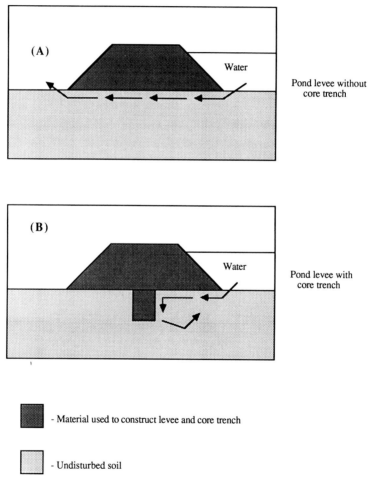

Figure 41. Arrows show how water can follow the base of a levee in a pond without a core trench *(A)* and eventually cause seepage. When a core trench is constructed, filled, and compacted with the same material as the levee *(B)*, seepage is blocked at the location of the core trench.

Figure 42. Levees between ponds should be wide enough to allow passage of vehicles associated with harvesting and feeding.

Trenches dug to contain drain lines serve the same purpose as core trenches and are constructed in the same way, except that a pipeline is placed in the trench before backfilling and compacting is undertaken. Core trenches are not required when ponds are equipped with liners.

The levee on at least one side of every pond should be sufficiently wide to drive on, thus providing vehicular access for harvesting, feeding, and other activities (Figure 42). Ideally, it should be possible to drive a vehicle completely around each pond. This is particularly useful if it is necessary to distribute feed evenly over the water; however, it is more expensive than constructing narrow levees. All levees should be sufficiently wide to accommodate a lawn mower. Extremely narrow levees are also more likely to erode than wider ones.

Most levees are constructed with slopes of 1:2 or 1:3 (for every unit of levee height, the base of the levee is two or three units wide). Levees with slopes of 1:1 have been constructed, but are difficult to build with a bulldozer because the banks are so steep.[60] Steep banks are also more subject to erosion than those with more gentle slopes. Bulldozer operators prefer 1:3 slopes, primarily because the equipment is not required to work at such steep angles, but as the steepness of the banks decreases, the amount of shallow water in the completed ponds increases, thereby increasing the area of pond bottom suitable for the development of rooted aquatic vegetation.

The pond bottom should slope toward the drain (discussed in the following section). Typical pond bottom slopes are about 1% (1 m drop in elevation for every 100 m of linear distance).

---

[60] A 1:1 slope leads to levees with a 45° angle on the sides.

As earth is moved into position to form pond levees, it should be compacted frequently. Compaction is achieved by running the bulldozer over the levee material from side to side and from base to top and back down the other side to the base. If the process is repeated every time 20 to 30 cm of material are added to the levee, uniform compaction can be achieved. Final grading of the top of each levee is most commonly accomplished with a motor grader.

Once the levees have been completed and the surfaces that are exposed to the atmosphere are covered with topsoil, those exposed areas should be seeded with grass. The grass will help control erosion and keep the levees from becoming quagmires of mud following rains. Road levees can be merely seeded, but since they are often subjected to frequent traffic, they should be graveled to keep them from eroding.

Drains and Inflow Lines

Drains are typically placed at one end of each pond. To conserve on plumbing and the associated expense, drainage systems are usually laid out so that the drain ends of parallel series of ponds share the same effluent course. Water exiting a pond may enter an underground pipe, or it may flow into an open ditch. As the number of ponds emptying into a common drain pipe increases, the diameter of the pipe should also be increased.

Drain structures vary considerably with respect to complexity. The most simple type of drain is a standpipe that protrudes from the pond bottom or is installed outside the pond in cases where drainage is into a ditch. The lower end of the standpipe is screwed into an elbow, the other end of which is joined to a pipe that passes through the levee. By turning the elbow with its attached standpipe, the culturist can control the level of water in the pond (Figure 43). When fully upright, the top of the standpipe should be lower than the top of the levee to prevent the pond from overflowing. When the standpipe is laid down against the pond bottom (or against the drainage ditch bottom in the case of external standpipes), the pond can be completely drained. Standpipe drains are normally constructed from metal rather than plastic pipe, though both have been used effectively. The threads on metal pipe tend to bind, and it may become difficult to adjust the elevation of the standpipe. PVC standpipes, on the other hand, have a tendency to tilt away from the position in which they are set unless they are tied into position.

The presence of a pipe through a levee provides another means by which water can seep along a discontinuity layer. Antiseep collars that are constructed of concrete, sheet metal (Figure 44), or some other suitable material and are placed within the levee will block seepage in much the same manner as a core trench.

While tilt-over standpipes are inexpensive and effective, they may not be an appropriate choice in large ponds where drain lines need to be of significantly greater diameter those that are satisfactory in small ponds. Structures of various complexities have been designed in conjunction with drains. In many cases gate valves are used in conjunction with drain lines in structures that are known as kettles or monks (Figure 45). Another option is to use dam boards placed in slots along the

Figure 43. Pond with internal standpipe that can be moved on an elbow to control water depth. The standpipe shown here is at about a 45° angle to the pond bottom.

Figure 44. An antiseep collar placed around the drainline that passes through each pond levee will prevent water from following along between the pipe and the surrounding soil and causing a leak in the levee.

Figure 45. Pond with concrete kettle drain system featuring a stairway for access, a harvest basin, gate valve drain, slots for screens, and a source of inflowing water. (Photography by Meryl Broussard).

side of the kettle (Figures 46 and 47). Water level is controlled by the number of boards put in place. Slots are often placed in the walls of kettles that have valved drains (Figure 45) to accommodate screens that will keep the aquaculture animals from escaping when the pond is drained.

Kettles may be relatively simple structures (Figure 47), or they may have stairways and catch basins designed into them (Figure 45). A harvest basin is usually a concreted area in front of the kettle that has an elevation several centimeters below that of the pond bottom. As the pond is drained, the animals under culture may accumulate in the harvest basin, where they can be easily removed with a dip net or fish pump. During summer, temperature can rise precipitously in the shallow water of a harvest basin, so a source of new water is a desirable feature. Thus, many pond systems are designed so that the inflow line is associated with the drain (Figures 45 to 47). When a pond is full and new water is being flowed through, which may be necessary in conjunction with resolving certain water quality problems, it is generally more desirable to have the inflow and outflow at opposite ends of the pond. Some aquaculture facilities, therefore, have inflow lines at both ends of each pond.

In most cases inflow water is delivered through pipes, and each pond should have a valve to control the rate of flow. If it is not under pressure, inflow water can also be delivered through pipes that are merely stoppered with some type of appropriate plug (Figure 48) or incoming water may be delivered through open channels (Figure 49).

The sizes of inflow and drain pipes used in various parts of an aquaculture facil-

Figure 46. Concrete kettle in which dam boards are placed in slots to control water level. In this case (a facility in the Philippines), earth was packed between two sets of dam boards to prevent seepage. Many types of wood will become waterlogged and seepage is not a problem. Note that the inflow water is being passed through saran cloth to filter out undesirable organisms.

ity should be sufficiently large to handle the required amount of water, but since plumbing is expensive, the pipes should not be greatly oversized. The larger the diameter of the pipes and valves used, the greater the cost per item of material purchased. For example, metal valves for 10 cm pipes may cost in the neighborhood of $20 each, whereas a metal valve for a 20 cm pipe can cost $250. Clearly, a 20 cm inflow line should not be used if a 10 cm line is sufficient.[61]

Other Considerations

If water for an aquaculture facility is from an unreliable source, it may be necessary to construct a holding reservoir to provide makeup water when the primary supply is unavailable. For example, many aquatic farms around the world rely on water supplied through irrigation canals, the flow through which is controlled in most instances by those who are concerned about supplying water at the proper time and in the right amount for irrigating terrestrial plants. The water may be unavailable when fields lie fallow, or a rupture in a canal, perhaps many kilometers from a fish

---

[61] It is important to note that the amount of water that can be carried through two 10 cm pipes is not the same as that which will pass through a 20 cm pipe. Recall that the area of a circle is πr^2. For a 10 cm diameter pipe (5 cm radius), the area is 3.14 x 25 = 78.5 cm^2, while for a 20 cm pipe (10 cm radius), it is 3.14 x 100 = 315 cm^2. Thus, the larger pipeline has four times as much capacity (314/78.5 = 4).

Figure 47. View of the front of a kettle showing placement of the dam boards.

farm, could cause shutdown of the supply for a period of hours, days, or even weeks. Knowing the frequency and duration of such potential shutdowns and projecting the continuous and intermittent water needs of the aquatic farm will provide the aquaculturist with the information needed to design a holding reservoir of the proper size.

Windbreaks, in the form of rows of trees such as poplars or hedges, may help the aquaculturist avoid levee erosion when placed upwind of the farm. Deciduous plants should not be placed in such close proximity to ponds that falling leaves will enter the water as they will contribute to organic loading, which will increase the oxygen demand in the ponds. Roots can also weaken levees. Windbreaks constructed from snow fence might be more desirable for use on fish farms.

Current practice in some types of aquaculture (e.g., catfish farming) is to allow a pond to remain in continuous operation through successive growing seasons. Fish may be selectively harvested (see Chapter 9) and restocked at intervals to keep them in production. As organic matter accumulates, production tends to decline, so that

Figure 48. Low-pressure water flowing into a pond in Jamaica. To stop the flow, the fish farmer place a wooden plug in the end of the pipe.

even such ponds should be drained and the bottoms allowed to dry out every few years. Some aquaculturists drain and dry their ponds annually and may disk the pond bottoms and reshape the sides with a bulldozer or tractor blade each year. Disking exposes organic matter to the atmosphere and helps speed oxidation. Shaping the pond levees can repair areas where minor erosion has occurred during the growing season.

OTHER TYPES OF WATER SYSTEMS

The basic types of culture systems used by aquaculturists around the world are covered in the above sections. Throughout the world one sees familiar types of facilities, each of which involves some variation on the themes presented above. Various modifications may be necessary to accommodate the particular needs of a given species, but there are relatively few dramatic departures from the basic types of culture systems: ponds, raceways or tanks, rafts or longlines, and cages or net pens. Bottom culture of oysters is an activity that does not utilize containment but might be likened more to hunting and gathering than to controlled production. Notions about raising fish in space stations are a variation on closed-system aquaculture, and hanging oysters or containers of fish from oil drilling platforms is merely another form of more traditional string and cage culture.

Crawfish farming, which uses a modified pond culture technique, deserves a few words. Crawfish have been farmed for many years in Louisiana, and a significant

Figure 49. A group of ponds in China to which water is delivered in open channels.

industry has developed since about 1980 in Texas. Other states have become interested not only in native crawfish species but also in exotics. Two species, the red swamp crawfish *(Procambarus clarkii)* and the white river crawfish *(Procambarus acutus),* currently dominate the aquaculture scene. The red swamp crawfish has historically been the major crop of most producers (LaCaze 1976).

The largest source of wild crawfish in the United States is the Atchafalaya Basin in Louisiana. Commercial crawfish fishermen in Louisiana harvest some 45,000,000 kg of crawfish annually, with about 40% coming from the Atchafalaya Basin and the other 60% from culture ponds (de la Bretonne 1988). As pond technology has developed, the effective harvest season has increased from 3 to 8 months annually.

Three types of ponds used in Louisiana have been described by LaCaze (1976) and de la Bretonne (1977). Wooded areas are sometimes impounded, or open ponds used exclusively for crawfish production are constructed. The third method involves rotation of crawfish and rice, which may require modifying rice fields (increasing the levee height from 15 cm for rice to 30 to 75 cm for crawfish). When rice planting and crawfish stocking are properly timed, both crops can be harvested, with the rice stubble serving as a forage base for the crawfish (Landau 1992). Wild rice, millet, alligator weed, and various other plants have also been provided as crawfish food.

Brood crayfish are stocked in flooded ponds during late May or June. LaCaze (1976) recommended stocking males and females in similar numbers at a size of about 20 to 30 per kilogram. Adults should be stocked at between 20 and 100 kg/ha depending on the amount of cover available and on whether or not crawfish are already present in the pond.

During the early summer, the ponds are drained and the adult crawfish burrow into the sediments, where their young are produced. The ponds are flooded again in the early fall, and the adults and young leave the burrows to forage. By late fall,

Figure 50. Crawfish are captured in traps, such as the pillow trap shown here being emptied into a tub.

Figure 51. A crawfish harvesting machine. As the machine moves through a pond, each trap is picked up and dumped and a new baited trap is put in place.

harvesting can begin (Landau 1992). Crawfish are harvested with various types of traps (Figure 50). The traps may be set from a boat or from wheeled machines designed specifically for the purpose (Figure 51).

Recent innovations in crawfish farming have involved certain types of modifications in pond design and the development of artificial bait to replace the trash fish that have traditionally been used to bait traps. While the techniques for crawfish culture are not the same as those used with respect to other types of aquaculture products, they are, once again, variations on a standard theme: pond culture.

LITERATURE CITED

Arthur, J. W., and D. I. Mount. 1975. Toxicity of a disinfected effluent to aquatic life. pp. 778–786, In: First International Symposium for Fish Culture. American Fisheries Society, Fish Culture Section, Bethesda, Maryland.

Balderston, W. L., and J. McN. Sieburth. 1976. Nitrate removal in closed-system aquaculture by columnar denitrification. Appl. Environ. Microbiol., 1976: 808–818.

Bardach, J. E., J. H. Ryther, and W. O. McLarney. 1972. Aquaculture. Wiley-Interscience, New York. 868 p.

Bender, J. 1984. An integrated system of aquaculture, vegetable production and solar heating in an urban environment. Aquacult. Eng., 3: 141–152.

Blasiola, G. C. 1984. Freshwater Mar. Aquar., 7(5): 10–12.

Boyd, C. E. 1968a. Evaluation of some common aquatic weeds as possible foodstuffs. Hyacinth Control J., 7: 26–27.

Boyd, C. E. 1968b. Fresh-water plants: a potential source of protein. Econ. Bot., 22: 359–368.

Boyd, C. E. 1990. Water quality ponds for aquaculture. Agricultural Experiment Station, Auburn University, Auburn, Alabama. 482 p.

Brannon, E., and G. Klontz. 1989. The Idaho aquaculture industry. Northwest Environ. J., 5: 23–35.

Buss, K., D. R. Graff, and E. R. Miller. 1970. Trout culture in vertical units. Prog. Fish-Cult., 32: 187–191.

Cadwallader, M. W., and C. B. Springer. 1992. Liner materials for water containment. World Aquacult., 23(2): 59–61.

Chamberlain, G., and K. Strawn. 1977. Submerged cage culture of fish in supersaturated thermal effluent. Proc. World Maricult. Soc., 8: 625–645.

Chen, J.-C., Y. Y. Ting, J.-N. Lin, and M.-N. Lin. 1990. Lethal effects of ammonia and nitrate on *Penaeus chinensis* juveniles. Mar. Biol., 107: 427–431.

Chien, Y.-H., and Y.-H. Chou. 1989. The effects of residual chlorine, temperatures and salinities on the development of Pacific oyster (*Crassostrea gigas*). Asian Fish. Sci., 3: 69–84.

Colberg, P. J., and A. J. Lingg. 1978. Effect of ozonation on microbial fish pathogens, ammonia, nitrate, nitrite, and BOD in simulated reuse hatchery water. J. Fish. Res. Board Can., 34: 1290–1296.

Colt, J. E., and D. A. Armstrong. 1981. Nitrogen toxicity to crustaceans, fish, and molluscs. pp. 34–47, In: L. J. Allen and E. C. Kinney (Eds.). Proceedings of the Bio-Engineering Symposium for Fish Culture. Fish Culture Section Publication No. 1. American Fisheries Society, Washington, D.C.

Ctiokeleuchai, P., and Duangswasdi, M. 1981. Acute toxicity of chlorine to sea bass and giant prawn. Thai Fish. Gaz., 34(5): 493–498.

Davis, J. T. 1977. Design of water reuse facilities for warm water fish culture. Ph.D. Dissertation, Texas A&M University, College Station. 100 p.

de la Bretonne, L. W. 1988. Commercial crawfish cultivation practices. J. Shellfish Res., 7: 210.

de la Bretonne, L. W., Jr. 1977. A review of crawfish culture in Louisiana. Proc. World Maricult. Soc., 8: 265–269.

DeManche, J. M., P. L. Donaghay, W. P. Breese, and L. F. Small. 1975. Residual toxicity of ozonated seawater to oyster larvae. Sea Grant College Program Publ. ORESUT-003, Oregon State University, Corvallis.

Dupree, H. K. 1981. An overview of the various techniques to control infectious diseases in water supplies and in water reuse aquaculture systems. pp. 83–89, In: L. J. Allen and E. C. Kinney (Eds.). Procceedings of the Bio-Engineering Symposium for Fish Culture. Fish Culture Section Publication No. 1. American Fisheries Society, Washington, D.C.

Fuss, J. T. 1983. Evaluation of a heat pump for an aquacultural application. Prog. Fish-Cult., 45: 121–123.

Gratzek, J. B., J. P. Gilbert, A. L. Lohr, E. B. Shotts, Jr., and J. Brown. 1983. Ultraviolet light control of *Ichthyophthirius multifiliis* Fouquet in a closed fish culture recirculation system. J. Fish Dis., 6: 145–153.

Hall, L. W., Jr., D. T. Burton, and L. B. Richardson. 1981. Comparison of ozone and chlorine toxicity to the developmental stages of striped bass, *Morone saxatilis*. Can. J. Fish. Aquat. Sci., 38: 752–757.

Hess, W. J. 1981. Performance ratings for submerged nitrification biofilters: development of a design calculation procedure. pp. 63–70, In: L. J. Allen and E. C. Kinney (Eds.). Procceedings of the Bio-Engineering Symposium for Fish Culture. Fish Culture Section Publication No. 1. American Fisheries Society, Washington, D.C.

Hoergensen, S. E. 1985. Application of ion exchange as biofilter for complete recirculation in aquaculture. Vatten, 41: 110–114.

Horsch, C. M. 1984. The use of a natural zeolite as both an ion-exchange and biological filter media at Eagle Creek National Fish Hatchery. Salmonid, 7: 12–15.

Hunter, C. J., and W. E. Farr. 1970. Large floating structure for holding adult Pacific salmon (*Oncorhynchus* spp.), J. Fish. Res. Board Can., 27: 947–950.

Klontz, G. W. 1981. Recent advances in evaluating biofilter performance. pp. 90–91, In: L. J. Allen and E. C. Kinney (Eds.). Procceedings of the Bio-Engineering Symposium for Fish Culture. Fish Culture Section Publication No. 1. American Fisheries Society, Washington, D.C.

LaCaze, C. G. 1976. Crawfish farming. Revised edition. Louisiana Wildlife and Fish Commission Fishery Bulletin No. 7, Baton Rouge. 27 p.

Landau, M. 1992. Introduction to aquaculture. John Wiley & Sons, New York. 440 p.

Layton, R. F. 1972. Analytical methods for ozone in water and wastewater applications. pp. 15–18, In: F. L. Evans (Ed.). Ozone in water and wastewater treatment. Ann Arbor Science Publishers, Ann Arbor, Michigan.

Lewis, W. M., and G. L. Buynak. 1976. Evaluation of a revolving plate type biofilter for use in recirculated fish production and holding units. Trans. Am. Fish. Soc., 105: 704–708.

Lewis, W. M., J. H. Yopp, A. M. Brandenburg, and K. D. Schnoor. 1981. On the maintenance of water quality for closed fish production systems by means of hydroponically grown vegetable crops. pp. 121–129, In: Vol. 1, Proceedings of the World Symposium on Aquaculture in Heated Effluents and Recirculated Systems, Stravanger, Norway, May 28–30, 1980, Heenemann, Berlin.

Lewis, W. M., J. H. Yopp, H. L. Schramm, Jr., and A. M. Brandenburg. 1978. Use of hydroponics to maintain quality of recirculated water in a fish culture system. Trans. Am. Fish. Soc., 107: 92–99.

Liao, P. B., and R. D. Mayo. 1974. Intensified fish culture combining water reconditioning with pollution abatement. Aquaculture, 3: 61–85.

Lucchetti, G. L., and G. A. Gray. 1988. Water reuse systems: a review of principal components. Prog. Fish-Cult., 50: 1–6.

Maruyama, T., K. Ochiai, A. Miura, and T. Yoshida. 1988. Effects of chloramine on the growth of *Porphyra yezoensis* (Rhodophyta). Bull. Jpn. Soc., Sci. Fish., 54: 1829–1834.

McMurtry, M. R., P. V. Nelson, D. C. Sanders, and L. Hodges. 1986. Sand culture of vegetables using recirculated aquacultural effluents. Appl. Agric. Res., 5: 280–284.

Mattice, J. S., S. C. Tsai, and M. B. Burch. 1981. Comparative toxicity of phyochlorous acid and hypochlorite ions to mosquitofish. Trans. Am. Fish. Soc., 110: 519–525.

Meade, T. L. 1974. The technology of closed system culture of salmonids. Marine Technology Report No. 30. University of Rhode Island Sea Grant Publication, Narragansett. 30 p.

Miller, G. E., and G. S. Libey. 1986. Evaluation of three biological filters suitable for aquaculture. J. World Maricult. Soc., 16: 158–168.

Milne, P. H. 1976. Engineering and the economics of aquaculture. J. Fish. Res. Board Can., 33: 888–898.

Muir, P. R., D. C. Sutton, and L. Owens. 1991. Nitrate toxicity to *Penaeus monodon* protozoea. Mar. Biol. 108: 67–71.

Murray, K. R., M. G. Poxton, B. T. Linfoot, and D. W. Watret. 1981. The design and performance of low pressure air lift pumps in a closed marine recirculation system. pp. 413–428, In: Tiews, K. (Ed.). Aquaculture in heated effluents and recirculation systems. Schriften der Bundesforschungsanstalt fuer Fischeri, Vols. 16–17.

Naegel, L. C. A. 1977. Combined production of fish and plants in recirculating water. Aquaculture, 10: 17–24.

Nijhof, M., and J. Bovendeur. 1990. Fixed film nitrification characteristics in sea-water recirculation fish culture systems. Aquaculture, 87: 133–143.

Novotny, A. J. 1969. The future of marine aquaculture in energy systems. Proc. Annu. Conf. West. Assoc. State Game Fish Comm., 49: 169–183.

Novotny, A. J. 1972. Fish and shellfish farming in coastal waters. Fish News (Books), London. 208 p.

Novotny, A. J. 1975. Net-pen culture of Pacific salmon in marine waters. Mar. Fish. Rev., 37: 36–47.

O'Farrell, T. P., F. P. Frauson, A. F. Cassel, and D. F. Bishop. 1972. Nitrogen removal by ammonia stripping. J. Water Pollut. Control Fed., 44: 1527–1535.

Paller, M. H., and W. M. Lewis. 1982. Reciprocating biofilter for water reuse in aquaculture. Aquacult. Eng., 1: 139–151.

Paller, M. H., and W. M. Lewis. 1988. Use of ozone and fluidized-bed biofilters for increased ammonia removal and fish loading rates. Prog. Fish-Cult., 50: 141–147.

Parker, N. C., and B. A. Simco. 1973. Evaluation of recirculating systems for the culture of channel catfish. Proc. Southeast. Assoc. Game Fish Comm., 27: 474–487.

Pierce, B. 1980. Water reuse aquaculture systems in two greenhouses in northern Vermont. Proc. World Maricult. Soc., 11: 118–127.

Piper, R. G., I. B. McElwain, L. E. Orme, J. P. McCraren, L. G. Fowler, and J. R. Leonard. 1982. Fish hatchery management. U.S. Fish and Wildlife Service, Washington, D.C. 517 p.

Poxton, M. G., K. R. Murray, B. T. Linfoot, and A. B. W. Pooley. 1981. The design and performance of biological filters in an experimental mariculture facility. Aquaculture in Heated Effluents and Recirculation Systems. pp. 369–382, In: K. Tiews (Ed.). Schriften der Bundesforschungsanstalt fuer Fishcheri, Vol. 16–17.

Rakocy, J. E. 1989a. Hydroponic lettuce production in a recirculating fish culture system. Univ. Virgin Islands Agricult. Exp. Stn., Island Perspectives, 3: 4–10.

Rakocy, J. E. 1989b. Vegetable hydroponics and fish culture—a productive interface. World Aquacult., 20(3): 42–47.

Rakocy, J. E., and R. Allison. 1979. Evaluation of a closed recirculating system for the culture of tilapia and aquatic macrophytes. pp. 296–307, In: L. J. Allen and E. C. Kinney (Eds.). Proceedings of the Bio-Engineering Symposium for Fish Culture. Fish Culture Section Publicaton No. 1. American Fisheries Society, Washington, D.C.

Rakocy, J. E., J. A. Hargreaves, and D. S. Bailey. 1989. Effects of hydroponic vegetable production on water quality in a closed recirculating system. J. World Aquacult. Soc., 20: 64A.

Reeves, J. G. 1972. Nitrogen removal: a literature review. J. Water Pollut. Control Fed., 44: 1895–1908.

Scott, K. R., and L. Allard. 1983. High-flowrate water recirculation system incorporating a hydrocyclone prefilter for rearing fish. Prog. Fish-Cult., 45: 148–153.

Scott, K. R., and L. Allard. 1984. A four-tank recirculation system with a hydrocyclone prefilter and a single water reconditioning unit. Prog. Fish-Cult., 46: 254–261.

Simpson, H. J., and M. Pedini. 1985. Brackishwater aquaculture in the tropics: the problem of acid sulfate soils. FAO Fishery Circular No. 791. 32 p.

Slone, W. J., D. B. Jester, and P. R. Turner. 1981. A closed vertical raceway fish cultural system containing clinoptilolite as an ammonia stripper. pp. 104–115, In: L. J. Allen and E. C. Kinney (Eds.). Proceedings of the Bio-Engineering Symposium for Fish Culture. Fish Culture Section Publication No. 1. American Fisheries Society, Washington, D.C.

Smith, C. E., R. G. Piper, and H. Tisher. 1981. The use of clinoptilolite and ion exchange as a method of ammonia removal in fish culture systems. Bozeman Informational Leaflet No. 4. U.S. Fish and Wildlife Service, Fish Cultural Development Center, Bozeman, Montana.

Speece, R. 1973. Trout metabolism characteristics and the rational design of nitrification facilities for water reuse in hatcheries. Trans. Am. Fish. Soc., 102: 323–334.

Stickney, R. R. 1988. The culture of macroscopic algae. World Aquacult., 19(3): 54–58.

Stickney, R. R. 1990. Controversies in salmon aquaculture and projections for the future of the aquaculture industry. pp. 455–461, In: Vol. 2, Proceedings of the Fourth Pacific Congress on Marine Science and Technology, Tokyo, Japan, July 16–20. PACON 90, Tokyo.

Stickney, R. R., and N. K. Person. 1985. An efficient heating method for recirculating water systems. Prog. Fish-Cult., 47: 71–73.

Sutterlin, A. M., C. Y. Couturier, and T. Devereaux. 1984. A recirculating system using ozone for the culture of Atlantic salmon. Prog. Fish-Cult., 46: 239–244.

Sutton, R. J., and W. M. Lewis. 1982. Further observations on a fish production system that incorporates hydroponically grown plants. Prog. Fish-Cult., 44: 55–59.

van Rign, J., and G. Rivera. 1990. Aerobic and anaerobic biofiltration in an aquaculture unit—nitrite accumulation as a result of nitrification and denitrification. Aquacult. Eng., 9: 217–234.

Vlasenko, M. I. 1969. Ultraviolet rays as a method for the control of diseases of fish eggs and young fishes. Probl. Ichthyol., 9: 697–705.

Wedemeyer, G. A., N. C. Nelson, and W. T. Yasutake. 1979. Potentials and limits for the use of ozone as a fish disease control agent. Ozone. Sci. Eng., 1: 295–318.

Wetzel, R. G. 1975. Limnology. W. B. Saunders, Philadelphia. 743 p.

Wheaton, F. W. 1977. Aquaculture engineering. Wiley-Interscience, New York. 708 p.

Wilde, E. W., R. J. Soracco, L. A. Mayack, R. L. Shealy, T. L. Boradwell, and R. F. Steffen. 1983. Comparison of chlorine and chlorine dioxide toxicity to fathead minnows and bluegill. Water Res., 17: 1327–1331.

Wren, S. W. 1984. Comparison of hydroponic crop production techniques in a recirculating fish culture system. M. S. Thesis, Texas A&M University, College Station.

Yunker, M. P., and E. D. Scura. 1985. An improved strategy for building brackishwater culture ponds with iron pyrite soils in mangrove swamps. p. 171, In: Y. Taki, J. H. Primavera, and J. A. Llobrera (Eds.). Proceedings of the First International Conference on the Culture of Penaeid Prawns/Shrimps, Iloilo City, Philippines, December 4-7, 1984.

Zweig, R. D. 1986. An integrated fish culture hydroponic vegetable production system. Aquacult. Mag., May/June: 34–40.

4 Nonconservative Aspects of Water Quality

GENERAL CONSIDERATIONS

Hundreds of thousands of chemical substances can be found in association with water. They range from such common ions as sodium and chloride, which account for much of the salinity in seawater, to trace elements such as cadmium, copper, and gold, and include various naturally occurring and manufactured organic compounds like petroleum fractions, pesticides, and polychlorinated biphenyls (PCBs). Water supplies intended for use in aquaculture may, as we have seen, contain naturally occurring substances like carbon dioxide, hydrogen sulfide, and iron at levels that make the water unsuitable for use unless the undesirable chemical is reduced to a safe level or completely removed. Some water sources are contaminated with toxic substances that may make them entirely unsuitable for aquaculture. In the majority of instances where water of suitable quantity is available, its quality is within acceptable limits. Routine chemical analyses can be obtained rather inexpensively, but if an array of evaluations for trace elements and organic chemicals is required, the analyses can run into the thousands of dollars.

While virtually every known chemical compound has been recovered from water, unless a water source is aberrant in some respect (unique to the particular area in which the water is found, or contaminated with toxic chemicals), there are a relatively small number of important water quality parameters of primary interest and concern to the aquaculturist. Of those, some may require frequent monitoring, while others may need to be determined only once, or at intervals of weeks, months, or even years to ensure that the values remain within acceptable limits.

The most important aspects of water quality—including some, like light, of a nonchemical nature—are considered in this and the following chapter. Various parameters and their importance are discussed, but there is little mention of synergisms, that is, the response of aquaculture organisms to two or more interacting water quality variables. The reason that synergisms are not discussed is because very little information is known about them. The aquaculturist should be aware, however, that the additive effect of two or more acceptable, though suboptimal, water quality conditions may create a stressful or even lethal condition.

This chapter discusses properties of water that can be affected by biological activity. Such properties are termed "nonconservative." In Chapter 5, those properties of water and water systems that are not affected by biological activity (conservative properties) are considered.

Among the nonconservative properties of water that are of interest and importance to aquaculturists are the levels of such nutrients as the various forms of nitrogen and phosphorus, dissolved oxygen (DO), and pH. Each of those variables is affected by organisms, often in association with photosynthetic activity by plants. Therefore, before discussing these and other important nonconservative properties of water, we will consider the processes that are responsible, in part, for controlling how some of them fluctuate temporally.

PHOTOSYNTHESIS AND PRIMARY PRODUCTION

Various types of primary producers can be found in the water systems associated with aquaculture facilities. Among them, depending on the type of water used, are aquatic macrophytes,[62] benthic algae (periphyton),[63] phytoplankton, filamentous algae, and autotrophic bacteria. In addition, terrestrial plants, including grasses, may grow out over the surface of a pond. Photosynthesis, the process by which carbon dioxide is converted into sugar, is a process undertaken by each of these types of organisms. A simplified equation for photosynthesis is:

$$CO_2 + H_2O \xrightarrow[\text{chlorophyll}]{\text{light}} CH_2O + O_2$$

Photosynthesis occurs in the presence of light and chlorophyll and requires the presence of certain enzymes. The sugar that is produced is biochemically converted within the cells of the autotrophic organism into the various other chemicals required for growth and metabolism (e.g., proteins, lipids, and complex carbohydrates).

Most photosynthetic organisms absorb light of wavelengths ranging from about 350 to 700 nm, though autotrophic bacteria often absorb light at up to 900 nm. This appears to be related to the capacity of the bacteria to utilize chemical energy as well as light energy in the elaboration of new biomass (Steemann Nielsen 1975).

For the majority of plants, photosynthesis occurs in response to wavelengths of light approximately within the range visible to humans. When light enters water, it becomes increasingly dispersed or attenuated as it penetrates more and more deeply because of absorption and scattering. The vertical attenuation coefficient of light in water as presented by Steemann Nielsen (1975) can be determined from the following formula:

---

[62] Macrophytes are defined as vascular plants and seaweeds. Included are rooted and floating freshwater plants along with kelp and other macroscopic algae.

[63] The periphyton community is made up of algal species that are attached to substrates. These include single-cell as well as filamentous algae. The *Aufwuchs* community includes both the plants and animals that are attached to, but do not penetrate, substrates in aquatic environments.

$$E_z = E_0 e^{-kz}$$

where E_z is the irradiance on a horizontal plane at a depth of z meters, E_0 is the irradiance just below the surface of the water, e is the natural logarithm, and k is the attenuation coefficient. From the form of the above equation it is apparent that light intensity decreases rapidly with increasing depth.

As light penetrates water, both the quality and quantity of the light at any given depth are affected. Long wavelengths (the red end of the spectrum) are absorbed first, with the shorter ones (the blue end of the spectrum) finally being absorbed at greater depths. Many marine species are bright red in color but are often difficult to see because that color is not transmitted under the low light levels at which many of those species reside.

The depth to which light of any given wavelength penetrates in water depends in part on the amount of suspended particulate matter present. Dissolved organic matter, suspended sediment particles (particularly silt and clay), and suspended organic matter (including living organisms such as phytoplankton and zooplankton) can exert a significant influence on light penetration. Certain inland and coastal waters that contain high levels of organic acids can turn the water a color similar to that of tea, for example. Plankton blooms may have a transitory but distinct and sometimes very important impact on photosynthesis by limiting light penetration. Phytoplankton blooms may, in fact, become self-limiting by greatly restricting light penetration and thereby reducing the volume of water in which photosynthesis can occur.

Since aquaculture is usually practiced in relatively shallow water and with species that seem to be relatively unaffected by its intensity, there is often little or no effort on the part of the culturist to control light. There are, however, instances when light control is warranted or even necessary. Some species, for example, spawn at depths below or near the maximum depth of light penetration. Certain species of penaeid shrimp are one example. Inducing such species to spawn in captivity has often been dependent on removal of one of the eyestalks (eyestalk ablation).[64] Wurts and Stickney (1984) suggested that attempts at spawning *Penaeus setiferus* in the laboratory, even when light was severely restricted, had been conducted at intensities well above those on the natural spawning grounds. That theory was supported by later studies conducted by George Chamberlain at Texas A&M University.[65]

It is generally accepted that the compensation depth—the depth at which plant photosynthesis and respiration are equivalent—coincides with the depth at which light intensity is 1% of that at the water surface. Even in the turbid waters that exist in many culture ponds, light intensity at the bottom typically exceeds 1% of the incident level. Even when this is not the case, there is usually sufficient light penetration to support significant amounts of photosynthetic activity.

Photosynthesis is dependent on light, but the rate at which it occurs will vary

---

[64] The hormones that control gamete development and the spawning of shrimp are located on the eyestalk, and their activation may be influenced by photoperiod, light intensity, or both.
[65] George Chamberlain, personal communication.

considerably as a function of water temperature. In general, the rate of chemical reactions increase with increasing temperature. This is certainly true of biochemical reactions that occur over the temperature range tolerated by species in which the reactions are occurring. Thus, the rate of photosynthesis can be expected to increase with increasing water temperature, all other factors being equal. In reality, blooms in primary productivity often occur during the spring and fall when the water is warming or cooling but not during periods of peak temperature for the year. The fall and winter blooms are related not to temperature or light penetration but to releases of required nutrients.

At least three forms of chlorophyll occur in green plants: chlorophylls *a, b,* and *c.* Chlorophyll *a* appears to be the most important in terms of energy absorption. Other photosensitive plant pigments include certain carotenoids (carotene, lutein, fucoxanthin, peridinin) and biliproteins (phycoerythrin and phycocyanin). Steemann Nielsen (1975) indicated that some pigments other than chlorophyll appear able to transfer the energy they absorb to the type of chlorophyll best able to utilize that energy in carbon fixation. Carotenoids other than those listed above appear to have little or no function in photosynthesis.

Importance of Primary Production to Aquaculture

Assuming that the culture species of interest is an animal rather than a plant, primary productivity may or may not be of immediate importance to the aquaculturist. As we have seen, culturists who employ closed water systems commonly attempt to restrict the presence of plants within their systems except in the case of utilizing a chamber containing photosynthetic organisms to assist with water purification. At the other extreme are mollusc culturists who may be completely dependent on the presence of high concentrations of algae in the water to feed their clams, oysters, scallops, or mussels. Phytoplankton culture to provide food for the larvae of aquaculture species, or to feed zooplanktonic organisms that are used for larval and juvenile feeds, are important to some aquaculturists. The production of phytoplankton as a food source is discussed in more detail in Chapter 6.

For most aquaculture strategies, macrophyte and filamentous algal growth in the culture chambers should be discouraged and the growth of phytoplankton encouraged.[66] Control of rooted plants can be effected to some extent by having pond levees with the proper side slopes, thereby limiting the amount of shallow water present in which such plants can become established. Establishment and maintenance of phytoplankton blooms through fertilization will also help control the development of other forms of aquatic vegetation. (Fertilization is discussed below.) Establishment and maintenance of a phytoplankton bloom may not only be neces-

---

[66] There are always exceptions, of course. The first one that comes to mind in this case is associated with the production of grass carp *(Ctenopharyngodon idella),* which feed on aquatic plants, though they are quite particular about the types of plants they readily consume. Another exception involves the culture of milkfish *(Chanos chanos).* The growth of dense mats of filamentous algae is encouraged in milkfish ponds since the fish feed upon that material.

Figure 52. A fish culture pond in which mats of algae have floated to the surface.

sary to provide food for larval and juvenile aquaculture species but also will reduce the level of light penetration, thereby limiting the establishment of undesirable plants.

While macrophytes can be found in both the freshwater and marine environments, the majority of the problems associated with them occur in freshwater culture systems, and most of those problems are associated with pond culture. The consequences of failure to control undesirable types of aquatic plants go well beyond their unsightliness (Figures 52 and 53) and the fact that they remove nutrients that might otherwise support phytoplankton.

Water temperature may be influenced to some extent by the presence of plants in culture ponds, especially when clear water is present as a result of limited phytoplankton productivity. Water tends to warm more slowly and lose heat more rapidly when it is clear. Sunlight can penetrate more deeply in clear water, thereby increasing the area that can be colonized by rooted plants. Such plants can spread from the margins and may eventually cover the pond bottom.

Heavy growth of rooted aquatic macrophytes and filamentous algae can entrap culture animals, immobilizing them and increasingly their vulnerability to attack by predatory aquatic insects, snakes, turtles, and so forth. Access to food may also be restricted in weed-choked ponds, and the wasted feed will add nutrients that go toward promoting additional plant growth.

Floating aquatic plants such as duckweed (*Lemna* sp.) can completely cover the surface of a culture pond (Figure 54), thereby limiting not only light but also the diffusion of oxygen into the water. Dissolved oxygen (DO) levels can be greatly reduced as a result. In the tilapia pond shown in Figure 54, the morning DO concentration reached 0.0 mg/l where the duckweed cover was complete. Tilapia have the ability to obtain oxygen even when the water is depleted by passing surface water

Figure 53. A pond in the Philippines in which rooted plants have taken over.

Figure 54. Duckweed (*Lemna* sp.) completely covers the surface of this small tilapia pond.

over their gills. Even in an oxygen-depleted body of water, the surface film, which is in contact with the atmosphere, will be saturated with oxygen. Tilapia are one of the few aquaculture organisms that can efficiently obtain oxygen from that surface film. In the case of the duckweed-covered pond, even the tilapia were frustrated in their attempt to obtain oxygen, and the entire population was asphyxiated.

An additional and significant problem associated with the presence of macrophytes and filamentous algae is that ponds supporting such growth are extremely difficult to sample or harvest. Seines are very difficult to pull through stands of vegetation and tend to roll up. As a consequence, the animals being harvested often have little difficulty avoiding the net. Those that are captured are difficult to separate from the plants and may die before they can be found, extricated, and placed back into the water.

Decomposition of aquatic vegetation that dies can greatly increase the biochemical oxygen demand in a culture pond and may lead to such low DO levels that the culture species is stressed or killed. Stressed animals that survive an oxygen depletion may develop disease problems as a result of their exposure to the degradation in water quality.

Aquatic plants have been evaluated as feeds for livestock (e.g., see Baily 1965, Lange 1965, Boyd 1968a, b, Culley and Epps 1973, Lizaman et al. 1988). The digestibility of 40 species of aquatic plants by livestock was determined by Boyd and McGinty (1981). While many species appear to be suitable as livestock feeds, their extremely high moisture content makes transport of the wet material economically unfeasible. Drying of the plants prior to shipment may also be a problem from an economic point of view, so in most cases it would appear as though the livestock should be produced in close proximity to the source of the aquatic vegetation.

Vegetation-choked lakes are better sources of aquatic plants for livestock feed than are aquaculture ponds with moderate amounts of plant growth. Most aquaculturists are more concerned about eliminating aquatic plants than they are interested in producing them. Exceptions exist in the case of such desirable plants as decorative water lilies and aquarium plants, along with edible aquatic plants such as watercress and water chestnuts.

Measuring Primary Productivity

The typical commercial aquaculturist does not need to measure primary productivity[67] and is, in reality, concerned with the subject only in terms of controlling nuisance vegetation and perhaps in maintaining a suitable level of DO in the culture system. Exceptions exist, of course, as alluded to previously. Persons involved in maintaining larval fishes and invertebrates may be particularly interested in the level of primary productivity, since it is the plants (usually phytoplankton) that form the base of the food web on which the aquaculture species rely. Those involved in growing molluscs will also need to monitor phytoplankton productivity because in most cases the animals being cultured utilize that resource as their source of food.

---

[67] Primary productivity is defined as the rate at which plant tissue is elaborated.

Whether directly involved in the establishment or maintenance of an algae bloom or not, aquaculturists will have encounters with aquatic plants and should have some familiarity with the importance of photosynthesis on certain water quality variables as well as an understanding of how primary productivity can be measured. Various techniques have been developed. Some of the more common ones are described in this section.

With respect to rooted aquatic macrophytes and floating plants, annual productivity can be determined by measuring the dry biomass per unit area of pond bottom at the end of the growing period.[68] Macrophytes, while subject to some consumption by invertebrates, typically die back during the winter and return when the water warms in the spring. A better estimate of annual production can be made by comparing samples taken at the beginning of the growing season with those at the end. Significant errors can be introduced if the population of herbivores present is high.

The technique of measuring primary production by determining biomass at one or two times during the year does not work for phytoplankton, filamentous algae, or periphyton, all of which are present in natural waters throughout the year. Their reproductive cycles, growth rates, and the rates at which they die or are consumed by zooplankton and other organisms make their communities highly dynamic. Different species may dominate these plant communities at different times of the year, and predictability of species succession is generally not possible.

An indication of periphyton production can be obtained by placing clean substrates (e.g., glass microscope slides, glass plates, glass rods, or sheets of plastic) in the water at selected depths in the photic zone and determining the rate at which plant material accumulates with time. The determination is usually based on ash-free dry weight (the difference in weight of the material after being dried at between 80 and 105°C, and then ashed at 550 to 600°C). One problem with this technique is that the substrate will attract various types of animals as well as plants. While it is possible to remove the larger insects from substrates placed in fresh water and the tunicates, bryozoans, and so forth from substrates placed in salt water, complete removal of such animals as protozoans and various kinds of larvae is not generally feasible. Thus, such measurements are actually of the *Aufwuchs* community rather than only of the plant component of that community.

Phytoplankton productivity has been measured indirectly in a number of ways. One of the first was the use of the light and dark bottle technique (Gaarder and Gran 1927) in which identical samples of water containing phytoplankton[69] are placed in paired bottles: one transparent and the other opaque. The DO concentration in the initial water sample is determined; then each bottle is suspended at a preselected depth in the water body from which the water came or in a chamber designed to simulate natural conditions. The bottles are allowed to incubate during daylight—

---

[68] The process is simple if plant density is uniform. In most cases it is not uniform because of variable pond depths and other factors, so a series of samples would have to be taken from various locations and a mean value established for the pond.

[69] No attempt is made to remove bacteria and other microorganisms from such samples, so there can be some influence on the results of metabolic activities by other organisms.

usually for several hours—after which the DO concentration is again determined. Following incubation, DO in the transparent (light) bottle should be higher than at the outset of the experiment as a result of photosynthetic activity during incubation (as long as the bottles were suspended above the compensation depth). The DO in the opaque (dark) bottle should be lower than in the initial sample because of respiration.[70] The amount of oxygen produced in the light bottle is added to that consumed in the dark bottle. An indication of gross primary production can be obtained by determining the theoretical amount of carbon fixed using the relationships between oxygen and carbon shown in the photosynthesis equation.

The light and dark bottle technique is considered to be crude as compared with more modern techniques. One of those that has been widely used since it was first described by Steemann Nielsen (1951) involves the use of radioactive carbon-14 (^{14}C). The technique involves the use of light and dark bottles as previously described. A known amount of ^{14}C-labeled carbon is injected into each bottle and the amount of the isotope that is absorbed by the phytoplankton during the incubation period is determined. By knowing the ratio of the ^{14}C to the other carbon isotopes, the photosynthesis equation can once again be used to develop an indication of the amount of primary productivity. Following incubation the samples are filtered through a membrane filter (0.45 μm), and the radioactivity is counted in a liquid scintillation counter to determine how much of the radioisotope was incorporated into the phytoplankton cells.

The carbon-14 technique has been modified over the years by a number of investigators who have introduced correction factors for various interferences or enhancements of apparent carbon uptake. For example, it has been learned that ^{12}C is absorbed by plants at a slower rate than ^{14}C (Steemann Nielsen 1952); therefore, a correction factor is required when the total amount of carbon fixed is calculated. In many cases the formula presented by Saunders et al. (1962) is utilized for determining photosynthetic production by the ^{14}C technique.

Apparent carbon fixation in the dark bottle may be the result of absorption or adsorption, or it may indicate that chemoautotrophic bacteria are present in the sample. In any case, the dark fixation is subtracted from that which occurs in the light. Despite the various problems associated with the ^{14}C technique, it has been widely used.

Determination of the total amount of chlorophyll in a water sample is another technique that has been used to provide an estimate of primary productivity. Procedures for extracting chlorophyll from plant material as well as formulas for calculating the levels of various pigments through spectrophotometric analysis of the extract have been available for many years (Richards and Thompson 1952). Ryther (1956) found that although both chlorophyll content and the rate of photosynthesis are variable, the rate of photosynthesis per unit of chlorophyll is relatively constant, permitting some indication of primary productivity to be obtained through chlorophyll analysis.

---

[70]Respiration in the dark bottle is due to both the plants and animals. Plants, like animals, continuously respire, but when plants are actively photosynthesizing, they commonly produce oxygen more rapidly than they consume it, so the net result is an increase (which only occurs in the light bottle).

Chlorophyll fluoresces when exposed to certain wavelengths of light, and the degree of fluorescence relates to the amount of chlorophyll present, and thus is related to primary productivity. Fluorometers are available that provide a continuous reading of the extent of fluorescence in water pumped through them, thereby allowing investigators to obtain rapid indications of productivity without having to collect and extract water samples, though discrete samples can also be run on most fluorometers. Instantaneous sampling with such an instrument in the field saves time and money and provides immediate results. While not used routinely in conjunction with commercial aquaculture ventures, the technique could be useful for measuring phytoplankton bloom development in ponds and the status of algae cultures in hatcheries that maintain such cultures to feed larvae and other stages of aquatic animals. Algae can also be enumerated in Coulter counters.

There are other techniques by which primary production can be measured. Some, such as remote sensing, require sophisticated equipment (even the use of spacecraft or satellites) and are even more indirect than those described above. Perhaps the most commonly used technique, and by far the simplest, is the Secchi disc. A Secchi disc is a flat, round sheet of metal, wood, or plastic that is commonly about the size of a dinner plate. The surface of the disc is usually divided into pie-shaped wedges (usually about six in number) of alternating black and white. A string is attached to the center of the disc, and the apparatus is lowered into the water column. A measurement of the string is taken at the depth where the disc disappears from sight, and that depth becomes the Secchi disc reading. The higher the turbidity of the water, the lower the reading will be. As is described below, Secchi disc readings are useful in evaluating of the establishment and maintenance of phytoplankton blooms in ponds. Secchi discs are available commercially, or they can be fabricated by the culturist.

PLANT NUTRIENTS

In order to grow properly, plants require certain levels of available nutrients. While a variety of such chemicals is required, the ones that typically become limiting are nitrogen and phosphorus. Silicon can become limiting to diatoms, which have a skeleton (test) that is comprised largely of that element. If the aquaculturist is attempting to maintain a continuous phytoplankton culture in the laboratory for use in feeding a particular type of animal, or trying to culture the algae as an end product, a complete nutrient medium will be required. Pond fertilization for establishment of a plankton bloom usually requires only the addition of nitrogen and phosphorus.

Nitrogen

Nitrogen is required by all living organisms, being an important component of protein and other required chemical substances. Nitrogen is taken up by plants primarily in the form of nitrate (NO_3^-) ions. Animals satisfy their nitrogen requirements through the intake of food. Nitrogenous wastes are excreted by animals in several

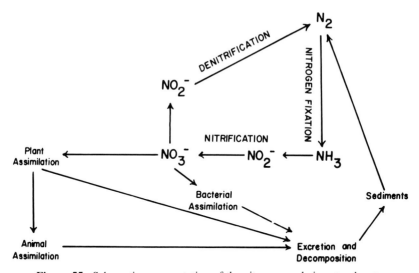

Figure 55. Schematic representation of the nitrogen cycle in natural waters.

forms: ammonia, creatine, creatinine, free amino acids, urea, and uric acid. Nitrogenous compounds are also released during the bacteriological decomposition of plant and animal matter. The primary source of nitrogen from aquaculture animals is in the form of ammonia (NH_3). We have already seen that aerobic bacteria in the genus *Nitrosomonas* are responsible for nitrifying ammonia to nitrite (NO_2-), while *Nitrobacter* bacteria are responsible for the step from nitrite to nitrate. The nitrification reactions are critical to efficient biofilter operation, but they also occur in ponds and other types of culture systems.

Other bacteria are able to denitrify nitrate and convert it to elemental nitrogen (N_2) gas. According to Meade (1976) denitrifying bacteria can be found in such genera as *Pseudomonas, Achromobacter, Bacillus, Micrococcus,* and *Corynebacterium.* Energy for the reduction reactions involved may come from certain carbohydrates and alcohols. Recirculating water systems have been designed that employ special chambers to promote these reactions by supplying the proper substrate. A simplifed nitrogen cycle is shown in Figure 55.

In pond environments there is little concern over the accumulation of nitrate because primary producers in the system generally remove it from the water nearly as rapidly as it is produced. In closed systems—including some hatchery systems where susceptible species such as shrimp are produced (Chen et al. 1990a, b, Muir et al. 1991)—nitrate levels may sometimes become sufficiently high to produce stress or mortality. Various types of water systems that receive high levels of organic or inorganic fertilization can also exhibit high nitrate concentrations.

While nitrogen is available to plants in the form of nitrate, it apparently must be reduced to ammonia once again before it can be absorbed into plant tissues (Fogg 1972). The reaction appears to be light catalyzed and to proceed as follows:

$$NO_{2^-} + H_3O^+ \xrightarrow[\text{light}]{} NH_3 + 2O_2$$

Nitrite and ammonia are both toxic to aquatic animals at much lower concentrations than is nitrate. Rare in natural waters because it is an intermediate that is quickly transformed by bacteria to nitrate, nitrite sometimes occurs in high concentrations in aquaculture systems. Historically, the problem has been largely found in flowing water systems and has been resolved through the incorporation of efficient biofiltration or suitable exchange rates with new water. In recent years, nitrite toxicity has occurred in ponds when very high densities of animals are being maintained. Catfish farmers in Mississippi, for example, have experienced nitrite toxicity during the late summer or early fall, when fish biomass is at the highest level of the year, the water is warm, and the feeding rate is extremely high.

Lees (1952) reported that since the growth of *Nitrobacter* is inhibited in the presence of ammonia, nitrite may become concentrated in biofilters until the ammonia concentration is greatly reduced. Once the biofilter begins to operate efficiently, nitrite usually ceases to be a problem unless some change occurs that results in destruction of the *Nitrobacter*. This can happen when chemicals are used to treat for diseases or if the system becomes anaerobic.

Nitrite toxicity in fish was reviewed by Lewis and Morris (1986). Their review included both species of aquaculture interest and others. The authors indicated that salmonids were among the most sensitive fishes tested. Results from some of the studies that have been conducted with species of aquaculture importance are presented in Table 12.

Experiments to determine the LC_{50} of a chemical on an aquatic species over a discrete time period (usually not more then 96 hr) are called acute studies. Long-term or chronic exposure to much lower levels of a toxicant like nitrite might be lethal or could cause pathological changes. Thus, safe levels are often considered to be some fraction (e.g., $\frac{1}{10}$ or even $\frac{1}{100}$) of the acute toxicity level. Most studies report the LC_{50} value, though some, like Chen and Chin (1988a), have developed what they consider to be a safe level of exposure from the LC_{50} data.

Effects of nitrite on the eggs, alevins, and fry of Atlantic salmon *(Salmo salar)* by Williams and Eddy (1989) have shown that the early development stages can tolerate very high levels (24 hr LC_{50} values were 3,276 mg/l for eggs, and 2,940 mg/l for early alevins, decreasing to 121.8 mg/l as the alevins developed). However, exposure of eggs to as little as 14 mg/l of nitrite in either fresh or brackish water delayed hatching and had measureable effects on the cardiovascular system.

The results of studies on the same species can vary considerably as shown in Table 12 for such species as *Anguilla anguilla, Ictalurus punctatus*, and *Penaeus monodon*. The duration of the studies is one factor, as is the size of the animals tested. As indicated by Russo (1984), pH, chloride concentration, and calcium concentration also affect nitrite toxicity. The study of Almendras (1987) with eels and milkfish *(Chanos chanos)* demonstrated that salinity can have a significant influence on the tolerance of fish to nitrite. Tomasso and Carmichael (1991) found that different strains of channel catfish responded differently to nitrite, with some strains being nitrite-susceptible and others being nitrite-resistant.

TABLE 12. Nitrite Toxicity for Selected Species of Fishes and Invertebrates of Aquaculture Importance

Species	Type of Trial	Lethal Level (mg/l)	Citation
		Fishes	
Anguilla anguilla	96 hr	84–974[a]	Saroglia et al. (1981)
Chanos chanos	48 hr LC_{50}	12,675[b]	Almendras (1987)
Clarias batrachus	48 hr LC_{50}	35.6	Duangsawasdi and Sripumun (1981)
	48 hr LC_{50}	15.8, 35.6[c]	Sripumun and Momsiri (1982)
Clarias lazera	96 hr LC_{50}	28, 32[c]	Hilmy et al. (1987)
Ctenophyaryngodon idella	96 hr LC_{50}	4.62	Wang and Hu (1989)
Dicentrarchus labrax	96 hr LC_{50}	154–274[d]	Saroglia et al. (1981)
Ictalurus punctatus	24 hr LC_{50}	33.8	Konikoff (1975)
	48 hr LC_{50}	28.8	Konikoff (1975)
	72 hr LC_{50}	27.3	Konikoff (1975)
	96 hr LC_{50}	24.8	Konikoff (1975)
	96 hr LC_{50}	7.1 ± 1.9	Palachek and Tomasso (1984)
Micropterus salmoides	96 hr LC_{50}	140.2 ± 8.1	Palachek and Tomasso (1984)
Morone saxatilis	24 hr LC_{50}	163	Mazik et al. (1991)
Tilapia aurea	96 hr LC_{50}	16.2 ± 2.3	Palachek and Tomasso (1984)
		Invertebrates	
Penaeus chinensis	24 hr LC_{50}	339	Chen et al. (1990a)
	96 hr LC_{50}	37.7	Chen et al. (1990a)
	120 hr LC_{50}	29.2	Chen et al. (1990a)
	144 hr LC_{50}	27.0	Chen et al. (1990a)
	192 hr LC_{50}	23.0	Chen et al. (1990a)
Penaeus monodon	96 hr LC_{50}	1.36[e]	Chen and Chin (1988)
	96 hr LC_{50}	0.11[e]	Chen and Chin (1988)
	24 hr LC_{50}	218[f]	Chen et al. (1990b)
	48 hr LC_{50}	193[f]	Chen et al. (1990b)
	96 hr LC_{50}	171[f]	Chen et al. (1990b)
	144 hr LC_{50}	140[f]	Chen et al. (1990b)
	192 hr LC_{50}	128[f]	Chen et al. (1990b)
	240 hr LC_{50}	106[f]	Chen et al. (1990b)

[a] Several values were obtained over a salinity range of from 0 to 36 parts per thousand. Tolerance to nitrite increased with increasing salinity. The values shown are for salinities of 0 (84 mg/l) and 36 parts per thousand (974 mg/l).

[b] The low values was obtained in fresh water, the higher one in 16 parts per thousand salinity.

[c] Two sizes of fish were tested.

[d] Trials were run at temperatures ranging from 17 to 27°C. Toxicity decreased with increasing temperature.

[e] The value represents what the authors considered to be a safe level for postlarvae (1.36 mg/l) and nauplii (0.11 mg/l)

[f] Experiments were run on animals of 91 ± 8 mm in length.

Greater toxicity has been found in exposure of *Penaeus monodon* to mixtures of ammonia and nitrate than from exposure to either chemical alone. The synergistic effects of the two forms of nitrogen became apparent after 96 hr of exposure (Chen and Chin 1988b).

In fish, nitrite combines with hemoglobin in the blood to produce methemoglobin and a condition known as methemoglobinemia. Hemoglobin that has been converted to methemoglobin is unable to carry oxygen, so affected animals are asphyxiated. Suspected incidences of nitrite toxicity can be quickly confirmed if the culturist sacrifices a fish and examines the blood. Nitrite is believed to enter the blood in conjunction with chloride/bicarbonate exchange. Eddy and Williams (1987) concluded that fish such as salmonids thats have high chloride uptake rates are much more susceptible to nitrite toxicity than fish such as carp that have low chloride uptake rates. In the common carp *(Cyprinus carpio)* there is a high correlation between chloride level and nitrite toxicity (Hasan and Macintosh 1986b). On the other hand, Tomasso (1986) found that while chloride ion increased the tolerance of channel catfish for nitrite, there was no such response in largemouth bass *(Micropterus salmoides)*.

In fish with methemoglobinemia the blood will be chocolate brown in color. In the case of channel catfish, affected fish will rest on the bottom of the culture chamber and will swim erratically for up to one minute immediately before dying. They die with their mouths open and their opercles closed (Konikoff 1975).

Exposures to sublethal concentrations of nitrite can cause pathology in fish. Hemolytic anemia was reported in the sea bass *(Dicentrarchus labrax)* by Scarano et al. (1984). Arillo et al. (1984) concluded that liver hypoxia was the cause of mortalities in rainbow trout exposed to nitrite. Gill hypertophy has been reported in rainbow trout (Gaino et al. 1984) and tilapia of various species and their hybrids (Lightner et al. 1988) exposed to sublethal levels of nitrite.

Catfish farmers have found that the addition of 25 mg/l of salt for each 1 mg/l of nitrite present is an effective treatment for methemoglobinemia (Anonymous 1980). The treatment does not remove the nitrite from the water.

Blanco and Meade (1980) reported that steelhead trout *(Oncorhynchus mykiss)* demonstrated increased tolerance to nitrite when the dietary level of vitamin C (ascorbic acid) was increased. The authors speculated that ascorbic acid acted to reduce methemoglobin to hemoglobin, but also indicated that the vitamin has a protective effect against stress in fish, which may have played a role.

Ammonia is one of the variables in water that are of primary concern to aquaculturists. Other forms of nitrogenous waste are relatively unimportant in most cases, so it is the ammonia level in water that has generally been monitored. As we have seen, nitrite can also be a critical factor, so it is also sometimes closely watched. Colorimetric tests that can be conducted virtually anywhere have been developed and are in wide use.

Fishes excrete most of their nitrogenous waste through the gills in the form of ammonium ion, NH_4^+ (Hochachka 1969). Ammonium ion accounts for as much as 60 to 90% of the total nitrogen excreted (H. W. Smith 1929, Wood 1958, Fromm 1963). In addition to the ionized form (NH_4^+), un-ionized ammonia (NH_3) occurs

TABLE 13. Percentage of Total Ammonia in the Un-ionized (NH_3) Form for a Few Temperatures and pH Values (Source: Emerson et al. 1975)

Temperature	pH				
(°C)	6.5	7.0	7.5	8.0	8.5
16	0.1	0.3	0.9	2.9	8.5
18	0.1	0.3	1.1	3.3	9.8
20	0.1	0.4	1.2	3.8	11.2
22	0.1	0.5	1.4	4.4	12.7
24	0.2	0.5	1.7	5.0	14.4
26	0.2	0.6	1.9	5.8	16.2
28	0.2	0.7	2.2	6.6	18.2
30	0.3	0.8	2.5	7.5	20.3

in water. The toxicity of ammonia to aquatic organisms in primarily associated with the level of un-ionized ammonia (Chipman 1934, Wuhrmann et al. 1947, Wuhrmann and Woker 1948, Downing and Merkens 1955, Merkens and Downing 1957, Lloyd 1961), whereas the ionized form appears to be relatively harmless (Tabata 1962).

Both ionized and un-ionized ammonia can occur together, but the ratio between them is dependent on temperature, pH, DO, carbon dioxide concentration, bicarbonate alkalinity, and salinity (H. W. Smith 1929, Lloyd and Herbert 1960, Lloyd 1961, Brown 1968, Emerson et al. 1975, Thurston et al. 1981a). Un-ionized ammonia increases relative to ionized ammonia with increasing temperature and pH (Table 13) but decreases as carbon dioxide increases and in hard and saline waters (Emerson et al. 1975, Thurston et al. 1981a). Long-term exposure of aquatic animals to elevated ammonia levels can result in reduced growth, impaired stamina (Burrows 1964), gill abnormalities, and ultimately death.

Ammonia electrodes used in conjunction with a pH meter and colorimetric tests for ammonia can provide the aquaculturist with a total ammonia value, which is satisfactory under most circumstances. Seawater interferes with the colorimetric technique for ammonia determination, so such samples should be distilled prior to testing. Ammonia concentration in the distillate is not changed, but the chemicals causing the interference are not passed in the condensed steam. Tables such as those produced by Emerson et al. (1975) will provide the culturist with a means of determining that actual level of un-ionized ammonia.

Studies of ammonia toxicity have been conducted on a number of species of aquaculture interest and under a variety of conditions. Table 14 provides an indication of how such factors as pH, life-cycle stage, salinity, and the form in which ammonia is added for purposes of the bioassay influence the results. In general, coldwater fishes are less tolerant of ammonia than are warmwater species (Table 14). Calamari et al. (1981) proposed a water quality standard of 0.02 mg NH_3 for rainbow trout. Haywood (1983), who reviewed the effects of ammonia on teleost

TABLE 14. Ammonia Toxicity for Selected Species of Fishes and Invertebrates of Aquaculture Importance

Species and Conditions	Type of Trial	Lethal Level (mg/l)	Citation
		Fishes	
Anguilla japonica			
pH = 5	24 hr LC$_{50}$	2844[a]	Yamagata and Niwa (1982)
pH = 7	24 hr LC$_{50}$	820[a]	Yamagata and Niwa (1982)
pH = 9	24 hr LC$_{50}$	16.8[a]	Yamagata and Niwa (1982)
Chanos chanos	4 hr LC$_{50}$	1.89[b]	Cruz (1981)
	48 hr LC$_{50}$	1.46[b]	Cruz (1981)
	72 hr LC$_{50}$	1.25[b]	Cruz (1981)
	96 hr LC$_{50}$	1.12[b]	Cruz (1981)
Clarias batrachus	48 hr LC$_{50}$	15.78[b]	Sripumun and Somsiri (1982)
Ctenopharyngodon idella			
26 days old	96 hr LC$_{50}$	0.57[b]	Zhou et al. (1986)
47 days old	96 hr LC$_{50}$	1.61[b]	Zhou et al. (1986)
125 days old	96 hr LC$_{50}$	1.68[b]	Zhou et al. (1986)
47 days old	48 hr LC$_{50}$	1.73[b]	Zhou et al. (1986)
60 days old	48 hr LC$_{50}$	2.05[b]	Zhou et al. (1986)
125 days old	48 hr LC$_{50}$	2.14[b]	Zhou et al. (1986)
Cyprinus carpio			
206 mg fry	48 hr LC$_{50}$	1.87[b]	Hasan and Macintosh (1986a)
206 mg fry	96 hr LC$_{50}$	1.84[b]	Hasan and Macintosh (1986a)
206 mg fry	168 hr LC$_{50}$	1.78[b]	Hasan and Macintosh (1986a)
299 mg fry	48 hr LC$_{50}$	1.76[b]	Hasan and Macintosh (1986a)
299 mg fry	96 hr LC$_{50}$	1.74[b]	Hasan and Macintosh (1986a)
299 mg fry	168 hr LC$_{50}$	1.68[b]	Hasan and Macintosh (1986a)
Ictalurus punctatus	24 hr LC$_{50}$	2.77[b]	Robinette (1976)
pH 8.8[d]	24 hr LC$_{50}$	1.91[b]	Sheehan and Lewis (1986)
pH 8.0[d]	24 hr LC$_{50}$	1.45[b]	Sheehan and Lewis (1986)
pH 7.2[d]	24 hr LC$_{50}$	1.04[b]	Sheehan and Lewis (1986)
pH 6.0[d]	24 hr LC$_{50}$	0.74[b]	Sheehan and Lewis (1986)
pH 8.8[e]	24 hr LC$_{50}$	2.24[b]	Sheehan and Lewis (1986)
pH 8.0[e]	24 hr LC$_{50}$	1.75[b]	Sheehan and Lewis (1986)
pH. 7.2[e]	24 hr LC$_{50}$	1.16[b]	Sheehan and Lewis (1986)
pH 6.0[e]	24 hr LC$_{50}$	0.81[b]	Sheehan and Lewis (1986)
	96 hr LC$_{50}$	1.5–3.1[b]	Ruffier et al. (1981)
Micropterus salmoides	96 hr LC$_{50}$	0.7–1.2[b]	Ruffier et al. (1981)
Oncorhynchus mykiss			
Eggs to hatch	96 hr LC$_{50}$	0.49[b]	Calamari et al. (1981)
70-day-old fry	96 hr LC$_{50}$	0.16[b]	Calamari et al. (1981)
Fingerlings	96 hr LC$_{50}$	0.44[b]	Calamari et al. (1981)
	96 hr LC$_{50}$	0.3b	Ruffier et al. (1981)

TABLE 14. (*continued*)

Species and Conditions	Type of Trial	Lethal Level (mg/l)	Citation
		Fishes	
O. tshawytscha parr			
Freshwater	24 hr LC_{50}	0.36[b]	Harader and Allen (1983)
9.6 ppt salinity[c]	24 hr LC_{50}	2.2[b]	Harader and Allen (1983)
Sparus aurata	96 hr LC_{50}	23.7[a]	Wajsbrot et al. (1991)
Tilapia aurea	48 hr LC_{50}	2.40[b]	Redner and Stickney (1979)
Tilapia mossambica	*48 hr LC_{50}*	6.6[b]	Daud et al. (1988)
× *T. nilotica*	72 hr LC_{50}	4.07[b]	Daud et al. (1988)
hybrid	96 hr LC_{50}	2.88[b]	Daud et al. (1988)
		Invertebrates	
Mercenaria mercenaria			
4 mm long	30 day LC_{50}	20.0[a]	Stevens (1982)
6 mm long	30 day LC_{50}	28.0[a]	Stevens (1982)
10 mm long	30 day LC_{50}	34.5[a]	Stevens (1982)
Penaeus chinensis	24 hr LC_{50}	3.29[b]	Chen et al. (1990a)
	48 hr LC_{50}	2.10[b]	Chen et al. (1990a)
	90 hr LC_{50}	1.53[b]	Chen et al. (1990a)
	120 hr LC_{50}	1.44[b]	Chen et al. (1990a)
Penaeus monodon	96 hr LC_{50}	1.69[b]	Allan, et al. (1990)
Penaeus simisulcatus	96 hr LC_{50}	23.7[a]	Wajsbrot et al. (1990)

[a] Reported as mg/l total ammonia
[b] Reported as mg/l unionized ammonia
[c] ppt is parts per thousand
[d] ammonium chloride
[e] ammonium sulfate

fishes, recommended maximum total ammonia exposure levels of 1.0 mg/l for salmonids and 2.5 mg/l for other freshwater and marine fishes.

Thurston et al. (1981b) found that rainbow trout and cutthroat trout *(Salmo clarki)* were more tolerant to constantly elevated ammonia levels than to fluctuating concentrations. Thurston and Russo (1983) indicated that the median tolerance limit for un-ionized ammonia ranged from 0.16 to 1.1 mg/l in rainbow trout, with susceptibility to ammonia decreasing as the fish developed from the sac fry to juvenile stage. They also found that toxicity decreased as temperature increased over the range of 12 to 19°C. Redner and Stickney (1979) found that *Tilapia aurea* could develop an increased tolerance to ammonia when exposed to sublethal levels in advance of bioassays.

Both fish (Yamagata and Niwa 1982, Sheehan and Lewis 1986) and shrimp (Chen and Chin 1989) are more tolerant of elevated ammonia with increasing pH.

Shrimp appear to be more sensitive to ammonia in the periods just before, during, and after ecdysis[71] (Wajsbrot et al. 1990, Lin et al. 1991). Further, the toxicity of ammonia is enhanced as DO concentration is reduced (Allan et al. 1990). The relationship between ammonia toxicity and DO has also been shown in gilthead seabream (Wajsbrot et al. 1991).

Sublethal concentrations of ammonia can cause histological changes in fish (Flis 1968, Smart 1976, Redner and Stickney 1979, Calamari et al. 1981, Thurston et al. 1984), and will also affect growth (Robinette 1976, Sadler 1981). Gill hyperplasia is a common sign of chronic ammonia toxicity (Smith and Piper 1975, Smart 1976).

Phosphorus

Phosphorus is a required nutrient for both plants and animals. It is often present in only minute concentrations in natural waters, though it reportedly can range from about 0.01 to 200 mg/l (Wetzel 1975). Phosphorus tends to be the first limiting nutrient in natural fresh waters, while nitrogen may be the first limiting nutrient in the marine environment. Phosphorus can be stored in plants, so when it is released in large amounts, such as during the collapse of a phytoplankton bloom, it will quickly cycle into other primary producers and be removed from solution.

In flowing water systems, with the exception of closed or partially recirculating systems, phosphorus levels do not tend to be very high within the system because of the rapid water exchange rates. There may be some increase above ambient levels in a raceway, cage, or net-pen system due to the continuous input of feed and production of feces. Even slight increases can support luxurient growths of primary producers.

Ponds and recirculating systems are excellent reservoirs for nutrient accumulations. As is the case with nitrogen compounds, phosphorus may be elevated in such systems, though fertilization is often required to stimulate blooms of phytoplankton. Fertilization is discussed in the following section.

The carbon/nitrogen/phosphorus ratio required by most phytoplankton species is near 106:16:1, indicating the potential of even trace levels of phosphorus to influence primary productivity in the presence of sufficient concentrations of the other two elements. Since phytoplankton blooms are often encouraged by aquaculturists, especially those engaged in pond culture, it is important to maintain sufficient levels of phosphorus to encourage the desired type of productivity. If, on the other hand, improper management leads to the collapse of a phytoplankton bloom, the release phosphate may be used to support an undesirable filamentous algae or macrophyte population.

Phosphorus (measured as phosphate) can be determined in a fairly crude way (to within a few tenths of a part per million) with a colorimetric technique. More precision can be obtained with sophisticated equipment, but precise measurements (down to micrograms per liter, for example) of phosphorus are usually not required

---

[71] Shrimp and other crustaceans grow by shedding their exoskeletons. Ecdysis is also known as molting.

by aquaculturists. Toxicity from high levels of phosphorus have not been reported by aquaculturists, but levels of several parts per million have been observed when ponds were heavily loaded with organic fertilizer. In at least one instance, the level of phosphorus was about 5 mg/l, and the pond apparently became nitrogen-limited as a bloom of blue-green algae (capable of fixing nitrogen) occurred that was readily consumed by the tilapia in the pond.[72]

FERTILIZATION

Fertilizers are often added to ponds, particularly in the spring, to encourage the development of phytoplankton blooms. Inorganic fertilizers are commonly employed in the United States, but around the world organic fertilization is probably more common.

Spring fertilization with nitrates and phosphates is usually initiated to provide a food base for early life stages of aquaculture animals. Fertilization will stimulate the development of a phytoplankton bloom, which will support a concomitant zooplankton bloom. Larvae, fry, and juvenile stages of various aquaculture animals require living food, and even those species that will accept prepared feeds at first feeding or soon thereafter will commonly benefit from the availability of natural food sources in the water. In addition to forming the base of a natural food web, phytoplankton blooms increase the turbidity in ponds and reduce light penetration, thereby discouraging the development of attached filamentous algae and rooted macrophytes.

Fertilization is not always recommened. Some soils contain relatively high levels of nutrients that may leach into pond water and support adequate blooms of phytoplankton. Incoming water, especially runoff water from land and from a surface source, may also contain sufficient supplies of nutrients to establish and support a phytoplankton bloom. As the growing season progresses, feeding rates increase and the nutrient level in ponds tends to increase because of the increased production of nutrient-rich wastes from the aquaculture species. Exogenous fertilization can be terminated when background nutrient levels remain sufficient to maintain the bloom. Fertilization has been used in both freshwater and marine ponds.

Inorganic Fertilization

Phytoplankton blooms can usually be induced in ponds by adding approximately 50 kg/ha of 16-20-4 fertilizer.[73] In ponds constructed where the soil contains sufficient potassium, good results can be obtained with 16-20-0. The optimum rate of fertilization varies to some extent from region to region because of inherent differences in soil chemistry and the concentrations of nutrients in the water.

---

[72] Observations by the author.

[73] The numbers refer to the percentages of nitrogen, phosphorus, and potassium (N-P-K) in the fertilizer. Other mixtures can be used. For example, 8-10-2 could be applied at 100 kg/ha.

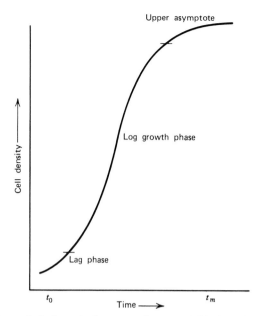

Figure 56. Theoretical phytoplankton growth curve. Initiation of a bloom (t_o) may be induced by application of fertilizer. Once the bloom has reached the desired level (t_m), it can be maintained by period fertilizer additions.

In warmwater ponds such as those used for the rearing of channel catfish, fertilization is usually initiated in the spring when the water temperature reaches 15 to 18°C. Fertilizing such ponds when the temperature is higher may lead to blooms of filamentous algae. The phytoplankton bloom should be maintained until the culture animals are well established on prepared feeds.

Fertilizer applications should be repeated at 10 to 14 day intervals until a Secchi disc reading of approximately 30 cm is obtained. The depth of the reading will vary to some extent depending on time of day (because of the angle of the sun on the water) and the amount of cloud cover. If a Secchi disc is not available, the measurement can be made by determining the depth at which the culturist's hand disappears when lowered into the water at right angles to the forearm. Proper fertilization has been achieved if the hand disappears when the arm is submerged to the elbow.

The phytoplankton community may respond to fertilization by logarithmic growth after a lag phase that usually lasts only a few days (Figure 56). The resulting phytoplankton bloom will reach a maximum when some factor becomes limiting. If steps are not taken to either eliminate the limiting factor (allowing for continued expansion of the bloom) or maintain the population at the asymptotic level, the population may quickly die in what is known as the crash of a bloom. The limiting factor may be a nutrient or could even be insufficient light caused by self-shading, since the turbidity of the water increases because of the high cell density. In aquaculture ponds, it is desirable to maintain the bloom at the 30 cm Sechhi disc level previously described. This can be done by applying small amounts of fertilizer periodically, though in most instances blooms are maintained with the nutrients being continuously added in feed and through fecal waste from the aquaculture animals.

The establishment of a proper phytoplankton bloom may require three to five fertilizer applications. If a bloom has not been established after five applications, less desirable plants may be taking over the system.

If a bloom crashes, the stored nutrients will be rapidly released and can lead to the establishment of a secondary bloom of a different species. When the initial fertilization protocol or the release of nutrients following the crash of a phytoplankton bloom results in the rapid growth of filamentous algae or undesirable macrophytes, it may be necessary to treat the pond with a herbicide and begin the process once again. Destruction of the undesirable plants will result in the release of nutrients, so the culturist should not excessively fertilize in an attempt to establish a phytoplankton bloom after herbiciding a pond.

Overfertilization can also lead to impaired water quality due to excessive phytoplankton production. As discussed in the section on dissolved oxygen (later in this chapter), primary producers are a major reason for diel oxygen fluctuations in aquaculture systems.

In general, well-managed culture ponds should not contain rooted plants or large mats of filamentous algae during any portion of the year. Exceptions do occur, however. For example, millet and rice are often grown in association with ponds used for crawfish production. In Asia, milkfish have been successfully grown in ponds where thick *Aufwuchs* mats called lab-lab are encouraged. They provide food for the fish (Pillay 1972).

Unispecific algae cultures of known species can be produced as food for aquaculture organisms. As described in Chapter 6, the practice is common with respect to mollusc hatcheries. In ponds, the culturist who fertilizes to stimulate a phytoplankton bloom has little or no control over the species of algae that will be produced, though green algae will usually bloom when the proper ratio of nutrients is provided. There is never any need to introduce algae into a pond, even if the water body is new or has been drained and thoroughly dried prior to filling and fertilization. Algae spores and cells are cosmopolitan in water and are transported by animals such as birds as well as by the wind. They can also be carried into ponds by surface runoff and with the introduction of water from surface supplies.

Organic Fertilization

A significant problem associated with the rearing of terrestrial livestock is the disposal of liquid and solid wastes. In addition to dealing with the problem of high volumes of organic wastes, there may be severe odor and fly problems associated with livestock production operations such as poultry houses and cattle and swine feedlots. As is true in the case of human sewage treatment, the wastes from terrestrial animals can be deposited in sewage lagoons, where degradation occurs. The high levels of nutrients released from animal waste oxidation can support the growth

Figure 57. Ducks living in association with culture ponds, such as this one in Nepal, provide organic fertilizer, which encourages production of planktonic and benthic food organisms for the fish.

of extensive phytoplankton blooms and lead to high levels of secondary productivity in the form of zooplankton and benthos.

Organic fertilizers can be derived from sewage lagoons as described above, or they may be added directly to aquaculture ponds. The practice of rearing ducks, swine, cattle, and other livestock adjacent to or over ponds is common in many countries, particularly in Asia (Figure 57). In some areas, untreated human sewage (night soil) is used as fertilizer in ponds. Treated domestic sewage was evaluated as a source of nutrients for various marine species in the 1970s (Huguenin 1975, De-Boer et al. 1977), and chlorinated sewage effluent has more recently been evaluated in conjunction with the production of tilapia (Turner et al. 1986a), mullet, and penaeid shrimp (Turner et al. 1986b). In a properly fertilized pond, sufficient natural food may be produced that the need for providing prepared feeds is obviated. In these studies no human pathogens were detected within the flesh of the animals produced. Elevations in the levels of copper and zinc did not appear to be attributable to the sewage, and no accumulations of chlorinated hydrocarbons was found.

Research on the use of animal wastes as fertilizer in aquaculture ponds has been conducted in various parts of the world. Outside of Asia, some of the early work was conducted in Israel by Schroeder (1974) and in the United States by Stickney et al. (1977b), Buck et al. (1978), and Stickney and Hesby (1978).

The question of whether poultry waste could be used as an enhancer of tilapia production was posed by L. O. Rowland in the mid-1970s and led to the initial research at the Aquaculture Research Center of Texas A&M University (Stickney et al. 1977a). Following the first study, in which poultry manure was used as a feed

Figure 58. Poultry wastes deposited in a pond can promote a phytoplankton bloom that will support certain types of fish, in this case *Tilapia aurea*.

ingredient, the emphasis changed to one in which the manure was used as a fertilizer deposited in ponds. At the same time, the waste from swine was also evaluated (Stickney et al. 1977b). Most of the studies were conducted with growing-finishing pigs or laying hens reared immediately adjacent to or directly over ponds (Figure 58). The studies ultimately developed stocking recommendations of 50 growing-finishing hogs per hectare (Stickney et al. 1979) and 4,000 laying hens per hectare (Burns and Stickney 1980) as a means of promoting growth in tilapia that were not provided with prepared feed. Some refinement of the recommendation for tilapia was made by McGeachin and Stickney (1982). They found that between 70 and 140 kg/ha/day of dry poultry manure would produce good growth in *T. aurea*. The recommendation was made that the lower end of that range should be applied to waters of high alkalinity and indicated that the higher end of the range might be appropriate for waters of low alkalinity. Broussard et al. (1983) recommended the application of dried poultry manure at a rate of 100 kg/ha/day (3,000 kg/ha/month) for the production of *T. nilotica* fingerlings in the Philippines, and Fagbenro and Sydenham (1988) reported good growth and survival of the walking catfish *Clarias ishariensis* in ponds fertilized with 90 kg/ha/day of dry poultry sweepings (30 to 40% manure).

Animal wastes are not commonly used as organic fertilizers in the United States, and the practice may not become widespread, particularly with respect to the production of fish produced for human consumption, because of concerns for public health associated with potential pathogens in manures. Pathogenic bacteria can frequently be isolated from treated sewage effluent, but they do not generally appear in the tissues of aquaculture animals exposed to the waste. The bacteria can occur on the surface or in the intestines of aquaculture species and may pose a health threat to individuals involved in handling and processing the animals.

Manure may have a place in United States sportfish production, however. Lanoiselee et al. (1986) showed that swine manure added to ponds at up to 15 kg/ha/day of dry matter enhanced the productivity of young northern pike *(Esox lucius)*.

While usually applied in freshwater situations, manure has also been used to increase tilapia production in brackish-water ponds (Tamse et al. 1985). Other organic fertilizers such as plant meals and grasses have been used in conjunction with the rearing of certain species. For example, freshwater hatcheries have used such fertilizers as hay, alfalfa, rice bran, and cottonseed meal for striped bass production (Bonn et al. 1976, Geiger 1983, Ludwig and Tackett 1991). Ludwig and Tackett (1991) found that larger striped bass fingerlings were produced when rice bran was used as a fertilizer than when ponds were fertilized with cottonseed meal. Cottonseed meal has been used in conjunction with saltwater ponds used to rear the early stages of red drum, spotted seatrout, and striped bass (Colura et al. 1976, Colura and Matlock 1984). Such fertilizers often promote zooplankton blooms upon which larval and postlarval aquaculture animals feed.

Combination Fertilization

Combinations of inorganic and organic fertilizers are used by some aquaculturists since the blends seem to promote a wider variety of both autotrophic and heterotrophic organisms (Geiger 1983). In the United States, the practice has been applied to the production of sportfish. Red drum culturists in Texas produce fingerlings for stocking coastal waters to enhance the recreational fishery. In recent years, red drum hatcheries have converted from organic fertilization with cottonseed meal alone to a schedule that employs both organic and inorganic (phosphoric acid and ammonium nitrate) fertilizers (Porter and Maciorowski 1984, Colura 1987). Combinations of granular (diammonium phosphate) or liquid (phosphoric acid and ammonium nitrate) inorganic fertilizer and cottonseed meal have been used to fertilize striped bass and smallmouth bass rearing ponds (Farquhar 1987). Broussard et al. (1983) recommended the addition of 3000 kg/ha/mo of dried chicken manure and 100 kg/ha/month of 16-20-0 inorganic fertilizer to tilapia fingerling production ponds in the Philippines.

CONTROL OF AQUATIC VEGETATION

Depending on the control method selected, it may be necessary to determine the species of plant or plants causing a particular problem in advance of treatment. Various publications have been developed to help aquaculturists, as well as the sport fishers, farmers, and ranchers determine what species of plants are present and the proper control procedures to eradicate them. Since aquatic plants vary regionally because of climate, specialized publications for a particular area are desirable. Such publications, when available, can often be obtained through country Cooperative Extension Service offices in the United States. General publications that will aid in the identification of aquatic plants include those of Eyles and Robertson (1944), Muenscher (1944), G. M. Smith (1950), Fassett (1960), Weldon et al. (1969), Applied Biochemists, Inc. (1976), Seagrave (1988), and Cook (1990).

In culture ponds, filamentous algae may grow as mats associated with the bottom, in conjunction with other types of aquatic vegetation, on the surface of the water, or suspended in the water column. In addition, various types of aquatic macrophytes may occur. Some species remain completely submerged, others are partly or mostly emergent. Emergent vegetation is most commonly associated with the shallow water at or near the shoreline. Finally, some species of aquatic macrophytes float at the surface.

There are three basic types of control that can be implemented by aquaculturists in their efforts to rid ponds of undesirable vegetation: mechanical, biological, and chemical. Each method has been widely used, and each can be effective. The control method selected may depend on the type of vegetation present, the extent of vegetative cover, local regulations, and other factors.

Mechanical Control

The objective of mechanical vegetation control is either to physically remove the offending plants or to alter the environment, thereby creating conditions that will discourage growth of the unwanted vegetation. When objectionable plants first appear, and before they become well established in a pond, harvesting the plants by hand may be the simplest and most cost-effective method of dealing with the problem. This approach is not practical in ponds where extensive areas of vegetation are growing. There are mechanical harvesting devices that will remove vegetation from large water bodies, but most of them are not suitable for use in aquaculture ponds, and regrowth typically requires periodic reharvesting. Studies have demonstrated that plant densities return to their initial levels within about 30 days of mechanical harvest (Mikol 1985, Shireman et al. 1986).

Plants like arrowhead (*Sagittaria* sp., shown in Figure 59), which is rooted, and water hyacinth *(Eichhornia crassipes),* which floats at the surface, can often be removed by hand during the early stages of colonization. Duckweed (*Lemna* sp.) may be skimmed off the water surface, but it is virtually impossible to harvest each of the very small individual plants, so mechanical harvesting usually has to be accompanied by some other technique.

Cattails *(Typha latifolia)* can sometimes be controlled by hand harvesting. The plants spread by sending out rhizomes beneath the surface of the substrate in addition to proliferating by seeds that are carried by the wind. During the spring and summer the spread of cattails by rhizomes can lead to colonization of extensive areas of shallow water within a few weeks (Figure 60). Thus it is often necessary to return to an area where the plants have been harvested to repeat the process since new plants will sprout from pieces of rhizome that remain in the soil. Again, an-

Figure 59. Arrowhead (*Sagitarria* sp.) is a rooted aquatic macrophyte that sends leaves to an often above-the-water surface.

Figure 60. Small patches of cattails *(Typha latifolia)* can be removed by hand if detected before they have an opportunity to spread. Once the rhizome network has developed, control becomes difficult.

Figure 61. Muskgrass or chara (*Chara* sp.), an alga that grows in clumps on pond bottoms.

other means of control (usually herbicide application) may be required in conjunction with mechanical harvest to rid a pond of cattails.

Some submerged plants that appear to be easy to remove by mechanical harvesting are, in reality, virtually impossible to control by that means. Bushy pondweed (*Naias* sp.), for example, can be gathered by hand, but the roots typically remain behind, and the plant will quickly regenerate. Hydrilla *(Hydrilla verticillata)* and watercress *(Nasturtium officinale)* are two among many other plants that are also difficult to control by mechanical means. Muskgrass or chara (*Chara* sp.) resembles bushy pondweed, at least grossly, though muskgrass is an alga. It grows in clumps (Figure 61), usually in association with the substrate. Removal of as much of the plant biomass as possible by hand before herbicide application may be desirable where heavy densities exist, since decaying plants will often lead to DO depletions, the extent of which depend largely on the amount of decaying biomass that is present.

One of the best ways to limit emergent vegetation in ponds is to avoid having extensive shallow areas. Proper pond design and construction should ensure that adequate slopes are constructed to limit the amount of shallow water in which vegetation can develop. Maintaining full ponds during the growing season, along with emptying, drying, and reshaping the slopes as necessary to maintain their design characteristics will also be helpful.

Some species, such as common carp *(Cyprinus carpio),* have a habit of rooting around in the substrate. That activity leads to increased turbidity, which can reduce colonization of rooted macrophytes by shading them out, but also causes damage to levees. Thus, the actions of carp and similarly behaving species can be considered a type of mechanical vegetation control, though not always a desirable one because of damage to levees. It may be necessary to reshape the levees of common carp ponds much more frequently than would be required for ponds containing species that do not display the rooting activity.

Lined ponds are being increasingly used in conjunction with sportfish hatcheries. Lined ponds are not only appropriate in regions where soil has poor water-holding ability (see Chapter 3), but also when small fish are being harvested. Losses of fish can be reduced when the pond bottom is plastic rather than mud. Another advantage of lined ponds is that rooted aquatic macrophyte growth is retarded or eliminated because there is little or no opportunity for the plants to become established in the absence of appropriate substrate.

Biological Control

The establishment and maintenance of a phytoplankton bloom in a pond is one form of biological plant control; that is, it is a means of using one organism to control another—in this case by using the increased turbidity of the plankton bloom to shade out rooted aquatic plants. The technique has been recognized for decades. Swingle (1947) indicated that a well-established phytoplankton bloom will shade out many of the rooted macrophytes in ponds deeper than 45 cm. The techniques associated with establishing a phytoplankton bloom was described earlier in this chapter.

Once undesirable vegetation has become established in a pond, it is too late to attempt to shade the plants out with a phytoplankton bloom. Adding fertilizer will often stimulate further growth of the undesirable plant species rather than leading to the growth of phytoplankton.

Various species of animals have been used to consume established undesirable plants and to prevent their establishment in the first place. Most animals recommended for use in biological plant control are fishes, though other organisms have also been employed. For example, the sea cow or manatee *(Trichechus manatus latirostris)* has been stocked in weed-infested waters within the state of Florida (Sguros et al. 1965). The application of manatees to aquaculture ponds is doubtful at best, though they may be useful to cage culturists working in irrigation ditches where natural populations of the large mammals occur. Large freshwater snails (Blackburn and Weldon 1965), and the South American flea beetle, *Agasicles* sp. (Anderson 1965), have been introduced into aquaculture systems for plant control purposes. Snails and insects usually cannot eliminate established plant populations, though they may have some utility in preventing their initial establishment. An additonal potential problem is that such biological control species cannot be easily contained in an aquaculture facility and may spread to surrounding waters where their activities may be considered undesirable.

An example of how an exotic introduction has led to a second such introduction in an attempt to control the first is associated with an Asian plant. Hydrilla *(Hydrilla verticillata)* is an introduced aquatic plant that has been known to grow in some 10 m of water in Lake Conroe, Texas (Martyn et al. 1986), and has choked many water bodies in various states leading to impaired fishing and even to the death of swimmers who become entangled in the plants. An aquatic moth *(Paraponyx diminutalis)*, the larvae of which feed on hydrilla, was apparently inadvertently introduced into Florida and Panama from Asia. While the moth may be useful in reducing hydrilla growth, its larvae also attack certain native plants (Buckingham and Bennett 1988), thereby perhaps obviating its positive impact.

Fish are more readily contained than some other biological weed control agents. While fish are still subject to escape, and few systems are foolproof, it is generally easier to retain fish than, for example, insects. As we shall see, there are also breeding techniques that can be used with fish to ensure that they will either not survive or that they will not reproduce in nature.

Perhaps the most widely used biological weed control agent, and one that has been highly controversial in the United States, is the grass carp, *Ctenopharyngodon idella*. The fish was first introduced into the United States from Malaysia in 1963 (Stevenson 1965) as a weed control agent since the fish appears to be a strict herbivore throughout most of its life. Experimental work with grass carp was initiated at the U.S. Fish and Wildlife Service's Fish Farming Experimental Station at Stuttgart, Arkansas, with that first introduction of 70 fish in November 1963. Auburn University also obtained a shipment of grass carp in 1963 and had spawned them for purposes of stocking research ponds by 1966 (Sills 1970). Their spread around the United States over the first 15 years of their presence in North America has been chronicled by Guillory and Gasaway (1978).

While Stanley et al. (1978) predicted poor spawning success of grass carp in nature, there was sufficient concern that at one point more than 30 states had outlawed stocking of the species. Some successful reproduction has been reported (discussed in Chapter 1), so obtaining permission to stock the fish as weed control agents in aquaculture ponds has often been difficult for fear of their eventual escape and proliferation in nature.

Reevaluation of the use of herbivorous fishes has come in more recent years after the techniques were perfected for developing hybrids (grass carp × bighead carp, *Aristichthys nobilis,* being the most common) and triploid fish. (Triploidy is discussed in Chapter 7.) Both techniques allow the stocking of sterile herbivorous fish. Many states have revised their policies with respect to grass carp stocking. In many instances, the aquaculturist must obtain certification that 100% of the fish being stocked are sterile.

The ability of grass carp to control vegetation was demonstrated in the United States within a few years after the fish were introduced (Avault 1965a, b, Mitzner 1978, Colle et al. 1978). In one of the largest experiments with grass carp, 270,000 of the fish were introduced into Lake Conroe, Texas, in 1981 and 1982. By October 1983, over 3,500 vegetated hectares had been cleared of hydrilla (Martyn et al. 1986) Other studies have made additional contributions regarding the food preferences of grass carp and its hybrids, evaluated appropriate stocking densities for various regions and situations, and compared grass carp performance with that of hybrids.

Many of the studies associated with grass carp efficacy in weed control have been conducted in natural or artificial lakes that are not used for aquaculture, but the results can certainly be applied to aquaculture ponds. Other studies have been conducted in the laboratory. Some of their results that could be useful to aquaculturists include the following:

- When stocked at 50 fish/ha, grass carp eliminated hydrilla for at least 6 yr. Other plant species were also eliminated or significantly reduced (Van Dyke et al. (1984). At 61 fish/ha, grass carp eliminated Illinois pondweed *(Potemogeton illinoeniss)* and significantly reduced Eurasian back milfoil *(Myriophyllum spicatum)*.

- Grass carp stocked at 110 per hectare (0.6 kg average weight) controlled hydrilla in southern Florida agricultural canals (Schramm and Jirka 1986).

- Hybrid grass carp were shown to have a feeding rate about one-third less than that of grass carp of similar size and a growth rate of only one-seventh of grass carp (Osborne 1982).

- Young et al. (1983) suggested that it might be necessary to stock twice as many hybrid grass carp to achieve the same level of vegetation control effected by grass carp. Hybrid grass carp preferred filamentous algae (primarily *Mougeotia* sp., *Oedogonium* sp., and *Spirogyra* sp.) and southern naiad *(Naias guadalupensis)* to coontail *(Ceratophyllum demersum)* and waterweed *(Elodea nuttallii)*.

- Hybrid grass carp were found to consume about as much plant material, with the same preferences, at 12 to 15°C as at 25 to 28°C (usually about six in number) (Cassani and Caton 1983). Of seven plant species evaluated, hybrids preferred duckweed *(Lemna gibba)* over the others, but also readily consumed chara, southern naiad, and sago pondweed *(Potemogeton pectinatus)*.

- Triploid grass carp stocked in California ponds at densities of 30 and 60 fish/ha showed the following preference for the plants present: sago pondweed > chara > Eurasian back milfoil (Pine and Anderson 1991).

- The introduction of triploid grass carp into a South Dakota lake at 49 fish/ha resulted in nearly complete elimination of chara within 2 yr in the presence of largemouth bass *(Micropterus salmoides)*. Grass carp were ineffective at 61 fish/ha in a lake containing northern pike *(Esox lucius)*, which apparently preyed upon the herbivorous fish (Bauer and Willis 1990).

- Weed control by grass carp has been shown to be more economical than mechanical harvesting or the use of herbicides (Shireman et al. 1986).

It is clear from the work accomplished to date that the number of carp to be stocked depends on the nature of the fish (normal, triploid, or hybrid), the types of plants present, plant density, and geographic region in which the problem occurs. For routine control of aquatic vegetation in aquaculture ponds, grass carp or hybrids can be stocked at relatively low densities (e.g., 15 fish/ha) if stocking occurs before a problem arises.

Various species of tilapia consume primary producers after attaining certain sizes. Some, such as *Tilapia aurea,* are efficient consumers of phytoplankton, while others such at *T. zillii,* have been used in irrigation canals to control higher aquatic plants. The food habits of various tilapia species have been elaborated by such authors as LeRoux (1956), McBay (1961), Yashouv and Chervinsky (1960, 1961), and Moriarty (1973). Silver carp *(Hypophthalmichthys molitrix)* have also been widely reported to consume phytoplankton, though recent information indicates that the species is actually quite omnivorous (Costa-Pierce 1992).

Chemical Control

Certain herbicides, including those that are used to control algae (algicides), are widely available, but their use involves certain risks and is strictly controlled, at least by the government in the United States, relative to their application in aquaculture. In cases where large amounts of plant material are killed with herbicides in a given pond, decaying vegetation can place a high demand on oxygen and may lead to DO depletions. Spot treatment—which might involve killing a portion of the aquatic vegetation in a pond and treating another portion after the first has decomposed—can be an effective way to avoid severe oxygen depletions, particularly if a contact chemical that does not dissolve in the water is employed. Pond DO should be carefully monitored (discussed later in this chapter) for several days following herbicide application to ensure that critically low levels (which might lead to mortality in the culture animals) do not occur or that, if they do, remedial action is taken. Carpenter and Greenlee (1981) developed a model to predict DO concentrations in unstratified lakes after herbicide applications.

Herbicides can be directly toxic to culture animals or to the organisms in the system upon which the target culture species feed. Virtually all herbicides are toxic to animals if present in sufficiently high concentration. The toxicity of a herbicide to nontarget organisms in a water body may be influenced by such factors as temperature, pH, alkalinity, humic substances,[74] and others (Seyrin-Reyssac 1990). In some cases the concentration of a herbicide required to control an aquatic plant may approach the toxic level for the target culture species, so it is critical that treatment levels are carefully calculated. The vulnerability of various culture species to herbicides can vary widely. Ramaprabhu and Ramachandran (1990) reviewed the effects of various herbicides on a variety of aquatic species and determined that 2,4-D was the safest of the chemicals evaluated. If information relative to a given culture species is not available on the chemical that is selected, a bioassay should be run to determine that the planned treatment level is not toxic to the culture animals. Such experiments can be conducted in aquaria or other small, aerated containers.

---

[74] These are typically large-molecular-weight organic acids found in certain types of soil and soluble in water.

TABLE 15. Herbicides Registered or Approved for Use in Aquaculture by the U.S. Food and Drug Administration or the U.S. Environmental Protection Agency (from Meyer and Schnick 1989)

Chemical	Trade Name(s)	Tolerance	Withdrawal Time and Comments
Acid blue and acid yellow	Aquashade[a]	Exempted	None required
Aluminum sulfate, calcium sulfate, boric acid	Clean-Flor lake cleaner[a]	Exempted	None required
Copper[b]	Algaetrol-76, Aquatrine, etc.	Exempted	Fish and shrimp may be immediately harvested; not for use on trout
Copper sulfate[a]	Agway copper sulfate	Exempted	None established
2,4-D[c]	Aquacide, AquaKleen, etc.	1 ppm in fish and shellfish	None established
Diquat dibromide[a]	AquaClear	0.1 ppm in fish and shellfish	None established
Endothall[c]	Aquathol, etc.	0.2 ppm in potable water	Three days
Fluridone[c]	Sonar	0.5 ppm in fish	None required; not for use in marine water or on crayfish
Glyphosate[c]	Rodeo	0.25 ppm in fish	None established
Potassium ricinoleate[b]	Solricin 135	Exempted	Four weeks
Simazine[a]	Aquazine	12 ppm in fish	None established
Xylene[c]	Xylene	Exempted	None established

[a] For algae and higher plants
[b] Algicide
[c] Not used on algae

All chemicals used in the United States with respect to treating fish or fish ponds for diseases, nuisance vegetation, and so forth must be approved for use by the U.S. Food and Drug Administration or the U.S. Environmental Protection Agency. As of 1989, the chemicals listed in Table 15 were approved for use in aquaculture under the limitations shown. Since the status of chemicals is constantly under review and subject to change, the information in Table 15 is meant to be illustrative only. Persons interested in applying the chemicals shown in the table, or other compounds, should carefully read the label on the product to determine approved usage and dosage. Some of the plants that have posed problems for aquaculturists and some of the chemicals that can be used to eradicate those plants are presented in Table 16.

Many herbicidal chemicals are dissolved in water and then applied to culture

TABLE 16. Partial List of Types of Aquatic Vegetation and Chemicals that Can Be Used to Control Them

Common Name	Genus	Herbicide
	Algae	
Filamentous algae	Several genera	Copper sulfate, Diquat
Muskgrass	*Chara*	endothall, Simazine
Nitella	*Nitella*	
	Macrophytes	
Elodea	*Elodea*	Diquat, endothal, 2,4-D
Coontail	*Ceratophyllum*	
Naiad	*Naias*	Diquat, endothall,
Pondweed	*Potamogeton*	Simazine, 2,4-D
Water milfoil	*Myriophyllum*	
Fanwort	*Cabomba*	Endothall, Simazine
Arrowhead	*Sagittaria*	Endothall, 2,4-D
Bullrush	*Scirpus*	2,4-D
Rush	*Juncus*	
Water smartweed	*Polygonum*	
Water hyacinth	*Eichhornia*	
Cattail	*Typha*	Diquat
Duckweed	*Lemna*	Diquat, Simazine

ponds. In most cases, it is desirable to spread the chemical solution as evenly as possible over the pond surface. Mixing of the chemical with the pond water can be affected by pouring the solution into the wake of an outboard motor. The motor may be operated from a boat in large ponds or from a stationary mounting along the shore of a small pond. Circulation of the pond with aerators such as paddlewheels (described later in this chapter) during herbicide application is another method for achieving thorough mixing.

Endothall can be dissolved in water or purchased adsorbed to sand particles that can be sprinkled directly on the offending vegetation. Since Endothall acts as a contact herbicide, placing it in direct contact with the plants allows the culturist to affect partial treatment and thereby control the amount of plant decomposition that is occurring at any given time. Glyphosate (Rodeo) has been shown to perform well against duckweed when sprayed directly on the plants but is ineffective when dissolved in the water (Lockhart et al. 1989).

THE CARBONATE BUFFER SYSTEM

Mixtures of weak acids and their salts are called buffers. Several buffer systems exist in water, including those associated with phosphate, borate (important in seawater), and carbonate. The last is usually the most significant buffer system in natural waters and is the system that is usually responsible for pH maintenance in aquatic ecosystems. Buffer solutions resist pH change because the ionization equilibrium between the weak acid and weak base changes in a manner that allows them to consume hydrogen ions (H^+) or hydroxide ions (OH^-) that are added to the system. The initial pH of an aquatic system will be controlled by the ionization constant that exists, which depends on the relative concentration of acid or base and its salt. Solutions at pH below 7.0 are acid, those above 7.0 are basic.

The carbonate buffer system involves a series of reversible chemical reactions. Carbon dioxide (which occurs in the atmosphere at a level of about 0.03%) is one source of carbon for the system. Carbon dioxide is dissolved in water through diffusion from the atmosphere and as a result of respiration as follows:

$$H_2O + CO_2 \leftrightarrows H_2CO_3 \qquad \text{(equation 1)}$$

Carbonic acid (H_2CO_3) dissociates in water to form bicarbonate ion (HCO_3^-) and hydrogen ion (H^+):

$$H_2CO_3 \leftrightarrows HCO_3^- + H^+ \qquad \text{(equation 2)}$$

Bicarbonate ion can further dissociate to produce another hydrogen ion and carbonate ion (CO_3^{2-}):

$$HCO_3^- \leftrightarrows H^+ + CO_3^{2-} \qquad \text{(equation 3)}$$

The carbonate buffer system resists changes in pH by releasing or absorbing hydrogen ions as necessary to maintain a steady state. In most cases fresh water has a pH of between 6.5 and 8.5, while seawater has a pH above 7.0.[75] The level of calcium carbonate and other buffering agents present in the system helps establish the equilibrium pH.

If a hydrogen ion (from respiration or the dissolution of atmospheric carbon dioxide) is added to a buffered water system, it will be captured by carbonate to form bicarbonate (moving equation 3 to the left). Thus, there will be no addition of free hydrogen ions to the system, and the pH will not be reduced. If, on the other hand, hydrogen ion is removed when the system is in equilibrium, such as when it combines with hydroxide (OH^-) to form water (H_2O), bicarbonate or carbonic acid can dissociate to replace the lost hydrogen ion (moving equation 2 or 3 to the right).

---

[75] The calcium carbonate that forms the exoskeletons of many marine mammals would dissolve under acid conditions. Coral reefs and oyster beds, as we know them, could not exist if salt water did not have a pH above 7.0.

Photosynthesis removes carbon dioxide from the system, thereby forcing equation 1 to the left and taking hydrogen ion out of the system. As long as there is a source of hydrogen ions from bicarbonate or carbonic acid, the pH of the system will not rise.

The keys to the system are the carbonate and bicarbonate ions, which serve as sinks and sources of hydrogen ions, respectively. If all of the carbonate present is converted to bicarbonate, calcium carbonate (if present) will dissociate to provide additional carbonate ions:

$$CaCO_3^- \leftrightarrows Ca^+ + CO_3^{2-} \qquad \text{(equation 4)}$$

The reaction in equation 4 may not keep up with demand in some instances, or the pool of calcium carbonate may be insufficient to meet the demand. In such instances, the pH may fall dramatically as hydrogen ions are added to the system. Similarly, if carbon dioxide is removed to the extent that the bicarbonate pool is exhausted, the pH may rise dramatically. In ponds with high levels of primary productivity, pH may actually increase significantly during the daytime as a result of photosynthesis. Conversely, at night when photosynthesis ceases, respiration by both plants and animals adds carbon dioxide to the system and the pH falls. In poorly buffered systems, the diel change can actually amount to several pH units (e.g., from less than 6.0 to over 10.0).[76] Such changes can be detrimental or lethal to aquatic animals, but they can be avoided if the system is well buffered. This can be accomplished by maintaining sufficiently high alkalinity (discussed in Chapter 5). Figure 62 shows how pH tends to change over a 24 hr period as a function of photosynthesis and respiration.

Bicarbonate ion appears to be an important element in the evolution of oxygen during photosynthesis (Stemler and Govindjee 1973) and sometimes can be substituted for carbon dioxide in the photosynthetic process. Bicarbonate usually serves as the carbon source in photosynthesis only when its concentration is at least 10 times that of free carbon dioxide (Wetzel 1975). This occurs frequently, at least in the marine environment. Steemann Nielsen (1975) indicated that about 1% of the total inorganic carbon in seawater is in the form of carbon dioxide, while 90% is bicarbonate and the remainder carbonate. When bicarbonate is utilized in photosytnesis, it is exchanged for hydroxide ions:

$$HCO_3^- \quad \rightarrow \quad CO_2 + OH^-$$

At sufficiently high pH, calcium carbonate crystals will form on the leaves of aquatic macrophytes:

---

[76] Since the pH scale is logarithmic, a 1 unit change represents an order of magnitude increase or decrease in the amount of hydrogen ion present. Thus, a change in pH from 6.0 to 10.0 represents four orders of magnitude.

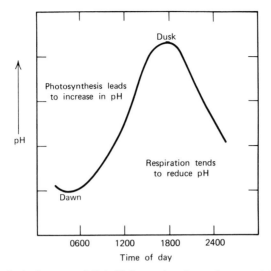

Figure 62. Hypothetical pattern of diel pH fluctuations in a culture pond. In a well-buffered pond the vertical scale might be fractions of a pH unit, while in poorly buffered systems each hashmark could indicate a full pH unit.

$$OH^- + HCO_3^- \rightarrow H_2O + CO_3^{2-}$$

and

$$CO_3^{2-} = +Ca^{2+} \rightarrow CaCO_3 \downarrow$$

In alkaline fresh waters it is often possible to create a white cloud of calcium carbonate by disturbing rooted macrophytes that have been photosynthesizing in this manner. The utilization of bicarbonate ions in photosynthesis may be more advantageous to macrophytic vegetation than to algae (Wetzel 1975), and in aquaculture, the growth of macrophytes is usually discouraged.

DISSOLVED OXYGEN

The level of DO available to the animals in an aquaculture system is perhaps the most critical among the water quality variables that are routinely monitored. If a sufficient level of DO is not constantly maintained, animals will become stressed, after which they may not eat well for some period of time and their susceptibility to disease can increase dramatically. In the worst instance the animals may be killed outright. Even slight reductions in DO below the minimum desirable levels can lead to reduced growth rates and suboptimal food conversion efficiencies (FCE). DO depletions can actually be exacerbated when the culturist provides food to animals that are not eating properly since waste feed will decompose and increase the oxygen demand on the system.

A few aquaculture organisms such as walking catfish (*Clarias* sp.) and many species of tilapia (*Tilapia* sp.) are able to tolerate low DO without apparent consequence, but most are not. Aquaculturists need to frequently measure the level of DO in their water system, should be familiar with the minimum DO level tolerated by the species with which they are working, and must be prepared to take remedial action when a DO deficiency is detected.

Sources, Sinks, and Acceptable Levels

Oxygen is dissolved in water by diffusion from the atmosphere and through its release into the water as a byproduct of photosynthesis by aquatic plants. Diffusion from the atmosphere is aided when turbulence occurs, such as during windy weather, since more water surface area is placed in contact with the atmosphere at any given time than when the surface of the water is calm. The greater the surface area of water in contact with the atmosphere, the more diffusion that will occur.

When air or pure oxygen is bubbled into a water body, the action serves to increase the amount of water surface that is in contact with the gas. The thousands of small bubbles produced per second from an air stone in an aquarium provide a large amount of surface area. The transfer of oxgyen from the bubbles to the water is by diffusion.

All aquaculturists recognize that respiration is a normal physiological function of animals and that the removal of oxygen from water through respiratory activity can greatly impact the DO level. Many do not realize, however, that plants also respire continuously. During daylight and above the compensation depth, photosynthesis should produce more oxygen than is being consumed by both plant and animal respiration. In some instances, oxygen production will be so great that the water becomes supersaturated. Oxygen may then be lost to the atmosphere. At night, there is no oxygen being produced, so only diffusion is operating to replace the oxygen being removed through plant and animal respiration. Usually, the respiratory demand during the night is not sufficient to reduce the DO to critical levels, but under some circumstances that can happen. Figure 63 shows a hypothetical diel oxygen curve.

Oxygen can also be removed from water as a result of certain inorganic chemical reactions (chemical oxygen demand) and through the decomposition of organic matter by microorganisms. The requirement for oxygen by the latter process plus that associated with plant and animal respiration comprise the biochemical oxygen demand (BOD). The BOD test is an empiracal one that has been standardized for use in many types of water (APHA 1989). The BOD in aquaculture systems can become important when large amounts of aquatic vegetation are decaying (e.g., after a pond has been treated with a herbicide) or when dead animals are allowed to decompose in the water system. A high BOD can trigger an oxygen depletion.

As a general rule, if the DO concentration is equal to, or in excess of 5 mg/l, conditions relative to this parameter should be acceptable for aquatic organisms (Wheaton 1977). Some species can tolerate DO levels well below 5 mg/l with little or no resultant stress or impact on growth, but none of the species currently being

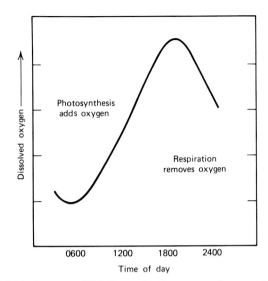

Figure 63. Hypothetical pattern of DO fluctuation in an aquaculture pond that has high rates of photosynthesis and respiration. The vertical scale hashmarks could be in milligrams per liter, or in fractions or multiples of 1 mg/l.

reared by commercial aquaculturists seem to require a level above 5 mg/l. Still, many culturists prefer to see DO levels at or near saturation at all times. Several species of tilapia are able to survive well at DO concentrations as low as 1 mg/l (Uchida and King 1962, Denzer 1968) and will continue to grow rapidly even if exposed to periods of such low levels on a daily basis (Stickney et al. 1977a). Other species may suffer severe stress or die if exposed to such low levels, though most will perform well at 4 mg/l and may survive for extended periods at 3 mg/l. While most fish species can tolerate 1 to 2 mg/l for brief periods, death is common if exposure to those levels exceeds a few hours (Piper et al. 1982). Salmonid culturists begin to worry when DO levels fall below 5 mg/l, while catfish farmers usually find levels of 3 mg/l or higher acceptable.

The amount of oxygen that can be dissolved in water under ambient equilibrium conditions is called the saturation concentration. DO solubility is dependent on various factors, with temperature, salinity, and altitude being the most important. The DO saturation concentration is reduced as each of the other factors increases. The solubility of oxygen in waters of various temperatures and salinities at one atmosphere (1 atm, sea level) are presented in Table 17. Even at high temperatures and salinities, the DO saturation value exceeds 5 mg/l at sea level; thus unless some other factor leads to an oxygen depletion, the water utilized in most aquaculture facilities should have the capacity to hold sufficient oxygen to support the species being reared. A mariculture facility located at high altitutude (e.g., a recirculating system located in the mountains) could face a chronic oxygen problem because of the inability to maintain the desired level at saturation unless supplemental oxygen is provided.

TABLE 17. Solubility of Oxygen in Water (mg/l[a] at 1 atm) with Varying Temperature and Salinity (Weiss 1970)

Temperature (°C)	Salinity (ppt)[b]				
	0	10	20	30	35
10	11.3	10.6	9.9	9.3	9.0
12	10.7	10.1	9.5	8.9	8.7
14	10.3	9.7	9.1	8.6	8.3
16	9.9	9.3	8.7	8.2	8.0
18	9.5	8.9	8.4	7.9	7.7
20	9.1	8.6	8.1	7.6	7.4
22	8.7	8.2	7.8	7.3	7.1
24	8.4	7.9	7.5	7.1	6.9
26	8.1	7.7	7.2	6.8	6.6
28	7.8	7.4	7.0	6.6	6.4
30	7.6	7.1	6.8	6.4	6.2
32	7.3	6.9	6.5	6.2	6.0
34	7.0	6.7	6.2	6.0	5.8

[a] Converted from ml/l using the following relationship: mg/l = 0.6998 mg/l.
[b] ppt = parts per thousand

Supersaturated oxygen conditions can occur, as previously indicated, when the rate of photosynthesis by aquatic plants is high. Levels of DO as high as 30 mg/l are not uncommon under some conditions. Supersaturated oxygen is not usually a problem in aquaculture.

Measuring Dissolved Oxygen

The aquaculturist should routinely monitor DO to ensure that depletions in the water system are not occurring. Measurements should be made daily, preferably at or before dawn, particularly in culture chambers where potential oxygen depletions are anticipated. Details regarding the causes and steps that can be taken to alleviate problems are presented in the next two sections of this chapter.

DO concentration can be measured by Winkler titration (APHA 1989) or with an oxygen meter (Figure 64). Both methods can accurately determine DO to within 0.1 mg/l. Oxygen meters provide readings within seconds, while the titration procedure requires a few minutes, the availability of several chemicals, and utilizes some items of glassware including a burette for titration.[77] Oxygen meters are quite reliable. Maintenance of models used by scientists usually involves infrequent battery replacement and periodic replacement of the membrane covering the end of the probe, along with replacement of the fluid within the probe.

---

[77] Some commercial test kits utilize an eye dropper for titration, with each drop of titrant added to effect a color change in the solution being equivalent to 1 mg/l of DO. For most purposes, knowing the DO concentration to within ± 1 mg/l is sufficient.

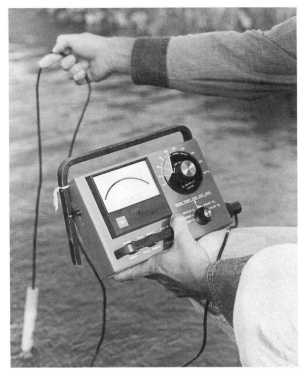

Figure 64. A portable DO meter can be used to make rapid determinations with good precision. The meter shown can be adjusted for altitude, temperature, and salinity.

Selection of a DO meter should be made with consideration of the type of environment in which it is to be used. For freshwater culture systems, temperature and altitude compensation will be required, while mariculturists will want to employ a DO meter that is salinity compensated as well. For DO meters that do not have built-in compensation features for each of the desired parameters, conversion tables are generally provided by the manufacturer.

A good-quality DO meter will cost a few hundred dollars, while titration kits are less expensive but require replacement chemicals periodically. On balance, DO meters save time, and the need to replace chemicals and broken glassware may make the meter the more economical choice in the long run.

Causes of Oxygen Depletions

Many aquatic species will provide the aquaculturist with an indication that an oxygen depletion is occurring by a change in their behavior. Feeding activity will often be reduced or cease completely (though that can occur in response to other problems, or it may be an aspect of normal behavior). When DO becomes critically low, many fishes will rise to the water surface and appear to gulp air. Crawfish may climb up on emergent vegetation to place their gills in contact with the atmosphere. Shrimp will often move into shallow water when the DO level is low. Shrimp farmers often check their ponds at night by shining a flashlight into the water. If shrimp are concentrated in the shallow water, the oxygen may be low. However, if the shrimp move quickly away from the light, the situation is typically not critical. When the animals ignore the light, then the problem may be severe.[78]

In ponds, sufficient oxygen is usually produced by photosynthesis during daylight and absorbed from the atmosphere at all times to maintain DO above the critical level. Flowing water systems may require continuous aeration in order to maintain the desired DO level. As biomass and the amount of feed being put into the culture chambers increase during the growing season, maintenance of the required DO level may become increasingly difficult. In ponds, the problem is exacerbated by high primary production rates because of the added respiratory demand at night. Temperature also plays in important role, not only because of the direct effect it has on the solubility of oxygen in water, but also because as temperature increases so does the metabolic rate of the organisms in the system.

In temperate climates, DO problems in ponds are fairly predictable. During spring, primary production may be high, but animal biomass tends to be relatively low and the water temperature is cool, so DO depletions tend not to be a problem. In the fall, water temperature cools once again, and primary production may show another, somewhat lower peak than that observed in the spring. Biomass may be reduced from the summer high due to harvesting, feed rate may be reduced (see Chapter 6), and the result is that DO problems are not too common. During winter, temperature is low, feeding rate may be reduced or feed may not even be offered, biomass may be further reduced due to harvesting, and primary productivity is also slowed because of temperature. Unless a pond is ice-covered and subjected to winterkill from oxygen depletions, problems with low DO should not occur.

Summer is the season when most factors work against the maintenance of high DO levels in ponds. Temperature is at its maximum, so the solubility of oxygen is at its lowest. Biomass and feeding rates are typically approaching maximum for the year, and while primary production may have declined considerably from the spring high, crashes in phytoplankton blooms may occur leading to greatly increased BOD as the cells decompose. Respiration by all the organisms in the pond is at its highest level, and the decomposer community is placing a considerable demand on the available oxygen.

Declines in phytoplankton are also associated with periods of cloudy weather that reduce the amount of available sunlight and restrict the photic zone. During cloudy periods the water temperature might cool somewhat, thereby increasing the solubility of oxygen, but the aquaculturist should be aware that cloudy weather can trigger the crash of a phytoplankton bloom.

Typically, the lowest level of DO that occurs in a pond over any 24 hr period will coincide with dawn because respiration will take place in the absence of photo-

---

[78] George Chamberlain, personal communication.

synthesis from about dusk the day before. Throughout the day, oxygen should increase as photosynthesis occurs. In a system at equilibrium, the diel pattern in DO may lead to a minimum level that is acceptable (5. 0 mg/l or more) at dawn. At about dusk, the level may be a few milligrams per liter higher, and the pattern will continue with little change from day to day (repeating the one shown in Figure 63). Ponds do not tend to be at equilibrium, however. As we have already seen, the photosynthetic rate can change significantly with temperature, the level of nutrients present, and as a function of light level (which is influenced by cloud cover). At the same time the demand on oxygen from respiration is constantly being altered as biomass changes. These and other factors tend to lead to a net change in DO concentration from one day to the next. That change may be either upward or downward and varies from one pond to the next.

By monitoring pond DO early in the morning, preferably before or at dawn, the culturist will be able to predict in advance when a problem may be imminent. To the delight of aquaculturists, particularly those with large numbers of ponds, few ponds respond in exactly the same way even when they are stocked at the same rate and subjected to the same management practices. It has been said that no two ponds are alike, and that tends to be true with respect to DO curves. Thus, a culturist may see critically low DO levels in one or two ponds on a given day, but it is rare when a large percentage of ponds on any one facility experience depletion problems.[79]

In aquaculture facilities other than ponds, primary productivity is not usually a major factor, though cages and netpens are sometimes an exception.[80] For tanks and raceways, whether in a flowthrough or recirculating configuration, oxygen depletions often occur in response to increasing respiratory demands as biomass increases. Failure of a biofilter can quickly lead to DO problems in a recirculating system, as can a reduction or loss of water input in any flowing system. Routine determination of DO can be accomplished at any time of the day in systems that are not subjected to diel fluctuations related to the photosynthetic cycle.

Overcoming Oxygen Depletions

When a DO depletion is detected, immediate action should be initiated to restore DO to a safe level. Several means can be used to accomplish that restoration. When it is available, the addition of large amounts of well-oxygenated new water to a culture system will quickly raise the DO level. Since new water is often at a premium and if it comes from a well may contain little or no oxygen, other means of aeration are commonly used.

Any technique that will increase the amount of water surface area in contact with

---

[79] As with most facts about aquaculture, there are exceptions. If a farmer were to apply herbicide to several weed-infested ponds, for example, each of those might experience a DO crisis on the same day as a result of the increased BOD from decaying vegetation. Even in that case it is likely that the problem would appear over a series of days since even the decomposition rate would not be the same in each pond.
[80] Toxic algae blooms have been a serious problem for net-pen culturists in some locations (e.g., Washington State and Norway), but those problems are not generally associated with oxygen depletions.

the atmosphere will increase the amount of oxygen in the exposed water. Aeration with compressors or air blowers is typically used in small systems. While such systems can sometimes be used in ponds, they are often not practical. Other ways of getting oxygen into water include aeration with compressed air, bottled oxygen, and liquid oxygen. Again, while commonly employed in conjunction with tank and raceway systems, those sources of oxygen are not usually used on ponds or with cages and netpens. Aquaculture aeration systems and their efficiencies have been reviewed by Boyd and Watten (1989).

An excellent means of aerating ponds is to get the water within the pond to circulate. By providing a means of continuously bringing deep water to the surface, the entire water volume can be put into motion. Splashing water does increase the amount of surface area in contact with the atmosphere, but it may not completely mix the pond. Some commercial aeration devices draw water from beneath them and throw it up in the air, where it becomes oxygenated. The water then falls back into the pond and is recycled through the aerator. The net result is that there is an improved DO level in the immediate vicinity of the aeration device but not elsewhere in the pond.

Figure 65. A series of ponds at an aquaculture research complex in Israel in which each pond is equipped with a paddlewheel aerator.

Figure 66. Paddlewheels used in the United States often operate off the power takeoff of a tractor, making it easy for the aquaculturist to move the aerator from pond to pond.

Total pond mixing can be accomplished with paddlewheel aerators (Figure 65), which are typically operated by electric motors or from the power takeoff of a tractor (Figure 66). Tractor-driven paddlewheel aerators are commonly used in the United States in instances where stocking levels are not so high that each pond in a complex will require aeration on a given day. Most farmers only have one or two such aerators available (no more than the available number of tractors) that can be quickly moved from pond to pond as the need arises. If very high densities are stocked and the farmer knows that DO problems will be chronic in a number of ponds, it may be necessary to provide an aerator for each, in which case electric devices are preferred. Engle (1989) reported that when a pond requires less than 250 hr of aeration a year, a tractor-driven paddlewheel is more efficient from an economic standpoint than an electric one, but that the reverse is true when more than 250 hr of aeration are required; pond size has an influence on efficiency as well, with floating electric paddlewheels being the more efficient choice in large ponds.

Paddlewheel aerators do not need to splash a great deal of water to be effective (Figure 67). Their primary purpose is to turn over the pond, though the splashing action may provide some psychological benefit since the farmer can clearly see that something is happening. Boyd and Watten (1989) discussed a number of types of paddlewheel designs along with other methods for aerating ponds.

When an oxygen depletion is anticipated, preventative measures can be taken. Paddlewheel or other types of aerators can be operated through the night. Alternatively, or in addition, oxygen-rich new water can be added to flush the pond. Both the addition of new water and aeration are often used by the catfish farmers in

Figure 67. Agitation is not always necessary for good aeration. The slowly moving paddlewheel in this photograph at a research laboratory in Alabama did not splash the water, but was effective at turning over the pond and efficiently maintaining the DO level in the water.

Mississippi during the period of the year when biomass and feeding rates are at their highest levels and water quality is difficult to maintain. The addition of new water reduces the amount of suspended organic material by flushing it from the pond. As a result, the BOD can be significantly reduced.[81]

Culture animals should not be fed when an oxygen depletion is anticipated or has recently occurred. The stress associated with oxygen depletions often causes the animals to reject feed, so the ration will only serve to increase the BOD when it decomposes. Stress that does not result in death of the animals may make them susceptible to diseases. Epizootics may occur as soon as 24 hr after an oxygen depletion or as long as 2 wk later. The culturist should be aware of that time frame and watch the animals closely. If a disease is detected, early treatment may be more effective than if a full-blown epizootic develops before the culturist is aware of the problem.

Potassium permanganate ($KMnO_4$), a strong oxidizing agent, has long been used by fish culturists at the rate of about 2 to 3 mg/l in instances where oxygen levels are low on the theory that the oxidation of organic matter will lower the BOD and chemical oxygen demand (COD). Tucker and Boyd (1977) found that 2 mg/l of $KMnO_4$ killed over 99% of the gram-negative bacteria present and a considerable percentage of the gram-positive bacteria. In addition to the oxidizing activity, potassium permanganate will release free oxygen directly to the water.

---

[81] Flushing water of degraded quality from ponds may not be possible in the future as increasingly stringent effluent standards are imposed.

Boyd (1990) reported on studies that had been conducted to determine if potassium permanganate was effective and concluded that ponds treated with the chemical early in the morning actually recover more slowly than those that are not treated. He also examined the chemistry of permanganate ion in water and found that free oxygen is released; however, the number of milligrams per liter of potassium permanganate that would be required to release 1 mg/l of O_2 is at least 6.58 mg/l. In order to increase DO in a pond, the amount of $KMnO_4$ needed to oxidize the organic matter present would have to be supplemented by over 6 mg/l for each milligram per liter of O_2 increase required. Thus, before any meaningful increase in DO could be obtained, the level of potassium permanganate required would be toxic to fish.

Unlike ponds, the oxygen dynamics in raceways are highly predictable. Because of the rapid turnover rate that exists in most raceway systems, there is little opportunity for phytoplankton and other photosynthetic organisms to become established. As indicated by Boyd and Watten (1989), the sources of oxygen in raceways are associated with the incoming water and supplemental aeration. The amount of oxygen used by fish in respiration can be related by a proportionality constant (K_1) that correlates with the grams of oxygen consumed per kilogram of feed offered. Boyd and Watten (1989) reviewed the literature and found the K_1 is slightly over 200 g/kg in channel catfish and from 200 to 220 g/kg for salmonids. These authors cautioned that those values are means. Oxygen demand increases dramatically during feeding and for a period thereafter because of the increased metabolic rate associated with feeding activity and digestion.

Hydrogen peroxide (H_2O_2) can be used to generate oxygen gas and has been evaluated as a source of oxygen in conjunction with live hauling. Innes Taylor and Ross (1988) described a technique by which hydrogen peroxide was used to generate oxygen in a container that was separate from the fish hauling tank. They found no toxicity in *Tilapia nilotica* exposed to a concentration of 0.15 ml/l of 12% H_2O_2.

Pure oxygen in the form of compressed gas or liquid oxygen (LOX) is increasingly finding use in aquaculture systems, particularly high-density tank and linear raceway systems. LOX is often available within easy trucking distance of even remotely located fish farms in the United States. It is generally more economical to have LOX delivered than to manufacture it on site.[82] By continuously injecting oxygen into a raceway system, much higher standing crops can be supported with a given water flow rate than with the same system operating without such supplementation.

LITERATURE CITED

Allan, G. L., G. B. Maguire, and S. J. Hopkins. 1990. Acute and chronic toxicity of ammonia to juvenile *Metapenaeus macleayi* and *Penaeus monodon* and the influence of dissolved-oxygen levels. Aquaculture, 91: 265–280.

---

[82] Ronald Mayo, personal communication.

Almendras, J. M. E. 1987. Acute nitrite toxicity and methemoglobinemia in juvenile milk-fish (*Chanos chanos* Forsskal). Aquaculture, 61: 33–40.

Anderson, W. H. 1965. Search for insects in South America that feed on aquatic weeds. Proc. South. Weed Conf., 18: 586–587.

Anonymous. 1980. Brown blood disease continues to cause problems in catfish. pp. 1–2, In: For fish farmers. Mississippi State University Cooperative Extension Service, Mississippi State.

APHA. 1989. Standard methods. 17th edition. American Public Health Association, Washington, D.C. 1,467 p.

Applied Biochemists, Inc. 1976. How to identify and control water weeds and algae. Applied Biochemists, Inc., Mequon, Wisconsin. 64 p.

Arillo, A., E. Gaino, C. Margiocco, P. Mensi, and G. Schenone. 1984. Biochemical and ultrastructural effects of nitrite in rainbow trout: liver hypoxia as the root of the acute toxicity mechanism. Environ. Res., 34: 135–154.

Avault, J. W., Jr. 1965a. Biological weed control with herbivorous fish. Proc. South. Weed Conf., 18: 590–591.

Avault, J. W., Jr. 1965b. Preliminary studies with grass carp for aquatic weed control. Prog. Fish-Cult., 27: 207–209.

Baily, T. A. 1965. Commercial possibilities of dehydrated aquatic plants. Proc. South. Weed Conf., 18: 543–551.

Bauer, D. L., and D. W. Willis. 1990. Effects of triploid grass carp on aquatic vegetation in two South Dakota lakes. Lake Reservoir Manage., 6: 175–180.

Blackburn, R. D., and L. W. Weldon. 1965. A fresh-water snail as a weed control agent. Proc. South. Weed Conf., 18: 589–591.

Blanco, O., and T. Meade. 1980. Effect of dietary ascorbic acid on the susceptibility of steelhead trout (*Salmo gairdneri*) to nitrite toxicity. Rev. Biol. Trop., 28: 91–107.

Bonn, E. W., W. M. Bailey, J. D. Bayless, K. E. Erickson, and R. E. Stevens. 1976. Guidelines for striped bass culture. Striped Bass Committee of the Southern Division, American Fisheries Society, Bethesda, Maryland. 103 p.

Boyd, C. E. 1968a. Evaluation of some common aquatic weeds as possible feedstuffs. Hyacinth Control J., 7: 26–27.

Boyd, C. E. 1968b. Fresh-water plants: a potential source of protein. Econ. Bot., 22: 359–368.

Boyd, C. E. 1990. Water quality in ponds for aquaculture. Agricultural Experiment Station, Auburn University, Auburn, Alabama. 482 p.

Boyd, C. E., and B. S. McGinty. 1981. Percentage digestible dry matter and crude protein in dried aquatic weeds. Econ. Bot., 35: 296–299.

Boyd, C. E., and B. J. Watten. 1989. Aeration systems in aquaculture. Rev. Aquat. Sci., 1: 425–472.

Broussard, M. C., Jr., R. Reyes, and F. Raguindin. 1983. Evaluation of hatchery management schemes for large scale production of *Oreochromis niloticus* fingerings in Central Luzon, Philippines. pp. 414–424, In: L. Fishelson and Z. Yaron (compilers). International Symposium on Tilapia in Aquaculture, Nazareth, Israel, May 8–13. Tel Aviv University Press, Tel Aviv.

Brown, V. M. 1968. The calculation of the acute toxicity of mixtures of poisons to rainbow trout. Water Res., 2: 723–733.

Buck, D. H., R. J. Baur, and C. R. Rose. 1989. Utilization of swine manure in a polyculture of Asian and North American fishes. Trans. Am. Fish. Soc., 107: 216–222.

Buckingham, G. R., and C. A. Bennett. 1988. Laboratory host range of *Paraponyx diminutalis* (Lepidoptera: Pyralidae), and Asian aquatic moth adventive in Florida and Panama on *Hydrilla verticillata* (Hydrocharitaceae). Environ. Entomol., 18: 526–530.

Burns, R. P., and R. R. Stickney. 1980. Growth of *Tilapia aurea* in ponds receiving poultry wastes. Aquaculture, 20: 117–121.

Burrows, R. E. 1964. Effects of accumulated excretory products on hatchery-reared salmonids. U.S. Bureau of Sport Fisheries and Wildlife Resources Report No. 66, Washington, D.C. 12 p.

Calamari, D., R. Marchetti, and G. Vailati. 1981. Effects of long-term exposure to ammonia on the developmental stages of rainbow trout (*Salmo gairdneri* Richardson). Rapp. P.-V. Reun. Ciem., 178: 81–85.

Carpenter, S. R., and J. K. Greenlee. 1981. Lake deoxygenation after herbicide use: a simulation model analysis. Aquat. Bot., 2: 173–186.

Cassani, J. R., and W. E. Caton. 1983. Feeding behavior of yearling and older hybrid grass carp. J. Fish Biol., 22: 35–41.

Chen, J.-C., and T.-S. Chin. 1988a. Acute toxicity of nitrite to tiger prawn, *Penaeus monodon*, larvae. Aquaculture, 69: 253–262.

Chen, J.-C., and T.-S. Chin. 1988b. Joint action of ammonia and nitrite on tiger prawn *Penaeus monodon*. J. World Aquacult. Soc., 19: 143–148.

Chen, J.-C., and T.-S. Chin. 1989. Effect of ammonia at different pH levels on *Penaeus monodon* postlarvae. Asian Fish. Sci., 2: 233–238.

Chen, J.-C., Y. Y. Ting, J.-N. Lin, and M.-N. Lin. 1990a. Lethal effects of ammonia and nitrate on *Penaeus chinensis* juveniles. Mar. Biol., 107: 427–431.

Chen, J.-C., P.-C. Liu, and S.-C. Lei. 1990b. Toxicities of ammonia and nitrite to *Penaeus monodon* adolescents. Aquaculture, 89: 127–137.

Chipman, W. A., Jr. 1934. The role of pH in determining the toxicity of ammonium compounds. Ph.D. Dissertation, University of Missouri, Columbia. 153 p.

Colle, D. E., J. V. Shireman, and R. W. Rottmann. 1978. Food selection by grass carp fingerlings in a vegetated pond. Trans. Am. Fish. Soc., 107: 149–152.

Colura, R. L. 1987. Saltwater pond fertilization. pp.51–53, In: G. W. Chamberlain, R. J. Miget, and M. G. Haby (Eds.). Manual on red drum aquaculture, Vol. 3. Texas Agricultural Extension Service, Corpus Christi.

Colura, R. L., and G. C. Matlock. 1984. Comparison of zooplankton production in brackish water ponds fertilized with cottonseed meal or chicken manure. Proc. Tex. Chapt., Am. Fish. Soc., 6: 68–83.

Colura, R. L., B. T. Hysmith, and R. E. Stevens. 1976. Fingerling production of striped bass *(Morone saxatilis),* spotted seatrout *(Cynoscion nebulosus),* and red drum *(Sciaenops ocellatus)* in saltwater ponds. Proc. World Maricult. Soc., 7: 79–92.

Cook, C. D. K. 1990. Aquatic plant book. SPB Academic Publ., The Hague, Netherlands. 228 p.

Costa-Pierce, B. A. 1992. Review oif the spawning requirements and feeding ecology of silver carp *(Hypophthalmichthys molitrix)* and reevaluation of its use in fisheries and aquaculture. Rev. Aquat. Sci., 6: 257–273.

Culley, D. D., Jr., and E. A. Epps. 1973. Use of duckweed for waste treatment and animal feed. J. Water Pollut. Control Fed., 45: 337–347.

Cruz, E. R. 1981. Acute toxicity of un-ionized ammonia to milkfish *(Chanos chanos)* fingerlings. Fish. Res. J. Philipp., 6: 33–38.

Daud, S. K., D. Hasbollah, and A. T. Law. 1988. Effects of un-ionized ammonia on red tilapia *(Oreochromis mossambicus/O. niloticus* hybrid). pp. 411–413, In: R. S. V. Pullin, T. Bhukaswan, K. Tonguthai, and J. L. Maclean (Eds.). The Second International Symposium on Tilapia in Aquaculture, Bangkok, Thailand, March 16–20, 1987. ICLARM Conf. Proc. No. 15, International Center for Living Aquatic Resources Management, Manila, Philippines.

DeBoer, J. A., B. E. Lapointe, and J. H. Ryther. 1977. Preliminary studies on a combined seaweed mariculture–tertiary waste treatment system. Proc. World Maricult. Soc., 8: 401–406.

Denzer, H. W. 1968. Studies on the physiology of young *Tilapia*. FAO Fish. Rep., 44: 357–366.

Downing, K. M., and J. C. Merkens. 1955. The influence of dissolved-oxygen concentration on the toxicity of un-ionized ammonia to rainbow trout *(Salmo gairdneri* Richardson). Ann. Appl. Biol., 43: 243–246.

Duangsawasdi, M., and C. Sripumun. 1981. Acute toxicities of ammonia and nitrite to *Clarias batrachus* and their interaction to chlorides. National Inland Fisheries Institute, Bangkok, Thailand. 20 p.

Eddy, F. B., and E. M. Williams. 1987. Nitrite and freshwater fish. Chem. Ecol., 3: 1–38.

Emerson, K., R. C. Russo, R. E. Lund, and R. V. Thurston. 1975. Aqueous ammonia equilibrium calculations: effect of pH and temperature. J. Fish. Res. Board Can., 32: 2379–2383.

Engle, C. R. 1989. An economic comparison of aeration devices for aquaculture ponds. Aquacult. Eng., 8: 193–207.

Eyles, D. E., and J. L. Robertson. 1944. A guide and key to the aquatic plants of the southeastern United States. U.S. Public Health Service Bulletin No. 286, Washington, D.C. 151 p.

Fagbenro, O. A., and D. H. J. Sydenham. 1988. Evaluation of *Clarias isheriensis* (Sydenham) under semi-intensive management in ponds. Aquaculture, 74: 287–291.

Farquhar, B. W. 1987. Comparison of granular and liquid inorganic fertilizers used in striped bass and smallmouth bass rearing ponds. Prog. Fish-Cult., 48: 21–28.

Fassett, N. C. 1960. A manual of aquatic plants. University of Wisconsin Press, Madison. 405 p.

Flis, J. 1968. Anatomicohistopathological changes induced in carp *(Cyprinus carpio* L.) by ammonia water. I. Effects of toxic concentrations. II. Effects of sub-toxic concentrations. Acta Hydrobiol., 10: 205–238.

Fogg, G. E. 1972. Photosynthesis. Elsevier, New York. 116 p.

Fromm, R. O. 1963. Studies on the renal and extra-renal excretion in a freshwater teleost, *Salmo gairdneri*. Comp. Biochem. Physiol., 10: 121–128.

Gaarder, T. and J. J. Gran. 1927. Investigations of the production of plankton in the Oslo Fjord. Rapp. P.-V. Cons. Int. Explor. Mer, 42: 1–48.

Gaino, E., A. Arillo, and P. Mensi. 1984. Involvement of the gill chloride cells of trout under acute nitrite intoxication. Comp. Biochem. Physiol., 77**A**: 611–617.

Geiger, J. G. 1983. Zooplankton production and manipulation in striped bass rearing ponds. Aquaculture, 35: 331–351.

Guillory, V., and R. D. Gasaway. 1978. Zoogeography of the grass carp in the United States. Trans. Am. Fish. Soc., 107: 105–112.

Harader, R. R., Jr., and G. H. Allen. 1983. Ammonia toxicity to chinook salmon parr: reduction in saline water. Trans. Am. Fish. Soc., 112: 834–837.

Hasan, M. R., and D. J. Macintosh. 1986a. Acute toxicity of ammonia to common carp fry. Aquaculture, 54: 97–107.

Hasan, M. R., and D. J. Macintosh. 1986b. Effect of chloride concentration on the acute toxicity of nitrite to carp, *Cyprinus carpio* L., fry. Aquacult. Fish. Manage., 17: 19–30.

Haywood, G. P. 1983. Ammonia toxicity in teleost fishes: a review. Canadian Technical Reports in Fisheries and Aquatic Sciences, No. 1177. Canadian Department of Fisheries and Oceans, Ottawa. 39 p.

Hilmy, A. M., N. A. El-Domiaty, and K. Wershana. 1987. Acute and chronic troxicity of nitrite to *Clarias lazera*. Comp. Biochem. Physiol., 86**C**: 247–253.

Hochachka, P. W. 1969. Intermediary metabolism in fishes. pp. 351–389, In: W. S. Hoar and D. J. Randall (Eds.). Fish Physiology, Vol. 1. Academic Press, New York.

Huguenin, J. E. 1975. Development of a marine aquaculture research complex. Aquaculture, 5: 135–150.

Innes Taylor, N., and L. G. Ross. 1988. The use of hydrogen peroxide as a source of oxygen for the transportation of live fish. Aquaculture, 70: 183–192.

Konikoff, M. 1975. Toxicity of nitrite to channel catfish. Prog. Fish-Cult., 37: 96–98.

Lange, S. R. 1965. The control of aquatic plants by commercial harvesting, processing and marketing. Proc. South. Weed Conf., 18: 536–542.

Lanoiselee, B., R. Billard, and G. de Montalembert. 1986. Organic fertilization of nursery ponds for pike *(Esox lucius)*. Aquaculture, 54: 141–147.

Lees, H. 1952. The biochemistry of the nitrifying organisms. 1. The ammonia-oxidizing system of *Nitrosomonas*. Biochem. J., 52: 134–139.

LeRoux, P. J. 1956. Feeding habits of the young of four species of *Tilapia*. S. Afr. J. Sci., 53: 96–100.

Lewis, W. M., Jr., and D. P. Morris. 1986. Toxicity of nitrite to fish: a review. Trans. Am. Fish. Soc., 115: 183–195.

Lightner, D., R. Redman, L. Mohney, G. Dickenson, and K. Fitzsimmons. 1988. Major diseases encountered in controlled environment culture of tilapias in fresh-and brackishwater over a three-year period in Arizona. pp. 111–116, In: R. S. V. Pullin, T. Bhukaswan, K. Tonguthai, and J. L. Maclean (Eds.). The Second International Symposium on Tilapia in Aquaculture, Bangkok, Thailand, March 16–20, 1987. ICLARM Conf. Proc. No. 15, International Center for Living Aquatic Resources Management, Manila, Philippines.

Lin, H. P., G. Charmantier, and J.-P. Trilles. 1991. Toxicite de l'ammonium au cours du development postembryonnaire de *Penaeus japonicus* (Crustacea, Decapoda). Effects sur l'osmoregulation. C. R. Acad. Sci., Ser. III Sci. Vie, 312: 99–105.

Lizaman, L. C., J. E. Marion, and L. R. McDowell. 1988. Utilization of aquatic plants

Elodea canadensis and *Hydrilla verticillata* in broiler chick diets. Anim. Feed Sci. Technol., 20: 155–161.

Lloyd, R. 1961. The toxicity of ammonia to rainbow trout (*Salmo gairdneri* Richardson). Water & Waste Treat. J., 8: 278–279.

Lloyd, R., and D. W. M. Herbert. 1960. The influence of carbon dioxide on the toxicity of un-ionized ammonia to rainbow trout (*Salmo gairdnerii* Richardson). Ann. Appl. Biol., 48: 399–404.

Lockhart, L., B. N. Billeck, and C. L. Baron. 1989. Bioassays with a floating aquatic plant *(Lemna minor)* for effects of sprayed and dissolved glyphosate. pp. 353–359, In: M. Munawar, G. Dixon, C. I. Mayfield, T. Reynoldson, and M. H. Sadar (Eds.). Environmental bioassay techniques and their application. Hydrobiologia No. 188–189.

Ludwig, G. M., and D. L. Tackett. 1991. Effects of using rice bran and cottonseed meal as organic fertilizers on water quality, plankton, and growth and yield of striped bass, *Morone saxatilis,* fingerlings in ponds. J. Appl. Aquacult., 1: 79–94.

Martyn, R. D., R. L. Noble, P. W. Bettoli, and R. C. Maggio. 1986. Mapping aquatic weeds with aerial infrared photography and evaluating their control by grass carp. J. Aquat. Plant Manage., 24: 46–56.

Mazik, P. M., M. L. Hinman, D. A. Winkelmann, S. J. Klaine, B. A. Simco, and N. C. Parker. 1991. Influence of nitrite and chloride concentrations on survival and hematological profiles of striped bass. Trans. Am. Fish. Soc., 120: 247–254.

McBay, L. G. 1961. The biology of *Tilapia nilotica* Linnaeus. Proc. Southeast. Assoc. Game Fish. Comm., 15: 208–218.

McGeachin, R. B., and R. R. Stickney. 1982. Manuring rates for production of blue tilapia in simulated sewage lagoons receiving laying hen waste. Prog. Fish-Cult., 44: 25–28.

Meade, T. L. 1976. Closed system salmonid culture in the United States. Marine Memorandum No. 40. Marine Advisory Service, University of Rhode Island, Kingston. 16 p.

Merkens, J. C., and K. M. Downing. 1957. The effect of tension of dissolved oxygen on the toxicity of un-ionized ammonia to several species of fish. Ann. App. Biol., 45: 521–527.

Meyer, F. P., and R. A. Schnick. 1989. A review of chemicals used for the control of fish diseases. Rev. Aquat. Sci., 1: 693–710.

Mikol, G. F. 1985. Effects of harvesting on aquatic vegetation and juvenile fish populations of Saratoga Lake, New York. J. Aquat. Plant Manage., 23: 59–63.

Mitzner, L. 1978. Evaluation of biological control of nuisance aquatic vegetation by grass carp. Trans. Am. Fish. Soc., 107: 135–145.

Moriarty, D. J. W. 1973. The physiology of digestion of blue-green algae in the cichlid fish *Tilapia nilotica.* J. Zool. (Lond.), 171: 25–39.

Muenscher, W. C. 1944. Aquatic plants of the United States. Comstock Publishing Co., Ithaca, New York. 374 p.

Muir, P. R., D. C. Sutton, and L. Owens. 1991. Nitrate toxicity to *Penaeus monodon* protozoea. Mar. Biol., 108: 67–71.

Osborne, J. A. 1982. The potential of the hybrid grass carp as a weed control agent. J. Freshwater Ecol., 1: 353–360.

Palachek, R. M., and J. R. Tomasso. 1984. Toxicity of nitrite to channel catfish *(Ictalurus punctatus),* tilapia *(Tilapia aurea),* and largemouth bass *(Micropterus salmoides).* Evidence for a nitrite exclusion mechanism. Can. J. Fish. Aquat. Sci., 41: 1739–1744.

Pillay, T. V. R. 1972. Coastal aquaculture in the Indo-Pacific region. Fishing News (Books), London. 497 p.

Pine, R. T., and L. W. J. Anderson. 1991. Effect of triploid grass carp on submersed aquatic plants in northern California ponds. Calif. Fish Game, 77: 27–35.

Piper, R. G., I. B. McElwain, L. E. Orme, J. P. McCraren, L. G. Fowler, and J. R. Leonard. 1982. Fish hatchery management. U.S. Fish and Wildlife Service, Washington, D.C. 517 p.

Porter, C. W., and A. F. Maciorowski. 1984. Spotted seatrout fingerling production in salt-water ponds. J. World Maricult. Soc., 15: 222–232.

Ramaprabhu, T., and Ramachandran, V. 1990. Observations on the effects and persistence of herbicides in aquaculture. Comp. Physiol. Ecol., 8: 223–233.

Redner, B. D., and R. R. Stickney. 1979. Acclimation to ammonia by *Tilapia aurea*. Trans. Am. Fish. Soc., 108: 383–388.

Richards, F. A., and T. C. Thompson. 1952. The estimation and characterization of plankton populations by pigment analyses. II. A spectrophotometric method for the estimation of plankton pigments. J. Mar. Res., 11: 156–172.

Robinette, H. R. 1976. Effect of selected sublethal levels of ammonia on the growth of channel catfish *(Ictalurus punctatus)*. Prog. Fish-Cult., 38: 26–29.

Ruffier, P. J., W. C. Boyle, and J. Kleinschmidt. 1981. Short-term acute bioassays to evaluate ammonia toxicity and effluent standards. J. Water Pollut. Control Fed., 53: 367–377.

Russo, R. C. 1984. Ammonia, nitrite, and nitrate. pp. 455–474, In: G. M. Rand and S. R. Petrocelli (Eds.). The fundamentals of aquatic toxicology: methods and applications. Hemisphere, Washington.

Ryther, J. H. 1956. The measurement of primary productivity. Limnol. Oceanogr., 1: 72–84.

Sadler, K. 1981. The toxicity of ammonia to the European eel (*Anguilla anguilla* L.). Aqua-culture, 28: 173–181.

Saroglia, M. G., G. Scarano, and E. Tibaldi. 1981. Acute toxicity of nitrite to sea bass *(Dicentrarchus labrax)* and European eel *(Anguilla anguilla)*. J. World Maricult. Soc., 12: 121–126.

Saunders, G. W., F. B. Trama, and R. W. Bachmann. 1962. Evaluation of a modified C-14 technique for shipboard estimation of photosynthesis in large lakes. Great Lakes Re-sources Division Publication No. 8. University of Michgan, Ann Arbor.

Scarano, G., M. G. Saroglia, R. H. Gray, and E. Tibaldi. 1984. Hematological responses of sea bass *Dicentrarchus labrax* to sublethal nitrite exposures. Trans. Am. Fish. Soc., 113: 360–364.

Schramm, H., Jr., and K. J. Jirka. 1986. Evaluation of methods for capturing grass carp in agricultural canals. J. Aquat. Plant Manage., 24: 57–59.

Schroeder, G. 1974. Use of cowshed manure in fish ponds. Bamidgeh, 26: 84–96.

Seagrave, C. 1988. Aquatic weed control. Fishing News, Farnham, Surrey, England. 160 p.

Seyrin-Reyssac, J. 1990. The use of copper sulphate and simazine in fish ponds. Effects on the fauna influencing fish production. Bull. Ecol., 21: 89–95.

Sguros, P. L., T. Monku, and C. Philipps. 1965. Observations and techniques of the Florida manatee—reticent but superb weed control agent. Proc. South. Weed Conf., 18: 588.

Sheehan, R. J., and W. M. Lewis. 1986. Influence of pH and ammonia salts on ammonia toxicity and water balance in young channel catfish. Trans. Am. Fish. Soc., 115: 891–899.

Shireman, J. V., D. E. Colle, and D. E. Canfield, Jr. 1986. Efficacy and cost of aquatic weed control in small ponds. Water Resour. Bull., 22: 43–48.

Sills, J. 1970. A review of herbiorous fish for weed control. Prog. Fish-Cult., 32: 158–161.

Smart, G. 1976. The effect of ammonia exposure on gill structure of the rainbow trout *(Salmo gairdneri)*. J. Fish Biol., 8: 471–475.

Smith, C. E., and R. G. Piper. 1975. Lesions associated with chronic exposure to ammonia. pp. 497–514, In: W. E. Ribelin and G. Migaka (Eds.). The pathology of fishes. University of Wisconsin Press, Madison.

Smith, G. M. 1950. Fresh-water algae of the United States. McGraw-Hill, New York. 145 p.

Smith, H. W. 1929. The excretion of ammonia and urea by the gills of fish. J. Biol. Chem., 81: 727–742.

Sripumun, C., and C. Somsiri. 1982. Acute toxicity of ammonia and nitrite to the walking catfish. Thai Fish. Gaz., 35: 373–378.

Stanley, J. G., W. W. Miley II, and D. L. Sutton. 1978. Reproductive requirements and likelihood for naturalization of escaped grass carp in the United States. Trans. Am. Fish. Soc., 107: 119–128.

Steemann Nielsen, E. 1951. Measurement of the production of organic matter in the sea by means of C-14. Nature, 167: 684–685.

Steemann Nielsen, E. 1952. The use of radioactive carbon (C-14) for measuring organic production in the sea. J. Cons., 18: 117–140.

Steemann Nielsen, E. 1975. Marine photosynthesis. Elsevier Oceanography Series, 13. Elsevier Scientific Publications, Amsterdam. 140 p.

Stemler, A., and Govindjee. 1973. Bicarbonate ion as a critical factor in photosynthetic oxygen evolution. Plant Physiol., 53: 119–123.

Stevens, F. S. 1982. Sensitivity of juvenile hard clams *(Mercenaria mercanaria)* to ammonia. J. Shellfish Res., 2: 107.

Stevenson, J. 1965. Observations on grass carp in Arkansas. Prog. Fish-Cult., 27: 203–206.

Stickney, R. R., and J. H. Hesby. 1978. Tilapia culture in ponds receiving swine waste. pp. 90–101, In: R. O. Smitherman, W. L. Shelton, and J. H. Grover (Eds.). Culture of Exotic Fishes Symposium Proceedings. Fish Culture Section, American Fisheries Society, Bethesda, Maryland.

Stickney, R. R., H. B. Simmons, and L. O. Rowland. 1977a. Growth responses of *Tilapia aurea* to feed supplemented with dried poultry waste. Tex. Acad. Sci., 29: 93–99.

Stickney, R. R., L. O. Rowland, and J. H. Hesby. 1977b. Water quality–*Tilapia aurea* interactions in ponds receiving swine and poultry wastes. Proc. World Maricult. Soc., 8: 55–71.

Stickney, R. R., J. H. Hesby, R. B. McGeachin, and W. A. Isbell. 1979. Growth of *Tilapia nilotica* in ponds with differing histories of organic fertilization. Aquaculture, 17: 189–194.

Swingle, H. S. 1947. Management of farm fish ponds. Alabama Agricultural Experiment Station Bulletin No. 254, Auburn University, Auburn, Alabama. 30 p.

Tabata, K. 1962. Toxicity of ammonia to aquatic animals with reference to the effect of pH and carbon dioxide. Bull. Tokai Reg. Fish. Res. Lab., 34: 67–74.

Tamse, A. F., N. R. Fortes, L. C. Catedrilla, and J. E. H. Yuseco. 1985. The effect of using piggery wastes in brackishwater fishpond on fish production. Fish. J. Coll. Fish. Univ. Philippines Visayas, 1: 69–76.

Thurston, R. V., and R. C. Russo. 1983. Acute toxicity of ammonia to rainbow trout. Trans. Am. Fish. Soc., 112: 696–704.

Thurston, R. V., R. C. Russo, and G. A. Vinogradov. 1981a. Ammonia toxicity to fishes. Effect of pH on the toxicity of un-ionized ammonia species. Environ. Sci. Technol., 15: 837–840.

Thurston, R. V., C. Chakoumakos, and R. C. Russo. 1981b. Effect of fluctuating exposures on the acute toxicity of ammonia to rainbow trout *(Salmo gairdneri)* and cutthroat trout *(S. clarki)*. Water Res., 15: 911–917.

Thurston, R. V., R. C. Russo, R. J. Luedtke, C. E. Smith, E. L. Meyn, C. Chakoumakos, K. C. Wang, and C. J. D. Brown. 1984. Chronic toxicity of ammonia to rainbow trout. Trans. Am. Fish. Soc., 113: 56–73.

Tomasso, J. R. 1986. Comparative toxicity of nitrite to freshwater fishes. Aquat. Toxicol.(N.Y.), 8: 129–137.

Tomasso, J. R., and G. J. Carmichael. 1991. Differential resistance among channel catfish strains and intraspecific hybrids to environmental nitrite. J. Aquat. Anim. Health, 3: 51–54.

Tucker, C. S., and C. E. Boyd. 1977. Relationships between potassium permanganate treatment and water quality. Trans. Am. Fish. Soc., 106: 481–488.

Turner, J. W. D., R. R. Sibbald, and J. Hamans. 1986a. Chlorinated secondary domestic sewage effluent as a fertilizer for marine aquaculture. 1. Tilapia culture. Aquaculture, 53: 133–143.

Turner, J. W. D., R. R. Sibbald, and J. Hamans. 1986b. Chlorinated secondary domestic sewage effluent as a fertilizer for marine aquaculture. 3. Assessments of bacterial and viral quality and accumulation of heavy metals and chlorinated pesticides in cultured fish and prawns. Aquaculture, 53: 157–188.ß

Uchida, R. M., and J. E. King. 1962. Tank culture of *Tilapia*. Fish. Bull., 62: 21–25.

Van Dyke, J. M., A. J. Leslie, Jr., and J. E. Nall. 1984. The effects of the grass carp on the aquatic macrophytes of four Florida lakes. J. Aquat. Plant Manage., 22: 87–95.

Wajsbrot, N., A. Gasith, M. D. Krom, and T. M. Samocha. 1990. Effect of dissolved oxygen and the molt stage on the acute toxicity of ammonia to juvenile green tiger prawn *Penaeus semisulcatus*. Environ. Toxicol. Chem., 9: 497–504.

Wajsbrot, N., A. Gasith, M. D. Krom, and D. M. Popper. 1991. Acute toxicity of ammonia to juvenile gilthead seabream *Sparus aurata* under reduced oxygen levels. Aquaculture, 92: 277–288.

Wang, H., and D. Hu. 1989. Toxicity of nitrite to grass carp *(Ctenopharyngodon idellus)* in ponds and its way of prevention. J. Fish. China, 13: 207–214.

Weiss, R. F. 1970. The solubility of nitrogen, oxygen, and argon in water and seawater. Deep-Sea Res., 17: 721–735.

Weldon, L. W., R. D. Blackburn, and D. S. Harrison. 1969. Common aquatic weeds. U.S. Department of Agriculture Handbook 352, Washington, D.C. 43 p.

Wetzel, R. G. 1975. Limnology. Saunders, Philadelphia. 743 p.

Wheaton, F. W. 1977. Aquaculture engineering. Wiley-Interscience, New York. 708 p.

Williams, E. M., and F. B. Eddy. 1989. Effect of nitrite on the embryonic development of Atlantic salmon *(Salmo salar)*. Can. J. Fish. Aquat. Sci., 46: 1726–1729.

Wood, J. D. 1958. Nitrogen excretion in some marine teleosts. Can. J. Biochem. Physiol., 36: 1237–1242.

Wuhrmann, K., and H. Woker. 1948. Experimentelle Untersuchungen über die Ammoniak- und Blausaurevergiftung. Schweiz. Z. Hydrol., 11: 210–214.

Wuhrmann, K., F. Zehender, and H. Woker. 1947. Über die fischereibiologische Bedeutung des Ammonium- und Ammiakgehaltes fliessender Gewässer. Vierteljahrsschr. Naturforsch. Ges. Zür., 92: 198–204.

Wurts, W. A., and R. R. Stickney. 1984. An hypothesis on the light requirements for spawning penaeid shrimp, with emphasis on *Penaeus setiferus*. Aquaculture, 41: 93–98.

Yamagata, Y., and M. Niwa. 1982. Acute and chronic toxicity of ammonia to eel *Anguilla japonica*. Bull. Jpn. Soc., Sci. Fish., 48: 171–176.

Yashouv, A., and J. Chervinsky. 1960. Evaluation of various food items in the diet of *Tilapia nilotica*. Bamidgeh, 92: 198–204.

Yashouv, A., and J. Chervinsky. 1961. The food of *Tilapia nilotica* in ponds of the fish culture reserach station at Dor. Bamidgeh, 13: 33–39.

Young, L. M., J. P. Monaghan, Jr., and R. C. Heidinger. 1983. Food preferences, food intake, and growth of the F_1 hybrid of grass carp (females) $\times$ bighead carp (males). Trans. Am. Fish. Soc., 112: 661–664.

Zhou, Y.-X., F.-Y. Zhang, and R.-Z. Zhou. 1986. The acute and subacute toxicity of ammonia to grass carp *(Ctenopharyngodon iella)*. Acta Hydrobiol. Sin., 10: 32–38.

5 Conservative Aspects of Water Quality and Physical Aspects of the Culture Environment

GENERAL CONSIDERATIONS

Conservative water quality properties are those that are not affected in any significant way by the activities of organisms living in the water. Such factors as temperature, salinity, alkalinity, and hardness are modified in some cases by physical or chemical processes, but are not perturbed to any extent by biological activity except at the microcosmic level in some instances. For example, metabolic activity in organisms releases some heat energy into the environment. Very precise measurement of the water immediately in contact with the animal involved could perhaps demonstrate a minute change in temperature, but the effect would be localized and trivial. The oceans, lakes, rivers, and culture chambers around the world are not thermally altered as a result of the dissipation of such heat energy into them.

Physical aspects of the culture environment that may have to be considered, depending on the type of water system and the species involved, include stocking density, physical space requirements of individual organisms, substrate requirements, light intensity, and photoperiod. The one physical factor that is of universal interest is temperature, and while it is technically possible to control that parameter and sometimes necessary to do so, it can be extremely expensive.

In most parts of the world where aquaculture is practiced there is at least some seasonal fluctuation in water temperature. That fluctuation is reduced toward the equator and the poles and is greatest in temperate regions. Tropical aquaculturists and those who rear fish at high latitudes can benefit from relatively constant annual temperatures, though in the polar regions the water has temperatures that are suboptimum even for coldwater species. While adjustments to temperature may be expensive, alteration of background levels of such things as hardness, alkalinity, suspended solids, and light are generally feasible at reasonable cost. This chapter outlines the physical conservative chemical factors considered most important by aquaculturists and discusses their influence on culture organisms and how they can be controlled.

TEMPERATURE

Thermal Requirements

Selection of an aquaculture species usually takes temperature into consideration, particularly when the site has been selected and the aquaculturist knows the type of water that will be used in the facility. Water temperature may fluctuate significantly in temperature on a seasonal basis, depending on the source of the water being used (see the later section on seasonal patterns in water temperature). Some culture species can survive a broad range of temperatures, while others cannot. Each species has a relatively narrow temperature range within which growth is optimum. Catfish, tilapia, striped bass, and trout provide good examples.

The original range of occurrence of the channel catfish,[83] for example, was from the Great Lakes region and prairie provinces of Canada to the Gulf states (Wellborn and Tucker 1985). The species can thus survive water temperatures approaching freezing as well as those that rise above 30°C. Growth rate differs significantly from the North to the South, with the optimum temperature lying within the range of 26 to 30°C (Kilambi et al. 1970, Andrews and Stickney 1972).[84] Temperatures in that range may never be reached at the northern end of the range, and several years might be required for a fish to reach marketable size (about 450 g) in high latitudes. In the southern United States, market size is generally reached in 18 months. If the temperature can be maintained within the optimum range at all times, the length of time from egg to market can be reduced by another 8 to 10 months. Based on its optimum temperature, channel catfish are considered to be warmwater fish (Table 18).

Tilapia are tropical fishes that can tolerate temperatures above those at which many warmwater species succumb. Tilapia are not tolerant of cold water, however. Death generally occurs when the water falls below 10 to 12°C (Chimits 1957, McBay 1961, Avault and Shell 1968). Growth is generally poor below about 20°C and disease epizootics, extremely rare when the water is around 30°C, become very common when the temperature approaches the lower lethal range. Additional information on tilapia temperature tolerance and the overwintering of tilapia in temperate climates was summarized by Stickney (1986).

Rainbow trout represent a coldwater species that, according to Piper et al. (1982) has a temperature tolerance range of about 1°C to nearly 26°C. While rainbow trout and other salmonids can survive in relatively warm water, the optimum temperature range of rainbows is about 10 to 16°C. Growth is retarded at higher and lower temperatures.

Striped bass can be considered a midrange species with respect to temperature tolerance. According to Piper et al. (1982), striped bass survive a temperature range of about 2 to 32°C, but have an optimum range of 13 to 24°C. Other species that

---

[83] Channel catfish have been stocked and cultured outside of their original range, as is true of many other species.

[84] Piper et al. (1982) placed the optimum temperature range for channel catfish at approximately 21 to 30°C.

TABLE 18. Classification of Selected Aquaculture Species as a Function of Temperature Required for Optimum Growth[a]

Optimum Temperature	Name
Coldwater Invertebrate	Pacific oyster *(Crassostrea gigas)*
Coldwater Fishes	Chum salmon *(Oncorhynchus keta)*
	Coho salmon *(Oncorhynchus kisutch)*
	Pink salmon *(Oncorhynchus gorbuscha)*
	Rainbow trout *(Oncorhynchus mykiss,* formerly *Salmo gairdneri)*
	Chinook salmon *(Oncorhynchus tshawytscha)*
	Plaice *(Pleuronectes platessa)*
	Atlantic salmon *(Salmo salar)*
	Brown trout *(Salmo trutta)*
	Sole *(Solea solea)*
Midrange	Northern pike *(Esox lucius)*
	Muskellunge *(Esox masquinongy)*
	Striped bass *(Morone saxatilis)*
	Yellow perch *(Perca flavescens)*
	Walleye *(Stizostedion vitreum vitreum)*
Warmwater Invertebrates	American oyster *(Crassostrea virginica)*
	Freshwater shrimp *(Macrobrachium rosenbergii)*
	Northern quahog *(Mercenaria mercenaria)*
	Southern quahog *(Mercenaria campechiensis)*
	Blue mussel *(Mytilus edulis)*
	Kuruma shrimp *(Penaeus japonicus)*
	Tiger shrimp *(Penaeus monodon)*
	Blue shrimp *(Penaeus stylirostris)*
Warmwater Fishes	Bighead carp *(Aristichthys nobilis)*
	Goldfish *(Carassius auratus)*
	Milkfish *(Chanos chanos)*
	Mud carp *(Cirrhina molitorella)*
	Walking catfish *(Clarias batrachus)*
	Grass carp *(Ctenopharyngodon idella)*
	Common carp *(Cyprinus carpio)*
	Sea bass *(Dicentrarchus labrax)*
	Silver carp *(Hypophthalmichthys molitrix)*
	Bigmouth buffalo *(Ictiobus bubalus)*
	Blue catfish *(Ictalurus furcatus)*
	Channel catfish *(Ictalurus punctatus)*
	Red drum *(Sciaenops ocellatus)*
	Yellowtail *(Seriola quinqueradiata)*
	Rabbitfish *(Siganus* spp.)
	Gilthead sea bream *(Sparus aurata)*
	Blue tilapia *(Tilapia aurea)*
	Mossambique tilapia *(Tilapia mossambica)*
	Nile tilapia *(Tilapia nilotica)*

[a] Warmwater refers to species with temperature optima at or above 25°C, coldwater refers to those with temperature optima below 20°C, and midrange species have optima that lie generally between the extremes, though there may be overlap by some species placed within any of the three groups.

are often considered to be midrange in terms of their temperature optima, such as northern pike, muskellunge, and walleye, may actually be better classified as cold-water species according to the optimum ranges presented by Piper et al. (1982).

Table 18 presents a list of some of the commonly reared culture species and an indication of their temperature preferences. In general, warmwater fish have a thermal optimum around 25°C, midrange species grow best around 20°C, and coldwater fish are most productive at cooler temperatures (often around 15°C). The temperature range at which spawning occurs is often below the optimum range for best growth, particularly in species native to temperate areas.

Temperature-tolerant coldwater, warmwater, and midrange species may coexist in water bodies that lie in close proximity or even in the same water body. For example, the deep waters of reservoirs that feature primarily warmwater species may have sufficiently cold water at depth that coldwater species can survive under the thermocline or in proximity to cold springs during summer. Latitude, altitude, and water source are all factors that affect the suitability of a given area for various species of aquaculture interest. The relatively low latitudes in which the southern United States occur are generally conducive for the production of warmwater fishes, yet at the higher elevations in many southern states it is possible to produce midrange and even coldwater fishes. Most southern states have trout hatcheries in them. Northern climates where midrange and coldwater species predominate will have native populations of some warmwater fishes (e.g., channel catfish and largemouth bass), and aquaculture of warmwater fishes may be possible in geothermal water or the effluent of power plants virtually anywhere.

Virtually all aquaculture candidates are poikilothermic;[85] thus, their metabolic rates are determined by ambient water temperature. For reasons that are largely unknown but undoubtedly entail genetic differences in enzyme systems among the various species, temperature tolerances and optima vary greatly. As a general rule, a coldwater or midrange species may attain a maximum size and weight that are as large or larger than those of a warmwater species, but the growth rate of the latter will often be more rapid because of the higher metabolic rate.

There has been interest expressed in recent years in genetically engineering fish to tolerate or even grow more rapidly in water temperatures outside of the normal range. For example, through the insertion of appropriate genes it might be possible to engineer a trout or salmon that can tolerate and grow well at 30°C or a tilapia that will not succumb to disease and experience mortality as the water temperature falls below 15°C. The latter would involve insertion of a so-called antifreeze gene. Certain polypeptides are present in the blood plasma of fishes that exhibit antifreeze protection. While presence of the antifreeze gene sounds impressive, and is certainly of importance to aquaculturists in certain regions where slightly more tolerance to winter minimum temperatures could mean the difference between success and devastating mortality, presence of the polypeptides that allow fish to live at unusually cold temperatures may impart less than an additional 1°C advantage. Adult cod *(Gadus morhua),* for example, freeze at about − 1.2°C as compared with

---

[85] The more common term is cold-blooded.

juveniles that can tolerate $-1.55°C$. That is compared with fish like halibut and salmon, which lack the antifreeze polypeptide and generally freeze at -0.7 to $-0.9°C$ (King et al. 1989), though Pacific halibut juveniles have been found to survive temperatures below $-1.0°C$ (Goff et al. 1989).

Some success has been obtained in not only the transfer of a winter flounder *(Pseudopleuronectes americanus)* antifreeze gene into Atlantic salmon *(Salmo salar)*, but also the expression of that gene in a cross between a transgenic male and a normal female salmon (Fletcher et al. 1990). Better tolerance to low temperature could come from the transfer to Atlantic salmon of genes from other species that have higher concentrations of the antifreeze polypeptides (Ewart and Fletcher 1989). These kinds of alterations may soon become routine, but are they desirable?

One of the certainties associated with the practice of aquaculture is that unless strict quarantine restrictions are imposed, culture animals will eventually escape into the natural environment. Restrictions have already been imposed on genetically engineered animals that ensure to the extent possible that the fish or their progeny do not escape into the wild. So far, transgenic animals of aquaculture interest have only been developed and evaluated in a research environment. If they are released for general aquaculture use, it will be very difficult to maintain the same stringent controls that are possible at research institutions.

It is difficult to know what the impact on native populations of warmwater trout or coldwater tilapia would be, but studies have shown that tilapia would compete with other species for nesting sites and would also compete with various species for food. At present, the threat of tilapia to native fish populations is greatly reduced since all but a very few locations in the United States are too cold in winter for tilapia to survive. While the aquaculturist might benefit from a cold-tolerant tilapia, native populations might suffer. Consideration of the welfare of native populations should supersede the interest of the aquaculturist in instances where the consequences could alter the natural ecological balance of a region.

The Influence of Genetics and Age on Response to Temperature

While the tendency is to indicate that a species will respond to temperature stress in a predictable manner, the fact is that genetics can play a significant role and that different stocks of animals may respond differently to the same stressor. For example, McGeer, et al. (1991) found that coho salmon from six hatcheries responded differently to thermal increases of 1°C/hr as confirmed by measurements of changes in the critical thermal maximum tolerated by the fish. Similarly, Beacham and Withler (1991) determined that a southern population of chinook salmon was better able to survive a challenge with high temperature than was a more northern population (the southern stock would experience higher summer maximum water temperatures than the northern stock). The response was affected by fish size, with larger (and presumably older) fish surviving the temperature challenge better than smaller fish.

The ability to tolerate low temperatures by Black Sea golden gray mullet *(Liza aurata)* increases with age until the fish reach sexual maturity, after which it decreases again (Shekk et al. 1990). The relationship between age and temperature tolerance has not been well documented for most aquaculture species.

Genetic influences on the cold tolerance of tilapia have also been demonstrated. For example, *Tilapia nilotica* from strains originating in Egypt, the Ivory Coast, and Ghana showed different lower lethal temperatures ranging from 10.0°C for the Egyptian strain to 14.1°C for fish from Ghana (Khater and Smitherman 1988). The Ivory Coast strain was intermediate (12.2°C). The responses to low temperature were correlated with the geographic origins of the three strains.

Response to Seasonal Patterns in Water Temperature

The temperature of surface water bodies typically fluctuates to one extent or another on a seasonal basis. The range of annual fluctuation tends to be maximized in temperate regions and minimized in Arctic and tropical ones. There are locations in temperate regions where temperatures approach the optimum for various culture species throughout much of the year. Surface waters in southern Florida and extreme southern Texas, for example, can support good growth of warmwater species throughout most of the year. Temperatures are typically sufficiently warm to support tilapia survival during all but the most exceptionally cold winters. The climate in Hawaii is tropical and will support year-round growth of warmwater species. By contrast, the waters of Puget Sound in Washington, and those off upper New England are sufficiently cold to support year-round salmon production. Parts of Alaska are subarctic and will also support year round salmon growth.

As we have already observed, there are instances wherein water that is unusually cold or atypically warm for a given region occur. There are also places where water temperature does not fluctuate to any extent seasonally. Cold springs may reduce the extent of spring and summer warming in some surface waters, while geothermal water may be used to produce warmwater species throughout the year in regions where even summer water temperatures might otherwise be suboptimum. Facilities producing tilapia and channel catfish in parts of Idaho where geothermal water is available are good examples. Similarly, and also in Idaho, is the production of rainbow trout in the Hagerman Valley where underground rivers of virtually constant temperature provide ideal year-round growing conditions.

Water that has had its temperature altered by the activities of humans can also sometimes be used for aquaculture. The best example is the warm water associated with fossil-fuel and nuclear power plants, but many other industries produce heated effluents that may be suitable for use in aquaculture, either directly, or through heat exchange with water in the culture chambers.

Most species that are being successfully produced by commercial aquaculturists reach the market in less than 2 yr, with 1 yr or less for growout being even more desirable. Tilapia can be grown to market size (about 500 g) in under a year in tropical regions or when geothermal or artificially heated water is available. Channel catfish can also be reared to the same size within a year if water temperature is constantly maintained within the optimum range for the species. In most parts of the United States where catfish are reared in waters that are subject to seasonal temperature fluctuations, growout requires about 18 months. In tropical regions it is easy to obtain two crops of penaeid shrimp annually, while in temperate regions, one crop per year is typical.

Some species of commercial importance require two or more years to reach market size. Pacific oysters and Atlantic salmon are examples, with 3 yr production cycles from egg to market being typical. The culture of Atlantic halibut is in its infancy, and those involved in the enterprise anticipate a 3 yr growout requirement for the production of 5 kg fish, which are considered to be market size for that species.[86] One of the reasons for slow growth in halibut is related to their requirement for cold water. Not only do they require cold water for proper growth, they are able survive at temperatures that would be too cold for many species.

At least modest reductions in the time required to produce marketable species in commercial culture and those of suitable stocking size in the recreational fish production arena may be possible as a result of improved nutrition, genetic manipulation, and aggressive culture system management. Research has already been initiated with respect to the introduction of growth hormone into fish. Yet, given the small initial size of the larvae and fry of aquatic species of aquaculture interest, there will be limits to how fast marketable animals can be produced. We are not likely to see fish or oysters reach the market in a few weeks as broiler chickens do, but we can anticipate some reduction in the growout period as technology advances. Maintenance of optimum temperature will be one key to reducing the period required for growout.

For most species the maximum temperature experienced during the year is not within the optimum temperature range for most rapid growth. At temperatures both above and below the optimum range, growth rate decreases. Metabolic rate is reduced when the water temperature is below the optimum range and increases when the temperature rises above that range. Low metabolism means reduced feed intake and slower growth. At temperatures above the optimum range, the animals' feed consumption rate may increase to accommodate the higher metabolic rate. Growth does not increase, and there is an economic impact associated with meeting the energy demand by providing more food for no additional increase in rate of gain. As temperature approaches the upper tolerance point for a given species, metabolic rate begins to fall as various systems fail. Feeding activity declines, and ultimately death occurs.

Given a range of temperature, most species will select one that is within the optimum for growth,[87] unless some mitigating factor is operating. For example, a fish that is approaching sexual maturity may select a temperature suitable for spawning rather than one that is optimum for growth. In nature, aquatic animals can often make such selections because water bodies tend to not have uniform temperatures throughout. Thermal stratification and inflowing warm or cold springs provide op-

---

[86] Rolf Engelsen, personal communication.

[87] According to William Neill (personal communication), certain some species of tilapia will go to the warmest portion of a temperature gradient, even if that temperature is lethal. The instinct to seek warm water apparently is greater than that for survival in at least some species.

portunities for aquatic organisms to be somewhat selective in terms of temperature. In water systems in which temperature is controlled in some way, conditions may be more uniform.

Aquaculture ponds tend to have zones of temperature. Even a pond that is less than 2 m deep may partially stratify, and the shallow water around the edges of culture ponds is often considerably warmer during the day than is water in the middle, particularly on sunny days during summer.

Synergisms with Temperature

The influence of temperature on growth, metabolism, disease resistance, and survival of aquaculture animals may be influenced by various other parameters. While many of these synergisms have not been evaluated in any detail, and none have been examined across a broad range of aquaculture species, there is some information documenting at least a few of the factors that can interact with temperature and influence performance of aquaculture species.

Zale and Gregory (1989) examined the cold tolerance of *Tilapia aurea* at salinities ranging from 0 to 35 parts per thousand (ppt) and determined that fish that were isosmotic with the medium[88] survived lower temperatures than when the external medium was at a higher or lower salinity. One implication of the relationship between temperature tolerance and salinity is that the species could be expected to have an extended range of distribution into estuarine waters as compared with freshwater or marine waters (see the section on salinity below for further explanation).

Various other chemical parameters associated with the water can influence the temperature tolerance of aquaculture species. The relationship between temperature and oxygen has already been discussed in Chapter 4, as has the relationship between temperature and ammonia toxicity. The ability of channel catfish to tolerate high temperature is not only reduced in the presence of ammonia (the un-ionized percentage of which increases with increasing temperature), but also when the fish are exposed to elevated nitrite levels. After exposure of catfish to 1.4 mg of nitrite for 24 hr at 20°C, the maximum temperature tolerated was 35.9°C as compared with 38.0°C for control fish (Watenpaugh et al. 1985).

Effects of Temperature on Development

Egg and embryo development are heavily influenced by temperature. A minimum temperature, below which larval development does not occur, has been identified for most species of aquaculture interest and importance. As the temperature increases above the minimum, development rate also increases. However, once the temperature at which eggs or developing larvae are exposed to temperatures rises above a maximum, which varies from species to species, abnormal development occurs and the rate of mortality increases until it becomes total as the upper temper-

---

[88] The salt concentration within the body of the fish and in the external medium was the same, in this case 11.6 ppt.

ature limit is reached. The upper limit for proper development of eggs and larvae may not be as high as the temperature maximum that can be tolerated by later stages in the life history of the animals. A series of abnormalities can occur when temperature becomes too high during egg and larvae development. Some of the common ones in fishes are unusual numbers of vertebrae and tail abnormalities. The proper temperature ranges for normal development of some aquaculture species are presented in Chapter 7.

A special condition related to development, temperature, and other variables in anadromous salmonids is smoltification, a physiological process that allows the fish to make the transition from fresh water to the ocean. Since the process is related to that transition, discussion of smoltification is taken up in the section on salinity in this chapter.

Thermal Shock and Tempering

Thermal shock results when aquatic animals are exposed to rapid changes in ambient water temperature. The stress associated with such shock may weaken the immune system thereby reducing the resistance of the animals to disease, or if severe enough, temperature shock can lead to death. The diel changes in temperature that occur as a result of daytime warming and nighttime cooling of the water within a culture chamber are usually small enough that no thermal shock occurs. An exception might be a small static tank or pond that is only partially filled. In those cases daytime warming could be significant.

Thermal shock is most commonly associated with harvesting, live hauling, and transfer of aquatic animals from one water body to another. When aquatic animals are harvested at times of the year when air and water temperatures are disparate, there can be a considerable amount of thermal stress that occurs during the harvesting process. When a pond is being harvested, the degree of temperature change can be significantly enhanced if the water level is greatly reduced. Total pond harvesting often involves seining followed by reduction in water volume, reseining, and so forth. Rather small volumes of water associated with harvest basins or the immediate vicinity of the drain in ponds without harvest basins may contain surprisingly large numbers of animals that avoided the seine. While those animals are being collected, the water can warm or cool considerably, depending on season of the year. Heating of small volumes of water has been a more significant problem than cooling. In warm climates during summer or fall, the water temperature can rise several degrees in only a few minutes. The resulting stress may be acceptable if the animals are being hauled a short distance for processing but can be devastating when the affected fish or invertebrates are being moved for restocking. This type of thermal stress may lead directly to mortality or reduce the disease resistance of affected animals.

In addition to thermal stress associated with the final stages of pond harvest, there is often significant reduction in dissolved oxygen (DO). Low DO occurs because the water can carry less oxygen at saturation when it is heated and because of increased metabolism in the animals that are being harvested. The problem can be alleviated by adding copious amounts of new, oxygen-rich pond water to the drain area during the latter stages of harvesting.

Water in hauling trucks should be at approximately the same temperature as the water from which the aquatic animals are collected. In many instances water from the culture chamber being harvested is used to fill the hauling tanks, thereby assuring a minimal initial change in temperature in most cases.

Whether a hauling tank or some other type of container is used for transporting fish (see Chapter 9), the temperature of the water is subject to change with time. Insulated boxes and hauling tanks are helpful in maintaining temperature, but it may be necessary to add ice to keep water within acceptable limits during summer if the boxes or hauling tanks are exposed to high temperature and direct sunlight.

Once live hauling has been completed, and before the animals are stocked, the temperature in the transportation tank and receiving water should be compared. If the difference in temperature is more than about 2°C, the animals should be gradually tempered. For animals transported in tanks, common practice is to slowly introduce water from the receiving site to the tank until the temperature within the tank equilibrates with that of the receiving water body. Tempering should not exceed a rate of more than about 5°C/hr and in all instances the process should be completed within 10 to 12 hr. Care should be taken to maintain a high level of DO at all times during the tempering process.

When fish are hauled in small containers such as plastic pails, plastic bags, or others that readily conduct heat, the containers can be floated in the receiving water body until the temperature of the water within the container equilibrates. This technique should not be used if a dramatic difference in initial temperature (5°C or more) exists since the rate of tempering may be too rapid and, therefore, stressful.

Rapid changes in temperature and exposure of aquatic animals to inappropriate temperatures sometimes cannot be avoided. While there continues to be little experimental evidence available, it appears less stressful to aquatic animals when temperature is changed toward the thermal optimum of the species rather than away from the optimum. Temperatures above and below the optimum can lead to increased rates of metabolism and other physiological changes that can be detrimental. Whether tempering can be conducted more rapidly when the temperature change is toward the thermal optimum has not been well researched, but there is at least some circumstantial evidence to support that approach. When fish are exposed to heat stress in an almost completely drained pond during summer, rapid reduction in the temperature by adding cool water is often an effective means of avoiding subsequent mortality even though the change in temperature may occur very rapidly.

Aquaculture animals can generally be handled without damage or severe stress in cool weather, but great care should be taken when handling them in the summer, particularly in temperate and tropical latitudes. Handling often leads to injury, and the high metabolic rate of aquatic animals in the summer may make them more active when caught, leading to an increase in the incidence of self-inflicted injury. Also, bacterial activity is higher in warm as compared with cold water; thus wounds are more likely to become infected during the summer than in the winter. Animals that have been exposed to water temperatures above their normal optimum may be more subjected to parasitic and bacterial epizootics than those that have not, especially when exposure to unusually high temperatures is coupled with handling.

It is generally a good idea to avoid handling aquatic animals during the summer insofar as possible. When it becomes necessary to handle animals during hot weather, they should be caught early in the morning when the water has reached its coolest temperature of the day. Care should be taken to ensure that a high level of DO is maintained at all times. The animals should be handled gently and returned to the water as quickly as possible. If the animals are to be weighed, the operation should be accomplished in pre-weighed, water-filled containers; to avoid damage, animal density in the weighing containers should be as low as practicable.

Overwintering

Most aquaculture in conducted in systems designed to operate under ambient water temperature conditions. Many aquaculture systems in temperate climates can vary in temperature from 4°C to over 30°C, and there have been very few attempts to alter the normal pattern.[89] Heating or chilling production units has typically been economically unviable in commercial systems, though both heating and cooling have been used in conjunction with research laboratories, in hatcheries, and when fish are being held over the winter.

Maintenance of certain minimum temperatures through artificial heating during the winter is often necessary in conjunction with tropical species being reared in temperate climates. Various species of tilapia, along with certain species of penaeid and freshwater shrimp require supplemental heat during winter. Overwintering can be accomplished by maintenance of proper air temperatures in buildings used to house such species in static or recirculating systems. In flowthrough systems, water heating can lead to significant and even staggering energy costs.

Overwintering is typically practiced in conjunction with the maintenance of the broodstock, though young animals may also be held during winter for subsequent stocking and growout the following spring and summer. It is generally not necessary or even desirable to maintain the overwintering temperature at the optimum for the species involved since the goal is to hold the animals, not have them grow. Thus, overwintering temperatures should be sufficiently low to slow the metabolic rate of the animals without stressing them and reducing their resistance to disease. For species such as tilapia, an overwintering temperature of around 20°C may be appropriate. Fish at that temperature can be expected to eat a sufficient amount of food to maintain their body weight and can be expected to survive well.

The number of overwintered broodstock of species with high fecundity that spawn during spring need not be very high, so modest facilities can be used to house sufficient adults for fairly large growout operations. Species vary in their ability to accept crowding, but for most species being reared commercially today, density limitations are usually not a factor as long as suitable water quality is maintained.

---

[89] Notable exceptions are the use of heated effluents and geothermal water.

When a source of inexpensive heated water is not available, heated buildings, including greenhouses, are typically used for overwintering. Most commonly, the aquaculture animals are maintained in circular or linear raceways, though ponds covered by greenhouses may be practical in low temperate regions where soil temperatures remain fairly high throughout the winter. A greenhouse system developed for overwintering tilapia in Texas by Chervinski and Stickney (1981) maintained the water within a range of 15 to 22°C from late October until April 1 (138 days).

Stickney and Person (1985) examined various ways of heating water before designing a heat exchanger used in conjunction with a recirculating water system. Such systems remove thermal energy from water that is of any initial temperature above freezing, thereby producing both a heat source that can be used to warm water and a source of cold water (the incoming water from which heat has been removed). Utilizing such a system could provide water suitable for the maintenance of both warmwater and coldwater species if the temperature of the incoming water falls between the optima of the two types of animals.

SALINITY

Salinity has been defined as the total amount of solid material in grams contained in 1 kg of seawater when all the carbonate has been converted to oxide, the bromine and iodine replaced by chlorine, and all organic matter completely oxidized (Sverdrup et al. 1942). The following empirical relationship can be used to determine salinity:

$$\text{Salinity} = 0.03 + 1.805 \times \text{chlorinity}$$

Chlorinity is defined as the total amount of chlorine, bromine, and iodine, in grams contained in 1 kg of seawater, assuming that the bromine and iodine have been replaced by chlorine. To make the determination, a water sample is titrated using silver nitrate (resulting in the precipitation of silver chloride). The titration is standardized with seawater of known chlorinity (standard seawater or Copenhagen seawater).[90]

Salinity Measurement and Naturally Occurring Levels

Titration is one of the methods that is still sometimes used for salinity measurement, though others that require less time and are simpler to make include the measurement of conductivity, density, and refractive index, each of which can be transformed into salinity.

Of the methods most commonly in use, the least expensive involves determina-

---

[90] Standard seawater is taken from the North Atlantic Ocean and is sealed in glass vials that are marked with the chlorinity of the sample. Interestingly, the source of the standard seawater is from a location where salinity is somewhat lower than for the world's ocean as a whole.

TABLE 19. Classification of Aquatic Environments on the Basis of Salinity Ranges (after Hedgpeth 1957) [a]

Classification of Environment	Salinity Range (ppt)
Freshwater	≤0.5
Oligohaline	0.5–3.0
Mesohaline	3.0–16.5
Polyhaline	16.5–30.0
Marine	>30.0

[a] Waters with salinities greater than about 40 ppt are referred to as brines or hypersaline waters.

tion of the density of a water sample with a hydrometer. With the reading from the hydrometer taken at the appropriate temperature (20°C), density can be rapidly converted to salinity by employing tables that have been developed for that purpose. The major drawback of the technique involves the fact that hydrometers are made from glass. They are quite fragile and are often broken even when great care is taken.

Refractometry is the simplest and most rapid means of measuring salinity. Hand-held refractometers have glass optics associated with them. They can be damaged or broken if dropped, but they are much more resistant to breakage than hydrometers; however, refractometers are also considerably more expensive (a few hundred dollars compared with a few dollars for a hydrometer). A typical refractometer looks like a pocket telescope and requires only a drop of water in order to provide the user with a measurement of salinity at a level of accuracy that is appropriate for aquaculture applications. While most refractometers once measured refractive index in density or units called brix (which could be converted by use of published tables into salinity), the demand became sufficiently high that instruments are now available that read salinity directly.

Conductivity meters can be used to measure salinity with a high degree of accuracy and precision. Many are quite expensive and are more commonly used by researchers than practicing aquaculturists.

Salinity is reported in parts per thousand (ppt, o/oo, or ''per mille''). One ppt is $\frac{1}{10}$ of 1%.[91] The refractometer is generally accurate to within ±0.5 ppt, whereas other methods of salinity determination may be accurate to much less than 0.1 ppt. Aquaculturists are usually not concerned with changes in salinity of less than 1 or even up to a few ppt, so the precision associated with some of the other methods is not required.

The salinity of fresh water is usually less than 0.5 ppt (Table 19). Humans can recognize salt in water at a level of about 2 ppt. The salinity of seawater varies with

---

[91] The abbreviation ppt can also indicate parts per trillion, which is commonly seen in conjunction with toxicity studies. In this book ppt is used only in conjunction with salinity as measured in parts per thousand.

distance from freshwater inputs, including glacial and iceberg melt; depth (surface waters on which rain has fallen will have lower salinity than underlying water layers); and other factors. The open ocean averages about 35 ppt (Pearse and Gunter 1957), although there is some variability from region to region. Salinity is not generally a consideration in freshwater aquaculture but is certainly an important parameter for mariculturists.

Salinity and Its Control in Mariculture Systems

Open ocean mariculture is in its infancy. Most marine aquaculture systems are located on shore or adjacent thereto, often in embayments where facilities are at least somewhat protected from storms. Salinities associated with mariculture operations may be relatively constant or they can vary, often to a considerable extent. Varying salinities below full-strength seawater occur in conjunction with mariculture operations located in estuaries.

As defined by Pritchard (1967), an estuary is a semienclosed coastal water body with unrestricted access to the open ocean, where salt water is measurably diluted by freshwater runoff. Technically, the definition is actually for positive estuaries where freshwater input exceeds evaporation, thereby causing dilution of seawater and a reduction in salinity. Negative estuaries are those in which evaporation exceeds freshwater inflow and hypersaline conditions occur. A neutral estuary would be one in which freshwater inflow and evaporation are equivalent.[92]

Examples of neutral estuaries do not come readily to mind. Most estuaries are of the positive type, but hypersaline lagoons do occur. One is the Laguna Madre in Texas, where salinity often exceeds 40 ppt and salinities of 50 to 80 ppt are not uncommon (Pearse and Gunter 1957). Since many aquaculture species are intolerant of hypersaline waters, fresh water for dilution will be required during periods when excessive salinities occur.

Euryhaline animals are adapted to the conditions that exist in estuaries. While each species may have an optimum salinity range that is relatively narrow, those that are euryhaline can often survive and grow well even when exposed to widely fluctuating salinities. For stenohaline species, a narrow salinity range is required. Proper siting or the use of recirculation technology are means for ensuring that salinity is maintained within the proper range for stenohaline species.

Most aquaculture facilities are operated without salinity adjustment except in the case where ambient salinity of incoming water is excessive. Freshwater dilution can be used to maintain the desired salinity, assuming that a source of freshwater is available. Salinity can easily be increased by adding salt to water (either in the form of sodium chloride or as a mixture of sodium chloride and other chemical salts that in the aggregate simulate the composition of seawater). Adding salts to water may not add inordinately to the expense associated with operating a recirculating water system, but routine salt addition to flowthrough systems may be economically un-

---

[92] The term ''estuary'' in this text refers to the positive type unless preceded by either ''negative'' or ''neutral.''

feasible. Obviously, as the amount of salinity increase required or the amount of water to which salt is to be added increases, the cost will escalate.

In static systems (e.g., ponds) and in recirculating systems, evaporation will lead to increased salinity. Adding saltwater to replace evaporative losses will reduce the salinity somewhat, but because there has been no net loss of total salt (water lost to the atmosphere during evaporation leaves its salt behind), there will be a continual net increase in salinity. Over time, this could lead to a situation where the salinity reaches a level that leads to stress and ultimately death of the culture species. To avoid the tendency of such systems to increase in salinity, freshwater should be added to replace evaporative losses.

Freshwater animals can typically tolerate at least modest levels of salinity (often up to about 10 ppt). In some instances, exposure of freshwater aquaculture species to low levels of salinity can actually be beneficial. For example, some disease organisms are intolerant of even low levels of salt, so exposure of animals that might be affected by such diseases may be protected when reared in low-salinity water (see Chapter 8).

Osmoregulation

The blood and tissue fluid salt concentrations of most fishes differ somewhat from the salinity of the surrounding water, while invertebrates have internal salt concentrations that mirror the environments in which they live. The ability to maintain a constant internal salt concentration in the face of external salinities that are higher, lower, or fluctuating is known as osmoregulation.

In the case of freshwater fishes, internal salt concentration tends to be in the range of 10 to 12 ppt, which is hypertonic to the surrounding medium (typically less than 1 ppt). Fishes living in the open ocean maintain internal salt concentrations that are hypotonic to the external medium. Estuarine species could maintain a constant relationship between their internal salt concentration and that of the external medium by maintaining themselves within a water mass of a selected salinity and following that water mass as it moves due to tidal flow. In reality, estuarine fishes tend to be exposed to variable external salt concentrations as they move about and are exposed to tide-induced salinity changes.

While crustaceans and other invertebrates have blood salt concentrations isotonic with the surrounding water, or very nearly so, osmoregulation is still required. Invertebrate blood ionic composition is not identical with seawater (Lockwood 1967), so physiological processes must occur to control and maintain that blood chemistry.

Osmoregulation involves the selective absorption of ions, and in some cases the selective removal of salts through the gills. In marine teleosts, the kidney is responsible for the excretion of salt. Because osmoregulation is a metabolic process, it requires energy. The more salt the body of an aquatic animal must take in or remove in order for it to maintain its internal salt concentration, the more energy is required. Energy used for osmoregulation may be used at the expense of growth (though there are many other metabolic energy losses in addition to osmoregulation). This does

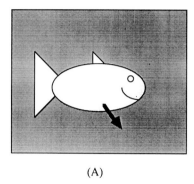

(A)

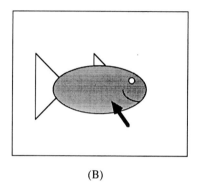

(B)

Figure 68. The skin of a fish acts like a semipermiable membrane. *(A)* In marine fishes osmosis tends to cause the fish to lose water, while *(B)* in freshwater fishes the same process leads to continuous uptake of water into the tissues. The arrows show the direction of water movement into or out of the fish.

not mean that maintaining aquatic animals in isotonic conditions will lead to optimum growth. Many stenohaline marine fishes will die when placed in water that is isotonic with their blood, as will many freshwater fishes. Those species have become physiologically adapted to salinities significantly different from their internal salt concentrations. As a general rule, aquatic species should be reared in waters of salinities similar to those in which the animals occur in nature. Again, euryhaline species may be broadly adaptable, and many will grow well over a broad range of salinity.

As shown in Figure 68, water will move through a semipermeable membrane (in this case the skin of a fish) from the region of lower ionic concentration to that of higher concentration in a process called osmosis. The movement of the water through the semipermeable membrane will continue until the salt concentration on both sides of the membrane is the same. Thus, because of osmosis, freshwater fishes continually gain water from the surrounding environment, while the water in the tissues of marine fishes is continually being lost to the sea. In the first instance, the freshwater fish must keep the external medium from lowering its internal salt concentration toward zero, while in the second, osmosis charges the marine fish with the onerous task of diluting the ocean to a salinity of about 10 ppt.

A marine fish (Figure 68A) is continuously losing water to the environment. In order to replace that water, the fish must drink copious amounts of seawater, thereby increasing the amount of salt they ingest. The kidneys and gills actively remove ions that are absorbed. At the same time, the water is retained. Thus, marine fish produce small quantities of urine that is highly concentrated with salts. A review of the literature by Conte (1969) indicated that marine teleosts ingest water at the rate of from 0.3 to 1.5% of body weight per hour. Estuarine fishes could, perhaps, find places where the salinity is isotonic with the fluids in their bodies, but in reality, there does not seem to be evidence to support the theory that the fish seek out and remain in such waters.

Since water is continuously entering the body of freshwater fish (Figure 68B), they do not drink. Fresh water contains only low levels of ions, so freshwater fish depend upon the salts in their food as a primary source of minerals, though some are actively transported from the water into the fish by means of the gills. Freshwater fish must continuously excrete water. They accomplish that by producing copious amounts of very dilute urine.

In the embryos and larvae of fish that do not have functional kidneys and gills, osmoregulation is thought to be controlled by so-called chloride cells. These are mitochondria-rich cells that have been shown to transport salt in at least one important aquaculture species, *Tilapia mossambica* (Foskett and Scheffey 1982). Chloride cells become established during embryonic development and may disappear after the yolk sac has been absorbed (Alderdice 1988). The presence and density of chloride cells relates to the age, species of fish, and the salinity in which development occurs.

Salinity Tolerance

The salinity tolerance of several fish species used in aquaculture has been reviewed by Stickney (1991a). In that review, the need for taking temperature into consideration when evaluating salinity tolerance was highlighted, particularly with respect to developing embryos and larvae. Proper development requires that both salinity and temperature remain within certain limits, which will vary depending on the species under consideration. Later stages in the life cycle may also demonstrate differential responses to salinity as a function of temperature. For example, silver carp demonstrate maximum salinity tolerance within a temperature range of 18 to 22°C (Oertzen 1985). Such species as rainbow trout, which have anadromous strains, are able to withstand increased salinity, though the transfer of trout from fresh to salt water is facilitated if the proper temperature is maintained (Alexis et al. 1984).

Anadromous Species. Steelhead trout represent anadromous strains of the rainbow trout *(Oncorhynchus mykiss).* Many strains of rainbow trout are not anadromous, but still exhibit reasonable salinity tolerance. Salmon—both Atlantic salmon *(Salmo salar)* and the Pacific salmon species of culture interest *(Oncorhynchus* spp.)—can be reared through their life cycles in fresh water. The successful introduction of chinook and coho salmon into the Great Lakes is a clear demonstration of that fact. By not allowing anadromous salmonids to enter salt water at the time of smolting (see the section on smoltification below), the fishes can be readily adapted to fresh water.

Striped bass *(Morone saxatilis)* migrate to the sea as juveniles in nature but can be retained in freshwater culture environments. Crosses between striped bass and the nonanadromous white bass *(M. chrysops)* are euryhaline and can be reared over a fairly broad salinity range. Hybrids can survive for at least a week at 36 ppt (Wattendorf and Shafland 1982) and survive well at salinities of up to 28 ppt (Smith et al. 1986). Larval striped bass should be maintained in fresh or low-salinity water. Kane et al. (1990) found that mortality of larval striped bass was significantly higher at 3 ppt than at 2 ppt salinity due to sodium chloride. The authors indicated that in dilute seawater the larval survival of striped bass was enhanced at salinities from 0.5 to 10.0 ppt.

In the United States, the anadromous white sturgeon *(Acipenser transmontanus)* has received the attention of aquaculturists, particularly along the West Coast to which the species is native. Freshwater rearing is practiced, though the fish normally spawns in fresh water and spends most of its life in marine and coastal waters. A related species in Asia, the sevruga *(A. stellatus)* is a freshwater sturgeon of culture interest and is probably more typical of the genus. The sevruga can tolerate transfer to 10 ppt salinity but cannot survive more than a few hours when placed in 20 ppt water (Shikin and Lapina 1982).

Marine and Estuarine Species. Marine species that live offshore are typically stenohaline, while those living in estuaries are most commonly euryhaline. Among the invertebrates, for example, the American oyster *(Crassostrea virginica)* can adapt to a salinity range of 3 to 35 ppt (Pearse and Gunter 1957). Flavor and growth are adversely affected in oysters reared at low salinities.

Many molluscs of culture interest are fairly euryhaline, and a search of recent literature provides little insight into salinity requirements. Kuriakose (1980) reported that the brown mussel, *Perna indica,* tolerates salinities over a range of about 21 to 46 ppt. The pinto abalone, *Haliotis kamtschatkana,* was studied by Olsen (1983), who found that salinities of less than 23 ppt were detrimental to growth and survival.

Bivalve molluscs, unlike most aquaculture species, are able to avoid some undesirable water quality episodes by closing their valves and discontinuing the pumping of water. There is a limit on the time that the animals can remain isolated from the external environment, usually for a period measured in days. Drastic changes in salinity, such as those associated with a major storm event when freshwater inflow increases dramatically, can be tolerated by species that are able to isolate themselves from the external medium for a period sufficiently long that conditions return to normal.

Penaeid shrimp vary from species to species with respect to salinity tolerance, though most species are relatively euryhaline (Charmentier 1987). The same author indicated that the ability of penaeid shrimp to osmoregulate is affected at temperatures lower than about 18 to 20°C. The impact of reduced temperature on osmoregulatory capability was demonstrated in a study by Charmantier-Daures et al. (1988), who used postlarval *Penaeus japonicus* and *P. chinensis*. The animals were exposed in 96 hr LC_{50} studies to a range of salinities and temperatures. Low temperatures led to reduced salinity tolerance in *P. japonicus* (19.3 ppt at 10°C as compared with 5.4 ppt at 25°C) but had little effect on *P. chinensis* (11.3 ppt at 14°C and 11.1 ppt at 25°C). Mortalities were reportedly lowest when the salinity of the water was isosmotic with the hemolymph of the shrimp. Ogle et al. (1988) examined the ability of *P. vannamei* postlarvae to survive at low salinities. They obtained shrimp from various hatcheries that had obtained broodstock from two different countries and found that there was a significant influence of shrimp stock on salinity tolerance. In general, survival of 22-day-old larvae improved slightly when salinity was decreased from 32 ppt to 16 ppt and was much poorer at salinities of 8, 4, and 2 ppt. Younger larvae (8 days old) were less tolerant of low salinity than the 22-day-old animals. The mysis stage of *P. merguiensis* were unable to survive even 2 hr exposure to salinities of 10 ppt or less (Prasad et al. 1988).

Juvenile *P. indicus* were able to osmoregulate over a range of salinities from 3 to 26 ppt, while adults of the same species could osmoregulate over a range of 5 to 30 ppt according to Diwan and Laxminarayana (1989). By contrast, Robaina (1983) found best growth and survival of *P. brasiliensis* juveniles at 40 ppt. Chung (1980) had previously reported that postlarvae of the same species were concentrated over a salinity range of 5 to 28 ppt when provided with a salinity gradient chamber ranging from 0 to 80 ppt. The author acknowledged that *P. brasiliensis* are not found in estuarine salinities of 5 to 25 ppt in South Africa and that their presence in hypersaline Venezuelan waters is not explained by the results of that laboratory experiment.

Dalla Via (1986a) found that *P. japonicus* juveniles tolerated a salinity range of 25 to 40 ppt without mortality and found relatively good survival could be expected over a range of 10 to 55 ppt. Dalla Via (1986b) found that the free amino acids in *P. japonicus* decreased with decreasing salinity. Measurements of individual amino acids showed that those most intimately involved in osmoregulation are glycine, proline, and alanine. Extensive culture of the species is recommended within a salinity range of 30 to 40 ppt (Dalla Via 1986a, b).

Spawning and egg development represent critical life stages that can be affected by salinity. Choo (1987) found that normal eggs and healthy nauplii of *P. merguiensis* could be produced at salinities above 20 ppt. (The highest salinity evaluated was 28 ppt.) Few eggs were produced, and the resulting nauplii were weak and did not survive when the animals were maintained at 18 ppt. Larvae produced in salinities of 20 ppt or higher could subsequently be reared in 18 ppt water but not at a salinity of 15 ppt.

Species that undergo annual migrations between estuarine and offshore waters may require different salinities at various times of year. A good example is the southern flounder, *Paralichthys lethostigma,* which appears to be physiologically adapted to different salinities at different ages (Stickney and White 1973). The same is true of other species of realized or potential culture potential, such as certain members of the drum family (Sciaenidae). Jones and Strawn (1984) cultured sciaenids and striped mullet *(Mugil cephalus)* in cages at Baytown, Texas. They found that immature black drum *(Pogonias cromis)* and Atlantic croaker *(Micropogonias undulatus)* were tolerant of the low temperatures and rapidly declining salinities associated with winter conditions. Older fish, which normally would have migrated offshore before the onset of winter, did not survive the same conditions.

Red drum, *Sciaenops ocellatus,* are being cultured for both enhancement of sport fisheries and for direct human consumption. While the eggs and larvae of red drum appear to require salinities above 25 ppt, juveniles larger than about 4 cm can be reared in fresh water (Neill 1987), although survival is often poor during the winter. Red drum, like other species of sciaenids, would normally move to high-salinity waters in winter. Year-round culture in fresh water is possible if water hardness is sufficiently high (Neill 1987, Wurts and Stickney 1989). A hardness of at least 100 mg/l seems to be appropriate (hardness is discussed later in this chapter). Cold tolerance is also improved by rearing the fish in water of 5 to 10 ppt salinity, though the lower lethal limit for the species even under the most ideal salinity or hardness conditions is between 8 and 10°C (Neill 1987).

It may be less stressful on euryhaline animals that have changing salinity requirements if the culturist can accommodate the physiological requirements of the animals. Most aquaculture facilities are located in areas where only modest salinity changes, if any, can be effected; thus it is often necessary either to produce animals to market size before lethal stress is experienced or to select species that are not dependent on different salinities during various life stages. Exceptions are species with specialized salinity requirements as juveniles. Those requirements can be met in hatcheries where closed systems are utilized to maintain the desired salinity. If the animals become suitably euryhaline at a small size, they can be transferred to ambient salinity water, perhaps even fresh water, relatively early in their life history. Red drum and freshwater shrimp are examples of species for which that approach can be used.

The milkfish, *Chanos chanos,* is a euryhaline clupeid fish that can be grown in low salinity water. In Indonesia the fish is cultured in ponds that can range from 10 to 35 ppt (Landau 1992). One of the more interesting locations for milkfish culture is in the Philippines in an upper arm of Manila Bay known as Laguna de Bay. The bay is shallow, has very high primary productivity and low salinity (1 to 10 ppt), and is almost perfect for rearing milkfish in net pens. With the abundant phytoplankton that exists in Laguna de Bay, milkfish can be reared without supplemental feed or fertilization. Fry are collected by fishermen who net the small fish in the nearshore marine environment, after which they are transported to the net pens in Laguna de Bay.

Truly marine finfish species of aquaculture importance are far from numerous. The gilthead sea bream *(Sparus aurata)* and sea bass *(Dicentrarchus labrax)* are being reared in Europe and the Middle East, while red sea bream *(Chrysophrys major)* and yellowtail *(Seriola quinqueradiata)* culture has progressed well in Japan. There has been research conducted on rabbitfish *(Siganus* spp.), and various species of snappers and groupers, but little in the way of commercial culture of those fishes has developed to date. Some research has been conducted on dolphin or mahi-mahi, *Coryphaena hippurus*. The status of marine fishes with potential in the United States has been discussed by the National Research Council (1992). Expressions of interest have even been heard with respect to the culture of tuna and flying fish. Research in Norway and Scotland has led to the development of culture interest in Atlantic halibut *(Hippoglossus hippoglossus),* and preliminary work has been conducted with Pacific halibut *(H. stenolepis)* in Washington State and Canada (Liu et al. 1991, Stickney et al. 1991).

Freshwater Species. One of the more intriguing anecdotes about an aquaculture breakthrough involves the freshwater shrimp *Macrobrachium rosenbergii*. Dr. S.-W. Ling, a scientist with the Food and Agriculture Organization of the United Nations (FAO), reportedly became intrigued with *M. rosenbergii* while on an assignment in Southeast Asia during the 1950s (Hanson and Goodwin 1977). He attempted to rear the species through a complete life cycle but was frustrated at not being able to keep the larvae alive for more than a few days, though he had no trouble maturing and spawning the adults. There are various versions with minor variations on how the breakthrough occurred. One of the best, reportedly told by Dr. Ling himself and frequently repeated even though possibly of questionable validity (Landau 1992), is that the FAO scientist came to the conclusion that some element was missing from the water in which the larvae were being reared. He placed larvae in watch glasses and began adding various chemicals to the water. The result continued to be complete mortality, even though he apparently tried virtually all the chemicals available in his laboratory. One day his wife delivered his lunch. Out of desperation, Dr. Ling poured a few drops of soy sauce into one of the watch glasses. To his surprise, those larvae survived!

While we don't employ soy sauce in the culture of freshwater shrimp today, because of Dr. Ling's discovery and the realization that it was a component of the soy sauce that allowed the shrimp to survive, larval culture of the species is routine today. The elusive ingredient turned out to be sodium chloride. We now know that *M. rosenbergii* females migrate to estuarine areas to spawn and that the eggs require about 12 ppt salinity for proper development. The larvae move upstream into fresh water as they grow. One lesson that can be learned here is that knowledge of the life history of an animal may provide considerable insight into the requirements for its culture.

M. rosenbergii is neither the only freshwater shrimp species that requires exposure to salt water as larvae, nor the only species of culture interest. Another Asian freshwater shrimp, *M. nipponense,* is commercially important in China and Japan

(Wong 1987). Maximum survival and the highest percentage of metamorphosis from larvae to juveniles occurs over a salinity range of 7.5 to 12.5 ppt, though Wong (1987) reported that some survival has been observed in fresh water.

While the various species of tilapia that exist today are typically restricted to freshwater environments,[93] many of them are highly euryhaline. Included are several that are important to aquaculturists. Stickney (1986) reviewed the literature on salinity tolerance in tilapia and concluded that the most salt-tolerant species are *Tilapia zillii* (produced primarily as a weed control species) and *T. mossambica* (widely cultured as human food). *T. mossambica* has been reported from salinities of 60 up to 120 ppt (Potts et al. 1967, Assem and Hanke 1979, Whitfield and Blaber 1979). It is less desirable for food production than such species as *T. aurea* and *T. nilotica*. The latter two species have smaller heads and deeper bodies, thereby providing higher dress-out percentages. There are other drawbacks of *T. mossambica* as well, including maturation at a small size and dark pigmentation, which is considered to be undesirable in some markets. *T. aurea* can be adapted to seawater salinity, but growth and disease resistance may be reduced compared with freshwater culture (McGeachin and Wicklund 1987). As reviewed by Perschbacher (1992), *T. mossambica* can usually be introduced directly into seawater without tempering. Other species, such as *T. aurea* and *T. nilotica* require multistep increases in salinity over a period of 4 to 8 days until seawater salinity is reached.

Red hybrid tilapia, which have been developed by crossing various species (Galman and Avtalion 1983), have been developed in Florida, Israel, the Philippines, and Taiwan. Those hybrids that contain *T. mossambica* have demonstrated excellent salinity tolerance. Some may even perform better in brackish and full-strength seawater than in fresh water (Meriwether et al. 1984, Watanabe 1985). The Florida-strain red tilapia survive better if the fish are not introduced into seawater until they are sufficiently old or if they are spawned in saline water. Tolerance to salinity has been shown to improve significantly after the fish reach 40 days posthatch (Watanabe et al. 1990). Red tilapia of the Florida strain spawned in 18 ppt salinity water demonstrated significantly better growth than those spawned at 4 ppt when the fish were reared at either 18 or 36 ppt (Watanabe et al. 1989). When rearing temperatures fell below 25°C, fish spawned at 18 ppt showed better survival than those spawned at the lower temperature.

Tilapia eggs are typically incubated in fresh water, but developing embryos may be somewhat salinity-tolerant. Watanabe et al. (1985) reported that *T. nilotica* eggs exhibited the same salinity tolerance as 7- and 120-day-old fry (LC_{50} of 18.9 ppt). Six days after hatching, survival ranged from 85.5% with respect to eggs incubated in fresh water to 0% for those incubated at 32 ppt. Survival in excess of 80% was reported for eggs incubated at 5 and 10 ppt.

Many commonly cultured freshwater species can tolerate limited amounts of

---

[93] A population of *Tilapia hornorum* has been reported to occur naturally in brackish water (Philippart and Ruwet 1982). Populations of tilapia have become established in some saline waters (e.g., upper Tampa Bay, Florida, and Pearl Harbor, Hawaii), but those represent escapement from exotic introductions.

salinity. Unacclimated common carp *(Cyprinus carpio)* have a tolerance of 15 ppt (Payne 1983). Larval common carp have demonstrated improved growth and survival when salinities are increased from that of fresh water up to at least 3 ppt (Lam and Sharma 1985). Grass carp *(Ctenopharyngodon idella)* reportedly have a salinity tolerance of 14 ppt (Maceina and Shireman 1979, Kilambi and Zdinak 1980).

Silver carp *(Hypopthalmichthys molitrix)* have less salt tolerance than either common or grass carp. Chervinski (1977) and Oertzen (1985) indicated that short-term tolerance is 10 ppt for silver carp.

Various species of so-called Indian carps are being cultured on the Indian subcontinent. They include *Labeo rohita, Catla catla,* and *Cirrhina mrigala.* A series of papers on the salinity tolerance of those species demonstrated that the fishes can be cultured in brackish water (Saha et al. 1984a, b, c).

Channel catfish *(Ictalurus punctatus),* another species that is normally grown in fresh water, can tolerate salinities as high as 14 ppt and have been reared in brackish ponds at salinities as high as 11 ppt (Perry 1969, Perry and Avault 1968, 1969, 1971, Turner 1988). Blue *(I. furcatus)* and white catfish *(I. catus),* as well as hybrids among the three species of *Ictalurus,* can also tolerate about 14 ppt (Perry 1967, Allen and Avault 1969, Perry and Avault 1969, Stickney and Simco 1971). Growth and food conversion appears to be about the same for channel catfish reared in fresh water and at salinities up to about 6.5 ppt (Turner 1988).

The salinity tolerance of walking catfish *(Clarias* spp.) has not been extensively studied. Chervinski (1984) indicated that the tolerance of *C. lazera* is 9.5 ppt.

While the channel catfish industry in North America is found virtually entirely in fresh water, a small amount of salt can be beneficial. For example, the Ich parasite can be controlled in water containing 2 ppt salt (Johnson 1976).

Smoltification

The process by which anadromous salmonids adapt from fresh to salt water (smolitification) has been widely studied. The subject has been reviewed by Folmar and Dickhoff (1980). Smoltification, also known as parr-smolt transformation, is a physiological process that involves the fading of the dark pigmented bars (parr marks) on the sides of the young fish, leaving them silvery in color. As the transformation occurs, the fish become increasingly tolerant of salt water and will, in nature, begin their migration to the sea. Smoltification appears to be controlled by thyroid hormones (Dickhoff and Sullivan 1987).

Aquaculturists who are involved in the transfer of smolts from freshwater rearing facilities to marine net pens often use fish size in determining when the transfer should be made. There is ample support in the literature to relate fish size to smolting (reviewed by Hoar 1976, 1989). Age, which relates directly to fish size in most instances, is also important (Parry 1960). In addition, the onset of smoltification may be influenced by light and temperature. The literature on that subject is rather extensive. Examples are papers by Wagner (1974), Hoar (1976), Clarke et al. (1978), Johnston and Saunders (1981), and Duston et al. (1989).

With respect to the photoperiod aspect of light and its effect on smoltification, the use of increasing daylength throughout the rearing period (winter and spring) stimulates the parr-smolt transformation in Atlantic salmon, whereas exposure to continuous light (24 hr/day) decreases the ability of the fish to tolerate increased salinity (McCormick et al. 1987). Rearing tank color, which is also a light-related parameter, can also affect smoltification. Stefansson and Hansen (1989) found that Atlantic salmon smolted in gray tanks but did not find evidence of smoltification in fish of the same age in green tanks.

LIGHT

Light quality (spectral characteristics), quantity (intensity), and photoperiod (ratio of hours of light to hours of dark during a 24 h period)[94] are all important parameters that strongly influence plant growth. Light also influences animals in various ways. Some of the most important to aquaculturists relate to the influence of light on growth, the onset and control of reproduction, and in the use of salmon smoltification.

Light often does not operate alone in its effect on animal physiology. Typically, as in the case of the maturation process, there is a synergistic effect between light and temperature. For many animals that spawn in the spring, increasing daylength coupled with increasing temperature can trigger gonadal maturation and spawning. For fall spawners, reducing daylength and falling water temperatures may provide the maturation and spawning triggers. It is not possible to generalize, however, as some species take their primary cue from one or the other parameter and with the remaining one has apparently little influence. Since light is a major factor influencing reproduction in aquatic animals, that subject is discussed in Chapter 7.

Activity patterns in aquatic animals can be associated with light and darkness. Some species of shrimp, for example, are active during daylight, while others are nocturnal. The freshwater shrimp *Macrobrachium rosenbergii,* for example, demonstrated enhanced growth and survival when reared in a light/dark regime of 0:24 (Withyachumnarnkul et al. 1990).

Most aquaculture species are active during daylight, but if a species is active primarily at night, accommodations for feeding the animals in the dark may lead to better growth and food conversion efficiency (FCE). The effects of time of day on aquaculture production can be difficult to predict. For example, Noeske et al. (1981) and Noeske and Spieler (1984) found that the growth of goldfish *(Carassius auratus)* was influenced by the time of day that the fish were fed. When a 12:12 photoperiod was used, goldfish that were fed 18 h after the light was turned on (6 hr into the dark period) demonstrated the best growth. However, another cyprinid, the common carp *(Cyprinus carpio)* did not respond with differential growth when fed at various times and exposed to a 16:8 photoperiod (Noeske and Spieler 1984).

---

[94] If photoperiod is shown as 12:12, there will be 12 hr of light followed by 12 hr of darkness. Similarly a photoperiod of 8:16 would be characterized by 8 hr of light and 16 hr of darkness. In all cases, the sum of the light and dark periods should equal 24 hr.

The economics of controlling photoperiod must, of course, be taken into consideration. If the effect on growth and FCE is not sufficient to offset the additional energy costs associated with photoperiod control, it does not make sense to attempt to manipulate light in production facilities.

Control of light in outdoor rearing facilities is often impractical and unnecessary. Indoor water systems may or may not have windows that provide some natural light and an indication of natural photoperiod though artificial lights can be used to extend daylength or windows covered during part of the day to reduce daylength. In buildings without windows, photoperiod may coincide with when workers are present in the facility or it may be controlled with timers to maintain a set schedule. Ten to 12 hr of daylight are typically used when photoperiod is controlled.

Culturists who find the need to visit an indoor facility during nighttime when artificial lights are turned off should gradually increase the light level in the room so as to avoid startling the animals. Activating all the light switches at once can not only lead to a startle response that may be stressful; some individuals may jump out of their tanks or bump into the walls of culture chambers causing physical damage and perhaps, mortality, particularly for those which land on the floor and are not quickly discovered and returned to the water. Having lights on rheostatic controls that allow them to be manually or automatically brought up in intensity over a period of several minutes (simulating sunrise) is desirable. It might also be appropriate to have the lights dim slowly as well when they are being extinguished. If rheostatic control is not available, lights should be wired into several circuits, one of which should have low-wattage bulbs. The low-wattage bulbs can be turned on for a few minutes, followed by additional lights periodically until the desired light level is reached. When checking a facility during a period when the room lights are out, a flashlight with a red lens will often provide sufficient light without startling the animals.

Some organisms change color when exposed to different quantities or qualities of light. The bodies of channel catfish become generally darker in dim light (and when exposed to cold water) and lighter in bright light (and in warm water). Many flatfish species are able to adjust their upper body-surface pigmentation to match the background on which they are resting (Figure 69). The ability to match a background provides the fish with protection from visual predators and extends both to color and the mottling pattern.

Specific studies on the effects of photoperiod on the growth rates of aquaculture animals have become more widespread in recent years, but there is certainly room for additional research on the subject. In early studies on the effects of photoperiod on the growth of channel catfish *(Ictalurus punctatus),* Kilambi et al. (1970) reported a growth response in fry. However, Stickney and Andrews (1971) and Page and Andrews (1975) were unable to demonstrate growth differences in response to photoperiod in fingerlings of the same species. There was also no apparent growth response in catfish fingerlings exposed to light of different intensities (Page and Andrews 1975).

There is a considerable amount of literature available on the effects of light on control of spawning in rainbow trout. That subject is considered in Chapter 7.

Figure 69. A postlarval flounder (*Paralichthys* sp.) lying on the sandy bottom in a culture tank is able to mimic the substrate to the point that it is virtually invisible.

Research has demonstrated that performance in salmonids is affected by light, though the results have been somewhat variable due undoubtedly, in part, to the fact that different experimental designs and fish sizes were used in the various studies. With respect to Atlantic salmon, *Salmo salar,* Huse et al. (1988) reported no effect on growth, mortality, ectoparasite infestations, maturation, or the fouling of net pens when covers were used to reduce incident light levels. Saunders and Harmon (1988), Villarreal et al. (1988), Saunders and Harmon (1990), and Stefansson et al. (1990) reported a positive influence of certain types of adjusted photoperiods (continuous light or extended periods of light and simulation of different seasons of the year out of phase with the actual photoperiod) on the growth of young Atlantic salmon. The development of salinity tolerance in juvenile Atlantic salmon appears to be influenced by photoperiod (Stefansson et al. 1992). Manipulating the photoperiod has also been shown to enhance seawater survival of coho salmon, *Oncorhynchus kisutch* (Corley-Smith 1989).

Photoperiod has been shown by Woiwode and Adelman 1991to have an influence on the growth of hybrid striped bass. Juveniles exposed to springtime photoperiod grew significantly better than those exposed to autumn photoperiod.

There is some published information to indicate that the percentage of normal Atlantic halibut *(Hippoglossus hippoglossus)* eggs surviving to hatching is improved if the eggs are incubated in the dark (Bolla and Holmefjord (1988), but more work needs to be conducted. Exposure to light during yolk sac absorption in halibut does not appear to be a problem so long as the intensity is not too high. Skiftesvik et al. (1990) found that yolk sac nutrient utilization was lower in Atlantic halibut larvae reared under 1,000 lux than in those reared between 1 and 10 lux. Studies with gilthead sea bream *(Sparus aurata)* showed that highest survival from the time of hatching to 20 days of age occurred under continuous light, while a 12:12 photoperiod was more desirable for older larvae (Tandler and Helps 1985).

Exposure to certain types of light reportedly has an adverse effect on developing fish eggs, particularly those of salmonids. Piper et al. (1982) indicated that the most detrimental light wavelengths for developing trout were in the visible and violet-blue range. They recommended the use of pink fluorescent tubes (which emit light in the yellow to reddish-orange range) rather than white fluorescent lights in hatcheries. They also recommended covering eggs and keeping them away from direct light. The same authors indicated that fish embryos subjected to bright artificial light before the eye pigments form can experience high rates of mortality, with surviving embryos often being deformed and growing poorly.

Harmful effects of certain types of light radiation have been reported for a number of salmonids. For example, Bell and Hoar (1950) reported that ultraviolet radiation caused premature hatching, retarded growth, and other abnormalities, including high mortality, in sockeye salmon *(Oncorhynchus nerka)*. However, the rearing of sockeye fry in continuous darkness led to significantly smaller fish, which experienced higher mortality that those reared for 12 hr/day under fluorescent lighting (Bilton 1972).

In a more recent study, Dey and Damkaer (1990) examined the effects of various types of light on eggs, alevins, and fry of chinook salmon *(O. tshawytscha)* and came to the following conclusions:

- Ultraviolet light at energy levels above 10^{-3} W/m^2 is harmful to each of the three life stages when exposure is 8 hr/day.
- Visible light at energy levels above 10^{-2} W/m^2 is harmful to eggs exposed for 8 hr/day.
- Chinook salmon eggs are more susceptible to damage from both UV and visible light than are alevins or fry.
- Semidarkness or the use of pink fluorescent bulbs appear to support good fry growth.
- Viability and hatching of chinook salmon eggs does not seem to be affected by exposure to 50 to 100 kJ/m^2 of light daily from incandescent bulbs, filtered pink fluorescent bulbs, or conventional white fluorescent bulbs.

Insufficient light levels can impair the ability of sight feeding fish to capture their prey as has been demonstrated for larval striped bass (Chesney 1989). At the same time, some shading to keep young fish from being exposed to direct sunlight may also be beneficial as has been shown for hybrid striped bass fry (Rees and Cook 1982). At this time there is not enough information on a sufficient number of species and their life stages to make specific recommendations.

Swim bladder inflation has been a significant problem in some aquaculture species. Many laboratories have attempted to find the reason why there is sometimes a high percentage of failure in swim bladder inflation. The problem has been a significant one for culturists of striped bass, *Morone saxatilis* (reviewed by Kerby 1986, 1992). Swim bladder inflation typically occurs in striped bass between the fifth and seventh day after hatching and coincides with the onset of feeding. In groups of fish that do not achieve proper swim bladder inflation, mortality can be quite high. While the cause of the problem in striped bass has yet to be determined, maintenance of the fish in well-aerated, turbulent water appears to reduce the problem. However, there may also be an effect of light as well. Weppe and Joassard (1986) found that certain light intensities seemed to enhance swim bladder inflation in sea bass *(Dicentrarchus labrax)*. They found a light intensity of 70 lux to be the most favorable to proper inflation.

The hormones that induce maturation and spawning in penaeid shrimp are located in the eyestalk, and it has long been the practice of shrimp hatchery personnel to ablate (sever) one or both eyestalks of females to induce maturation and spawning.[95] Low light levels are also typically employed during the process. Wurts and Stickney (1984) calculated the probable light intensity on the spawning grounds of *Penaeus setiferus* in the Gulf of Mexico shrimp and compared that value with light levels used in shrimp hatcheries. The authors found that hatchery light levels were typically orders of magnitude more intense than those at which the shrimp spawned in nature and proposed the use of much lower light intensities for controlled spawning. At about the same time the hypothesis was developed, George Chamberlain (personal communication), working at Texas A&M University in a carefully controlled culture environment, found that by maintaining adult *P. stylorostris* in virtual darkness, spawning could be induced without eyestalk ablation.

Relatively little work has been conducted with regard to the effects of light color (wavelength) on the performance of aquaculture species. Emmerson et al. (1983) found that *P. indicus* would spawn under dim blue or green light but not under dim white light. The same was not found to be true for *P. monodon*. Unablated shrimp exposed to blue, red, or white light failed to fully mature, whereas ablated shrimp did mature and spawn under all light conditions. The differences in response of the two species could be related to differences in light intensity between the two studies, differences in the depth at which the shrimp spawn in nature, or to other factors.

As a general principle, it should not be necessary to mutilate an animal to induce maturation and spawning. A better approach is to determine the conditions that exist on natural spawning grounds and try to duplicate those in the hatchery.

An important and interesting light-related effect has been reported in conjunction with the production of brine shrimp (*Artemia* sp.), an animal that is widely used as a first feed for a variety of aquaculture species that require living food (see Chapter 6 for a more complete discussion). Van der Linden et al. (1985) found that a certain amount of light energy is required for proper hatching of brine shrimp cysts. Wavelengths between 400 and 600 nm were the most effective in promoting hatching.

---

[95] An alternative approach is to slice open the eye and squeeze out the contents.

SUSPENDED SOLIDS

Suspended solids are small pieces of particulate matter, larger than 0.45 mm, found in the water column. Particles smaller than the indicated size are considered to be dissolved in the water.[96] Suspended solids are made up of sediment particles (fine sand, silt, clay), organic material (detritus composed of plant and animal remains); waste feed particles; as well as bacteria, fungi, phytoplankton, and other microorganisms. Each of these materials contributes to water turbidity.

The higher the concentration of suspended solids in the water, the more turbid the water becomes. If turbidity becomes too high as a result of the presence of suspended inorganic solids, primary productivity may be reduced because of shading. Similarly, when a phytoplankton bloom becomes well established, self-shading can occur and lead to the eventual crash of the bloom (which can occur as a result of nutrient limitation as well, or because of a combination of the two factors). Shading, whether because of inorganic or organic turbidity, can be beneficial in retarding the growth of filamentous algae and rooted aquatic plants. There is a distinct disadvantage in having water with high levels of inorganic turbidity in instances wherein the culturist is attempting to establish a plankton bloom in order to feed young animals.

The early feeding stages of many fishes depend on phytoplankton. Included are tilapia and milkfish. Many molluscs consume phytoplanktonic food throughout their lives (e.g., clams, mussels, oysters, scallops), and the phytoplankton will support large zooplankton communities that are used as food by a variety of vertebrate and invertebrate species. The green water technique of rearing the larval and postlarval stages of shrimp and marine fish is common employed around the world. Green water contains both phytoplankton and zooplankton. In some cases the aquaculture animals only eat the zooplankton, but the presence of phytoplankton seems to enhance survival. That may be associated with some beneficial impacts on water quality, or it may relate to the fact that even carnivores ingest some phytoplankton cells and may receive an unknown nutritional benefit.

Sediment particles composed largely of silt and clay (commonly called mud) may become suspended in the water column as a result of currents or wind mixing, or they may be washed into the water system from land runoff. In some cases, if conditions are sufficiently severe, sand may also become suspended in the water, though it will settle rapidly unless there is continuous turbulent mixing. Suspended solids levels in large ponds, bays, rivers, and so forth may vary considerably with time as a function of weather conditions, especially with respect to wind mixing and precipitation in the watershed.

Inorganic particles can have detrimental effects on aquatic organisms (Cairns 1967). The mechanical action of such particles can lead to clogging of the gills or the irritation of gill filaments and other membranes. If a great deal of suspended particulate matter is introduced into an aquaculture pond (e.g., as a result of mud-

---

[96] The selection of 0.45 mm to separate particulates from dissolved material is based on the use of membrane filters with pore spaces that are 0.45mm in size.

laden precipitation runoff), subsequent settling of the material may lead to the burial of eggs, larvae, fry, and even the juveniles of benthic organisms. Sessile species such as oysters can be suffocated through direct burial.

Sedimentation of the nests of various fishes is not an uncommon event in nature. Heavy precipitation may cause silting in of the nests of trout, catfish, and bass, for example. Spawning mats used in conjunction with the production of minnows, goldfish, and in recent years, largemouth bass, are subject to siltation with resulting egg losses when turbidity is high. Siltation in streams where salmon nests (called redds) are located can clog the gravel in which the eggs are incubating, leading to a reduction in, or the cessation of water flow through the gravel. This will lead to oxygen deprivation and death of the developing embryos. The controlled spawning and indoor incubation of many species reared by aquaculturists provides protection against exposure to high turbidity in most instances, particularly if well water is used. When surface waters are the source of incubation water, increased turbidity can impact animals being held in the hatchery unless the water is filtered before use.

Suspended particles provide an enormous amount of surface area for colonization by bacteria and fungi (Cairns 1967). In addition, such particles can absorb or adsorb dissolved chemicals, including trace metals and nutrients such as phosphate. Thus, fertilization may be less effective when the water is turbid, not only because of light reduction but also because the nutrients become unavailable since they are tied up by the particulate matter.

Diel temperature swings tend to be less severe in turbid water than in water that is clear. At the same time, sediment-laden water will warm more rapidly than clear water. It also holds heat better, so it does not cool down as rapidly at night.

Aquaculturists generally exercise little control over suspended solids in incoming pond water. In addition, water entering outdoor raceways is not often filtered. When necessary, turbidity can be reduced by passing the water through a reservoir where residence time is sufficiently long and water movement slow enough to allow the material to settle before the water is placed into the culture chambers. A problem with this type of system is that the reservoir may eventually become filled with sediment. As it fills, the residence time and effectiveness in sediment removal are reduced.

Mechanical filters can effectively remove suspended material (see Chapter 3). Filters require periodic backflushing so that required flow rates can be maintained.

Suspended inorganic material can cause direct damage to culture animals and will, of course, sediment out in culture chambers if not removed prior to its introduction. Culture ponds may require reworking to maintain the depth as sediments accumulate. Raceways can also become filled with sediments. Reduction in the volume of a culture chamber means that fewer animals can be produced. In addition, harvesting can be impaired. Turbidity also makes observation of the culture animals more difficult.

Even with the potential problems that can occur in conjunction with high levels of turbidity, there are some benefits that can accrue when there is at least some turbidity present. Many species, including catfish and carp, are commonly found in turbid water and can tolerate and perform well even when turbidity levels are high.

In addition to the shading effect of turbidity on unwanted vegetation, limiting light penetration may be beneficial to the aquaculture animals. Some fish may actually sunburn when exposed to direct sunlight through clear water. Anecdotal evidence for sunburn in trout has been found, particularly where the fish are raised at high altitudes where less atmospheric filtration of the light occurs. Shade cloths may be useful to reduce incident light when the water is very clear.

One of the first laws of pond aquaculture is that no two ponds are alike. Ponds that are identical in terms of construction, receive the same amount of water from the same source simultaneously, are stocked at the same time with the same number and species of animals, and are fertilized and fed identically will often differ in some significant way. Those differences will appear in terms of water quality, production rate, and the appearance of the water. Clarity and color will typically vary widely from one pond to another. Aerial photographs of pond facilities show these differences clearly. Further, during subsequent growing seasons the same pond will often have water of different colors and transparency levels.

In most cases some turbidity is not objectionable, and as indicated above, may even be desirable. However, if a phytoplankton bloom is required and the pond has a muddy appearance, it may be necessary to settle the suspended inorganic matter prior to fertilizing. There are two relatively simple ways in which this can be accomplished. The first involves spreading chopped hay over the pond surface. As the hay settles, it scrubs clay particles from the water, or at least it is supposed to accomplish that task. Experience has shown that hay is not always highly effective. The second method involves the use of a chemical to bind with the sediment particles and settle them out of suspension. One such chemical is gypsum ($CaSO4$). Gypsum has been found to reduce turbidity by from 6.7 to 62.5% when added to ponds at 250 to 500 mg/l (Wu and Boyd 1990). The treatment can be repeated at 7 to 10 day intervals, as necessary. Another chemical that will function in suspended sediment removal is alum. Boyd (1990) indicated that an effective coagulant is filter alum: $Al_2(SO_4)_3 \cdot 14H_2O$. Significant reductions in turbidity can be produced with 15 to 25 mg/l of alum. Since the reaction of alum in water produces acid, pH and alkalinity can be reduced. Every milligram per liter of added alum will reduce alkalinity by 0.5 mg/l. Boyd (1990) recommended incorporating a treatment of $Ca(OH)_2$ to maintain pH in soft water. For every 0.37 mg/l of $Ca(OH)_2$ added, the effect of 1 mg/l of alum can be counteracted.

Some animals can induce a certain amount of turbidity into otherwise clear water as a result of their feeding or reproductive activities. Examples of fishes that tend to increase turbidity are common carp *(Cyprinus carpio),* buffalo (*Ictiobus* spp.), and tilapia (*Tilapia* spp.). The levels of suspended solids induced by these types of animals are generally not detrimental to other organisms directly, though they may affect primary productivity through shading. Carp have a habit of rooting around pond bottoms, often working the levees just below the water surface. The activity not only increases turbidity but, as mentioned earlier, can cause damage to levees, including slumping. Reshaping the levees of carp ponds may be required between crops. Tilapia dig nests in the bottom during the spawning season. The nests are formed by fanning action of the sediments with the fins, which can cause an increase in turbidity. In muddy sediments, the fish may dig nests that are many centimeters deep.

Oysters such as the American oyster *(Crassostrea virginica)* feed most efficiently when the ratio of food to water volume is relatively low (Loosanoff and Tommers 1948). In turbid water the pumping rate of oysters is greatly reduced; thus growth rates can be affected. Many species of fish can survive extremely high levels of suspended solids for at least short periods. Wallen (1951) examined the effects of turbidity on 16 species of fish, including channel catfish and common carp, and observed no adverse effects until turbidity exceeded 20,000 mg/l. Most species were able to withstand turbidities up to 100,000 mg/l for a week or more. Since most turbidity levels can be measured in a few hundred milligrams per liter or less, aquaculturists usually do not have to concern themselves about direct mortality as a result of turbid water. Species that do not normally experience turbid waters in nature may be exceptions. The salmonids, for example, appear to be more susceptible to high turbidity than some of the other species mentioned.

For sight feeders, high levels of turbidity can interfere with the ability of the fish to capture their prey. Not only does increased turbidity directly interfere with the ability of a fish to identify food items, it also reduces light transmittance, which further compounds the problem. Reducing light directly and increasing turbidity by adding clay minerals to the water have both been shown to negatively impact feeding rates and growth with respect to striped bass larvae maintained in the laboratory (Chesney 1989).

Algae blooms have been implicated in mortalities that have occurred in salmonids reared in net pens over the past few years. Algae have been suspected of releasing toxins into the water (Harrell et al. 1986, Kent 1990) and have also been thought to directly cause mortality by causing mechanical injury to gills and other tissues (reviewed by Stickney 1991b). Algae species that have been implicated in mortalities include *Chaetoceros consulutus, Ceratium fusus, Gymnodinium splendens,* and *Heterosigma akashiwo.* The problem has become so significant in some places that a secondary industry has developed around disposal of mortalities collected from net pens (Egan 1990).

ALKALINITY

The capacity of a natural water system to resist changes in pH can be measured in terms of the amount of bicarbonate and carbonate ions that are available in the system. This measurement is called alkalinity. Chapter 4 described the effects of the carbonate buffer system on pH and discussed the importance of photosynthesis and respiration in the availability of carbonate and bicarbonate.

Alkalinity is measured by titrating water samples with dilute sulfuric acid to end points identified by color changes in the indicators phenolphthalein and methyl orange. Bicarbonate alkalinity is derived from the difference between carbonate (phenolphthalein) and total (methyl orange) alkalinity values (APHA 1989).

In aquaculture systems the alkalinity should generally be between 30 and 200 mg/l in fresh water, though water of higher and lower alkalinity has often been utilized successfully by culturists. The recommended minimum alkalinity for rearing striped bass *(Morone saxatilis),* for example, is 20 mg/l $CaCO_3$ (Hall 1991). Water of very low alkalinity has little capacity to resist pH changes and should be avoided under most circumstances. Low-alkalinity water may be suitable for intensive closed culture systems because a buffering agent is routinely recommended for inclusion in the design. The soils in ponds exchange ions with the water and can have a fairly profound effect on alkalinity until carbonates are depleted. It is possible to add calcium carbonate to ponds as a means of increasing alkalinity (Arce and Boyd 1975). A good discussion on liming ponds was developed by Boyd (1990). Boyd and Hollerman (1982) found that finely pulverized agricultural limestone was more effective than coarse limestone for increasing alkalinity, undoubtedly because of the increased surface area that would allow the finer material to dissolve more rapidly and more completely.

Water can be passed through beds of crushed limestone or oyster shell to increase alkalinity (and hardness). Introduction of the water into such a chamber from the bottom under pressure, or using high volumes of air introduced into the chamber from the bottom will keep the medium in motion. Limestone and oyster shell can quickly become covered with sediments or the growth of bacteria, thereby losing direct contact with the water and reducing or eliminating the rate of dissolution. When the medium is kept in motion, the particles are continuously scrubbed.

When alkalinity is extremely high, carbonates may precipitate on surfaces in the culture system (e.g., on the walls of plumbing and culture tanks). This is especially likely to happen when sufficient levels of calcium or magnesium ions are present in the water to react with the carbonate. The alkalinity of seawater is always fairly high because of the abundance of carbonates in marine sediments and dissolved in the water. However, it is generally not so high that precipitation of calcium carbonate occurs in culture chambers.

The pH of water affects the percentage of alkalinity contributed by carbonic acid, bicarbonates, and carbonates. Temperature and salinity also affect these relationships. For example, in seawater of pH 8.0 and 24°C temperature, slightly more than 8% of the alkalinity is in the form of carbonates, whereas in fresh water at the same pH and temperature, less than 0.5% of the alkalinity is represented by carbonate ions (Spotte 1970). In general, most of the alkalinity of fresh water and low-salinity estuarine waters may be attributed to bicarbonates.

HARDNESS

The concentration of divalent cations (primarily calcium and magnesium) present in the water determines its hardness, which is expressed in terms of milligrams per liter of calcium carbonate (APHA 1989). According to Landau (1992) the level of hardness, as measured in mg/l $CaCO_3$, can be defined as follows:

Soft water	0 to 55
Slightly hard water	56 to 100
Moderately hard water	101 to 200
Very hard water	201 to 500

To that list could be added extremely hard water of greater than 500 mg/l $CaCO_3$.

Although high alkalinity and high hardness often occur simultaneously, the two are actually independent, especially in fresh water. In salt water, the levels of both alkalinity and hardness are more predictable.

Salt water tends to be quite hard, while very soft (low hardness) fresh waters are not uncommon. Water that is soft may have low, intermediate, or high alkalinity. One well at the Aquaculture Research Center at Texas A&M University produced high-alkalinity water with a hardness that was sometimes measured at only 1 mg/l.[97]

Low hardness can adversely affect some aquaculture animals. For example, when euryhaline fishes such as red drum *(Sciaenops ocellatus)* are maintained in very soft fresh water, survival is poor (probably because of the inability of the animals to osmoregulate efficiently in a solution that is nearly devoid of divalent cations).

Excessive hardness can also be detrimental. The eggs of Atlantic salmon *(Salmo salar)*, rainbow trout *(Oncorhynchus mykiss)*, and brook trout *(Salvelinus fontinalis)* have been shown to suffer severe mortality when exposed to extremely hard water during the first few hours after fertilization (Ketola et al. 1988). Very hard water may be required by some species, however. Gonzal et al. (1987) found that excessive water absorption in silver carp *(Hypopthhalmichthys molitrix)* eggs incubated at a hardness of 100 to 200 mg/l $CaCO_3$ caused them to burst. Minimum water absorption occurred at 600 mg/l. The authors recommended incubating silver carp eggs in water of 300 to 500 mg/l hardness.

Vasquez et al. (1989) and Brown et al. (1991) found that very hard water depressed growth in *Macrobrachium rosenbergii*. There was no significantly negative impact seen from exposure of freshwater shrimp to soft water. Still, Vasquez et al. (1989) recommended hardness levels between 20 and 200 mg/l. That range would appear to be acceptable for rearing most freshwater species, though as indicated above, it may not always be an appropriate range for egg development.

When fish are transported from one facility to another, they are often exposed to significant differences in various aspects of water quality. One of those changes, temperature, has been previously discussed. Hardness is another. Mazik et al. (1991) found that if 10 ppt sodium chloride is added to the water of striped bass that have been transferred from a soft-water facility to one with hard water, the stress associated with the change in hardness can be reduced. Calcium chloride has been successfully used to improve survival of striped bass by directly increasing

---

[97] An advantage of that water source was that it was possible to determine the calcium and magnesium requirements of fishes reared in it. Deficiency signs often do not develop in fish fed diets containing little or no calcium or magnesium because the fish are able to absorb the ions from the water.

hardness. When $CaCl_2$ was used to increase hardness from the 10 mg/l that was present in hybrid striped bass rearing ponds to 70 to 200 mg/l $CaCO_3$ in holding and transportation tank water, survival ranged from 80 to 99% (Grizzle et al. (1985). That compared with a 16% survival rate for fish transported in water to which no additional calcium had been added. For general rearing, striped bass should be exposed to water with a hardness greater than 150 mg/l $CaCO_3$ (Hall 1991).

Hardness is not a consideration in marine culture systems since all natural salt waters have relatively high levels of divalent cations. Artificial sea salts are formulated to mimic seawater. If saltwater is produced solely from sodium chloride, it would not be sufficiently hard, and supplementation with calcium and/or magnesium (preferably both) would be recommended in proportions similar to those found in natural seawater.

In ponds, the hardness of the incoming water may be increased by the leaching of calcium and magnesium from the sediments, though this should not be relied on as an adequate source of those elements. Limestone can be used to increase hardness as well as alkalinity (Boyd 1990). If hardness alone is to be increased, lime (CaO), can be added. Slaked lime [$Ca(OH_2)$] should not be used because it will make the water basic (through the addition of hydroxyl ions), overcoming the buffer system in many cases. Calcium carbonate can be used but might not be effective in waters of high pH because the limestone would not dissolve rapidly. Gypsum can also be used to increase water hardness without affecting pH. No attempts should be made to increase hardness after the addition of phosphate fertilizers, since the calcium released will combine with the phosphate as insoluble $Ca(PO_4)_2$, making the fertilizer treatment ineffective.

The amounts of chemicals to be added to a particular body of water vary considerably with the initial water quality and desired final hardness. The culturist may begin by adding as much as 1,000 kg/ha; 200 to 500 kg/ha might be preferred and would be less expensive if effective. Hardness should be checked after a few days to determine the effectiveness of the treatment. Thereafter, additional treatments at the same rate or at higher or lower rates may be utilized until the desired hardness is obtained. Hardness can be readily determined through titration (APHA 1989).

Arce and Boyd (1975) examined the effects of adding limestone to soft-water ponds in Alabama at rates ranging from 4,300 to 4,900 kg/ha, after which the ponds were fertilized at 2 wk intervals for 13 applications with 45 kg/ha of 20-20-0 fertilizer. Total hardness (initially less than 10 mg/l) was increased fourfold, as was total alkalinity (initially less than 15 mg/l). Phytoplankton productivity was increased, as was the production of the largely herbivorous fish, *Tilapia aurea*, in ponds receiving calcium carbonate treatment.

Lime is sometimes used to sterilize the bottoms of dry ponds before refilling and stocking. A layer of lime is spread over the pond bottom and worked into the sediments with a disk. This technique is effective in killing undesirable benthic animals and may help control pathogenic organisms, along with adding to subsequent water hardness.

SUBSTRATE REQUIREMENTS

Many aquaculture species adapt as well to fiberglass, plastic, metal, concrete, or wooden culture tanks or raceways as they do to earthen ponds. In addition, many species can be reared without difficulty in cages or net pens with mesh bottoms. However, there are species grow better or have improved final product quality if they are reared over the proper type of substrate. Outdoor ponds are usually suitable or can be made suitable for species that require a particular type of substrate. Also, natural types of substrates, which would include mud, sand, shell-hash, or combinations of those constituents, can be placed into tanks or raceways when required.

Species that normally burrow may require some form of natural substrate in which to live, though in many instances performance is not significantly impacted when the animals are reared on an artificial surface such as the bottom of a fiberglass tank. In some instances, natural substrates provide invertebrates a hiding place during the period immediately after ecdysis when their exoskeletons are soft and the animals are susceptible to predation and cannibalism. Natural substrates into which such species can burrow are not always required. Artificial hiding places may be effectively employed to provide molting individuals protection from cannibalism. For example, among both penaeid and palaemonid shrimps of aquaculture interest *(Penaeus)* spp. and *Macrobrachium* spp., respectively), cannibalism is common. If the animals have a place in which to hide during ecdysis, their vulnerability is significantly reduced. Pieces of PVC pipe, fine-mesh screens hung either vertically or horizontally in a culture chamber, and other devices have been shown effective. For some species, neither natural substrates or artificial hiding places are sufficient under aquaculture conditions. A prime example is the American lobster, *Homarus americanus*. The only known way that lobsters can be reared in captivity is to provide them with separate culture chambers. Lobster condominiums have been constructed that provide individual housing for a number of animals. As the lobsters grow they are moved to increasingly larger condos until they reach market size. If lobsters are grown communally, the culturist may finish the growing season with one large individual, no matter how many were present when the animals were stocked as juveniles.

Oysters require firm substrates on which to attach or on which they can be supported if they are detached from the surface to which they first attached and grown individually (cultchless oysters). Sediments containing high percentages of silt and clay may not provide the proper support, and the oysters could sink and become asphyxiated. If not directly lethal, soft substrates may be sources of turbidity, which can lead to gill clogging and reduced pumping rates, both of which would impair growth. Oysters and clams can be reared in culture chambers with solid bottoms, in baskets that contain no sedimentary material, on strings, or on firm substrates in natural or pond environments. None of these methods of culture should damage the animals in any way, although each has its advantages and disadvantages in terms of suitability in a particular culture facility or available culture environment.

Fishes that normally swim within the water column, such as salmon and trout, would seem to be appropriate candidates for cage or net-pen culture, and that is

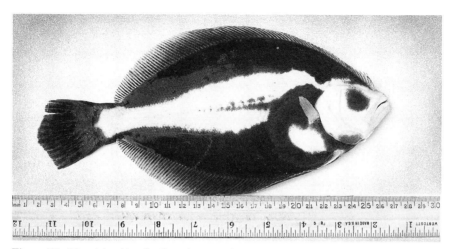

Figure 70. The right side of a flounder, *Paralichthys* sp. This side of the fish faces toward the substrate and is normally white in color. Ambicoloration can occur when flounders are reared on bare tank bottoms.

indeed the case. Catfish, which are often thought of as living in close proximity with the bottom, might be considered marginally suitable for cage or net-pen rearing, when in fact they adapt to such culture systems quite readily. More surprising is the fact that Pacific halibut *(Hippoglossus stenolepis)*—a species of flatfish which are commonly found not only living on but often partially buried in the sediment— will adapt to life in net pens (Stickney and Liu 1991). In that study, broodfish were initially held in circular tanks, to which they quickly adapted. In order to keep larger numbers of animals, a commercial salmon net-pen operator and National Marine Fisheries Service biologists became involved and provided salmon net pens for holding some of the halibut. The initial working assumption was that the net pens would have to be modified so the fish would have a hard substrate on which to rest. Modifying a net pen to support a solid bottom was difficult, so one was modified to allow the netting to reach the bottom of the water column at all tidal stages. Halibut introduced into that cage did not survive, though the reason for the mortality was not determined. It was then decided to put several adult halibut in a standard net pen. Surprisingly, the fish adapted well and survived for over a year without mortality.

Flounders of the genus *Paralichthys* have been shown to develop dark pigmentation on the underside[98] when cultured in fiberglass tanks without added substrate materials such as sand or shell-hash (Stickney and White 1975). This anomaly is shown in Figure 70. Ambicoloration does not affect the flavor or texture of the flesh, but consumers would probably reject fish that have developed pigment on the side which is normally facing the substrate. Since flounders are normally sold

---

[98] This condition is known as ambicoloration.

headed and gutted but with the skin and scales intact, marketing of such fish would be difficult. If the fish are able to partially bury in a natural substrate, the dark coloration on the bottom surface does not occur. In fact, providing fish with substrate once the problem has begun will lead to reversion to normal coloration. It appears as though the white underside of flounders acts in a manner similar to photographic film. When exposed to light, presumably reflecting from the bottom of the culture tank, the coloration develops.

The density of benthic animals in a culture chamber can sometimes be increased by providing artificial substrates. Platforms, netting, and so forth can provide additional substrate for benthic organisms such as shrimp (while also providing protection from cannibalism). Structures add to the costs involved in rearing the animals and would have to be removed prior to harvest. Flounders will associate with any type of unbroken surface, whether vertical, horizontal, or at an angle. While netting might not provide an appropriate vertical substrate for flounders, it might be useful if placed in the horizontal position off the bottom. Given a solid vertical surface, flounders form their bodies into a configuration that simulates a suction cup. They have been observed to hang on the walls of fiberglass tanks for hours without apparent difficulty. Rearing flounders on artificial substrates has the disadvantage of promoting ambicoloration (Figure 70).

DENSITY

Different species respond in a variety of ways to crowding. Some animals become aggressive at low densities and may show less aggression as the density is increased, while others may exhibit increasing levels of aggression or cannibalism as density is raised. For some species it is possible to successfully employ a range of densities. Examples are the rearing of penaeid shrimp and channel catfish both extensively and intensively. The proper density for optimum performance of each species under culture can be determined in terms of numbers animals per unit area or volume of water. An important criterion, of course, is to maintain suitable water quality no matter what density is ultimately selected.

As animal density is increased, competition for food may become severe and territorial behavior often breaks down. Channel catfish, trout, and many other species appear to coexist well under crowded conditions, but some species do not. It would seem to be to the advantage of the aquaculturist to select species for culture that accept crowding since aquaculture economics dictate getting the most production per unit area as possible.

Aggressive and cannibalistic species are, of course, being cultured. Survival of those species is often not as good as that obtained from species that are more socially compatible. We have already seen the desirability of providing hiding places for freshwater shrimp during the molting period and for raising lobsters individually. Marine shrimp are also cannibalistic, but the problem is not as severe and no unusual measures are taken, at least by pond culturists, to circumvent or even reduce the problem. Feeding practices can sometimes reduce both aggression and cannibalism. By providing several daily feedings, feeding throughout the pond, or both, the ability of each individual to obtain food without having to fight for it can be reduced to some extent.[99]

For species that tolerate crowding, culture chamber densities can reach almost incredible levels. Densities in neighborhood of 150 kg/m[3] have been maintained in flowing water systems, including those of the recirculating variety. When fish are present at such densities, they will literally fill the culture chamber and can do little more than keep station in the current. Lateral and vertical movement is greatly impaired, and such densities are atypical for most species. Cultured walking catfish may reach incredibly high densities, however, as those fishes are very tolerant of crowding as well as of poor water quality.

Filling the water column with pelagic species that are not negatively impacted by crowding is sometimes possible. For benthic species such as shrimp, crawfish, clams, oysters, and flatfish, the surface area available is sometimes a more important consideration than total water volume. Cannibalism can be expected to increase in shrimp ponds as the number of animals stocked per unit area is increased. When that happens, the proximity of individuals, and therefore their susceptibility to being discovered and consumed immediately after molting, increases. So long as the proper amount of food is made available, cannibalism may be controlled to some extent. Maintaining proper food levels can also be a problem with species that do not exhibit aggression and are not cannibalistic; e.g., clams and oysters. While such species can tolerate extreme crowding, competition for food can severely limit growth. Flounders and halibut seem to be highly tolerant of crowding. Those fishes will lie upon one another in a culture tank, though there is undoubtedly some limit to the how high the stack might become before some intolerance is demonstrated by individuals on or near the substrate.

Crowding can, of course, be a form of stress, thereby increasing the probability of a disease epizootic. Furthermore, disease transmission is enhanced when the affected animals are in close proximity to one another. Thus, it is very important to maintain excellent water quality and other conditions in facilities where high densities of culture animals are maintained.

Several studies conducted with salmon have demonstrated the effects of rearing conditions on physiological responses. In a study that reviewed previous work and reported on additional original research, Patiño et al. (1986) reported that crowding stress alone can have significant effects on the physiology of coho salmon. In addition, increased rearing density appears to reduce the return rate of Atlantic salmon (Hosmer et al. 1979).

The maximum density at which animals can be maintained is related to the size of the individuals in the population. It should be obvious that many more 1 g animals can be maintained per unit area of pond than 10 g or 100 g animals. For example, bass fry are often stocked at densities of 500,000/ha or more in ponds and

---

[99] An additional problem associated with rearing lobsters in individual containers is that the animals must be individually fed. That requires either a large expenditure on automatic feeding equipment or the use of people to hand-feed each lobster.

reared to a few grams. Their density may then be reduced to around 60,000/ha for further growout (reviewed by Simco et al. 1986, Williamson et al. 1992). Broodstock of the same species are typically held at densities of less than 2,500/ha. Similar figures could be presented for various other species. It is common practice to stock high densities of small animals and to reduce the density and distribute the animals more broadly into additional culture chambers as they grow. In many cases aquaculture animals are stocked at densities that will provide them with sufficient volume for growth throughout a given growing season.[100] They are then restocked for the subsequent growing season. The process for largemouth bass outlined above is one of many exceptions in that the fish are stocked at very high densities (over 100,000/ha) but are retained in ponds for only about a month. They are then either stocked into lakes and reservoirs to enhance sportfish populations or are grown at densities of several tens of thousands per hectare for a few more weeks or months before being stocked. Some of the variations in technique are discussed further in Chapter 7.

A laboratory study with *Tilapia aurea* and *T. mossambica* by Henderson-Arzapalo et al. (1980) led to the discovery of an autoimmune response in *T. mossambica*. Fish were stocked in circular culture tanks at various densities and reared over a period of 115 days. As the biomass of *T. mossambica* reached about 20 g/l (water turnover rate was 30 min in the tanks), mortalities began that were not associated with water quality or disease problems. The *T. aurea* showed no such response. Injection of *T. mossambica* with a concentrated extract of the water from affected tanks led to an allergic reaction. After several repetitions of similar experiments with the same result, the autoimmune response was lost from the afflicted tilapia population. Anecdotal information from individuals who have been involved in the aquarium fish production industry have indicated that similar responses to high density have been observed in other species.

PESTICIDES, HERBICIDES, AND TRACE METALS

The chronic and acute toxicity of various pesticides, herbicides, and trace metals on aquatic animals have been determined for a number of species, including many of aquaculture interest. As discussed in Chapter 3, the water utilized for aquaculture should be essentially free of herbicides and pesticides, and it should not contain high levels of such trace metals as cadmium, zinc, copper, silver, lead, and mercury. No attempt will be made to review the literature on these subjects with respect to the toxicity of individual chemicals on species of aquaculture interest. There are some general guidelines that aquaculturists should follow, however, relative to

---

[100] For fish such as largemouth bass the availability of natural food organisms (zooplankton for young fish and stocked forage fishes for older fingerlings) is a compounding factor. In some cases higher bass densities might be accommodated if sufficient prey density could be provided. With fish that are fed prepared rations, the food factor can be obviated assuming that even distribution of the feed over the pond area can be affected.

ensuring that the species under culture are not contaminated by such substances. Sources of contamination are many. Among them are the activities of the culturist with regard to the use of agricultural chemicals in the control of diseases and nuisance vegetation.

Regulations in the United States regarding the use of biocides (pesticides and herbicides) have been implemented to ensure the safe use of those compounds. Nations other than the United States may have more strict, or in many cases, very relaxed or nonexistent controls on the use of these chemicals. In many cases, chemicals designed for use in land-based agriculture have been applied to aquaculture systems without sufficient testing of the potential negative impacts those chemicals might have. An example is the use of Nuvan (an organophosphate pesticide) to control sea lice in salmon net pens in the United Kingdom (Ross 1989).

The goal of the aquaculturist should be to employ water with no traces of pesticide or herbicide contamination and with acceptable levels of trace metals. Water normally contains trace metals, and some of them are required for proper nutrition as discussed in Chapter 6. Their mere presence is no cause for alarm unless recommended maximum levels are exceeded. The U.S. Environmental Protection Agency has set standards (subject to change) for trace metals in United States waters. That agency (or the appropriate local parallel organization) should be consulted in case questions arise.

Trace metal contamination is most likely to occur when an aquifer is contaminated by recharge water into which significant amounts of metallic substances are introduced (e.g., from an industrial plant or municipal waste treatment plant). The analysis of water samples for trace metals is somewhat expensive (several dollars per element), but a complete water chemistry analysis, including analysis for trace metals, makes sense if there is any reason to believe that the water source could be contaminated. Such analyses need not be repeated unless there is some reason to suspect that the situation may have changed or if the level of some trace element is marginal and requires monitoring. In the latter instance, it would not be necessary to analyze samples for elements other than the one(s) of interest.

Trace metals can, of course, occur in runoff water entering an aquaculture facility. Various sources of such contamination can be present, some of which may not be readily apparent. One example is abandoned dump sites which have become a significant problem in many regions where surface and groundwater intrusion have led to contamination of municipal water supplies as well as water bodies used by livestock, wildlife, wild fish, and of course, aquaculture species. Another source of potential metal contamination is from vehicle exhaust. Madigesky et al. (1991) found that crawfish *(Procambarus clarkii)* in wetland areas adjacent to roadways had elevated levels of lead, cadmium, and aluminum compared with control populations from culture facilities.

The required levels of trace metals vary from one group of aquaculture animals to another. Copper, for example, is found in much higher concentration in crustaceans than in vertebrates because the blood pigment of those invertebrates is based on the copper-containing compound hemocyanin. The copper requirement for vertebrates is not the same as that of the crustaceans because the former have blood pigment containing iron in the hemoglobin molecule. Even so, a fish that feeds primarily on crustaceans should be expected to have a higher level of copper in its body than a species that is herbivorous, but that does not seem to be the case. Rather than biological magnification, there is often a diminution of copper levels from crustaceans to teleost fishes (Stickney et al. 1975). The difference between the biological magnification that occurs in conjunction with some pesticides and the situation with respect to trace metals (at least in the case of copper) may relate to the presence or absence of a biochemical pathway that can handle high levels of substances that are not required. Fish evolved in a world where crustaceans required high levels of body copper, so the ability to eliminate copper from their bodies would be necessary in order for them to avoid toxicity from ingesting crustaceans. It should be noted that fish are susceptible to elevated copper levels in water, as are present when copper-containing herbicides are applied in excessive concentrations.

Not all trace metals respond in the same way as copper when consumed in relatively high concentrations by fish. Mercury, lead, arsenic, and others are not required for normal metabolism in any animal and are stored by fish when ingested. If ingested in sufficient quantities, they can reach toxic levels. Cadmium, a metal that is toxic when present in high concentration, also has no metabolic function in animals, but its chemistry is very similar to that of zinc. Cadmium appears to compete with zinc for enzyme active sites, and as the ratio of zinc to cadmium decreases, the cadmium begins to outcompete zinc and eventually denatures the enzymes, leading to a disruption of normal metabolism.

Pesticides and herbicides can enter relatively shallow aquifers from surface recharge and could conceivably contaminate deep wells if the water table approached the surface at any point. Most incidents of pesticide and herbicide contamination are likely to occur in conjunction with the use of contaminated surface waters for aquaculture. Locating aquaculture facilities in close proximity to cropland can result in significant contamination problems, particularly in areas where aerial spraying is conducted. Cotton fields, for example, may be sprayed 10 or more times during a growing season. The wind can carry the spray significant distances.

Some agriculturists have employed aquatic organisms in crop rotation. For example, Arkansas farmers have rotated rice, soybeans, and catfish. Crawfish have been rotated with rice in Louisiana. Since most of the herbicides and pesticides in present use are short-lived organophosphates, rather than such chlorinated and long-lived hydrocarbons as DDT, problems with residues should not be a problem. However, if the land that is being used in crop rotation has been converted from farmland to a dedicated pond site that previously received applications of chlorinated hydrocarbons, the residual toxicity can be present for decades. The difficulties and expense involved in cleaning up such sites can be inordinate.

Long-lived pesticides, such as DDT, are known to undergo biological magnification, a process by which they enter the food web at a low trophic level (e.g., when ingested by a target insect) and increase in concentration as they pass through successive trophic levels. Many of the pesticides and herbicides in use today are short-lived—that is, they are biologically active for only a few days rather than years—so they do not biologically magnify. Trace metals sometimes exhibit the phenomenon of biological magnification, though often they do not.

Fish are highly susceptible to most pesticides, but crustaceans are even more vulnerable since they are more closely related to the target insect species than are fish. Herbicides may kill aquaculture animals directly, or they can lead to an increase in biochemical oxygen demand by destroying the primary producers in the system. The result may be an oxygen depletion that will also lead to high mortality unless countermeasures are taken (see Chapter 4).

The impact of pesticides on nontarget species is usually relatively insignificant in ponds and other confined culture systems, particularly when the chemicals become neutralized before any effluent water is released to the surrounding environment. In cage and net-pen culture, use of pesticides can be expected to have a more widespread influence than on only the target organism. For example, net-pen Atlantic salmon farmers in Norway and Canada have used pesticides such as Nuvan to treat for epizootics of sea lice *(Lepeophtharius salmonia)*. The chemical is quite toxic to various other invertebrates including crabs and lobsters. Caution in its use and the use of other similar compounds has been recommended (Egidius and Moester (1987), even though one study reported that while laboratory experiments confirmed the toxicity of Nuvan to various invertebrates, larval and juvenile lobsters *(Homarus americanus)* housed adjacent to treated netpens were not killed (Cusack and Johnson 1990).

Herbicides containing arsenic were once commonly employed. Samples of sediments and biota can remain elevated for decades after the soils have become contaminated (Tanner and Clayton 1990). Mercurial fungicides were once widely used in the United States but are no longer available because of the dangers posed to humans. Many algicides still employ copper as the active ingredient. Since copper is more toxic to algae and other types of plants than it is to fish, it can be used safely, but caution should be exercised in the calculation of treatment rates. An error could lead to the addition of sufficient copper to create a toxic situation, resulting in the destruction of the culture species.

Herbicides, like pesticides, can become concentrated in the water and soil and may then affect nontarget species, including fish and other animals. The manner in which herbicides concentrate and dissipate and their effects on plankton, invertebrates, and fishes were reviewed by Ramaprabhu and Ramachandran (1990). The authors concluded that 2,4–D appeared to be the least persistent and was considered to be generally harmless to fish and other animals. It must be stressed that in order to ensure that toxicity to nontarget species is minimized, the chemicals must be properly applied.

Water quality conditions can have a significant effect on the toxicity of a given pesticide or herbicide. Temperature, pH, hardness, and alkalinity can impact the solubility or toxicity of a chemical. Before applying any chemical substance, the label should be carefully read and the directions precisely followed. In the United States, an applicator's license must be obtained before an individual is allowed to handle and apply certain pesticides.

DISPOSAL OF AQUACULTURE EFFLUENT

Water is released from aquaculture systems of all types, sometimes constantly, sometimes occasionally. In virtually all cases, the effluent from aquaculture systems varies in quality from that of the influent water.

For a number of years the emphasis on obtaining permits for aquaculture facilities was largely associated with procuring water rights and, in some instances, permits for the rearing of exotic species. For a number of years there has been discussion about controlling aquaculture effluents in the United States, but the Environmental Protection Agency (EPA) had not developed guidelines and basically deferred to the states for the development of policies. There were some attempts to promulgate national standards for effluents from aquaculture in the 1970s, but nothing was forthcoming at a national level. There was even literature available indicating that the overall impact from fish hatchery discharges was "less than significant" (Harris 1981). That statement was made with respect primarily to government hatcheries. McLaughlin (1981) discussed methods that had been examined for removing pollutants from the discharges of U.S. Fish and Wildlife Service salmon hatcheries and indicated that settling appeared to be effective. The requirement to settle solids in ponds or to otherwise remove settleable solids has been imposed by various state agencies and became a commonplace feature of many governmental as well as private aquaculture facilities. While settling is relatively effective, the rate at which it occurs in ponds is highly variable; achieving solids concentrations of less than 6 mg/l is difficult (Henderson and Bromage 1988) and depends, in part, on maintaining a low flow rate (less than 1 m/min if possible). Sieving has also been used to remove solids from aquaculture effluents and may have a role, particularly in high-intensity systems (Maekinen et al. 1988).

An interesting alternative to effluent treatment is the production of organisms in the outflowing water. Plants and filter-feeding invertebrates can remove significant amounts of dissolved and suspended particulate matter, respectively from aquaculture effluent. Their use can result in the production of a secondary crop (a form of polyculture). For example, aquaculture system effluents have been used to raise the algae *Ulva* (Vandermeulen and Gordin 1990), clams (Vaughan 1988, Shpigel and Fridman 1990), and oysters (Lam and Wang 1989, Shpigel et al. 1989, Wang 1990).

As the number, size, and production capability per unit volume of water grew, increasing demands for regulation of aquaculture effluents were heard. By 1988, the discharge from aquaculture facilities of particulate matter, at least, was reported to represent "a serious pollution problem" (Stachey 1988). Research revealed that even government hatcheries were not exempt from having measurable impacts on receiving waters. During a summer low-flow period, Kandra (1991) measured increases in temperature, pH, suspended solids, ammonia, organic nitrogen, total phosphorus, and chemical oxygen demand in the portion of a stream that was receiving inflow from a salmon hatchery. Effects on downstream benthos were seen by Kandra (1991) in Washington State and by Munro et al. (1985) in British Columbia, Canada. A study in Norway by Bergheim et al. (1984) indicated that the cleaning of fish tanks on a fish farm led to increased concentrations of pollutants up to several thousand times those in the inlet water. Marked increases in receiving stream phosphate concentration was seen in conjunction with trout farms in England by Carr and Goulder (1990b). Increases in bacteria downstream of the trout farms was also seen (Carr and Goulder 1990a).

Interest in the impacts of effluents from fish hatcheries and commercial fish farms has not been limited to areas where coldwater facilities were depositing their runoff water into clear rivers and streams. Various states began developing effluent guidelines, with some of the strictest being in the north-central states. Stoppage of the construction of a state fish hatchery in Michigan[101] because the effluent that would be produced was calculated to contain a level of phosphorus that would lead to eutrophication[102] of receiving water bodies is one example of how aquaculture effluent standards can be applied. Since removal of nutrients requires tertiary treatment, which is very expensive,[103] attempts have been made in recent years to reduce nutrient inputs. One way in which that can be accomplished is to design feeds in which the contained nutrients are efficiently utilized and not excreted.

Edsall and Smith (1989) added clinoptilolite to the diets of coho salmon at 5 and 10% to determine if the zeolite would reduce ammonia level in the effluent. Growth rate of the fish was not affected, but there was no significant change in ammonia production.

Environmental issues stemming from mariculture operations were recently addressed by the National Academy of Science (National Research Council 1992). Included in that assessment was not only effluent impacts, but also impacts from the introduction of exotic species and the use of feed additives.

During the 1980s, controversies surrounding the establishment of net-pen salmon farms in the United States developed in the Pacific Northwest and along the coast of Maine. In Washington State, a long list of objections included charges of visual, noise, and water pollution; sterilization of the bottom with waste feed and feces; possible creation of resistant strains of marine bacteria because of the use of antibiotics; and the spread of diseases from cultured to wild fish. Each of these issues and others, including impacts on recreation and commercial fishing, was addressed in an Environmental Impact Statement (EIS) issued on behalf of the Washington Department of Fisheries in 1990 (Parametrix 1990). Various alternatives, including the notion of placing diapers around netpens to collect falling feed and feces, were considered in the EIS. Most of the objections were readily ad-

---

[101] Harry Westers, personal communication.

[102] Eutrophication is the process by which lakes age, moving from low productivity, nutrient-poor waters (oligotrophic) to waters of high productivity that are nutrient-rich (eutrophic). Eventually, the process leads to filling in of the lake basin and the creation of marshland. Increased inputs of phosphorus and nitrogen from activities associated with humans (including the raising of captive fishes) can accelerate the process by adding nutrients to waters that would otherwise be less eutrophic.

[103] Primary treatment involves settling suspended material, and is the process that has been accepted for the treatment of aquaculture wastes in the past. Secondary treatment involves the release of nutrients from organic matter in the water that are used for the production of bacteria and other organisms which can then be removed from the system as sewage sludge. To remove dissolved nutrients such as phosphorus requires the employment of higher technology, such as ion exchange resins, which will scrub the nutrients from the water.

dressable by proper siting of net pen facilities. In any event, because of the time and money involved in working through the permitting process, expansion of the salmon net-pen industry in the State of Washington has been stalled.

As the wastes from net-pens can impact the local environment, so too can effluents from various sources have a negative effect on the fishes being reared in those net pens and on shellfish beds. Sewage effluent can be particularly devastating to shellfish culturists (Young 1988) because the animals can concentrate pathogens, making them unfit for human consumption until they can be properly depurated.[104]

In 1989, the EPA determined that the effluents from fish farms fall under the National Pollution Discharge Elimination System (NPDES) and that permits would be required for all facilities that produce over 9,091 kg (20,000 lb) annually. The EPA presently follows state pollution control guidelines in the issuance of NPDES permits. Monitoring is generally required in conjunction with NPDES permits. Typically, permit holders are required to monitor biochemical oxygen demand and chemical oxygen demand, total ammonia, pH, total suspended solids, and settleable solids. Some states also require the monitoring of phosphate and nitrate. Since state requirements vary and the requirements for marine and freshwater aquaculture facilities in coastal states may also vary, the prospective aquaculturist would be best advised to contact the appropriate state agency. Most states now have a designated Aquaculture Coordinator. While there is no consistency with respect to which department that person might be employed in within a given state, many work through their state's Department of Agriculture, so that department would be a good place to begin.

LITERATURE CITED

Alderdice, D. F. 1988. Osmotic and ionic regulation in teleost eggs and larvae. pp. 163–251, In: W. S. Hoar and D. J. Randall (Eds.). Fish physiology Vol. 11, Part A. Academic Press, New York.

Alexis, M. N., E. Papaparaskeva-Papoutsoglou, and S. Papoutsoglou. 1984. Influence of acclimation temperature on the osmotic regulation and survival of rainbow trout *(Salmo gairdneri)* rapidly transferred from fresh water to salt water. Aquaculture, 40: 333–341.

Allen, K. O., and J. W. Avault, Jr. 1969. Effects of salinity on growth and survival of channel catfish, *Ictalurus punctatus*. Proc. Southeast. Assoc. Game Fish Comm., 23: 319–331.

APHA. 1989. Standard methods. 17th edition. American Public Health Association, Washington, D.C. 1,467 p.

Andrews, J. W., and R. R. Stickney. 1972. Interactions of feeding rates and environmental temperature on growth, food conversion, and body composition of channel catfish. Trans. Am. Fish. Soc., 101: 94–99.

Arce, R. G., and C. E. Boyd. 1975. Effects of agricultural limestone on water chemistry,

---

[104] Depuration is the elimination of a toxin or pathogen from a carrier organism and will be discussed in more detail in Chapter 9.

phytoplankton productivity, and fish production in ponds. Trans. Am. Fish. Soc., 104: 308–312.

Assem, H., and W. Hanke. 1979. Concentrations of carbohydrates during osmotic adjustment of the euryhaline teleost *Tilapia mossambica*. Comp. Biochem. Physiol., 64**A**: 5–16.

Avault, J. W., Jr., and E. W. Shell. 1968. Preliminary studies with the hybrid tilapia *Tilapia nilotica* × *Tilapia mossambica*. FAO Fish. Rep., 44: 237–242.

Beacham, T. D., and R. E. Withler. 1991. Genetic variation in mortality of chinook salmon, *Oncorhynchus tshawytscha* (Walbaum), challenged with high water temperatures. Aquacult. Fish. Manage., 22: 125–133.

Bell, G. M., and W. S. Hoar. 1950. Some effects of ultraviolet radiation on sockeye salmon eggs and alevins. Can. J. Res., 28: 35–43.

Bergheim, A. H. Hustveit, A. Kittelsen, and A. R. Selmer-Olsen. 1984. Estimated pollution loadings from Norwegian fish farms. 2. Investigations 1980–1981. Aquaculture, 36: 157–168.

Bilton, H. T. 1972. A comparison of body and scale growth of young sockeye salmon *(Oncorhynchus nerka)* reared under two light periods and in total darkness. Fisheries Research Board of Canada Technical Report No. 330. Fisheries Research Board of Canada, Ottawa.

Bolla, S., and I. Holmefjord. 1988. Effect of temperature and light on development of Atlantic halibut larvae. Aquaculture, 74: 355–358.

Boyd, C. E. 1990. Water quality in ponds for aquaculture. Agricultural Experiment Station, Auburn University, Auburn, Alabama. 482 p.

Boyd, C. E., and W. D. Hollerman. 1982. Influence of particle size of agricultural limestone on pond liming. Proc. Annu. Conf. Southeast. Assoc. Fish Wildl. Agencies, 36: 196–201.

Brown, J. H., J. F. Wickins, and M. H. MacLean. 1991. The effect of water hardness on growth and carapace mineralization of juvenile freshwater prawns, *Macrobrachium rosenbergii* de Man. Aquaculture, 95: 329–345.

Cairns, J., Jr. 1967. Suspended solids standards for the protection of aquatic organisms. Purdue Univ. Eng. Bull., 129: 16–27.

Carr, D. J., and R. Goulder. 1990a. Fish-farm effluents in rivers—1. Effects on bacterial populations and alkaline phosphatase activity. Water Res., 24: 631–638.

Carr, D. J., and R. Goulder. 1990b. Fish-farm effluents in rivers—2. Effects on inorganic nutrients, algae and the macrophyte *Ranunculus penicillatus*. Water Res., 24: 639–647.

Charmentier, G. 1987. L'osmoregulation chez les crevettes Penaeidae (Crustacea, Decapoda). pp. 179–186, In: L. Laubier (Ed.). 1976–1986: Dix ans de Recherche en Aquaculture. Oceanis (Documents Oceanographiques), 13(2): 179–186.

Charmantier-Daures, M., P. Thuet, G. Charmentier, and J.-P. Trilles. 1988. Tolerance a la salinite et osmoregulation chez les post-larves de *Penaeus japonicus* et *P. chinensis*. Effect de la temperature. Ressour. Vivantes Aquat., 1: 267–276.

Chervinski, J. 1977. Note on the adaptability of silver carp—*Hypophthalmichthys molitrix* (Val.) to various salinity concentrations. Aquaculture, 11: 179–182.

Chervinski, J. 1984. Salinity tolerance of young catfish, *Clarias lazera* (Burchell). J. Fish Biol., 25: 147–149.

Chervinski, J., and R. R. Stickney. 1981. Overwintering facilities for tilapia in Texas. Prog. Fish-Cult., 43: 20–21.

Chesney, E. J., Jr. 1989. Estimating the food requirements of striped bass larvae *Morone saxatilis:* effects of light, turbidity and turbulence. Mar. Ecol. Prog. Ser., 53: 191–200.

Chimits, P. 1957. The tilapias and their culture. A second review and bibliography. FAO Fish. Bull., 10: 1–24.

Choo, P. S. 1987. Effects of salinity on the spawning, egg-incubation and larviculture of *Penaeus merguiensis.* Fish. Bull. Dept Fish. (Malaysia), No. 52:1–9.

Chung, K. S. 1980. A note on salinity preference of *P. braziliensis.* Bull. Jpn. Soc. Sci. Fish., 46: 389.

Clarke, W. C., J. E. Shelbourn, and J. R. Brett. 1978. Growth and adaptation to seawater in 'underyearling' sockeye *(Oncorhynchus nerka)* and coho *(O. kisutch)* salmon subjected to regimes of constant or changing temperature and daylength. Can. J. Zool., 56: 2413–2421.

Conte, F. P. 1969. Salt secretion. pp. 241–292, In: W. S. Hoar and D. J. Randall (Eds.). Fish physiology, Vol. 1. Academic Press, New York.

Corley-Smith, G. E. 1989. Delayed photoperiod, seapen survival and growth rate of 0–age coho salmon *(Oncorhynchus kisutch),* under intensive culture conditions. Aquaculture, 82: 375–376.

Cusack, R., and G. Johnson. 1990. A study of dichlorvos (Nuvan: 2,2 dichloroethenyl dimethyl phosphate), a therapeutic agent for the treatment of salmonids infected with sea lice *(Lepeophtherius salmonis).* Aquaculture, 90: 101–112.

Dalla Via, G. J. 1986a. Salinity responses of the juvenile penaeid shrimp (prawn) *P. japonicus.* 2. Oxygen consumption and estimations of productivity. Aquaculture, 55: 297–305.

Dalla Via, G. J. 1986b. Salinity responses of the juvenile penaeid shrimp (prawn) *P. japonicus.* 2. Free amino acids. Aquaculture, 55: 307–316.

Dey, D. B., and D. M. Damkaer. 1990. Effects of spectral irradiance on the early development of chinook salmon. Prog. Fish-Cult., 52: 141–154.

Dickhoff, W. W., and C. V. Sullivan. 1987. Involvement of the thyroid gland in smoltification with special reference to metabolic and developmental processes. pp. 197–210, In: M. J. Dadswell, R. J. Klauda, C. M. Moffit, R. L. Saunders, R. A. Rulifson, and J. E. Cooper (Eds.). Common strategies of anadromous and catadromous fishes. American Fisheries Society, Bethesda, Maryland.

Diwan, A. D., and A. Laxminarayana. 1989. Osmoregulatory ability of *P. indicus* H. Milne-Edwards in relation to varying salinities. Proc. Indian Acad. Sci. Anim. Sci., 98: 105–111.

Duston, J., R. Saunders, P. Harmon, and D. Knox. 1989. Increase in photoperiod and temperature in winter advance completion of some aspects of smoltification in Atlantic salmon. Bull. Aquacult. Assoc. Can., 89(3): 19–21.

Edsall, D. A., and C. E. Smith. 1989. Effects of dietary clinoptilolite on levels of effluent ammonia from hatchery coho salmon. Prog. Fish-Cult., 51: 98–100.

Egan, B. D. 1990. All dredged up and no place to go (disposal of farmed salmon killed by algae blooms, west coast of Canada). Bull. Aquacult. Assoc. Can., 90(1): 7–15.

Egidius, E., and B. Moester. 1987. Effect of Neguvon and Nuvan treatment on crabs *(Cancer pagurus, C. (Carcinus) meanes),* lobster *(Homarus gammarus)* and blue mussel *(Mytilus edulis).* Aquaculture, 60: 165–168.

Emmerson, W. D., D. P. Hayes, and M. Ngonyame. 1983. Growth and maturation of *Penaeus indicus* under blue and green light. S. Afr. J. Zool., 18: 71–75.

Ewart, K. V., and G. L. Fletcher. 1989. Further study of the various antifreezes may help to optimise transfer of freeze resistance. Bull. Aqucult. Assoc. Can., 89(3): 25–27.

Fletcher, G. L., S.-J. Du, M. A. Shears, C. L. Hew, and P. L. Davies. 1990. Antifreeze and growth hormone gene transfer in Atlantic salmon. Bull. Aquacult. Assoc. Can., 90(1): 70–71.

Folmar, L. C., and W. W. Dickhoff. 1980. The parr-smolt transformation (smoltification) and seawater adaptation in salmonids. A review of selected literature. Aquaculture, 21: 1–17.

Foskett, J. K., and C. Scheffey. 1982. The chloride cell: definitive identification as a salt-secretory cell in teleosts. Science, 215: 164–166.

Galman, O. R., and R. R. Avtalion. 1983. A preliminary investigation of the characteristics of red tilapias from the Philippines and Taiwan. pp. 291–301, In: In: L. Fishelson and Z. Yaron (compilers). Proceedings, International Symposium on Tilapia in Aquaculture, Nazareth, Israel, May 8–13. Tel Aviv University Press, Tel Aviv.

Goff, G. P., D. A. Methven, and J. A. Brown. 1989. Low temperature tolerance of Atlantic halibut, *Hippoglossus hippoglossus*, at ambient ocean temperatures in Newfoundland. Bull. Aquacult. Assoc. Can., 89(3): 53–55.

Gonzal, A. C., E. V. Aralar, and J. M. F. Pavico. 1987. The effects of water hardness on the hatching and viability of silver carp *(Hypopthhalmichthys molitrix)* eggs. Aquaculture, 64: 111–118.

Grizzle, J. M., A. C. Mauldin, II, D. Young, and E. Henderson. 1985. Survival of juvenile striped bass *(Morone saxatilis)* and *Morone* hybrid bass *(Morone chrysops* × *Morone saxatilis)* increased by addition of calcium to soft water. Aquaculture, 46: 167–171.

Hall, L. W., Jr. 1991. A synthesis of water quality and contaminants data on early life stages of striped bass, *Morone saxatilis*. Rev. Aquat. Sci., 4: 261–288.

Hanson, J. A., and H. L. Goodwin. 1977. Shrimp and prawn farming in the western hemisphere. Dowden, Hutchinson & Ross, Stroudsburg, Pennsylvania. 439 p.

Harrell, L. W., R. A. Elston, T. M. Scott, and M. T. Wilkinson. 1986. A significant new systemic disease of net-pen reared chinook salmon *(Oncorhynchus tshawytscha)* brood stock. Aquaculture, 55: 249–262.

Harris, J. 1981. Federal regulation of fish hatchery effluent quality. pp. 157–161, In: L. J. Allen, and E. C. Kinney (Eds.). Proceedings of the Bio-engineering Symposium for Fish Culture. Fish Culture Section, American Fisheries Society, Bethesda, Maryland.

Hedgpeth, J. W. 1957. Classification of marine environments. pp. 17–27, In: J. W. Hedgpeth (Ed.). Treatise on marine ecology and paleoecology, Vol. 1. Memoir No. 69. Geological Society of America, New York.

Henderson, J. P., and N. R. Bromage. 1988. Optimising the removal of suspended solids from aquacultural effluents in settlement lakes. Aquacult. Eng., 7: 167–181.

Henderson-Arzapalo, A., R. R. Stickney, and D. H. Lewis. 1980. Immune hypersensitivity in intensively cultured *Tilapia* species. Trans. Am. Fish Soc., 109: 244–247.

Hoar, W. S. 1976. Smolt transformation: evolution, behavior, and physiology. J. Fish. Res. Board Can., 33: 1234–1252.

Hoar, W. S. 1989. The physiology of smolting salmonids. pp. 275–343, In: W. S. Hoar, and D. J. Randall (Eds.). The physiology of developing fish. Fish physiology, Vol. 11, Part B. Academic Press, New York.

Hosmer, M. J., J. G. Stanley, and R. W. Hatch. 1979. Effects of hatchery procedures on later return of Atlantic salmon to rivers in Maine. Prog. Fish-Cult., 41: 115–119.

Huse, I., A. Bjordal, A. Fernoe, and D. Furevik. 1988. The effect of shading in pen rearing of Atlantic salmon *(Salmo salar)*. ICES Council Meeting (Collected Papers). International Council for the Exploration of the Sea, Copenhagen. 11 p.

Johnson, S. K. 1976. Laboratory evaluation of several chemicals as preventatives of ich disease. pp. 91–96, In: Proceedings of the 1976 Fish Farming Conference and Annual Convention of the Catfish Farmers of Texas. Texas A&M University, College Station.

Johnston, C. E., and R. L. Saunders. 1981. Parr-smolt transformation of yearling Atlantic salmon *(Salmo salar)* at several rearing temperatures. Can. J. Fish. Aquat. Sci., 38: 1189–1198.

Jones, F. V., and K. Strawn. 1984. Survival of cage cultured black drum, Atlantic croaker, and striped mullet in upper estuarine water in relation to life history strategy. J. World Maricult. Soc., 14: 233–243.

Kandra, W. 1991. Quality of salmonid hatchery effluents during a summer low-flow season. Trans. Am. Fish. Soc., 120: 43–51.

Kane, A. S., R. O. Bennett, and E. B. May. 1990. Effect of hardness and salinity on survival of striped bass larvae. N. Am. J. Fish. Manage., 10: 67–71.

Kent, M. L. 1990. Netpen liver disease (NLD) of salmonid fishes reared in sea water: species susceptibility, recovery, and probable cause. Dis. Aquat. Org., 8: 21–28.

Kerby, J. H. 1986. Striped bass and striped bass hybrids. pp. 127–147, In: R. R. Stickney (Ed.). Culture of nonsalmonid freshwater fishes. CRC Press, Boca Raton, Florida.

Kerby, J. H. 1992. Striped bass and striped bass hybrids. pp. 251–306, In: R. R. Stickney (Ed.). Culture of nonsalmonid freshwater fishes (revised edition). CRC Press, Boca Raton, Florida.

Ketola, H. G., D. Longacre, A. Greulich, L. Phetterplace, and R. Lashomb. 1988. High calcium concentration in water increases mortality of salmon and trout eggs. Prog. Fish-Cult., 50: 129–135.

Khater, A. A., and R. O. Smitherman. 1988. Cold tolerance and growth of three strains of *Oreochromis niloticus*. pp. 215–218, In: R. S. V. Pullin, T. Bhukaswan, K. Tonguthai, and J. L. Maclean (Eds.). The Second International Symposium on Tilapia in Aquaculture, Bangkok, Thailand, March 16–20, 1987. ICLARM Conference Proceedings No. 15, International Center for Living Aquatic Resources Management, Manila, Philippines.

Kilambi, R. V., and Z. Zdinak. 1980. The effect of acclimation on the salinity tolerance of grass carp *Ctenopharyngodon idella* (Gur. and Val.). J. Fish Biol., 16: 171–175.

Kilambi, R. W., J. Noble, and C. E. Hoffman. 1970. Influence of temperature and photoperiod on growth, food consumption and food conversion efficiency of channel catfish. Proc. Southeast. Assoc. Game Fish Comm., 24: 519–531.

King, M. J., M. H. Kao, J. A. Brown, and G. L. Fletcher. 1989. Lethal freezing temperatures of fish: limitations to seapen culture in Atlantic Canada. Bull. Aquacult. Assoc. Can., 89(3): 47–49.

Kuriakose, P. S. 1980. Salinity tolerance of brown mussel *Perna indica*. Symp. Ser. Mar. Biol. Assoc. India, 6: 705.

Lam, C.-Y., and J.-K. Wang. 1989. The effects of feed water flow rate on the growth of aquaculture *Crassostrea virginica* in Hawaii. J. Shellfish Res., 8: 476.

Lam, T. J., and R. Sharma. 1985. Effects of salinity and thyroxine on larval survival, growth and development in the carp, *Cyprinus carpio*. Aquaculture, 44: 201–212.

Landau, M. 1992. Introduction to aquaculture. John Wiley & Sons, New York. 440 p.

Liu, H. W., R. R. Stickney, and S. D. Smith. 1991. A note on the artificial spawning of Pacific halibut. Prog. Fish-Cult., 53:189–192.

Lockwood, A. P. M. 1967. Aspects of the physiology of Crustacea. W. H. Freeman, San Francisco. 328 p.

Loosanoff, V. L., and F. D. Tommers. 1948. Effect of suspended silt and other substances on the rate of feeding of oysters. Science, 107: 69–70.

Maceina, M. J., and J. V. Shireman. 1979. Grass carp: effects of salinity on survival, weight loss, and muscle water content. Prog. Fish-Cult., 41: 69–73.

Madigesky, S. R., X. Alvarez-Hernandez, and J. Glass. 1991. Lead, cadmium, and aluminum accumulation in the red swamp crayfish *Procambarus clarkii* G. collected from roadside drainage ditches in Louisiana. Arch. Environ. Contam. Toxicol., 20: 253–259.

Maekinen, T., S. Lindgren, and P. Eskelinen. 1988. Sieving as an effluent treatment method for aquaculture. Aquacult. Eng., 7: 367–377.

Mazik, P. M., B. A. Simco, and N. C. Parker. 1991. Influence of water hardness and salts on survival and physiological characteristics of striped bass during and after transport. Trans. Am. Fish. Soc., 120: 121–126.

McBay, L. G. 1961. The biology of *Tilapia nilotica* Linnaeus. Proc. Southeast. Assoc. Game Fish Comm., 15: 208–218.

McCormick, S. D., R. L. Saunders, E. B. Henderson, and P. R. Harmon. 1987. Photoperiod control of parr-smolt transformation in Atlantic salmon *(Salmo salar):* changes in salinity tolerance, gill Na^+, K^+-ATPase activity, and plasma thyroid hormones. Can. J. Fish. Aquat. Sci., 44: 1452–1458.

McGeachin, R. B., and R. I. Wicklund. 1987. Growth of *Tilapia aurea* in seawater cages. J. World Aquacult. Soc., 18: 31–34.

McGeer, J. C., L. Baranyi, and G. K. Iwama. 1991. Physiological responses to challenge tests in six stocks of coho salmon *(Oncorhynchus kisutch)*. Can. J. Fish. Aquat. Sci., 48: 1761–1771.

McLaughlin, T. W. 1981. Hatchery effluent treatment—U.S. Fish and Wildlife Service. pp. 167–173, In: L. J. Allen and E. C. Kinney (Eds.). Proceedings of the Bio-engineering Symposium for Fish Culture. Fish Culture Section, American Fisheries Society, Bethesda, Maryland.

Meriwether, F. H., II, E. D. Scura, and W. Y. Okamura. 1984. Cage culture of red tilapia in prawn and shrimp ponds. J. World Maricult. Soc., 15: 254–265.

Munro, K. A., S. C. Samis, and M. D. Nassichuk. 1985. The effects of hatchery effluents on water chemistry, periphyton and benthic invertebrates of selected British Columbia streams. Canadian Manuscript Report of Fisheries and Aquatic Sciences, No. 1830. Department of Fisheries and Oceans, Ottawa. 220 p.

National Research Council. 1992. Marine aquaculture. National Academy Press, Washington, D.C. 290 p.

Neill, W. H. 1987. Environmental requirements of red drum. Section IV, pp. IV1–IV8, In:

G. W. Chamberlain, R. J. Miget, and M. G. Haby (Eds.). Manual on red drum aquaculture. Texas Agricultural Extension Service and Sea Grant College Program, Texas A&M University, College Station.

Noeske, T. A., and R. E. Spieler. 1984. Circadian feeding time affects growth of fish. Trans. Am. Fish. Soc., 113: 540–544.

Noeske, T. A., D. A. Erickson, and R. E. Spieler. 1981. The time-of-day goldfish receive a single daily meal affects growth. J. World Maricult. Soc., 12(2): 73–77.

Oertzen, J.-A. 1985. Resistance and capacity adaptation of juvenile silver carp, *Hypophthalmichthys molitrix* (Val.), to temperature and salinity. Aquaculture, 44: 321–332.

Ogle, J., K. Beaugez, and T. D. McIlwain. 1988. Survival of *Penaeus vannamei* postlarvae challenged with low-salinity water. J. Shellfish Res., 7: 172–173.

Olsen, S. 1983. Abalone and scallop culture in Puget Sound. J. Shellfish Res., 3: 113.

Page, J. W., and J. W. Andrews. 1975. Effects of light intensity and photoperiod on growth or normally pigmented and albino channel catfish. Prog. Fish-Cult., 37: 121–125.

Parry, G. 1960. The development of salinity tolerance in salmon, *Salmo salar* L. and some related species. J. Exp. Biol., 37: 425–434.

Parametrix, Inc. 1990. Final programmatic environmental impact statement: fish culture in floating net pens. Washington Department of Fisheries, Olympia. 161 p.

Patiño, R., C. B. Schreck, J. L. Banks, and W. S. Zaugg. 1986. Effects of rearing conditions on the developmental physiology of smolting coho salmon. Trans. Am. Fish. Soc., 115: 828–837.

Payne, A. I. 1983. Estuarine and salt tolerant tilapias. pp. 534–543, In: L. Fishelson and Z. Yaron (compilers). International Symposium on Tilapia in Aquaculture, Nazareth, Israel, May 8–13. Tel Aviv University Press, Tel Aviv.

Pearse, A. S., and G. Gunter. 1957. Salinity. pp. 129–157, In: J. W. Hedgpeth (Ed.). Treatise on marine ecology and paleoecology, Vol. 1. Memoir 69. Geological Society of America, New York.

Perry, W. G., Jr. 1967. Distribution and relative abundance of blue catfish, *Ictalurus furcatus,* and channel catfish, *Ictalurus punctatus,* with relation to salinity. Proc. Southeast. Assoc. Game Fish Comm., 21: 436–444.

Perry, W. G., Jr. 1969. Food habits of blue and channel catfish collected from a brackishwater habitat. Prog. Fish-Cult., 31: 47–50.

Perry, W. G., Jr., and J. W. Avault, Jr. 1968. Preliminary experiments on the culture of blue, channel, and white catfish in brackish water ponds. Proc. Southeast. Assoc. Game Fish. Comm., 22: 397–406.

Perry, W. G., Jr., and J. W. Avault, Jr. 1969. Culture of blue, channel and white catfish in brackish water ponds. Proc. Southeast. Assoc. Game Fish Comm., 23: 592–605.

Perry, W. G., Jr., and J. W. Avault, Jr. 1971. Polyculture studies with blue, white and channel catfish in brackish water ponds. Proc. Southeast. Assoc. Game Fish Comm., 25: 466–479.

Perschbacher, P. W. 1992. A review of seawater acclimation procedures for commercially important euryhaline tilapias. Asian Fish. Sci., 5: 241–248.

Philippart, J.-C., and J.-C. Ruwet. 1982. Ecology and distribution of tilapias. pp. 15–59, In: R. S. V. Pullin and R. H. Lowe-McConnell (Eds.). The biology and culture of tilapias. ICLARM Conference Proceedings Vol. 7, International Center for Living Aquatic Resources Management, Manila, Philippines.

Piper, R. G., I. B. McElwain, L. E. Orme, J. P. McCraren, L. G. Fowler, and J. R. Leonard. 1982. Fish hatchery management. U.S. Fish and Wildlife Service, Washington, D.C. 517 p.

Potts, W. I. M., M. A. Foster, P. P. Rady, and G. P. Howell. 1967. Sodium and water balance in the cichlid teleost *Tilapia mossambica*. J. Exp. Biol., 47: 461–470.

Prasad, C. V. N., P. N. Prasad, and B. Neelakantan. 1988. Salinity tolerance in the larvae (mysis 3) of banana prawn *Penaeus merguiensis* (DeMan). Seafood Export J., 20: 21–26.

Pritchard, D. W. 1967. What is an estuary: physical viewpoint. pp. 3–5, In: G. H. Lauff (Ed.). Estuaries. American Association for the Advancement of Science, Washington, D.C.

Ramaprabhu, T., and Ramachandran, V. 1990. Observations on the effects and persistence of herbicides in aquaculture. Comp. Physiol. Ecol., 8: 223–233.

Rees, R. A., and S. F. Cook. 1982. Effects of sunlight intensity on survival of striped bass × white bass fry. Proc. Annu. Conf. Southeast. Assoc. Fish Wildl. Agencies, 36: 83–94.

Robaina, G. O. 1983. Efectos de la salinidad y la temperatura en la sobrevivencia del camaron *Penaeus brasiliensis* Latreille (Crustacea, Decapoda, Penaeidea). Rev. Latinoam. Acuicult., 17: 25–37.

Ross, A. 1989. Nuvan use in salmon farming: the antithesis of the precautionary principle. Mar. Pollut. Bull., 20: 372–374.

Saha, K. C., D. N. Chakraborty, B. K. De, and S. Shakraborty. 1984a. Studies on the salinity tolerance of fry of Indian major carps in captivity. Indian J. Fish., 11: 247–248.

Saha, K. C., D. N. Chakraborty, A. Mahalanabish, D. K. Nag, G. C. Paul, and H. B. Dey. 1984b. Studies on the potentiality of brackish water fish farming at Junput sea coast, Contai, West Bengal. Indian J. Fish., 11: 249–255.

Saha, K. C., D. N. Chakraborty, B. K. Jana, B. K. De, J. N. Misra, B. K. Pal, and A. K. Talapatra. 1984c. Studies on the potentiality of brackish water fish farming along Alampore coast, West Bengal. Indian J. Fish., 11: 256–267.

Saunders, R. L., and P. R. Harmon. 1988. Extended daylength increases postsmolt growth of Atlantic salmon. World Aquacult., 19(4): 72–73.

Saunders, R. L., and P. R. Harmon. 1990. Influence of photoperiod on growth of juvenile Atlantic salmon and development of salinity tolerance during winter-spring. Trans. Am. Fish. Soc., 119: 689–697.

Shekk, P. B., N. I. Kulikova, and V. I. Rudenko. 1990. Age-related changes in reaction of the Black Sea golden gray mullet, *Liza aurata,* to low temperature. J. Ichthyol., 30: 132–147.

Shikin, Yu. N., and N. N. Lapina. 1982. Locomotor and feeding activity of juvenile sevruga, *Acipenser stellatis* (Acipenseridae), with increasing salinity. J. Ichthyol., 22: 138–142.

Shpigel, M., and B. Fridman. 1990. Propagation of the Manila clam *(Tapes semidecussatus)* in the effluent of fish aquaculture ponds in Eilat, Israel. Aquaculture, 90: 113–122.

Shpigel, M., J. J. Lee, and B. Soohoo. 1989. Fish-oyster polyculture in warm water marine ponds. J. Shellfish Res., 8: 481.

Simco, B. A., J. H. Williamson, G. J. Carmichael, and J. R. Tomasso. 1986. Centrarchids.

pp. 73–89, In: R. R. Stickney (Ed.). Culture of nonsalmonid freshwater fishes. CRC Press, Boca Raton, Florida.

Skiftesvik, A. B., I. Opstad, O. Begh, K. Pittman, and L. H. Skjolddal. 1990. Effects of light on the development, activity and mortality of halibut (*Hippoglossus hippoglossus* L.) yolk sac larvae. International Council for the Exploration of the Sea, Copenhagen. 15 p.

Smith, T. I. J., W. E. Jenkins, and R. W. Haggerty. 1986. Growth and survival of juvenile striped bass *(Morone saxatilis)* × white bass *(M. chrysops)* hybrids reared at different salinities. Proc. Annu. Conf. Southeast. Assoc. Fish Wildl. Agencies, 40: 143–151.

Spotte, S. H. 1970. Fish and invertebrate culture. John Wiley & Sons, New York. 145 p.

Stachey, D. 1988. Factors influencing the design of effluent quality control facilities for commercial aquaculture. p. 54, In: Proceedings of the Aquaculture International Congress and Exposition, Vancouver Trade and Convention Center, Vancouver, British Columbia, September 6–9.

Stefansson, S. O., and T. Hansen. 1989. Effects of tank colour on growth and smoltification of Atlantic salmon (*Salmo salar* L.). Aquaculture, 81: 379–386.

Stefansson, S. O., R. Nortvedt, T. J. Hansen, and G. L. Taranger. 1990. First feeding of Atlantic salmon, *Salmo salar* L., under different photoperiods and light intensities. Aquacult. Fish. Manage, 21: 435–441.

Stefansson, S. O., A. E. Berg, T. Hansen, and R. L. Saunders. 1992. The potential for development of salinity tolerance in underyearling Atlantic salmon *(Salmo salar)*. World Aquacult., 23(2): 52–55.

Stickney, R. R. 1986. Tilapia. pp. 57–72, In: R. R. Stickney (Ed.). Culture of nonsalmonid freshwater fishes. CRC Press, Boca Raton, Florida.

Stickney, R. R. 1991a. Effects of salinity on aquaculture production. pp. 105–132, In: D. E. Brune and J. R. Tomasso (Eds.). Aquaculture and water quality. Advances in World Aquaculture, Vol. 3. World Aquaculture Society, Baton Rouge, Louisiana.

Stickney, R. R. 1991b. Growout of Pacific salmon in net-pens. pp. 71–83, In:. R. R. Stickney (Ed.). Culture of salmonid fishes. CRC Press, Boca Raton, Florida.

Stickney, R. R., and J. W. Andrews. 1971. The influence of photoperiod on growth and food conversion of channel catfish. Prog. Fish-Cult., 33: 204–205.

Stickney, R. R., and H. W. Liu. 1991. Spawning and egg incubation of Pacific halibut. World Aquacult., 22(4): 46–48.

Stickney, R. R., and N. K. Person. 1985. An efficient heating method for recirculating water systems. Prog. Fish-Cult., 47: 71–73.

Stickney, R. R., and B. A. Simco. 1971. Salinity tolerance of catfish hybrids. Trans. Am. Fish. Soc., 100: 790–792.

Stickney, R. R., and D. B. White. 1973. Effects of salinity on growth of *Paralichthys lethostigma* postlarvae reared under aquaculture conditions. Proc. Southeast. Assoc. Game Fish Comm., 27: 532–540.

Stickney, R. R., and D. B. White. 1975. Ambicoloration in tank cultured flounder, *Paralichthys dentatus*. Trans. Am. Fish. Soc., 104: 158–160.

Stickney, R. R., H. L. Windom, D. B. White, and F. E. Taylor. 1975. Heavy-metal concentrations in selected Georgia estuarine organisms with comparative food-habit data. pp. 257–267, In: F. G. Howell, J. B. Gentry, and M. H. Smith (Eds.). Mineral cycling in southeastern ecosystems. U.S. Energy Research and Development Administration, CONF–740513, Springfield, Virginia.

Stickney, R. R., H. W. Liu, and S. D. Smith. 1991. Recent advances in halibut (*Hippoglossus* spp.) culture. pp. 9–13, In: R. S. Svrjcek (Ed.)., Marine Ranching: Proceedings of the 17th U.S.-Japan Meeting on Aquaculture; Ise, Mie Prefecture, Japan, October 16, 17, and 18, 1988. NOAA Tech. Rep. NMFS 102. U.S. Department of Commerce, Washington, D.C.

Sverdrup, H. N., M. W. Johnson, and R. H. Fleming. 1942. The oceans. Prentice-Hall, Englewood Cliffs, New Jersey. 1087 p.

Tandler, A., and S. Helps. 1985. The effects of photoperiod and water exchange on growth and survival of gilthead sea bream (*Sparus aurata,* Linnaeus; Sparidae) from hatching to metamorphosis in mass rearing systems. Aquaculture, 48: 71–82.

Tanner, C. C., and J. S. Clayton. 1990. Persistence of arsenic 24 years after sodium arsenite herbicide application to Lake Rotorua, Milton, New Zealand.

Turner, P. R. 1988. Growth of channel catfish in saline groundwaters of the Pecos Valley of New Mexico. New Mexico Water Resources Research Institute Report No. 231. New Mexico State University, Las Cruces. 54 p.

Van der Linden, A., R. Blust, and W. Decleir. 1985. The influence of light on the hatching of *Artemia* cysts (Anostraca: Branchiopoda: Crustacea). J. Exp. Mar. Biol. Ecol., 92: 207–214.

Vandermeulen, H., and H. Gordin. 1990. Ammonium uptake using *Ulva* (Chlorophyta) in intensive fishpond systems: mass culture and treatment of effluent. J. Appl. Phycol., 2: 363–374.

Vasquez, D. E., D. B. Rouse, and W. A. Rogers. 1989. Growth response of *Macrobrachium rosenbergii* to different levels of hardness. J. World Aquacult. Soc., 20: 90–92.

Vaughan, D. E. 1988. Clam culture: state of the art in Florida, U.S.A. J. Shellfish Res., 7: 546.

Villarreal, C. A., J. E. Thorpe, and M. S. Miles. 1988. Influence of photoperiod on growth changes in juvenile Atlantic salmon, *Salmo salar* L. J. Fish Biol., 33: 15–30.

Wagner, H. H. 1974. Photoperiod and temperature regulation of smolting in steelhead trout *(Salmo gairdneri).* Can. J. Zool., 52: 219–234.

Wallen, I. E. 1951. The direct effect of turbidity on fishes. Bull. Okla. Agric. Mech. Coll., 48: 1–27.

Wang, J.-K. 1990. Managing shrimp pond water to reduce discharge problems. Aquacult. Eng., 9: 61–73.

Watanabe, W. O. 1985. Experimental approaches to the saltwater culture of tilapias. pp. 3–5, In: ICLARM Newsletter, January. International Center for Living Aquatic Resources Management, Manila, Philippines.

Watanabe, W. O., C.-M. Kuo, and M.-C. Huang. 1985. Salinity tolerance of Nile tilapia fry *(Oreochromis niloticus),* spawned and hatched at various salinities. Aquaculture, 48: 159–176.

Watanabe, W. O., K. E. French, D. H. Ernst, B. Olla, and R. I. Wicklund. 1989. Salinity during early development influences growth and survival of Florida red tilapia in brackish and seawater. J. World Aquacult.. Soc., 20: 134–142.

Watanabe, W. O., L. J. Ellingson, B. L. Olla, D. H. Ernst, and R. I. Wicklund. 1990.

Salinity tolerance and seawater survival vary ontogenetically in Florida red tilapia. Aquaculture, 87: 311–321.

Watenpaugh, D. E., T. L. Beitinger, and D. W. Huey. 1985. Temperature tolerance of nitrite-exposed channel catfish. Trans. Am. Fish. Soc., 114: 274–278.

Wattendorf, R. J., and P. L. Shafland. 1982. Observations on salinity tolerances of striped bass × white bass hybrids in aquaria. Prog. Fish-Cult., 44: 148–149.

Wellborn, T. L., and C. S. Tucker. 1985. An overview of commercial catfish culture. pp. 1–12, In: C. S. Tucker (Ed.). Channel catfish culture. Elsevier, New York.

Weppe, M., and L. Joassard. 1986. Preliminary study: effects of light on swim-bladder's inflation of cultured seabass *(Dicentrarchus labrax)* larvae. pp. 379–380, In: C. P. Vivares, J.-R. Bonami, and E. Jaspers (Eds.). Pathology in marine aquaculture. Special Publication No. 9, European Aquaculture Society, Gent, Belgium.

Whitfield, A. K., and S. J. M. Blaber. 1979. The distribution of the freshwater cichlid *Sarotherodon mossambica* in estuarine systems. Environ. Biol. Fishes, 4: 77–81.

Williamson, J. H., G. J. Carmichael, K. G. Graves, B. A. Simco, and J. R. Tomasso. 1992. Centrarchids. pp. In: 145–197, R. R. Stickney (Ed.). Culture of nonsalmonid freshwater fishes (revised edition). CRC Press, Boca Raton, Florida.

Withyachumnarnkul, B., B. Poolsanguan, and W. Poolsanguan. 1990. Continuous darkness stimulates body growth of the juvenile giant freshwater prawn, *Macrobrachium rosenbergii* De man. Chronobiol. Int., 7: 93–97.

Woiwode, J. G., and I. R. Adelman. 1991. Effects of temperature, photoperiod, and ration size on growth of hybrid striped bass × white bass. Trans. Am. Fish. Soc., 120: 217–228.

Wong, J. T. Y. 1987. Responses to salinity in larvae of a freshwater shrimp, *Macrobrachium nipponense* (de Haan), from Hong Kong. Aquacult. Fish. Manage., 18: 203–207.

Wu, R., and C. E. Boyd. 1990. Evaluation of calcium sulfate for use in aquaculture ponds. Prog. Fish-Cult., 52: 26–31.

Wurts, W. A., and R.R. Stickney. 1984. An hypothesis on the light requirements for spawning penaeid shrimp, with emphasis on *Penaeus setiferus.* Aquaculture, 41: 93–98.

Wurts, W. A., and R. R. Stickney. 1989. Responses of red drum *(Sciaenops ocellatus)* to calcium and magnesium concentrations in fresh and salt water. Aquaculture, 76: 21–35.

Young, J. S. 1988. Impacts from sewage effluent on an open ocean shellfish farm. J. Shellfish Res., 7: 566.

Zale, A. V., and R. W. Gregory. 1989. Effect of salinity on cold tolerance of juvenile blue tilapias. Trans. Am. Fish. Soc., 118: 718–720.

6 Feeds, Nutrition, and Growth

FEEDING STRATEGIES AND FOOD REQUIREMENTS

At the time of first feeding, aquaculture animals are always quite small. Many require natural feeds at this time, and in a number of instances, those natural feeds must be alive. While it is theoretically possible to make a prepared feed[105] with a particle size appropriate for even the smallest aquaculture animal, in practice, fish nutritionists have been unsuccessful in getting some species to accept anything but live food. There are various reasons why that failure may have occurred, any one or combination of which may be valid. A prepared feed may not behave properly in the water (at least from the standpoint of the aquaculture animal) and will, therefore, not be recognized as food. Alternatively, the feed may not contain the proper odor to attract the target animal. Third, it may not be the proper color. Or, it may not have the proper texture.

Whether or not natural foods must initially be provided, many aquaculture species that ultimately become carnivores are herbivores or omnivores during the early stages of their life history.[106] It is often possible to train animals that prefer live or at least fresh natural food to accept feed pellets, though the process can be tedious and somewhat frustrating in some cases.

Many aquatic animals, including a number of species that are of great interest to aquaculturists, have extremely small eggs (see Chapter 7). As a consequence, the larvae that hatch from those eggs are also very small and tend to be difficult to feed in captivity. Some of the most successful aquaculture species hatch from relatively large eggs (e.g., tilapia, channel catfish, trout, and salmon). Those species are characterized by their willingness to accept prepared feeds from the onset of exogenous feeding.[107]

Once most aquaculture species reach the juvenile stage, the most costly aspect of their rearing is associated with feed. Aquatic plants, which may be provided

---

[105] Synonyms that are continually seen include manufactured feeds, artificial feeds, and formulated feeds.

[106] Some animals, like the filter feeding molluscs, are thought to be herbivorous for life though even oyster larvae have been found to consume protozoa and bacteria (Baldwin et al. 1989). The grass carp is perhaps the best example of a finfish that is exclusively herbivorous at fingerling size and larger. It tends to feed on zooplankton during the fry stage.

[107] For a period of days to weeks after hatching, larval fish are nourished from their yolk sacs (endogenous feeding). Once the yolk supply is exhausted, the fish must begin to ingest food they obtain from the environment (exogenous feeding).

along with a prepared nutrient medium if aquaculture animals are reared in confinement or are expected to obtain their nutrients from existing levels when reared in natural environments, are exceptions. Other exceptions are oysters, mussels, clams, and a few other molluscs raised in the natural environment. Such filter feeding animals generally depend on high concentrations of naturally occurring organisms (primarily phytoplankton). There have been at least some attempts to produce formulated feeds for molluscs, however. Castagna et al. (1984) tested several agricultural and fishery byproduct meals fed in slurries to clams *(Mercenaria mercenaria)*. Microencapsulated diets (discussed further below) have been fed by Chu et al. (1987) and Langdon (1989) to oyster larvae (*Crassostrea virginica* and *C. gigas,* respectively). The same type of diet was offered to sea scallops *(Placopecten magellanicus)* by Kean-Howie et al. (1989). Still, natural foods are relied upon at all life stages of most filter-feeding molluscs.

Such invertebrates as shrimp, along with most cultured fishes, will consume prepared feeds. As much or more than half the variable cost involved with rearing such animals can be ascribed to feed.

NATURAL FOODS FOR LARVAE AND FRY

Natural foods may be living organisms such as phytoplankton or zooplankton, or they may be whole or fresh products such as yeast, egg yolk, sea urchin eggs, clam necks, organ meats (heart, kidney), fresh fish, and shellfish. Such products may be ground or made into slurries. Larval aquaculture animals that will not accept a prepared feed at first feeding are often fed one or a combination of the above items, though ground animal material is also used to feed older aquaculture organisms in some instances. A first-feeding aquaculture animal may refuse prepared feed for any of a number of reasons, though in many instances refusal represents an inability of the animal to ingest the feed because of particle size.

The mouths of first-feeding larvae may be too small to accept conventional types of prepared feeds (the manufacture of which is discussed later in this chapter). When such feeds are ground finely enough that the particles can be ingested, the integrity of the feed is often disrupted. That is, the fine particles do not reflect the diet formulation but may, rather, be discrete feed ingredients, such as a piece of corn, soybean, or fish meal. It may be difficult to provide balanced nutrition from such products.

Other reasons why aquaculture species may not accept prepared feeds, either at all or only after training, typically involve problems with recognition. If a feed particle of the appropriate size does not behave, look, feel, or taste right to the aquaculture animal, that particle may be ignored or picked up and then rejected. Odor, flavor, texture, and color may be important. Also, the way a food particle acts can be critical. Many fish species, for example, are sight feeders and recognize potential prey by the behavior of the food item. A predatory fish may not recognize a floating or slowly sinking pellet as food. All of these characteristics of feeds are discussed in more detail below.

In cases where live food must be provided, the aquaculturist often becomes a polyculturist by default. While it is possible to purchase live foods in some instances, most culturists prefer to rear their own. If required to grow algae, such as might be the case in an oyster hatchery, the culturist has to maintain and manage at least two species: the algae and the oysters.[108] When zooplankton are required as a food item, the culturist will often be required to maintain at least three species: zooplankton, algae to feed the zooplankton, and the primary culture species. Each of the species has its own, often somewhat specialized needs with respect to culture conditions. Failure of any of the cultures can lead to disaster. The greater the number of species raised in conjunction with one another, the higher the likelihood of a disastrous failure.

Yeast

Yeast has been used as an ingredient in prepared feeds, directly as a primary food source for larval aquaculture animals, and as a feed for zooplankton that are raised to feed larval fishes and invertebrates. Yeast cultures can be enriched with certain nutrients, one of the most common being high-molecular-weight polyunsaturated fatty acids in the n–3 or $\omega 3$ family. Basically, many aquatic animals require dietary fatty acids from that family (see the section on lipids for details), but those fatty acids are not often present in sufficiently high quantity in the algae that are normally cultured as zooplankton food. Providing the proper types of oil in the yeast culture medium creates what the Japanese have called omega-yeast. Baker's yeast, brewer's yeast, and other types have been used in aquaculture.

Most of the zooplankters reared for use as live food for other aquaculture species are rotifers (mostly *Brachionus plicatilis*) and brine shrimp (*Artemia* sp.). Both are discussed in the following sections. Examples of enriching *B. plicatilis* by feeding it yeast that had been fortified with the desired fatty acids include studies by Kitajima et al. (1980a, b) and Rainuzzo et al. (1989). Japanese scientists have fed enriched rotifers were fed to larval ayu, *Plecoglossus altivelis,* and red sea bream, *Pagrus major* (Kitajima et al. 1980a and 1980b, respectively). Coutteau et al. (1990) found that baker's yeast led to poor growth in *Artemia* sp. unless the yeast cell walls were first removed by enzymatic treatment.

Yeast has served as the sole or a major ingredient in prepared diets for larval fishes, either as first feeds or when the fish are being converted from live to prepared feeds. Among the fishes that have been fed yeast are ayu (Shimma et al. 1980), common carp (Dabrowski et al. 1983, Shcherbina et al. 1987), and the sharptooth catfish, *Clarias gariepinus* (Hecht 1981). Yeasts have also been evaluated as supplements or replacements for algae in the feeding of postlarval penaeid shrimp (Aujero et al. 1985, Danakusumah et al 1985, Choo 1986).

---

[108] In many cases two or more species of algae are used, further complicating the picture.

Algae Culture

While the seaweed industry represents a significant aquaculture enterprise in its own right, this discussion of algae culture is limited to the production of algae for use in feeding animals being reared in captivity. Most of the algae reared for that purpose are single-celled phytoplankton species, though some colonial species have also been raised. Green algae and diatoms are the two most popular types.

The use of microalgae in mariculture as a food source for various types of animals has been reviewed by Brown et al. (1989). Individual algae genera are mentioned later in this chapter with regard to their use in feeding particular species of animals.

The basic apparatus used for the culture of algae involves chambers in which to grow the cultures, proper illumination to maintain photosynthetic growth in the logarithmic phase (see Figure 56), and the provision of a nutrient medium. Culture media are formulated to meet the specific requirements of the algae being produced. Diatoms, for example, have a silicon requirement that green algae do not, so the culture medium needs to accommodate that among other differences in nutrient requirements. Other factors that vary from one algal culture to another are such things as salinity and temperature. Aeration is generally provided to keep the cells in motion and ensure that all of them have equal exposure to the light. Carbon dioxide may be bubbled into the culture chambers since it is also required for photosynthesis.

Batch or continuous algae culture can be practiced. In batch culture the nutrient medium is introduced along with an inoculate of algae cells into the culture chamber. The algae are allowed to multiply for a specified period of time (usually a few days) after which the entire chamber is harvested. In continuous culture, a known quantity of nutrient solution is continuously or intermittently added while an equivalent amount of culture water is removed. The rate of flow through such systems is slow enough to allow cell division to maintain a consistently high algae density in the harvest water. Harvest densities are typically 10^6 cells/liter or higher, with orders of magnitude higher cell densities being readily obtained when small algae species are being cultured.

It is virtually impossible to keep contaminating species out of either batch or continuous algae cultures.[109] Since batch cultures are harvested after only a few days, the chance that the algae species of interest will be displaced to any extent by an invading species is remote. Continuous algae cultures should be routinely monitored, however. After some period of time competing species may become dominant. The culture should be discontinued before that happens. In all cases where a culture system has been harvested, the equipment should be thoroughly cleaned and, if possible, sterilized.

Since algae cultures tend to be invaded by other algal species and by bacteria

---

[109] Algae cells, or more likely their reproductive cysts, are carried in the atmosphere and cannot be readily controlled, particularly in operations where thousands of liters of algae are routinely being cultured.

and other microorganisms, algae culturists maintain axenic[110] stock cultures in incubators. Aseptic handling of the stock cultures is required to keep them from becoming contaminated. The nutrient broth prepared in conjunction with stock cultures is sterilized prior to use. Small vials of each algae species are kept in the incubator and are used to inoculate larger culture chambers. In some cases, there is an intermediate step from vials through carboys of several liters' volume before transfer into the final-growth chambers, which may contain hundreds or even thousands of liters in a large operation. If only modest amounts of algae are required, the final growout chambers may be of the size used in what would otherwise be the intermediate step. All the cultures, no matter what the size of the chambers in which they are being grown, require harvest and reinoculation at intervals. Typically, the cells are allowed to approach or reach an upper equilibirum density within a vial, the contents of the vial are added to a carboy containing nutrient medium,[111] the cells in the carboy are allowed to grow exponentially until they approach their upper equilibrium density and the final transfer is made to the larger growout tank. The process usually involves several culture chambers of each size and daily transfers.

Intermediate chambers may be glass carboys as mentioned, but in recent years plastic bags have become popular. The bags can be discarded after harvest, saving the time and labor expense involved in cleaning and sterilizing them. Bags of various sizes, from a few liters to tens of liters, have been successfully used.

Some culturists have used banks of fluorescent lights suspended above algae culture tanks or located on the sides of plastic bags and carboys. Lights of the type used to illuminate sports stadiums are also commonly seen algae culture facilities. Such lights are much more powerful than fluorescent bulbs.

Certain species of algae have become popular for the production of specfic types of larval aquaculture species or for feeding zooplankton used as live food. For example, algae in the genus *Tetraselmis* have been widely used in the production of the rotifer *Brachionus plicatilis* and also can serve as a food for larval penaeid shrimp (Okauchi 1988), though the level of high-molecular-weight fatty acids in rotifers cultured on *Tetraselmis tetrathele* was found to be insufficient for the rearing of fish larvae. Another popular alga is *Spirulina* sp. It has been widely used as a source of pigment in ornamental fish and also has been relied upon as an excellent source of various other nutrients. Its use in aquaculture has been reviewed by Hanson (1990). The list of algae that have been or are being grown by aquaculturists is a long one. In addition to the genera mentioned above, many have used *Isochrysis, Monochrysis, Chlorella,* and a number of other genera. Selection of an alga depends on whether the animal to which it will be fed is freshwater or marine, how large the animal is, and what its nutritional requirements are. The thermal regime of the species is another consideration. In general, coldwater algae are not cultured as food for warmwater species, and vice versa.

The so-called green water technique has been used successfully in conjunction with the culture of various larval fishes and shrimp. The technique involves the

---

[110] Axenic is defined as bacteria-free.

[111] Enough of the solution is retained to initiate a new stock culture.

maintenance of sufficient concentrations of phytoplanktonic algae in the culture tanks of the aquaculture target species to provide a green color. Even in the case of animals that are considered to be carnivorous in their early life stages, the presence of algae in the water has been found to enhance growth and survival in a number of species. The introduction of pond effluent into laboratory tanks in which shrimp were being reared led to a near doubling of the growth rate in a study by Leber and Pruder (1988). The conclusion from such results is that some unknown factor in the water stimulates growth and that the factor in question may be a product of phytoplankton algae.

The green water technique has been widely applied in the production of shrimp, including *Macrobrachium rosenbergii* (Aniello and Singh 1982). It has also been applied in the culture of such species as common carp, *Cyprinus carpio* (Kim and Lee 1981); bighead carp, *Aristichthys nobilis* (Fermin and Recometa 1988), and Atlantic halibut, *Hippoglossus hippoglossus* (Naess et al. 1990).

Once an algae culture has reached the desired cell concentration, it is harvested. This may consist of merely pouring the algae into the culture chamber in which the animals being fed are held. Alternatively, the algae can be concentrated and placed in frozen storage or immediately fed. Alage can be concentrated by passing the water in which they are cultured through various types of filter materials or by using continuous centrifuges. Commercial cream separators that are used by dairies provide a type of continuous centrifuge that can be used to concentrate algae. One commercial oyster firm in Washington State has used a cream separator to concentrate algae reared in culture tanks of about 10,000 l down to a volume of about 1 l. The concentrated algae can be immediately fed to oyster spat or frozen for subsequent use. When the frozen algae concentrate is placed in water, the cells disperse and are readily consumed by the larval oysters. Such frozen algae can also be used to support zooplankton. Some success has also been obtained by feeding spray-dried algae to various molluscs (Laing et al. 1990; Laing and Verdugo 1991).

Zooplankton Culture

Aquaculturists have attempted, sometimes quite successfully, to produce zooplankton food in considerable quantities from wild populations. Collection of wild zooplankton is tedious, sometimes expensive work and is usually not reliable since populations may crash leaving the culturist without a food source for larvae that cannot withstand long periods without proper nutrition. To circumvent the problems, unispecific cultures of zooplankton have been established.

The most simple way to provide zooplankton for larval fish food is to induce natural plankton populations to grow within the culture chamber. An example of that technique would be fertilization of a pond such as is done in conjunction with the rearing of channel catfish fry. Because the culturist has no control over the zooplankton species that develops, the technique is not widely used, particularly in the marine environment where the threat of producing predatory zooplankton is high. An exception involves the culture of red drum. Good success in the production of desirable zooplankton as food for red drum has been achieved with both organic and inorganic fertilization (reviewed by Colura 1987). Production of wild zooplankton has also been used in conjunction with the rearing of striped bass (Colura et al. 1976, Ludwig and Tackett, 1991) and spotted seatrout, *Cynoscion nebulosus* (Colura et al. 1976).

In theory, at least, virtually any zooplankton species could be cultured. In practice, only a few species have been successfully reared in large concentrations by people interested in feeding the zooplankton to fish or other animals being reared for food or recreational purposes. Among those zooplankton species are copepods such as *Eurytemora affinis, Acartia tonsa,* and *Tigriopus japonicus*; the rotifer *Brachionus plicatilus*; and the brine shrimp *Artemia* sp. The culture of *Brachionus* and *Artemia* has become so widespread among aquaculturists, and so much information has been generated on those organisms, that they can be thought of as being as domesticated as any other aquaculture species. Yet, as pointed out by Naas (1989), the nutritional quality of wild zooplankton is generally superior to rotifers and brine shrimp. Enrichment of both *Brachionus* and *Artemia* with such things as high-molecular-weight polyunsaturated fatty acids and other nutrients has helped improve the quality of those zooplankton species for use in the feeding of a variety of marine fishes. Studies demonstrating the value of enrichment to one or both of the subject zooplankters include those of Watanabe et al. (1980a, 1983), Van Ballaer et al. (1985), Sorgeloos et al. (1988), and Dhert et al. (1990).

Studies in Japan (Seikai 1985) have shown that, while rotifers and brine shrimp nauplii produced better growth and survival in flounder *(Paralichthys olivaceus)* than did wild zooplankton, the incidence of albinism was greatly reduced in fish fed wild zooplankton. From 97.2 to 100% of the fish on *Brachiolnus* and *Artemia* were albinos, compared with 0.1 to 1.8% of fish fed wild zooplankton.

The general strategy with respect to feeding fish larval zooplankton foods is to begin with rotifers, convert to brine shrimp nauplii as the fish reach sufficient size to ingest the *Artemia,* and then convert them to prepared feeds.

Some of the fishes that have been offered rotifers and/or brine shrimp during the early stages of their lives are sea bream *(Sparus aurata, Pagrus major,* and *Chrysophrys major),* porgies *(Acanthopagrus schlegeli* and *A. cuveieri)* grouper *(Epinephalus salmoides* and *E. tauvina),* seabass *(Lates calcarifer* and *Dicentrarchus labrax)* snook and tarpon snook *(Centropomus undecimalis* and *C. pectinatus),* dolphin *(Coryphaena hippurus),* rabbitfish *(Siganus* sp.), milkfish *(Chanos chanos),* Dover sole *(Solea solea),* turbot *(Scophthalmus maximus),* and flounders *(Paralichthys olivaceus).* More information on feeding those species can be found in papers by Pyen and Jo 1982, James et al. 1983, Fukusho 1985, Fukusho et al. 1985, Kungvankij et al. 1986, Tucker 1987, Dendrinos and Thorpe 1987, Sorgeloos et al. 1988, Gatesoupe et al. 1989, and Dhert et al. 1990). *Artemia* sp. are also widely used in the larval culture of both freshwater and marine shrimp.

Brachionus culture. Rotifers are often sufficiently small to be an acceptable first food for larval fishes and invertebrates. Cultures can be established quite easily. First, the culturist must obtain a sample of *Brachionus* with which to initiate the process. The stock culture (which can be obtained from biological supply houses or from another aquaculturist) is introduced into one or more culture tanks (often circular fiberglass tanks). An appropriate alga, or algae in conjunction with yeast (Planas and Estevez 1989), is grown to provide food for the rotifers. Some of the algae that have been used in conjunction with the culture of *Brachionus* include *Isochrysis galbana, Tetraselmis chuii, T. suecica, Chlorella* sp., *Dunaliella* sp., *Chlamydomonas* sp., *Platymonas* sp., *Skeletonema costatum, Cyclotella* sp., *Thalassiosira* sp., and *Johannesbaptistia* sp. (Chiang and Young 1978, Planas and Estevez 1989). The culture is allowed to develop for several days. Depending on the algae used, the rotifer population can increase 10 to 46 times within 120 hr (Chiang and Young 1978). McGeachin (1977) reared *Brachionus* in 28 ppt seawater at 26°C and fed the algae *Tetraselmis chuii*. He maintained algae cell density at 5,000 to 20,000 cells/ml and harvested 25 to 75% of the rotifers each day to maintain the culture in the log phase of growth. The rotifers can be harvested with a fine mesh net.

Artemia *culture*. One of the most popular zooplanktonic animals used in feeding larval aquaculture species is the brine shrimp, *Artemia* sp. The animal has been the focus of hundreds of research papers. General information on the biology and culture of brine shrimp can be found in such publications as Manzi et al. (1980), Sorgeloos (1980, 1986), Persoone et al. (1980a, b), and Sorgeloos et al. (1986).

Under certain conditions, brine shrimp females produce cysts, which are a resting stage, often erroneously referred to as eggs. In some cases the cysts, each measuring a fraction of a millimeter in diameter, are produced by the hundreds of millions. They can form large windrows along beaches or pondbanks where they are collected and vacuum-packaged in cans. The cysts will remain dormant but viable for many months.

Traditionally, there were two major sources of brine shrimp: Great Salt Lake in Utah, and San Francisco Bay, California. The sizes of nauplii produced from cysts collected at the two sites are slightly different, though that difference can be important in terms of acceptance by the target aquaculture animal. As the demand for brine shrimp by the aquaculture industry has increased, additional natural sources have been found and populations have been intentionally or accidentally established in various locations around the world.

Brine shrimp rearing is relatively simple and requires no elaborate equipment. If the newly hatched nauplii are used within 24 hr of hatching as food for other species, there is no need to maintain algae cultures to feed them. However, various studies have demonstrated that the food value of *Artemia* can be enriched by feeding them certain types of algae or yeast as previously discussed. The nauplii tend to be particularly deficient in high-molecular-weight polyunsaturated fatty acids.

Thousands of *Artemia* cysts will easily fit in a teaspoon, and millions of nauplii can be produced in a few liters of water. Hatching is possible over a range of salinities, with good results typically being obtained in water from 35 to 40 ppt at approximately room temperature (25°C). Natural and artificial seawater can be used, or the cysts can be hatched in a solution of sodium chloride. During incubation, the cysts should be constantly mixed into the water column. This can easily be accomplished by providing vigorous aeration (which also maintains the oxygen

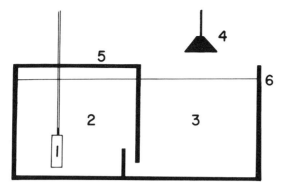

Figure 71. Schematic representation of an *Artemia* hatching chamber. Brine shrimp cysts are kept in suspension with an airstone *(1)* in the chamber on the left *(2)*. Following hatching, the cyst cases and unhatched cysts remain in the hatching chamber while the swimming nauplii are attracted to the chamber on the right *(3)* by a light suspended above the water *(4)*. A cover *(5)* over the hatching chamber prevents light from entering that compartment *(2)*. The water level *(6)* is the same on both sides of the partition. All dimensions are variable depending on the needs of the culturist.

level for the nauplii once they hatch). Hatching typically occurs within 24 to 36 hr after the cysts are placed in the water. If the cysts fail to begin hatching within that time period at the salinity range and temperature given, they can be considered unviable, and a new culture should be started.

Harvesting of hatched nauplii while at the same time leaving behind unhatched and dead cysts can be facilitated by taking advantage of the positive phototaxic response of the nauplii and the fact that the cyst cases float while unhatched cysts tend to sink. A light placed near the side of the *Artemia* culture chamber will attract the nauplii away from dead or unhatched cysts and cyst cases. Once the nauplii have been concentrated, they can be siphoned into another container or into a fine-mesh net. The nauplii can be fed directly to a culture species or first be provided with algae or yeast for a day or two to enrich them once they begin to accept feed. A simple, effective design for a brine shrimp hatching chamber is shown in Figure 71. That particular design simplifies the harvesting process.

Removal of the protective coating on *Artemia* cysts prior to initiating hatching, a process called decapsulation, has been found to enhance the viability of the resulting nauplii. The technique has been discussed Bruggeman et al. (1980).

PREPARED FEEDS

Prepared or manufactured feeds come in a variety of types and can contain an array of ingredients. They may float at the surface of the water, sink to the bottom, or be neutrally buoyant. They come in various colors, sizes, flavors, and textures. Some can be used to feed a variety of different species, while others are especially designed for a particular type of animal. Before any of them can begin to satisfy the nutritional requirements of any culture species, the animals must consume the diet. For some species, like catfish, tilapia, and trout, prepared feeds will be accepted at first feeding. Other species that initially require live foods may need to be trained to accept prepared feeds. That training is known as weaning.

Weaning

Weaning is quite simple in some cases and very difficult in others. The first experience of the author with weaning marine fishes came in the early 1970s during attempts to rear flounders of the genus *Paralichthys*. Since the technology for spawning adults had not yet been developed, we caught postlarvae of 10 to 11 mm long in a plankton net. The fish quickly adapted to small culture tanks with flowthrough water of intermediate salinity. They would not accept prepared feeds but readily consumed *Artemia* nauplii. Before the fish became too large to effectively capture brine shrimp,[112] we began to supplement the brine shrimp with pieces of finely cut frozen penaeid shrimp. After the transition from brine shrimp to frozen shrimp was made, we would supplement the frozen shrimp with pieces of freeze-dried shrimp. The freeze-dried shrimp would float, and the fish learned to feed at the surface. The frozen shrimp pieces were then removed from the diet, and floating feed pellets were introduced along with the freeze-dried shrimp. Gradually, as the fish began consuming the pellets, the freeze-dried shrimp was removed until only pellets were fed. The process took several weeks, and it became apparent that the conversion from shrimp to pellets was accelerated in any given tank once an individual fish learned to take the prepared feed. Other fish seemed to learn from that fish. Placing one or more trained fish (whether of the same or another species) into a tank of fish that are being weaned has been tried with varying success.

Weaning is a significant problem in a number of marine fishes with larvae that do not readily accept prepared feeds and for which the manufacture of uniform-quality feed (in terms of ingredient content) with extremely small particle sizes remains a technological problem. Various weaning programs have been developed. Some involve weaning directly from live to prepared feeds, while others use frozen animals (e.g., shrimp, polychaetes, and molluscs). When the conversion is made directly from live to prepared feeds, moist feeds are sometimes initially used because their soft texture is more akin to that of living organisms than to that of a dry feed particle. Attractants (discussed later in this chapter) may be added to prepared feeds to augment the process. Among the species for which weaning techniques have been developed are *Solea vulgaris* (Metailler et al. 1981), *Solea solea* (Fuchs 1982, Gatesoupe and Luquet 1982, Gatesoupe 1983, Person-Le Ruyet et al. 1983, Bromley and Sykes 1985), *Scophthalmus maximus* (Gatesoupe 1982, Bromley and Sykes 1985), *Dicentrarchus labrax* (Barahona-Fernandes 1982a, b, c), *Anguilla*

---

[112] Once the flounders reached a certain size, they were observed to take in *Artemia* nauplii through their mouths, but the brine shrimp were not retained by the gill rakers and would escape through the opercles of the fish.

anguilla (Kastelein 1983), *Chanos chanos* (Duray and Bagarinao 1984), *Clarias gariepinus* (Verreth and van Tongeren 1989, Haylor 1991), along with *Morone saxatilis* and *Morone* hybrids (Tuncer et al. 1990).

Many fishes and invertebrates are cannibalistic to a greater or lesser extent. Cannibalism incidence often increases when there is a great size disparity within a culture chamber. It can also increase during or after weaning as has been observed in such species as *C. gariepinus* (Haylor 1991) and *D. labrax* (Katavic et al. 1989).

Purified, Semipurified, and Practical Feeds. Of the three types of formulations that can be made, two are used exclusively for research (purified and semipurified feeds), while the third (practical feeds) is used both by researchers and producers. To understand the difference among the three types of formulations, it is best to discuss them in reverse order from how they are listed in the above heading.

Practical feeds are manufactured from natural and readily available ingredients that have not been subjected to high levels of alteration beyond grinding, pressing, and solvent extraction.[113] Among those ingredients are cereal grains, oilseed meals, fish meals, meat and meat byproducts, and animal and vegetable oils. To those ingredients may be added vitamin and mineral supplements and various nonnutritional ingredients (binders, color or flavor enhancers, attractants, and so forth). Vitamins, minerals, and the nonnutritional ingredients typically make up less than 5% of a practical diet. A typical practical diet might contain the following:

- Menhaden fish meal (primarily protein, but also some lipid and minerals)
- Soybean meal (protein, carbohydrate, lipid)
- Fish oil (lipid with trace amounts of other nutrients)
- Wheat (protein, carbohydrate, lipid)
- Vitamins and minerals

A semipurified research diet, the most common type of diet used in determining the nutritional requirements of an animal, has the majority of its ingredients in a form that is more highly refined than the ingredients in a practical diet. For example, casein (milk protein) and gelatin may be used as the sole protein sources. Casein is more than 95% protein and gelatin is also almost pure protein. Other ingredients would be similarly pure. A list of ingredients in a semipurified diet might look like the following:

- Casein and gelatin (protein)
- Corn starch (carbohydrate)
- Corn oil (lipid)
- Vitamins and minerals
- Carboxymethylcellulose or agar (binder)

---

[113] Fats are extracted from various feed ingredients either by mechanical pressure or through the use of solvents. Examples are the extraction of the oil before making fish meal and soybean meal.

If a nutritionist wants to examine, for example, the amino acid requirements of a fish, mixtures of purified amino acids might replace casein and gelatin in a number of different formulations (the other ingredients would be as shown above). Since there are about 20 amino acids, a group of diets could be made, each of which lacks one of them. Fish that performed poorly on any of the diets would presumably have a dietary requirement for the missing amino acid. Alternatively, if a nutritionist knows that a particular amino acid is present in casein at a level below that required by a certain fish species, a series of diets might be made up with casein as the primary protein source and different levels of the deficient amino acid added. By measuring fish growth and other performance characteristics it should be possible to determine the requirement for the deficient amino acid. Many fish culturists refer to this type of diet as a purified ration, when in reality it is semipurified.

A purified diet gets down to the very basics. Truly purified diets, which have rarely if ever been used in studies with fish, are manufactured exclusively from the most basic nutrients as follows:

- Mixture of individual amino acids
- A simple sugar such as glucose
- A mixture of individual fatty acids
- A mixture of individual vitamins and minerals
- Cellulose (undigestible fiber for filler)

The more purified the diet, the poorer the fish performance, in general. Fish tend to grow more slowly and convert food more poorly as the degree of purification increases. Palatability can be a major problem, which of course, translates into reduced performance.

Types of Practical Feeds

Microencapsulated Diets. One way to provide feed in extremely small particles that are uniform with respect to their ingredient makeup is to use the process of microencapsulation. Water is mixed with extremely finely ground ingredients to make a slurry that is then manufactured into microcapsules (many pharmaceutical capsules are filled with microencapsulated drugs). The slurry ends up being encased in proteinaceous membranes (the microcapsules), with the particles being as small as a few tens of microns in diameter. Microencapsulation is expensive though it has been employed to some extent in aquaculture.

The use of microencapsulated feeds in mollusc culture has already been mentioned. Microencapsulated diets have also been used with at least limited success as an initial food for some larval marine fishes or to train fish that are feeding on algae or zooplankton to accept prepared feeds (Appelbaum 1985). Several studies have investigated the use of microencapsulated diets in conjunction with the culture of penaeid shrimp, and the results have been encouraging (Jones 1985, Scura et al. 1985, Jones et al. 1987, Kurmaly et al. 1989). Microencapsulated diets have even

TABLE 20. A typical Oregon Moist Pellet Diet Formulation (Hardy 1989)

Ingredient	Percent in Diet
Wet fish	20.0
Herring meal	49.9
Wheat germ meal	10.0
Dried whey	8.0
Trace mineral mixture	0.1
Vitamin mixture	2.0
Fish oil	10.0

been used in conjuntion with the culture of *Brachionus plicatilis* (Teshima et al. 1981). Prolonged use leads to decreased rotifer population densities.

Moist and Semimoist Feeds. Moist feeds are formulated with significant percentages of such ingredients as whole fish or the frames remaining after fish have been processed. Dry ingredients, including such things as fish meal and various types of plant meal are mixed with the wet ingredients. A vitamin and mineral mix may also be added. The material is then mixed and made into pellets. The final feed is very soft and has a high water content. It is either fed to fish immediately or can be stored frozen. Because of the high moisture content, the feed will rapidly deteriorate if it is not frozen.

The commonly used moist diet used in the United States is the Oregon Moist Pellet (OMP), which is used in salmon culture (Hublou 1963), particularly in the Pacific Northwest. Modern OMP has a moisture content of about 32% according to Hardy (1989). A typical OMP formula is presented in Table 20. Net-pen culturists of sea bream and yellowtail in southern Japan feed them moist pellets that are manufactured immediately prior to being offered to the fish (Figure 72).

Because OMP and similar diets must be stored in a freezer, maintaining and moving them are expensive (both because they required frozen storage and because a significant amount of the feed weight is composed of water). In most cases OMP is manufactured and fed in relatively close proximity to the source of the fresh fish that goes into the product.

Because of their high water content, moist feeds must be provided at much higher levels than dry feeds of the same nutrient composition.[114] For example, shrimp may be fed at a few percent of body weight when a dry diet is used. However, when an unprocessed diet of 43% anchovy, 33% shrimp head, and 24% squid was fed to *Penaeus kerathurus,* 100% of biomass was required for animals 0.6 to 1.8 g, 60% for those from 1.8 to 4.3 g, and 40% for larger shrimp (Faranda et al. 1984).

---

[114] Water is considered a nutrient by many nutritionists. In this context the distinction between the feeding rates is made on the basis of whether or not water is present at a high level in the two feeds.

Figure 72. Feeding net-pen fish in Japan with soft pellets.

Semi-moist pellets that have about the consistency of hamburger have been developed. They are stabilized with preservatives to keep them from decomposing and then packed in airtight plastic bags. Some popular dogfoods are prepared in the same way. The quality of such feeds is excellent, but the cost is high because of the specialized processing and packaging required. Such feeds have enjoyed some success as starter feeds for fish fry but are too expensive to feed to fingerlings when cost is a consideration since the feed typically costs about $1.50 per kilogram.

Dry Feeds. The feeds that are least expensive, most easily stored, and most convenient to feed are dry feeds. They are manufactured entirely from dry ingedients[115] such as fish meal, soybean meal, wheat middlings, corn meal, and vitamins and minerals. Liquid fat may be added in the form of fish or oilseed oils.

The manufactured feed may be in the form of pellets, crumbles, or flakes. Any of those products may be produced in a form that floats or sinks. How dry feeds are manufactured is considered in a later section of this chapter.

Flaked feeds are widely used by fish hobbiests for the maintenance of aquarium fishes, including goldfish. The early stages of most commercially produced fish are fed granular feeds, called crumbles. Those feeds are prepared by grinding up larger pellets and passing the resulting pieces through standard sieves to separate them into known size groups as shown in Table 21. The size of the particles is increased as the fish grow. Eventually, the fish are converted from the largest crumble to

---

[115] "Dry" is a relative term. Most feed ingredients contain some moisture, and a typical manufactured feed will contain about 9 to 10% moisture.

TABLE 21. Size Designation, Size Range, and Sieve Size Used to Produce Crumbled Feeds (National Research Council 1981)

Size Designation	Size Range (mm)	Standard Sieve No.
0	0.420–0.595	40
1	0.595–0.841	30
2	0.841–1.19	20
3	1.19 –1.68	16
4	1.68 –2.38	12
5	2.38 –3.36	8

pellets. Only a few pellet sizes are used for feeding fish, with the largest generally being no more than 6.4 mm in diameter.

Feed Storage

All feeds, whether moist, semimoist, or dry, are susceptible to degradation with time. We have already seen that moist feeds should be stored in a freezer. Semimoist feeds will degrade once removed from their vacuum packs even though they are highly stabilized. Dry feeds are perhaps the most stable, but they are subject to degradation in a number of ways. If exposed to high humidity, dry feeds will take on water, increasing their susceptibility to attack by molds. Exposure to heat destroys certain vitamins and enhances degradation of lipids, leading to rancidity. Some vitamins are also destroyed when exposed to light.

To reduce the likelihood of developing problems with stored feed, it should be kept in a *cool, dry place*.[116] Most cool, dry places do not have high levels of light, but in the event your cool, dry place is brightly lit, the feed should also be protected from bright light. In addition, if two or more shipments of feed are on hand at the same time, the oldest feed (assuming it is still of good quality) should be used first. In general, feed should be used within 90 days of purchase. An aquaculturist should not buy feed a year's supply at a time because it will lose some of its nutritional quality, and could even become toxic, by the time the growing season ends. If the feed should become moldy, it should be destroyed. Under no circumstances should moldy feed by used since it may be highly toxic.[117]

Dry feed is typically available in 22.7 kg (50 lb) bags or in bulk. Bags can be stacked in a cool, dry storage room, while bulk feed is typically placed in feed storage bins (Figure 73). Feed bins are often placed out in the open, and while dry, they certainly cannot be considered cool when exposed to direct sunlight on a hot

[116] Cool does not mean cold. If the feed is used in a timely manner (within 90 days), normal room temperature, about 25°C, is acceptable.

[117] Moldy feeds may produce aflatoxins, chemicals that are highly toxic to animals that consume them. The natural tendency is to get rid of moldy feed by throwing it in a pond. That urge should be overcome because the result may be lethal to the animals inhabiting that water body.

Figure 73. Feed truck with screw auger that allows feed to be lifted from the truck into a large feed storage bin.

day. Most operators who use bulk feed storage bins receive feed deliveries once or more each week so there is typically no problem with freshness.

Commercial feed mills can produce tons of feed every hour and may operate virtually around the clock during the height of the growing season. Trains and trucks must resupply the feed bins with new ingredients as quickly as those ingredients are made into feed and delivered to customers (Figure 73).

At the other extreme of feed manufacturing is the preparation of small amounts of experimental feeds for research. A laboratory pellet mill is shown in Figure 74. It operates on the same principle as commercial-sized pellet mills and is useful when only a few hundred grams or a few kilograms of diet are required. A commercial feed mill may waste several tens of kilograms of feed before the equipment is properly adjusted to produce pellets of the desired quality. Since the ingredients in some experimental diets cost several dollars per kilogram, such wastage cannot be allowed.

Dry Feed Manufacture

Before a feed can be manufactured, the ingredients must be properly ground and mixed in the proportions called for in the feed formula. Mixers at large feed mills can handle tons of feed at a time. The ingredients are delivered to the mixer through pipes from storage bins. Each ingredient is automatically weighed before being dropped into the mixer. Once the ingredients have been thoroughly mixed, they are moved from the mixer to the appropriate manufacturing equipment where flakes, pellets, or crumbles are produced. The entire process is controlled from a central

Figure 74. A laboratory model pellet mill. Machines of this size can produce several kilograms per hour as compared with commercial feed mills that can produce tons per hour.

control booth (Figure 75). In modern feed mills the entire process is controlled by computers.

Double Drum Dried Feed. The paper-thin flakes that many of us have used to feed tropical aquarium fishes are manufactured from feed ingredients slurries that are processed on double drum dryers. The feed mixture, which is composed of very finely ground ingredients, is liquified through the addition of large percentages of water, and the liquid is pressed between two hot rollers (the drums), creating a thin, dry sheet of feed. The sheets are then broken up into the small flakes that are canned for sale in pet stores and other outlets. A can of flakes will typically contain particles with different colors. The different colored flakes are made from different formulations that are mixed together. Differences in formulations may be attributable to nothing more than the use of different food colors (red, green, yellow), with flakes that have a formula with no coloring added appearing brown.

Those flakes are very light. They tend to float on the surface of the water but will eventually sink. The process of their manufacture requires a great deal of heat energy, and a good deal of time is required to produce large quantities. Therefore,

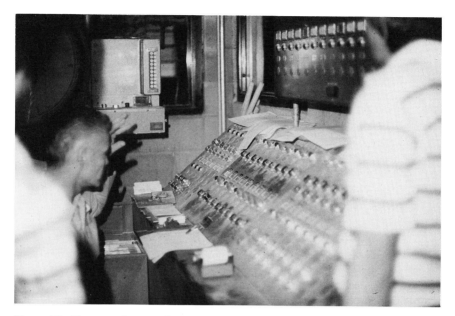

Figure 75. The control room of a commercial feed mill showing the console that handles the movement of ingredients from feed bins to mixers and pelleting equipment.

the process is quite expensive. If you have any question about the retail cost of flaked diets, just look at the price demanded for the typical ounce (30 g) of pet fish food, calculate the cost per ton, and compare that with the typical cost of a ton of commercial fish food pellets, which is typically less than $400.

Pressure Pellets. Pellet mills, such as that shown in Figure 74, are designed to force the feed mixture through holes in a steel cylinder or disc form known as a die. An auger delivers the feed to the die, and the pressure created from passage of the ingredients through the holes in the die causes the pellets to be formed. Dies tend to be only a few centimeters thick. Residence time of the feed in the die can be measured in fractions of a second. Each die has a number of openings. Depending on the size of the equipment and the size of the pellets being manufactured, a single die may have up to several hundred holes through it, each of which forms pellets when the equipment is in operation.

A knife cuts off the pellets to the desired length as they emerge from the die. Some heating occurs because of friction, and many feed mills expose the ingredients to live steam to heat them just before they enter the die. The steam also adds water to the pellets.

Pellet quality can be adversely affected if the lipid level in the formula is too high. When high levels of lipid are desired in the final feed, fat can be sprayed-on the finished pellets. Following passage through the mill and the addition of sprayed on fat, if required, the pellets are dried to remove excess moisture. The finished feed is then loaded into trucks for bulk shipment or placed in feed sacks.

Binders may be added to the feed formulation, as discussed below, to enhance the water stability of pressure pellets. As a general rule, a pressure pellet has been properly manufactured if it remains intact for a minimum of 10 min after being placed in water. That amount of time is sufficient to allow most species to consume the feed, though there are exceptions. Shrimp, for example, feed by nibbling off pieces of feed pellets and may spend hours each day in feeding activity. Pellets for animals that feed in that manner must be very water-stable. If the proper type and amount of binder is added, highly water-stable pressure pellets can be manufactured.

Pressure pellets sink when placed in water. That is an advantage for some species, such as shrimp, which will not take floating feeds. The relatively low amount of energy required in the manufacture of pressure pellets makes them less expensive than feeds produced in other ways. Also, the relatively low heat required in the process is less destructive on heat-labile ingredients than some other feed manufacturing processes.

In most cases crumbles are not manufactured from pressure pellets because when those pellets are reground they tend to disintegrate into particles the size of the original meals that were used in the manufacturing process.

Extruded Feeds. Extruded feeds are made by passing mixed ingredients through a long metal cylinder. The extruder barrel has one hole through it and may be 1 m or more in length. The opening is smaller than the size of the product that is produced in many cases and the opening at the exit end can have a variety of shapes. This can be verified by looking at the shapes used in pasta, dry dog and cat food, and breakfast cereals, all of which are made with extruders.

All along the barrel of the extruder are heating elements that can be individually adjusted. As the feed passes through the barrel, it can be exposed to a constant temperature or to fluctuating temperatures.

The feed ingredients are exposed to much higher pressures and temperatures in an extruder than in a pellet mill. The ingredients are literally cooked as they pass through the extruder barrel. When the material exits, it is cut into the proper length with a knife. In the case of fish-feed pellets, the material exiting the barrel expands as the pressure is released. Starches, which have become gelatinized during their passage through the extruder barrel, harden as they begin to cool, making the pellets highly water impervious and trapping air in the process. The result is a pellet that floats. By properly controlling the manufacturing process (keeping temperatures down and limiting dietary starch level), the pellets can also be made to sink or to be neutrally buoyant. Crumbles are made by grinding up extruded pellets that sink.

As is the case with pressure pellets, fat can be sprayed on extruded feeds after the pellets have been manufactured. Up to about 5% lipid can be added in that manner. Feed mill operators prefer to have the ingredients that pass through the extruder contain less than 10% total lipid.

Flakes can be made by passing extruded pellets through a roller mill immediately after they are manufactured. Flake breakfast cereals are made in that way.

Once the final product has been made, the feed is cooled and dried and is then bagged or placed into bulk storage containers or trucks. Bags holding feed with high lipid levels are usually lined with polyethylene to keep the fat from oozing through the paper.

Extruded feeds are more expensive to make than pressure pellets because the equipment is more expensive and the amount of energy required to run it is considerably higher. Very large motors are required to create the pressures needed to push the feed ingrdients through the extruder barrel and the energy required for heating the barrel is a significant expense. The difference in cost for pressure pelleted and extruded feeds is generally a few dollars a ton. That is not too significant for the small producer, but a farmer who feeds several tons a day can see a meaningful difference in annual feed cost between the two types of pellets.

The major advantage of floating extruded feeds to aquaculturists is that the fish farmer can observe feeding activity. We will see in a later section how to calculate feeding rates based on estimates of fish biomass and growth rate. Calculations of feeding rate can be avoided when floating feed is used. A farmer merely feeds what the animals will consume over a set period of time. If all the feed that is offered is gone before the allotted time, additional feed is provided until the fish become satiated.

The cooking process that occurs in the manufacture of extruded feeds actually enhances the digestibility of some ingredients. An additional advantage of floating feed is that the farmer can observe the health of the fish. Actively feeding animals tend to be healthy animals. If feeding behavior changes or if the fish cease to feed, it is readily apparent. When that happens, it is commonly a sign of stress caused by degradation in water quality or the onset of a disease.

The primary disadvantage of floating feeds is that heat-labile ingredients must be overfortified if they are to be maintained at the desired level in the final feed. Vitamin C, for example, may be present at only 20% of its initial level after passing through an extruder as a component of a feed formulation. That problem has been largely overcome in recent years with the development of heat-stabilized vitamin C compounds, as discussed in the vitamins section of this chapter.

ENERGY AND GROWTH

Metabolism is the result of all the chemical and energy transformations that occur within a living organism. All phases of metabolism require energy, which animals obtain from the food they ingest. Energy is expended for the maintenance of life and for growth and reproduction. Metabolism includes the storage of energy (anabolism) as fat, protein, and carbohydrate, and the transformation of those storage products into free energy (catabolism).

Food energy is measured in calories. In nutrition, reference to the calorie generally implies the large calorie or kilocalorie (kcal), which is equivalent to the amount of heat required to raise the temperature of 1 kg of water 1°C, as distinguished from the small or gram calorie (g·cal), which is the amount of heat required to raise the temperature of 1 g of water 1°C.

The immediate source of free energy in biochemical reactions is adenosine triphosphate (ATP). During anabolism, adenosine diphosphate (ADP) traps energy in phosphate bonds, forming ATP. Conversely, during catabolic reactions ATP is converted back to ADP and free energy is released.

The amount of energy expended for growth compared with that required for maintenance is dependent on the species of animal, its age, existing environmental conditions, composition of the diet, reproductive state, and other factors (all of which influence basal metabolic rate). Maintenance energy is a combination of basal metabolism and specific dynamic action (SDA). SDA is the amount of heat produced in addition to that of basal metabolism as a result of food ingestion (White et al. 1964). SDA is considerably higher for proteins than for carbohydrates or lipids.

Since metabolic rate in poikilothermic (so-called cold-blooded) animals is highly dependent on environmental temperature, energy requirements can be expected to vary both seasonally and diurnally as fluctuations occur in water temperature. In addition, small animals generally have higher metabolic rates than larger ones, and the growth rate of young animals is more rapid than that of older individuals. Based on the rate of oxygen consumption, the metabolic rate of fish increases shortly after feeding, indicating increased energy utilization as a function of digestion, absorption, and assimilation. Increased oxygen consumption associated with feeding activity has been shown for such aquaculture species as channel catfish (Andrews and Matsuda 1975); American eel, *Anguilla rostrata* (Gallagher and Matthews 1987); rainbow trout (Medland and Beamish 1985, Kaushik and Gomes 1988); largemouth bass (Tandler and Beamish 1981); and tiger prawn, *Penaeus esculentus* (Hewitt and Iving 1990). Ammonia production may also be increased in association with feeding (Kaushik and Gomes 1988, Hewitt and Iving 1990), though apparently not in all cases (Gallagher and Matthews 1987).

One of the goals of many aquaculturists is to produce a harvestable crop as rapidly as possible. Thus it is necessary to accelerate growth through water quality management, dietary manipulation, disease control, and in other ways. A formula for predicting short-term growth is as follows (Everhart et al. 1975):

$$W_t = W_0 e^{gt,}$$

where W_t is the weight at time t, W_0 is the initial weight, e is the natural logarithm, and g is the growth coefficient. The growth coefficient is determined as follows:

$$g = \ln\frac{W_1}{W_0}$$

The model generally holds fairly well for a growing season but cannot be extended to cover more than 1 yr. A modification of this formula can be used to calculate required feeding rates for maintenance of a desired growth rate of aquaculture animals. That subject is covered later in this chapter.

Optimum harvest size for many aquaculture animals occurs at or near the size where growth rate begins to decrease significantly. For example, channel catfish can reach a maximum weight in excess of 20 kg, but the most rapid growth occurs between hatching and the first 500 g. The time required to double the weight of a 500 g channel catfish may be as long as that needed to bring the fish from the egg to the first 500 g. This same general growth pattern is followed by many aquatic animals that live for several years, although the actual size at which growth rate begins to decline varies from species to species. Reduction in the rate of growth is often accompanied by a decrease in food conversion efficiency (FCE);[118] therefore, the costs of doubling the weight of fish with declining growth rates can increase considerably. If the animals have reached marketable size when the change in growth rate occurs, it often makes good economic sense to harvest them at that time.[119]

Part of the economics associated with marketing aquaculture animals, at least in temperate climates, also relates to the size attained by the animals at the end of the growing season. If they are marketable at the end of the growing season but might bring a better price if grown somewhat larger, the costs involved in overwintering the fish need to be taken into consideration. Those costs include water, labor, energy, and possibly food.[120]

Phillips (1972) discussed some of the factors that relate to energy requirements in fish. In general, the requirements for energy are somewhat greater for carnivores than herbivores because more energy is needed to eliminate the higher levels of nitrogenous wastes produced when animal protein is digested than those resulting when dietary protein comes primarily from plants. Similarly, animals that consume diets high in protein use more energy than do animals that consume low-protein diets. Diets with high mineral levels also tend to increase metabolic rates because of increased demands on osmoregulation to rid the body of excess salts.

Energy requirements increase during periods of gametogenesis and spawning. The energy drain due to reproduction may be so great that the adults die or are severely weakened after spawning (e.g., many salmonids). The quality of aquatic animals after spawning may be reduced to the point where they are not suitable for human consumption or at least are much lower in value than before spawning; salmonids and oysters[121] are examples. Most aquaculture species are harvested be-

---

[118] FCE and food conversion ratio (FCR), along with the calculations for them are discussed in more detail in a later section of this chapter.

[119] An exception would be when the value of a larger fish is higher per kilogram than that of a smaller, though marketable animal.

[120] A low level of winter feeding is appropriate for some species. Others may not be fed. In any case, overwintering weight loss must be replaced before additional growth can be obtained during the subsequent growing season.

[121] Sex-reversed fish and molluscs are exceptions, but since they do not spawn, they are excluded from the statements made here.

fore reaching adulthood and thus before food energy is diverted from growth to gamete production, which will lead to reduced growth and reduced FCE.

The metabolic rate of aquaculture species is controlled to a large extent by water temperature, but other physical and chemical conditions in the culture water can also influence energy utilization. Low levels of dissolved oxygen (DO) lead to increased respiration rates, at least to a point. When the DO becomes extremely low, some species compensate by reducing their rate of metabolism, including their respiration rate. Soft water may lead to increased metabolic rate because of stress related to insufficient levels of divalent cations. Increased energy expenditure is required to move those ions into the body against the concentration gradient. High ammonia levels increase metabolic rate, as does exposure to various organic pollutants. Under severe stress, metabolism may decrease as a means of conserving energy.

Energy requirements are increased in flowing water because the animals must swim more rapidly to maintain station. Exercise resulting from orientation with currents, while costing energy, may be important for maintaining muscle tone, which can translate into better quality in the final product. So, while aquaculturists usually attempt to divert as much food energy to growth as possible, there may be circumstances in which using energy for other purposes, like swimming, which might be considered wasteful to some culturists, make sense in terms of how the consumer perceives the product.

Not all of the energy present in the diet is used for growth, of course. In fact, not all dietary energy goes toward a combination of growth and metabolism. Before the energy can be metabolized or used for growth, it must be absorbed into the body of the animal. The amount of energy in the food is known as gross energy (GE). It can be determined by bomb calorimetry, which measures the amount of heat given off when a feed sample is burned in the presence of pure oxygen. Bomb calorimetry is also used to measure energy levels in feces and tissues.

The amount of energy that is actually available for metabolism and growth is called digestible energy (DE). DE is determined by subtracting GE in the diet from the energy present in the feces (FE):

$$DE = GE - FE.$$

DE is relatively easy to determine in most aquaculture species. If the animals are on prepared feed, pellet samples are readily available and fairly uniform in energy content. Feces can be collected, though that is not always a simple matter since they may dissolve rapidly in the water. A great deal of work has gone into development of methods for collecting fish and invertebrate feces, and at least some systems work fairly well with some species.

The physiological fuel value (PFV), also called the metabolizable energy (ME) value, of a food is the amount of energy that is absorbed by the animal after digestion. Because of physiological differeces, the ME of a given food item will vary from one species to another. In addition, ME can be affected by environmental conditions.

To determine ME, the amount of energy excreted by the gills through respiration (RE) and the energy in the urine (UE) must be subracted from DE according to the following formula:

$$ME = DE - (RE + UE)$$

Livestock feeds are usually formulated to meet the ME requirements of the animals. Feces are relatively easy to collect, RE can be measured by collecting respiratory gases and measuring the energy therein, and the urine can be collected without difficulty. To determine the RE of a fish (by collecting the combustible products excreted by the gills) is not so simple, nor is collecting fish urine. In fact, determining the RE and UE of fish is extremely difficult, so difficult, in fact, that most fish feeds, if formulated on the basis of energy, are based on DE rather than ME values. Several years ago, fish nutritionists often used ME values from poultry or swine for formulating fish feeds, but there is little reason to believe that the values are transferable from one species to another and that practice has been dropped.

The energy-containing components of foodstuffs are proteins, lipids, and carbohydrates. The GE levels contained in those food groups are 5.65, 9.40, and 4.15 kcal/g, respectively (White et al. 1964). By utilizing pure sources of the energy-bearing foodstuffs to feed a variety of animals, researchers have determined that the average ME for protein, carbohydrate, and lipid average 4, 4, and 9 kcal/g, respectively. Those values are virtually never obtained from practical-feed ingredients since most are not pure protein, lipid, or carbohydrate, nor are they 100% digestible. If the PFV of a diet is based on the 4, 4, 9 values for energy, there may be a considerable overestimate of the energy available in that diet. This may be more true of aquaculture diets than those formulated for terrestrial animals, at least with respect to some feed ingredients. For example, while the PFV of starch in livestock feed is nearly 4 kcal/g, channel catfish are able to extract only about 2.5 kcal/g from the same feedstuff (Wilson 1977b).

Not only should a feed have the appropriate level of DE, it should also have the proper amount of protein, and as we shall see in the following section, protein quality is as important or more important than quantity. Since the proper amounts of energy and protein both affect fish performance, the ratio of energy to protein (E:P) in the feed is an important consideration. The ratios have been reported as energy:protein and as protein:energy (P:E). E:P ratios are typically reported in units of kcal DE/g dietary protein, while P:E ratios have often been reported in mg protein/kcal DE or g protein/kcal DE (Table 22). Energy has also been reported in kilojoules per unit weight (e.g., kj/g), as was done in a study on *Penaeus semisulcatus* by El-Dakour (1986). Some standardization of reporting units would simplify comparisons from study to study and species to species.

Apparent optimum E:P ratio (or P:E ratio) can be affected by a number of factors. Included are fish size as seen for channel catfish in Table 22. Other factors

TABLE 22. Reportedly Optimum Protein-Energy and Energy-Protein Ratios for a Few Species of Aquaculture Interest

Species	Protein/energy Ratio or Energy/protein Ratio	Reference
Indian carp (Labeo rohita)	95 mg protein/kcal	Das et al. (1991)
Catla catla	124.8 mg protein/kcal	Singh and Bhanot (1987)
Silver carp (Hypopthhalmichthys molitrix)	118.4 mg protein/kcal	Singh (1990)
Eel (Anguilla anguilla) 40–120-g eels	100 mg protein/kcal	Degani et al. (1987)
Tilapia (Tilapia nilotica)	400 kcal/100 g protein	Yong et al. (1989)
Catfish (Ictalurus punctatus)		
General requirement	8 to 9 kcal/g protein	Robinson and Wilson (1985)
General requirement	11.4 kcal/g protein	Smith (1989)
Swim-up fry	126 mg protein/kcal	Winfree and Stickney (1984)
0.2-g fry	122 mg protein/kcal	Winfree and Stickney (1984)
1.7-g fry	117 mg protein/kcal	Winfree and Stickney (1984)
3–5-g fingerlings	100 mg protein/kcal	Winfree and Stickney (1984)
Crayfish (Procambarus clarkii)	120 mg protein/kcal	Hubbard et al. (1986)
Lobster (Homarus americanus)		
Eyestalk ablated	3.83 kcal/g protein	Koshio et al. (1990)
Unablated	3.63 kcal/g protein	Koshio et al. (1990)
Freshwater shrimp (Macrobrachium rosenbergii)	3.84 kcal/g protein	Gomez et al. (1988)
Shrimp (Penaeus monodon)	2.85–3.70 kcal/g pro.	Bautista (1986)
	4.13 kcal/g protein	Hajra et al. (1988)

that can influence the parameter include water temperature (Degani and Gallagher 1988) and dietary cellulose content (Yong et al. 1989).

An indirect means of looking at how effectively dietary protein is being used for growth is a measurement known as protein conversion efficiency. This parameter relates dietary protein with the amount of that protein that is converted into new tissue. Machiels (1987) evaluated several protein sources in feeds offered to the African catfish, *Clarias gariepinus*, and reported a protein conversion efficiency of 60%.

Most of the energy used by fish in nature comes from protein, which is required for growth. Energy for maintenance metabolism may be derived from lipids and carbohydrates. An ideal diet would be one in which all the lipid and carbohydrate energy was used to support metabolism, freeing up all the energy in protein for growth. That freeing up of protein energy is known as protein sparing. A properly formulated diet can produce at least some protein sparing. The concept of protein sparing by carbohydrate in salmonid diets has been discussed by Halver (1972). Stickney (1977) and Wilson (1977a) discussed the subject with respect to channel catfish. When lipids and carbohydrates are provided in excess, nutritional diseases may develop. For example, salmonids may store high levels of glycogen in the liver and exhibit signs of diabetes when fed too much carbohydrate, whereas high levels of dietary lipid can lead to fatty livers in a variety of animals. In channel catfish high levels of dietary lipid may give rise to the presence of excessive concentrations of visceral fat, which is lost during processing and decreases the dress-out percentage of the fish. When feeds containing the same amount of lipid are fed to channel catfish at different environmental temperatures, there is a trend toward increased lipid concentration with increasing temperature (Andrews and Stickney 1972).

Practical animal diets are composed of minerals, vitamins, water, and fiber in addition to the energy-bearing components. Researchers have made significant strides in the last several years in defining the nutritional requirements of a number of aquaculture species with respect to many dietary factors. The next five sections of this chapter provide background on proteins, lipids, carbohydrates, vitamins, and minerals and provide an indication of the known requirements of each with respect to some representative species.

PROTEIN

Proteins make up much of the structure of an animal's body, comprising the muscles and other connective tissues. They are also important as enzymes, which catalyze the thousands of biochemical reactions that are required to maintain life. Although, as we have seen, proteins can be metabolized for energy, one goal of aquaculture nutrition is to utilize as much dietary protein as possible for growth, allowing carbohydrates and lipids to provide metabolic energy.

Several types of proteins have been identified. Those that make up connective tissues and tendons in animals are called fibrous proteins. Conjugated proteins have another food group attached to them. Mucoproteins and glycoproteins are attached to carbohydrates, and lipoproteins are attached to lipids. Enzymes makes up another entire class of proteins.

Proteins are composed of long chains of amino acids. Each amino acid has the general formula

$$
\begin{array}{c}
\text{H} \\
| \\
\text{R---C---COOH} \\
| \\
\text{NH}_2
\end{array}
$$

where R is an organic radical or hydrogen.

In addition to carbon, hydrogen, oxygen, and nitrogen, three amino acids may contain sulfur: methionine, cystine, and cysteine. A particular protein molecule may contain thousands of amino acid residues, but there are only about 20 different amino acids involved (Figure 76).

Animals are able to synthesize some of their amino acids (nonessential amino acids) from carbohydrates, lipids, and some type of nitrogen compound, including

**TABLE 23. Qualitative Amino Acid Requirements of
Aquaculture Animals (Source: National Research
Council (1973)** [a]

Essental Amino Acids	Nonessential Amino Acids
Arginine	Alanine
Histidine	Aspartic acid
Isoleucine	Cystine
Leucine	Glutamic acid
Lysine	Glycine
Methionine	Proline
Phenylalanine	Serine
Threonine	Tyrosine
Tryptophan	
Valine	

[a] The National Research Council list was based on data from channel catfish and trout but appears to be universally applicable.

other nonessential amino acids. Animals require at least one sulfur-bearing amino acid in their diets, though if the required one is present in sufficient quantity, other sulfur-bearing amino acids can be synthesized from it.

Ten amino acids are required in the diets of animals. Those are the essential amino acids and may be universally required throughout the animal kingdom. Their essentiality has been demonstrated in each of the fishes studied to date (reviewed by Millikin 1982). The essential and nonessential amino acids are listed in Table 23.

Diets should be formulated to include the essential amino acids in the proper percentages of total protein, if those percentages are known, as is the case with at least a few species (Table 24). Limited information is available on other species. For example, the requirement of red drum for sulfur-bearing amino acids was estimated at 1.06% by Moon and Gatlin (1991). The same authors also determined that cystine was able to spare approximately 40% of the methionine requirement.[122] The lysine requirement for red drum has been placed at 1.55% of dry diet (Craig and Gatlin 1992). Information is lacking on the other essential amino acids with respect to red drum. Kim et al. (1992) reported that rainbow trout require lysine and arginine, respectively, at 3.71 and 4.03 percent of diet.

Cho et al. (1992) found that there may be some distinct differences within at least some essential amino acid requirements among closely related species. They reported that the arginine requirement for rainbow trout is somewhat less than that reported for salmon and called for additional studies to confirm the accepted value for that essential amino acid with respect to the salmonids.

Lacking precise information on the quantitative amino acid requirements of a species, aquaculture nutritionists will often formulate diets based on known or per-

[122] Substituting one amino acid to spare another is analogous to the concept, mentioned in the previous section, of sparing protein with dietary lipid or carbohydrate.

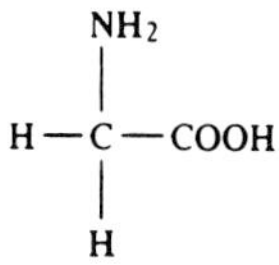

Glycine

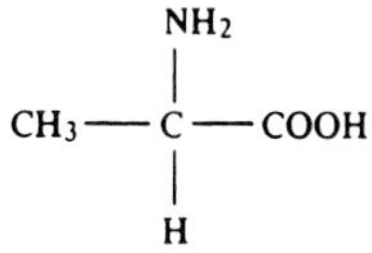

Alanine

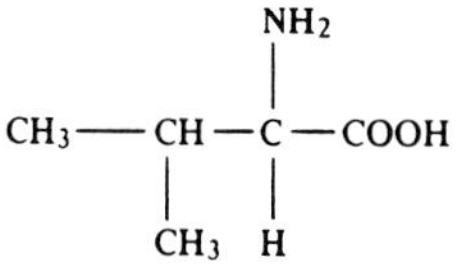

Valine

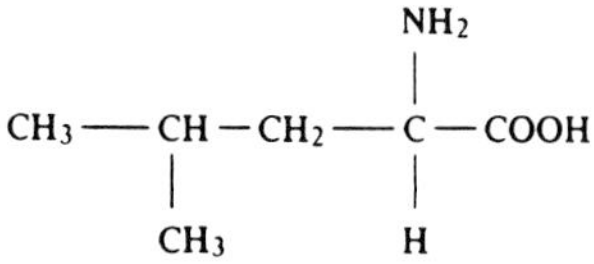

Leucine

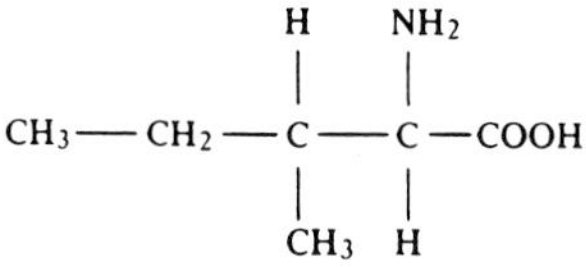

Isoleucine

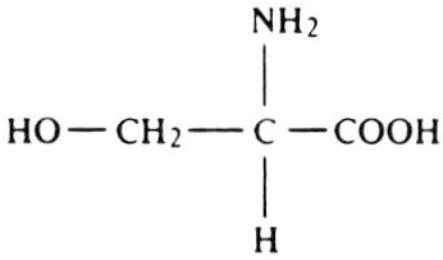

Serine

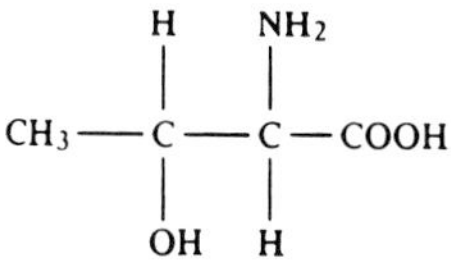

Threonine

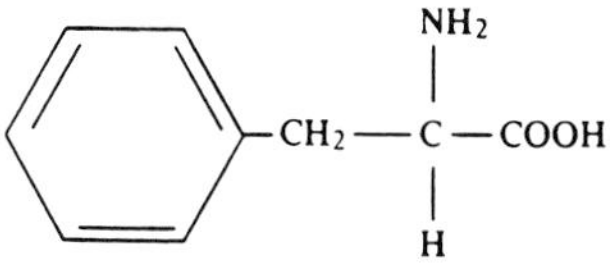

Phenylalanine

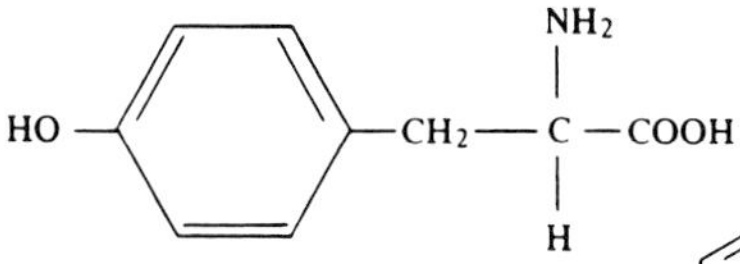

Tyrosine

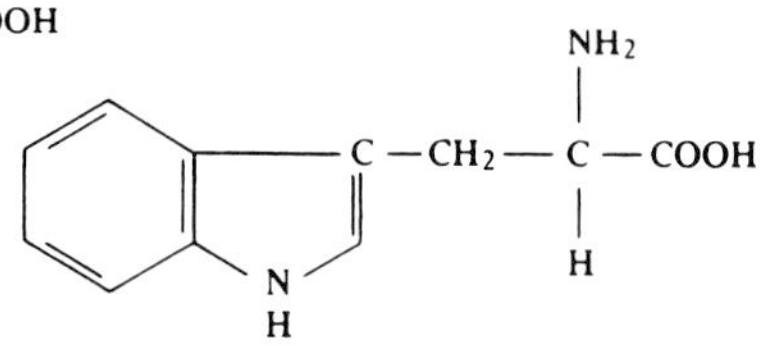

Tryptophan

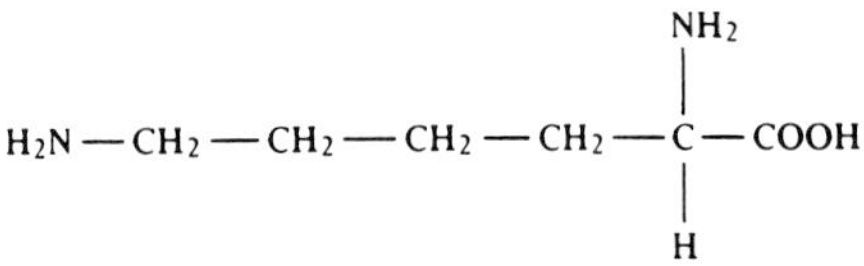

Lysine

Figure 76. Structural formulas of common amino acids.

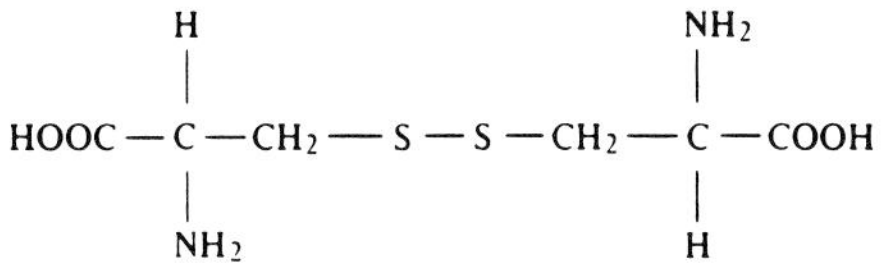

$$HOOC - \underset{\underset{NH_2}{|}}{\overset{\overset{H}{|}}{C}} - CH_2 - S - S - CH_2 - \underset{\underset{H}{|}}{\overset{\overset{NH_2}{|}}{C}} - COOH$$

Cystine

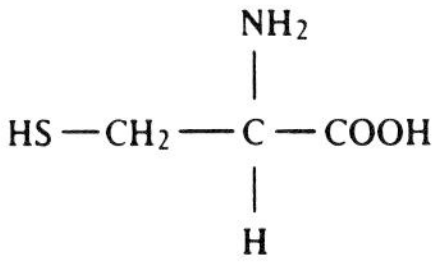

$$HS - CH_2 - \underset{\underset{H}{|}}{\overset{\overset{NH_2}{|}}{C}} - COOH$$

Cysteine

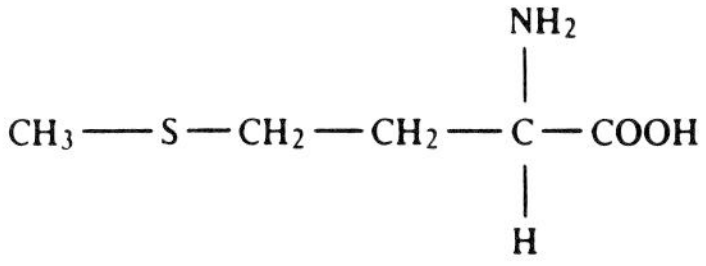

$$CH_3 - S - CH_2 - CH_2 - \underset{\underset{H}{|}}{\overset{\overset{NH_2}{|}}{C}} - COOH$$

Methionine

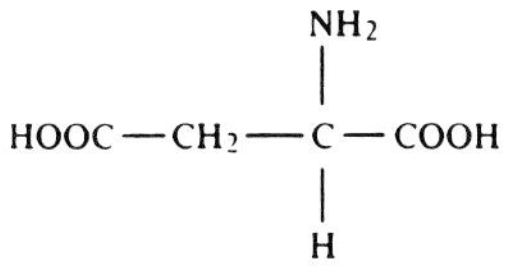

$$HOOC - CH_2 - \underset{\underset{H}{|}}{\overset{\overset{NH_2}{|}}{C}} - COOH$$

Aspartic acid

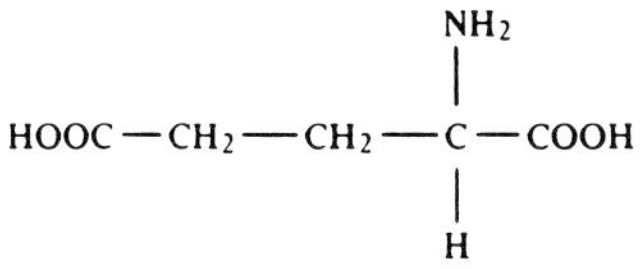

$$HOOC - CH_2 - CH_2 - \underset{\underset{H}{|}}{\overset{\overset{NH_2}{|}}{C}} - COOH$$

Glutamic acid

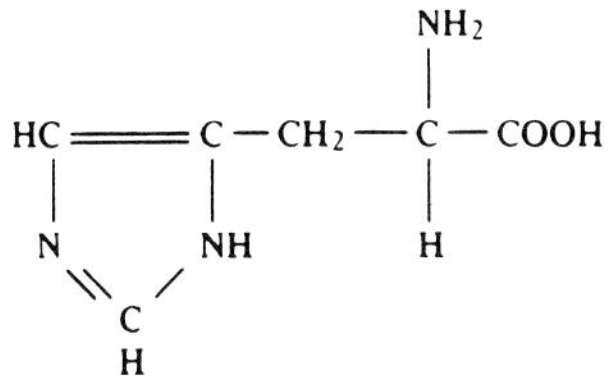

$$HC = C - CH_2 - \underset{\underset{H}{|}}{\overset{\overset{NH_2}{|}}{C}} - COOH$$

Histidine

$$H_2N - \underset{\underset{NH}{\|}}{C} - NH - CH_2 - CH_2 - CH_2 - \underset{\underset{H}{|}}{\overset{\overset{NH_2}{|}}{C}} - COOH$$

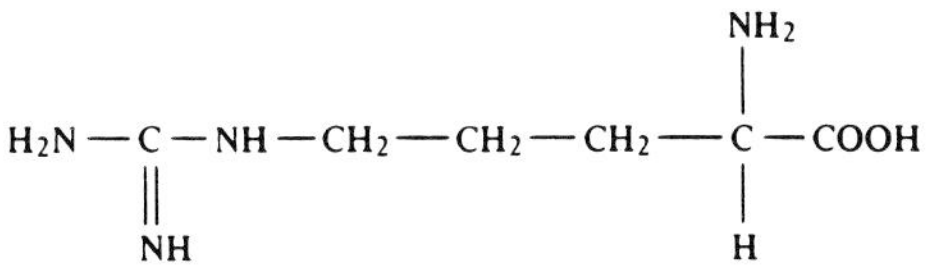

Arginine

Figure 76. (*continued*).

TABLE 24. Quantitative Amino Acid Requirements of Some Aquaculture Species Expressed as Percent of Protein

Amino Acid	Japanese Eel[a]	Common Carp[a]	Channel Catfish[a]	Chinook Salmon[a]	Milkfish[b]
Arginine	4.5	4.2	4.3	6.0	5.3
Histidine	2.1	2.1	1.5	1.8	2.0
Isoleucine	4.0	2.3	2.6	2.2	4.0
Leucine	5.3	3.4	3.5	3.9	5.1
Lysine	5.3	5.7	5.0	5.0	4.0
Methionine	5.0[c]	3.1[c]	2.3[c]	4.0[c]	2.5[d]
Phenylalanine[d]	5.8[e]	6.5[e]	5.0[e]	5.1[e]	4.2[f]
Threonine	4.0	3.9	2.0	2.2	4.5
Tryptophan	1.1	0.8	0.5	0.5	0.6
Valine	4.0	3.6	3.0	3.2	3.6

[a] National Research Council (1983)
[b] Borlongan and Coloso (1993)
[c] In the absence of cystine
[d] In the presence of 0.75% cystine
[e] In the absence of tyrosine
[f] In the presence of 1.0% tyrosine

ceived dietary protein requirement. However, since practical protein sources vary considerably in amino acid distribution, formulation on the basis of only protein percentage can lead to deficiencies in essential amino acids. For example, animal protein is required in virtually all aquaculture diets if the requirements for such amino acids as lysine, arginine, and methionine are to be met.[123]

Fish meal is the animal protein of choice in aquaculture diets. Various sources of fish meal are available, with the most commonly used fish meals in the United States being menhaden, herring, and anchoveta meal. The meals vary in proximate composition,[124] not only among species of fish used to manufacture the meal, but also from batch to batch within meals. Other animal protein sources such as poultry byproduct meal and meat and bone meal[125] can also be used in aquaculture diets when the alternative meals are available at competetive prices and in sufficient quantity.

Animal protein is the most expensive of the major ingredients in aquaculture

[123] Certain plant proteins used in the diets of at least some aquaculture species may be deficient in no more than a single amino acid, but it is generally true that animal protein is required at some level to fulfill all the amino acid requirements of aquatic animals.

[124] To determine the proximate composition of a feed ingredient such as fish meal, the meal is analyzed for moisture, protein, lipid, and ash (and the carbohydrate level by subtracting the total of those measurements from 100%).

[125] Meat and bone meal, a byproduct of the animal rendering industry, has been found to perform identically to fish meal when substituted on a one-to-one basis in practical diets (Robert R. Stickney, unpublished information). The source and quality of the meat and bone meal is critical for proper performance, as the product can vary considerably.

feeds. Nutritionists try to utilize as little animal protein as possible and depend on vegetable protein for the remainder. Meals made from soybeans, peanuts, wheat, corn, cottonseed, and rice are popular plant protein sources in aquaculture feeds. The major concern is maintenance of the proper amino acid balance, though for many species, information on amino acid requirements is lacking. In the absence of such information experiments can be conducted with a series of diets that feature increasing substitution of animal protein with plant protein. When those diets are fed to experimental fish for a period of several weeks, the maximum level of substitution should become apparent from growth and other measurements.

In many parts of the world, animal protein for use in aquaculture diets is a scarce commodity, may be of poor quality, and is high in price. Even the plant proteins that are typically found in great abundance in the United States (corn, wheat, and soybeans) are often difficult to obtain and expensive in developing nations. In the United States and around the world, the supply of fish meal and its price have fluctuated significantly over the years. The El Niño phenomenon[126] has led to population crashes in the anchovetas upon which the Peruvian fish meal industry is based, thereby affecting the cost of manufacturing fish feeds and enhancing the search for alternative protein sources. There have been numerous attempts to use locally available, and sometimes highly exotic, feedstuffs as protein sources in aquatic animal feeds. Examples of some of them are presented in Table 25. Some of the alternatives that have been evaluated represent agricultural wastes that have no use in livestock or human foods, and there seems to be a working hypothesis that aquatic species might thrive on such wastes. That hypothesis is often incorrect since aquatic animals typically require feedstuffs of a quality as high or higher than those required by mammals.

There may be instances where relatively small quantities of alternative proteins will serve as attractants in fish feeds. Murai et al. (1985) examined krill meal and earthworm meal at the 5% inclusion level, along with glycine, sucrose, and mussel extract in diets fed to *Penaeus monodon*. All of the attractants led to improved growth over a control diet that contained full-fat soybean meal (soybean meal made without extracting the oil), fish meal, and shrimp meal as protein sources.

Replacement of some fish meal or even a considerable percentage of it with nontraditional protein sources is often possible,[127] though the typical result is that as the rate of replacement increases, there is a decrease in performance. In many instances where plant protein is used for direct replacement of fish meal, the result is a diet with lower protein because fish meal tends to have a considerably higher protein level than the meals used for substitution. Some reduction in performance is also commonly seen when isonitrogenous diets are used; that is, when fish meal is replaced with sufficient plant protein to maintain the initial protein level. Differ-

[126] The phenomenon involves a crash of phytoplankton blooms off the west coast of South America during years when the winds do not blow in a direction that brings nutrient-rich deep water to the surface off Peru and Chile. Crashes of the phytoplankton blooms leads to drastic reductions in the production of the anchovetas that support the fish meal industry.

[127] When an alternative protein source does perform well in aquatic species, it will also tend to perform well as a livestock feed.

TABLE 25. Examples of Alternative Protein Sources That Have Been Evaluated for Use in Aquaculture Animal Feeds

Alternative Protein(s)	Reference
Paper-processing sludge	Orme and Lemm (1973)
Brewers single-cell protein	Windell et al. (1974)
Mixtures of blood meal and cattle rumen contents	Reece et al. (1975)
Egg-waste protein	Davis et al. (1976)
Coffee pulp	Bayne et al. (1976)
Dried poultry waste	Stickney et al. (1977)
Bird viscera flour	Pezzato et al. (1981)
Pito brewery waste (from sorghum fermentation)	Oduro-Boateng, and Bart-Plange (1988)
Industrial single-cell protein (Eurolysine Fodder Protein)	Davies and Wareham (1988)
Cabbage	Rajadevan and Schramm (1989)
Clitoria ternatea leaf (common Indian weed)	Raj (1989)
Kikuyu grass	Rajadevan and Schramm (1989)
Palm kernal meal	Ipinjolu et al. (1989)
Processed grass clippings	Bender et al. (1989)
Earthworms	Mossmann et al. (1990)
Insects	Reviewed by Ramos Elorduy (1990)
Various aquatic plants such as hornwort, duckweed and water hyacinth	Reviewed by Wee (1991)
Nontraditional terrestrial plant protein sources such as rapeseed, cassava, mustard, linseed, and sesame	Reviewed by Wee (1991)

ences in the levels of essential amino acids in the various protein sources account for some of the observed differences, though there may also be antinutritional factors present in plant proteins.

An example of a plant protein that contains an antinutritional factor is cottonseed meal. Normal cottonseeds have glands associated with them that contain a yellow pigment called gossypol. Free gossypol has been shown to have an adverse physiological effect on certain animals. Growth in channel catfish is depressed when the level of dietary glanded cottonseed meal exceeds 17.4% (Dorsa et al. 1982), though salmonids seem to tolerate somewhat higher levels (Wolf 1952, Fowler 1980). Glandless cottonseed meal has been developed in which the level of free gossypol is significantly lower than in glanded meal. Robinson et al. (1984a) found that glandless cottonseed meal did not depress growth in channel catfish when fed at the same level of glanded meal that did have a depressive effect on growth. Tilapia *(Tilapia aurea)* performance was poorer when the fish were fed diets containing either glanded or unglanded cottonseed meal than on diets containing peanut meal or soybean meal (Robinson et al. 1984b). Dietary gossypol level did not appear to influence the growth response of tilapia.

Not all of the alternative protein research has been aimed at finding plant proteins to replace some or all of the animal protein in aquaculture feeds. There have also been attempts to replace either expensive plant proteins with less expensive ones, or to replace unavailable plant proteins with those that are locally grown.[128] Using rice as a feed ingredient is logical in much of tropical Asia where rice is a staple. Similarly, peanut meal is readily available in the southern United States and can replace corn in catfish diets. In Canada, rapeseed (canola) meal has received a good deal of attention in recent years with respect to its potential for use in fish feeds. Davies et al. (1990) found that as much as 15% rapeseed could be used as soybean meal replacement in diets fed to *Tilapia mossambica* before performance was negatively affected. The overall nutritive value of rapeseed in salmonid and warmwater fish feeds has been reviewed by Higgs et al. (1988).

How a protein source is processed can have a significant bearing on digestibility and utilization. Heating may make a particular ingredient more or less digestible depending on the amount of heat employed and the duration of exposure. Many oilseed meals have the lipid component removed before the meal is manufactured. Examples are soybean and peanut meal. Experiments conducted with aquaculture species fed diets containing full-fat soybean meal, as compared with soybean meal from which the fat was extracted, have demonstrated improved performance in some instances and no response in others. Tacon et al. (1983) found that either extracted or full-fat soybean meal could replace up to 75% of the Peruvian fish meal in the diet of rainbow trout without affecting fish performance. Various other studies with full-fat soybean meal have also been conducted, including one with the shrimp *Penaeus vannamei* (Lim and Dominy 1992) in which the full-fat meal replaced solvent-extracted (defatted) soybean meal at various rates of replacement. No significant differences in performance or proximate composition were observed.

Digestibility can vary considerably from one protein source to another. Once again, the method of feed manufacture can influence the digestibility of an ingredient. The effects of both protein source and method of treatment on performance have been clearly demonstrated in channel catfish by Wilson and Poe (1985a) as shown in Table 26. Soybeans contain a trypsin enzyme inhibitor that has a limiting effect on fish growth, as demonstrated in such species as channel catfish (Wilson and Poe 1985b), carp (Viola et al. 1983), and tilapia (Wee and Shu 1989). Proper heating of soybeans will reduce or eliminate the problem. Trypsin inhibitor can be reduced even when the beans are heated only to 90 or 95°C (Abel et al. 1984), and exposing them to 100°C for 1 hr will eliminate the inhibitor (Wee and Shu 1989). Abel et al (1984) found that oil extracted from heated beans was more resistant to oxidation than oil from unheated beans.

Diets deficient in one or more amino acids can be supplied with those ingredients in purified form, as is commonly done in the livestock feeding industry. Purified amino acids are expensive, and, while there is experimental evidence that they are absorbed and utilized (Wilson 1989), there are few if any commercial diets being manufactured that incorporate them.

---

[128] Economics often enter the picture in both cases. Locally produced plants tend to be less expensive than those that are imported, for example.

TABLE 26. Average Apparent Digestibility of Various Protein Sources by Channel Catfish (Wilson and Poe 1985a)

Protein Source	Protein (%)	Digestibility (%)	
		In Extruded Feed	In Pelleted Feed
Corn grain	9.6	96	97
Cottonseed meal	41.3	80	83
Fish meal (menhaden)	61.1	76	85
Meat and bone meal	50.5	64	61
Peanut meal	48.9	76	74
Rice bran	12.7	78	73
Soybean meal	48.8	93	97
Wheat grain	12.6	88	92

Examination of the literature shows that the apparent protein requirement of fish and aquatic invertebrates of aquaculture interest varies considerably from one study to another. Reasons for differences in results can be attributed to ingredient source as already mentioned, though there are a number of other factors that can affect experimental results. For example, how feedstuffs are treated prior to manufacture and how the finished feed is stored can have a bearing on results. In some cases cooking may make a protein more highly digestible than if the ingredient has not been heated, yet in other cases heating may denature the protein and make it less available. The temperature at which an experiment is conducted and the initial size of the animals are other factors that can influence results, as can general animal health.

While there may well be differences in protein requirements from one species to another, one would anticipate animals within a genus to have similar requirements. That has certainly not been confirmed for marine shrimp in the genus *Penaeus,* where the apparent quantitative protein requirements vary over a considerable range (Table 27). While all the other problems that can influence results of a protein requirement study apply to shrimp, an additional problem is that shrimp do not eat whole pellets. Instead, they consume very small particles (a few microns in diameter). They ingest prepared feeds by breaking off pieces of a pellet. Consumption of a pellet requires a relatively long period. This means not only that some components may not be ingested (they could be discarded or lost as the pellet begins to soften and break up), but also that microorganisms may change the nutrient content before the feed particles are ingested. It is possible, for example, for microorganisms to produce essential amino acids and intact protein from other ingredients; thus, shrimp growth may be excellent even on a diet that is actually deficient as manufactured. There are few hard data on this subject, but given the range of results that have been published on marine and freshwater shrimp nutritional requirements [see New (1976, 1980) for extensive bibliographies], the theory seems supportable.

Information has been developed on the protein requirement of a number of aquaculture species (Table 28). Protein source can have a major influence on the apparent requirement because of differences in digestibility, amino acid balance, nonpro-

TABLE 27. Apparent Protein Requirements of Shrimp in the Genus Penaeus

Species	Apparent Protein Requirement (% of diet)	Reference
P. indicus	43[a]	Colvin (1976)
	up to 60	Sambasivam et al. (1982)
P. japonicus	≥40	Balazs et al. (1973)
	60	Deshimaru and Shegino (1972),
	54	Deshimaru and Kuroki (1974)
	52–57	Deshimaru and Yone (1978)
P. merguiensis	34–42	Sedgwick (1979)
P. monodon	up to 45.8[b]	Lee (1971)
	40[c]	Alava and Lim (1983)
P. orientalis	35.5–50.8[d]	Liang and Wenjuan (1986)
	44	Xu and Li (1988)
P. setiferus	30	Andrews et al. (1972)
	28–32	Shewbart et al. (1973)
P. stylirostris	44	Colvin and Brand (1977)

[a] Based on enhanced growth and food conversion efficiency up to that level; survival was better at 50%.
[b] Growth reportedly inceased with increasing dietary protein from casein or fish meal over a tested range of 18.3 to 45.8%.
[c] Best growth at 40%, but good growth also obtained at 30, 35, and 45% (range tested was 25 to 60%).
[d] Protein requirement increased with increasing body size.

tein dietary nutrients, E:P relationships, feeding rate, and other factors. Thus, the apparent requirement levels are not absolute. As an example of how apparent protein requirement can vary, Ogino (1980) reported that the optimum dietary protein level for both rainbow trout and common carp ranges from 35 to 50% depending on feeding rate. Kin et al. (1991) indicated that the conventionally accepted 40% protein requirement for rainbow trout includes 16% to meet energy needs. Growth would come from the remaining 24%.

The channel catfish industry currently employs mostly 32% protein feed, having converted in the past few years from 35%, though Brown and Robinson (1989) found that in some cases feeds with protein as low as 26% can perform well under certain circumstances. The result was based on the use of high-quality sources of protein, the proper balance of amino acids, the proper E:P, and culturing fish at low to moderate density (7,413 fish/ha in ponds). In contrast, Webster et al. (1992) reported that channel catfish reared in cages may require as much as 38% dietary protein for optimum performance.

LIPID

Lipids are defined as the portion of animal or plant tissue that can be extracted in such solvents as ether, chloroform, and benzene (White et al. 1964). Lipids occur

TABLE 28. Apparent Dietary Protein Requirements of Selected Aquaculture Species

Species	Apparent Protein Requirement (% of diet)	Reference
Acipsner baeri	36–42	Kaushik et al. (1989)
Anguilla japonica	44.5	Nose and Arai (1972)
Aristichthys nobilis fry	30	Santiago and Reyes (1991)
Catla catla	47	Singh and Bhanot (1988)
Channa micropeltes	52	Wee and Tacon (1982)
Chanos chanos	40	Lim et al. (1979)
Chrysophrys major	55	Yone (1976)
Clarias batrachus	30	Chatiyanwongse and Chuapoehuk (1982)
Ctenopharyngodon idella	41–43	Dabrowski (1977)
Cyprinus carpio	35–50	Ogino (1980)
Dicentrarchus labrax	40	Hidalgo and Alliot (1988)
Epinephelus salmoides	40–50	Teng et al. (1978)
Fugu rubripes	50	Kanazawa et al. (1980a)
Heteropneustes fossilis fry	27.73–35.43[a]	Akand et al. (1989)
Hypophthalmichthys molitrix	37–42	Singh (1990)
Ictalurus punctatus	32–36	Garling and Wilson (1976)
Labeo rohita	38	Mazid et al. (1987)
Micropterus dolomieui	45	Anderson et al. (1981)
Micropterus salmoides	40	Anderson et al. (1981)
Morone saxatilis	52	Berger and Halver (1987)
Mylopharyngodon piceus	41[b]	Yang et al. (1981)
Oncorhynchus keta	38–43[c]	Akiyama et al. (1981)
Oncorhynchus mykiss	35–50	Ogino (1980)
Plecoglossus altivelis	37	Arai and Nose (1983)
Sparus macrocephalus	45	Xu et al. (1991)
Tilapia aurea fry	56	Winfree and Stickney (1981)
Tilapia aurea fingerlings	34	Winfree and Stickney (1981)
Tilapia mossambica fry	40	Jauncey (1982)
Tilapia nilotica	25	Wang et al. (1985)
	35	Santiago et al. (1982)
Tilapia zillii	35	Mazid et al. (1979)
Homarus sp.	≤30	Norman-Boudreau and Conklin (1984)
Macrobrachium rosenbergii	13–25	Gomez et al. (1988)
Palaemon serratus	40	Forster and Beard (1973)
Penaeus aztecus	>40%	Venkataramiah et al. (1975)
	51.5%	Zein-Eldin and Corliss (1976)

[a] Range is based on levels of protein in test diets which included 0, 19.87, 23.9, 27.73, 31.33, 35.43, and 39.1%.

[b] Based on feeding trials with black carp that covered a protein range of 5–41% only. Thus, actual requirement could be higher.

[c] Optimum protein level was found to be 43% with 5% dietary fat and 38% with 10% fat.

as fatty acids, triglycerides (neutral fats), phospholipids, glycolipids, aliphatic alcohols and waxes, terpenes, and steroids. For many aquaculture animals the main source of dietary lipids in prepared feeds is triglycerides. Triglycerides are formed when three fatty acids are combined with glycerol:

$$
\begin{array}{llll}
\text{R—COO—CH} & \text{R—COOH} & & \text{CH}_2\text{OH} \\
\;\;\;\;\;\;| & & & \;\;\;\;| \\
\text{R}'\text{—COO—CH} + 3\text{H}_2\text{O} \leftrightarrow & \text{R}'\text{—COOH} & + & \text{HCOH} \\
\;\;\;\;\;\;| & & & \;\;\;\;| \\
\text{R}''\text{—COO—CH} & \text{R}''\text{—COOH} & & \text{CH}_2\text{OH} \\
\\
\;\;\;\;\text{triglyceride} & \;\;\;\text{free fatty acids} & & \text{glycerol}
\end{array}
$$

During digestion, certain enzymes cleave the fatty acids from the triglyceride molecule through hydrolysis. Following absorption into the blood, the free fatty acids and glycerol molecules may recombine to form new triglycerides. The three fatty acids associated with each glycerol molecule may be the same or different (as indicated by R, R′, and R″).

Saturated fatty acids are those that contain no double bonds, whereas unsaturated fatty acids may contain one (monounsaturated) or more (polyunsaturated) double bonds. All fatty acids are composed of carbon, hydrogen, and oxygen and are acyclic, unbranched molecules having an even number of carbon atoms. Polyunsaturated fatty acids are often abbreviated PUFA. The high-molecular-weight fatty acids such as eicosahexaenoic and docosahexaenoic acid (see Table 29) are sometimes referred to as HUFA.

The PUFA have been divided into three major families[129] which contain fatty acids that are required by one or more types of animal. The families are named for the fatty acid with the shortest chain length that is representative of each: oleic, linoleic, and linolenic acids. To simplify the classification of fatty acids and to aid in recognition of members of the three PUFA families, a shorthand system of nomenclature has been devised. Members of the oleic acid family are called n–9 fatty acids, and those in the linoleic and linolenic fatty acid families are referred to as n–6 and n–3 fatty acids, respectively.[130] Using shorthand notation, linolenic acid is abbreviated 18:3n–3 (Table 29), where the number before the colon indicates the number of carbon atoms in the chain (18), the number behind the colon indicates the number of double bonds (3), and the n–3 indicates that the first double bond is located three carbons from the methyl end of the molecule (between the third and fourth carbon atoms from the CH_3 end of the molecule). The list of fatty acids in

[129] A fourth family (n–7) is represented by only one fatty acid, palmitoleic acid, and does not represent a type of fatty acid required in aquatic organisms, though it is often present. The n–11 family also exists but is of limited importance in aquatic animal nutrition.

[130] Alternatively, the n–9, n–6, and n–3 families have been referred to as $\omega9$, $\omega6$, and $\omega3$, but the convention was changed several years ago (perhaps because the Greek letters are not available on many typewriters). The media and some scientists continue to use the ω (omega) system in referring to PUFA families. In particular, the media has touted the supposed health benefits of omega–3 fatty acids.

TABLE 29. Common Names, Structural Formulas, and Shorthand Notations for Representative Fatty Acids

Common Name	Structural Formula	Shorthand Notation
Caproic acid	$CH_3(CH_2)_4COOH$	6:0
Caprylic acid	$CH_3(CH_2)_6COOH$	8:0
Capric acid	$CH_3(CH_2)_8COOH$	10:0
Lauric acid	$CH_3(CH_2)_{10}COOH$	12:0
Myristic acid	$CH_3(CH_2)_{12}COOH$	14:0
Palmitic acid	$CH_3(CH_2)_{14}COOH$	16:0
Palmitoleic acid	$CH_3(CH_2)_5CH = CH(CH_2)_7COOH$	17:1n-7
Stearic acid	$CH_3(CH_2)COOH$	18:0
Oleic acid	$CH_3(CH_2)_7CH = CH(CH_2)_7COOH$	18:1n-9
Linoleic acid	$CH_3(CH_2)_4CH = CHCH_2CH = CH(CH_2)_7COOH$	18:2n-6
Linolenic acid	$CH_3(CH_2)CH = CHCH_2CH = CHCH_2CH = CH(CH_2)_7COOH$	18:3n-3
Arachidonic acid	$CH_3(CH_2)_4CH = CHCH_2CH = CHCH_2CH = CH(CH_2)CH = CH(CH_2)_3COOH$	20:4n-6
Docosahexaenoic acid	$CH_3(CH_2)CH = CHCH_2CH = CHCH_2CH = CHCH_2CH = CHCH_2CH = CHCH_2CH = CH(CH_2)_3COOH$	22:6n-3

Table 29 is far from complete, but it does cover the range of chain lengths generally found in fish.

Fish appear able to interconvert fatty acids within families, but like terrestrial animals, they are unable to convert from one family to another (Owen et al. 1975). Freshwater shrimp, *Macrobrachium rosenbergii,* are able to elongate and desaturate linolenic and linoleic acid (Reigh and Stickney 1989). For example, 18:3n–3 can be converted to 20:5n–3, 22:5n–3, and 22:6n–3, but 18:3n–3 cannot give rise to 18:2n–6 or other fatty acids in the linoleic acid family. The ability to elongate and desaturate, while present, varies in terms of efficiency from species to species. Long-chain n–3 and n–6 fatty acids often have greater essential fatty acid activity than the precursors by some animals and may not be efficiently produced from the 18 carbon precursors.

The melting point of lipids is proportional to their degree of unsaturation, except for the very low-molecular-weight saturated fatty acids, which are liquids at room temperature. Saturated fatty acids with 10 or more carbons are solid at room temperature (25°C) as are some naturally occuring fats made up of combinations of saturated and unsaturated fatty acids, but they have a high percentage of saturated fatty acids. Caprylic acid (8:0) has a melting point of 16°C and capric acid (10:0) has a melting point of 31.5°C. Beef tallow, high in 16:0, 18:0, and 18:1n–9, is solid at room temperature while oilseed oils, high in 18:1n–9 and 18:2n–6, are liquids under the same conditions. Examples of the fatty acid composition of a few typical lipids are presented in Table 30.

There has been a large number of studies on the lipid and fatty acid requirements of fishes and other aquaculture animals. Reviews on the lipid requirements of fishes have been published by Millikin (1982), Cowey and Sargent (1977), Castell (1979), Watanabe (1982), Greene and Selivonchick (1987), and Stickney and Hardy (1989). PUFA are required for membrane permeability and plasticity, enzyme activation, and for various other functions. Fishes differ in their PUFA requirements, with many having a requirement for n–3 or n–6 family fatty acids and even for n–9 fatty acids. Specific requirements vary from species to species and as a function of such environmental variables as temperature and salinity (Castell 1979).

In general, marine fish and salmonids typically require n–3 fatty acids, while nonanadromous freshwater fish may require lipids from the n–3 and/or n–6 families. Freshwater fish typically convert 18:3n–3 to HUFA better than marine fishes, so the latter may require a dietary source of HUFA (Yone 1982). The rainbow trout requirement for n–3 fatty acids can be met with 1% dietary linolenic acid (Castell et al. 1972a, b). Borlongan and Parazo (1991) found that sea bass *(Lates calcarifer)* grew better on a diet containing a 1:1 mixture of cod liver oil and soybean oil than on diets supplemented with either of those lipids alone. That implies that sea bass require fatty acids from both the n–3 and n–6 families. Freshwater shrimp also appear to require fatty acids from both families (Reigh and Stickney 1989).

Tilapia may have a requirement for n–6 or both n–3 and n–6 (Kanazawa et al. 1980b, Takeuchi et al. 1983, Stickney and McGeachin 1983), while grass carp selectively deposit n–6 fatty acids (Cai and Curtis 1989). Attempts to determine the fatty acid requirements of channel catfish have been somewhat inconclusive (re-

TABLE 30. Fatty Acid Composition of Some Dietary Lipids (from Stickney 1971)

Dietary Lipid	Fatty Acid	Percentage in Lipid
Coconut oil	8:0	3.1
	10:0	4.6
	12:0	27.1
	14:0	22.3
	16:0	17.9
	18:0	6.1
	18:1n-9	14.6
	18:2n-6	3.8
	Others	0.4
Beef tallow	14:0	2.9
	16:0	18.8
	16:1n-7	5.1
	18:0	10.9
	18:1n-9	57.5
	18:2n-6	2.1
	Others	2.5
Safflower oil	16:0	5.7
	16:1n-7	1.9
	18:0	1.9
	18:1n-9	12.4
	18:2n-6	72.5
	18:3n-3	2.9
	Others	2.7
Menhaden fish oil	14:0	5.1
	16:0	17.0
	16:1n-7	9.4
	18:0	3.2
	18:1n-9	16.8
	18:2n-6	2.5
	18:3n-3	3.1
	20:3n-9	0.2
	20:3n-6	0.8
	20:4n-3	2.0
	20:5n-3	17.2
	22:5n-3	2.9
	22:6n-3	13.2
	Others	6.6

viewed by Stickney and Hardy 1989). Studies by Satoh et al. (1989b) failed to resolve the issue but did show that n–3 highly unsaturated fatty acids were responsible for enhanced growth.

Diets that exceed 8 to 10% lipid are difficult to manufacture, though by spraying pellets with additional fat, the lipid content can be increased by 5% or more. Atlantic salmon diets are particularly high in lipids, while other fish, such as halibut, will not perform well when fed fatty diets. Experience has shown that Pacific halibut *(Hippoglossus stenolepis)* will reject feed that is high in lipid.[131] Fewer studies have been conducted with invertebrates than fishes. The lipid requirement of the white swamp crawfish appears to be 9% or higher (Davis and Robinson 1986).

With time and exposure to oxygen, the PUFA in a feed will begin to oxidize, forming peroxides. The result is rancid, and potentially harmful feed. At the least, fish fed rancid feed may grow poorly (Ketola et al. 1989). Gross changes in *Tilapia nilotica* fed rancid marine oil included lordosis, exophthalmia, and edema.[132] Histological changes also occurred (Soliman et al. 1983). Lipid oxidation is enhanced by dietary iron, which serves as a catalyst (Desjardins et al. 1987). Antioxidants should be added to the feed to protect the lipids from oxidation. Ethoxyquin is one chemical that is quite effective when added at a small percentage of the diet (e.g., 0.15%). In addition, highly unsaturated oils, such as fish oils, are typically stabilized with ethoxyquin when they are manufactured.

One dangerous side effect of lipid oxidation involves the production of heat. Feeds have actually become sufficiently hot due to lipid oxidation for spontaneous combustion to occur. The result could be destruction of a feed storage building from fire. Other antioxidants, such as vitamin E, are effective in retarding vitamin oxidation, but they do not appear to work as well at reducing oxidation of the lipid fraction of the feed.

Cholesterol is not required in the diet of finfish, but it should be added to formulated feeds offered to crustaceans. Studies with penaeid shrimp indicate that there is a dietary requirement of about 0.5% (Teshima and Kanazawa 1986, Teshima et al. 1989, Zhou and Wang 1991, Chen and Jenn 1991). Freshwater shrimp, such as *Macrobrachium rosenbergii,* may have a somewhat lower cholesterol requirement (Briggs et al. 1988). Phosphatidylcholine, a precursor of cholesterol, can be used to satisfy the cholesterol requirement in marine shrimp. Chen and Jenn (1991) found that the requirement of *P. penicillatus* for sterol could be met with 1.25% dietary phosphatidylcholine.

D'Abramo et al. (1984) found dietary cholesterol as low as 0.12% to be effective for the maintenance of lobster (*Homarus* sp.) growth, while Bordner et al. (1986) obtained good growth of lobsters on diets supplemented with 0.2% cholesterol. The cholesterol requirement of lobsters seems more similar to that of freshwater shrimp than to marine shrimp.

---

[131] Han Wu Liu, personal communication.
[132] Swelling of the tissues.

CARBOHYDRATE

Carbohydrates are, in terms of their chemical constituents, the most simple of the energy-containing food groups. They are made up only of the elements carbon, hydrogen, and oxygen. Sugars and starches are the forms in which carbohydrates appear in nature. More specifically, sugars can occur as simple sugars or monosaccharides (e.g., glucose, fructose, and galactose), or as compound sugars or disaccharides formed by the chemical union of two simple sugars. Examples of disaccharides are sucrose (glucose + fructose), maltose (glucose + glucose), and lactose (glucose + galactose). Complex sugars, known as polysaccharides, are a group of compounds with high molecular weights that are made up of long chains of simple sugars. The complex sugars make up such compounds as starches, cellulose, and hemicellulose.

Carbohydrates provide a significant amount of energy in mammals, but appear to be less useful energy sources in at least some aquaculture animals. Carbohydrates should be used in aquaculture feeds to the extent possible, however. Since carbohydrates represent the least expensive dietary source of energy, as the level of efficiently employed carbohydrate in the diet of an animal increases the need to employ protein as a source of energy is decreased.

The poor utilization of carbohydrates in some species may partly be due to the fact that carbohydrates tend to occur in diets as high-molecular-weight polysaccharides, which may be poorly digested by many aquatic species (National Research Council 1973). Many marine animals, for example, do not encounter such substances in nature and may not be equipped with the proper enzymes to digest them. The diets of fish such as rainbow trout, for example, typically contain low levels of carbohydrate. Trout have only a limited ability to digest starch (Hilton and Slinger 1983, Spannhof and Plantikow 1983) and tend to develop dangerously high levels of liver glycogen when fed diets with excessive carbohydrate levels (National Research Council 1973). Trout appear to absorb simple sugars more readily than complex carbohydrates. Channel catfish, on the other hand, handle starch well (Wilson 1977a, Wilson and Poe 1987), and diets for that species often contain about 40% starch.

Refstie and Austreng (1981) found that different families of rainbow trout demonstrated significant differences in terms of their responses to varying levels of dietary carbohydrates as demonstrated by growth, condition factor, proximate composition, liver weight and color, and dress-out percentage. However, those authors did not believe that selective breeding could be successfully employed to develop a strain of rainbow trout that could tolerate high levels of carbohydrate (their experimental feeds ranged from 15 to 49% carbohydrate in terms of energy) even though genetics is a factor in carbohydrate metabolism in trout.

There is a limited amount of information on carbohydrate utilization in the European eel, *Anguilla anguilla* (Degani et al. 1986, Degani and Levanon 1987, Degani and Viola 1987); striped bass, *Morone saxatilis* (Berger and Halver 1987); yellowtail, *Seriola quinqueradiata* (Shimeno et al. 1985, Furuichi et al. 1986); and red drum, *Sciaenops ocellatus* (Ellis and Reigh 1991). Dietary glucose seems to be less available to the yellowtail than is starch, a finding which also applies to the common carp, *Cypinus carpio,* and the red sea bream, *Chrysophrys major* (Furuichi et al. 1986). The European eel, on the other hand, is more similar to trout in that it has been shown to grow better on diets containing glucose at 20 or 30% than on diets containing corn starch as the carbohydrate source (Degani and Levanon 1987). The red drum utilizes dietary carbohydrate poorly compared with such fishes as the common carp and channel catfish (Ellis and Reigh 1991)

As seems always to be the case with research aimed at determining the nutritional requirements of aquatic animals, factors other than diet composition can affect results. One of those factors is feeding frequency. Tung and Shiau (1991) fed hybrid tilapia diets containing starch, dextrin, or glucose at 44% and found that the fish performed significantly better when fed six times as compared with twice daily. At the two-times-per-day feeding rate, fish fed the diet containing glucose grew poorly compared with those fed the more complex carbohydrates.

Some information on the utilization of carbohydrates by crustaceans has been developed. *Macrobrachium rosenbergii* has been found to grow poorly when fed diets supplemented with glucose but exhibited good growth when the dietary carbohydrate was in the form of starch (Gomez Diaz and Nakagawa 1990). Of the three sugars fed to *Penaeus monodon* by Alava and Pascual (1987), glucose was the poorest nutrient source. Trehalose and sucrose led to better weight gains, and diets containing 20% sugar outperformed those containing 30%.

Piedat-Pascual et al. (1983) reported that complete mortality occurred after 10 days in *P. monodon* juveniles fed diets containing 40% maltose or molasses. Various dietary starches supported growth in the shrimp, though there were significant differences among diets containing cassava starch, corn starch, or sago palm starch, with the palm starch being the best carbohydrate source. Catacutan (1991) fed *P. monodon* diets containing 40% crude protein at carbohydrate (gelatinized bread flour) levels of 5, 15, 25, and 35% and found no significant difference in growth, food conversion, or survival. While not significant, growth and food conversion were poorest in shrimp fed the 35% carbohydrate diet. Xu and Li (1988) indicated that the optimum dietary carbohydrate level for *P. orientalis* was 26%.

Fishes do not manufacture the enzyme cellulase, which breaks down cellulose, so the fiber contained in fish feeds is generally considered to have no nutritional value. Some fishes, such as channel catfish, may contain intestinal flora that produce cellulase (Stickney and Shumway 1974). Whether or not the cellulase activity is sufficient to provide significant levels of simpler carbohydrates that can be further digested or directly absorbed into the bloodstream of channel catfish has not been determined, though Fagbenro (1990) concluded that the presence of cellulase activity in the digestive tract of *Clarias isheriensis* may be responsible for the ability of that species of walking catfish to digest large quantities of blue-green algae (Cyanophyceae) in pond culture. The efficiency of the American oyster, *Crassostrea virginica,* to assimilate crude fiber has been placed at 3% (Crosby 1988, Langdon and Newell 1990). Xu and Li (1988) placed the optimum level of dietary fiber for *Pen-*

aeus orientalis at 4.5%, which would indicate that little or no nutrient value is contained in the fiber fraction. It is difficult to reduce the fiber level in prepared feeds much below that amount.

In general, fiber is kept to a minimum in commercial feeds. Sumagaysay and Chiu-Charn (1991) concluded that a low-protein, high-fiber (24%) diet could be economical for feeding milkfish, *Chanos chanos,* and indicated that the fiber was utilized as a direct or indirect energy source by that species. Mao et al. (1985) indicated that the daily requirement of grass carp, *Ctenopharyngodon idella,* for fiber is 15 to 20%, and Bonar et al. (1990) felt that the rate at which grass carp consume aquatic vegetation is related, in part, to the cellulose content of the plants. Shiloh and Viola (1973), on the other hand, concluded that common carp are unable to utilize dietary fiber as an energy source. Dietary fiber supplementation in the form of cellulose, agar, carrageenan, guar gum, or carboxymethylcellulose has led to decreased growth in tilapia[133] (Shiau and Kwok 1989).

Dietary levels of α-cellulose of 10 and 20% have resulted in growth depression in rainbow trout, *Oncorhynchus mykiss,* and to the conclusion that cellulose digestibility in that species is near zero (Hilton et al. 1983). In a study in which cassava and rice were sources of dietary fiber, protein digestibility was not influenced by high fiber level, though carbohydrate digestibility was poor (Ufodike and Matty 1986).

Since undigestible fiber adds bulk to a feed and may affect rate of passage through the gut, there are indirect effects that can influence performance when high levels of complex carbohydrates are incorporated into the diet. An animal must consume a higher volume of feed containing fiber to obtain the same amount of metabolizable energy available in a more highly digestible diet. Even if the fish are fed on an *ad libitum* basis they may not obtain sufficient food in a day to meet their protein or energy requirements.

VITAMINS

Vitamins are organic compounds that are required by at least some species in small quantitites for normal growth and health (White et al. 1964). Vitamins are considered to be catalytic since they take part in biochemical reactions but are not contained in the end products of those reactions. Most vitamins serve as coenzymes in biological systems. Animals are capable of producing their own enzymes biochemically but are unable to synthesize the vitamin coenzymes (or cofactors, which are mineral components of enzyme systems). All the vitamins known to be required by animals have been chemically identified (Figure 77) and can be synthetically produced in the laboratory.

Vitamins were initially assigned letters of the alphabet, and some of those origi-

---

[133] Agar, guar gum, and carboxymethylcellulose have been used as binders in aquaculture feeds and are considered to have no impact on growth when used at low levels. Binders are discussed in more detail later in this chapter.

nal designations are still in common usage. As the chemical structures of the vitamins became known, the names of the vitamins were changed to reflect the chemical nomenclature. The fat-soluble vitamins (A, D, E, and K) are still widely known
by their alphabetic designations. The water soluble vitamins (B complex and vitamin C) are becoming increasingly commonly known by their chemical names,
though cyanocobalamin is still often referred to as vitamin B_{12} and ascorbic acid is
known to most people as vitamin C.

A variety of signs of hypo- and hypervitaminosis [134] have been found in conjunction with aquatic animals. Most of the work to date has been conducted with finfish,
though invertebrates also have requirements for vitamins and may show deficiency
or toxicity signs. A summary of the types of hypo- and hypervitaminosis signs that
are commonly observed in fish in conjunction with each of the vitamins is presented
in Table 31.

The signs associated with vitamin deficiencies or excesses are not often exclusive
to nutritional problems. Many things can cause fish to lose their appetites and demonstrate poor growth. Exophthalmia may be a sign of gas bubble disease or the
presence of certain infectious diseases. Thus, when fish show one or more signs
associated with vitamin deficiency or excess, other causes of the problem need to
be considered as well.

When present in excess within the diet, water-soluble vitamins can be eliminated
in the urine. Thus, there can be problems associated with hypovitaminosis, but
excessive amounts in the diet are not harmful. Fat-soluble vitamins, on the other
hand, can lead to problems if present in either excessive or insufficient quantities in
the diet since they can be stored in body lipids.

The potency of vitamins can be expressed in any of four ways:

- International Units
  (IU)
  Vitamin activity is compared with an international stan-
  dard controlled by the Expert Committee on Biologi-
  cal Standardization of the World Health Organi-
  zation.
- United States Phar-
  macopoeia (USP)
  Vitamin activity is compared with standards maintained
  in the United States (IU and USP units are often iden-
  tical)
- International Chick
  Units (ICU)
  Vitamin activity is measured in terms of the response
  elicited in chickens.
- Weight
  Activity is shown as milligrams per kilogram of feed.

Most of the vitamin packages used in aquaculture feed formulations employ weight
as the primary measure of vitamin potency. That simplifies formulation since vitamin packages can be added to feed formulations as percentages of total diet. Some
of the fat-soluble vitamins are still typically added in IU or USP units. The vitamins
are provided in conjunction with a carrier, such as a feed grain, and are measured

---

[134] Hypovitaminosis leads to problems associated with vitamin deficiency, while hypervitaminosis problems occur when a vitamin is fed in excess of the animal's requirement.

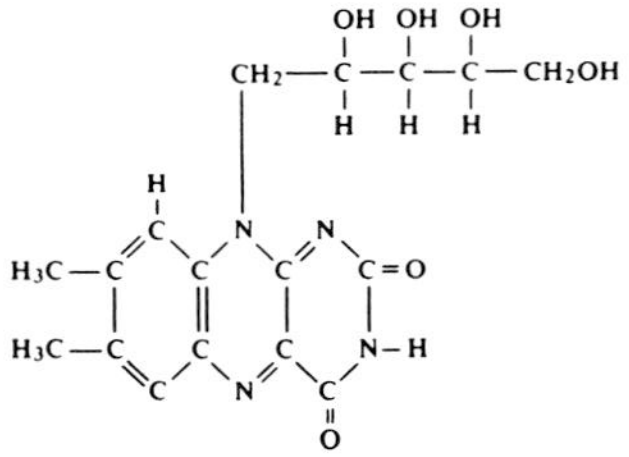

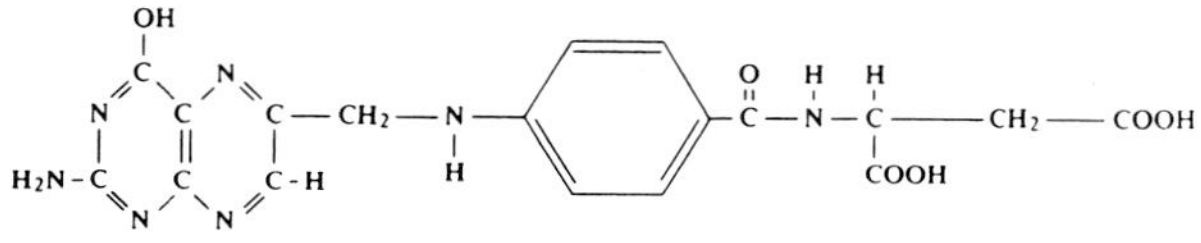

Folic acid

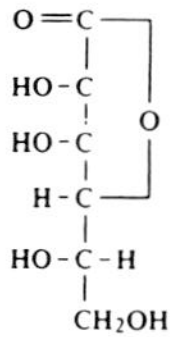

L-Ascorbic acid

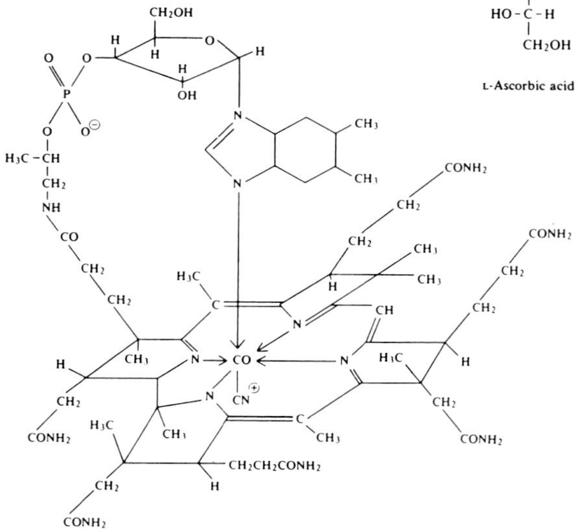

Cyanocobalamin

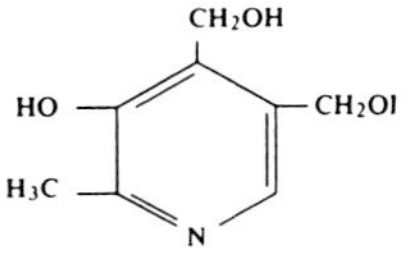

Riboflavin

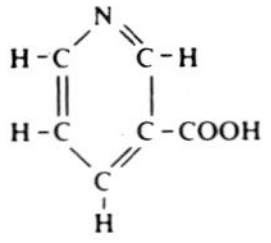

Pyridoxine

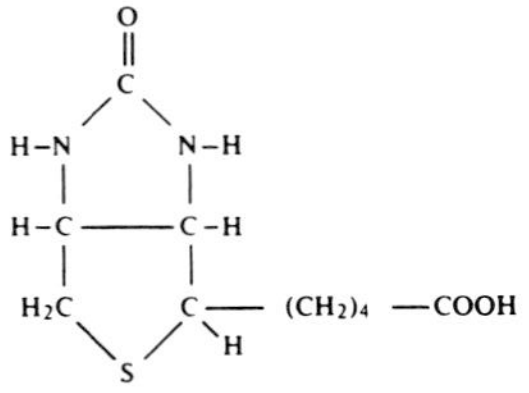

Niacin
(nicotinic acid)

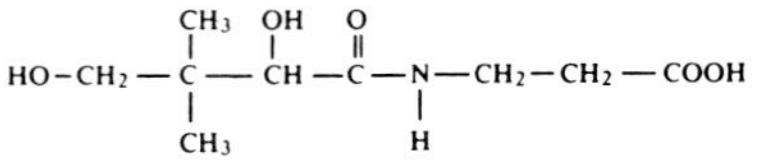

Pantothenic acid

Biotin

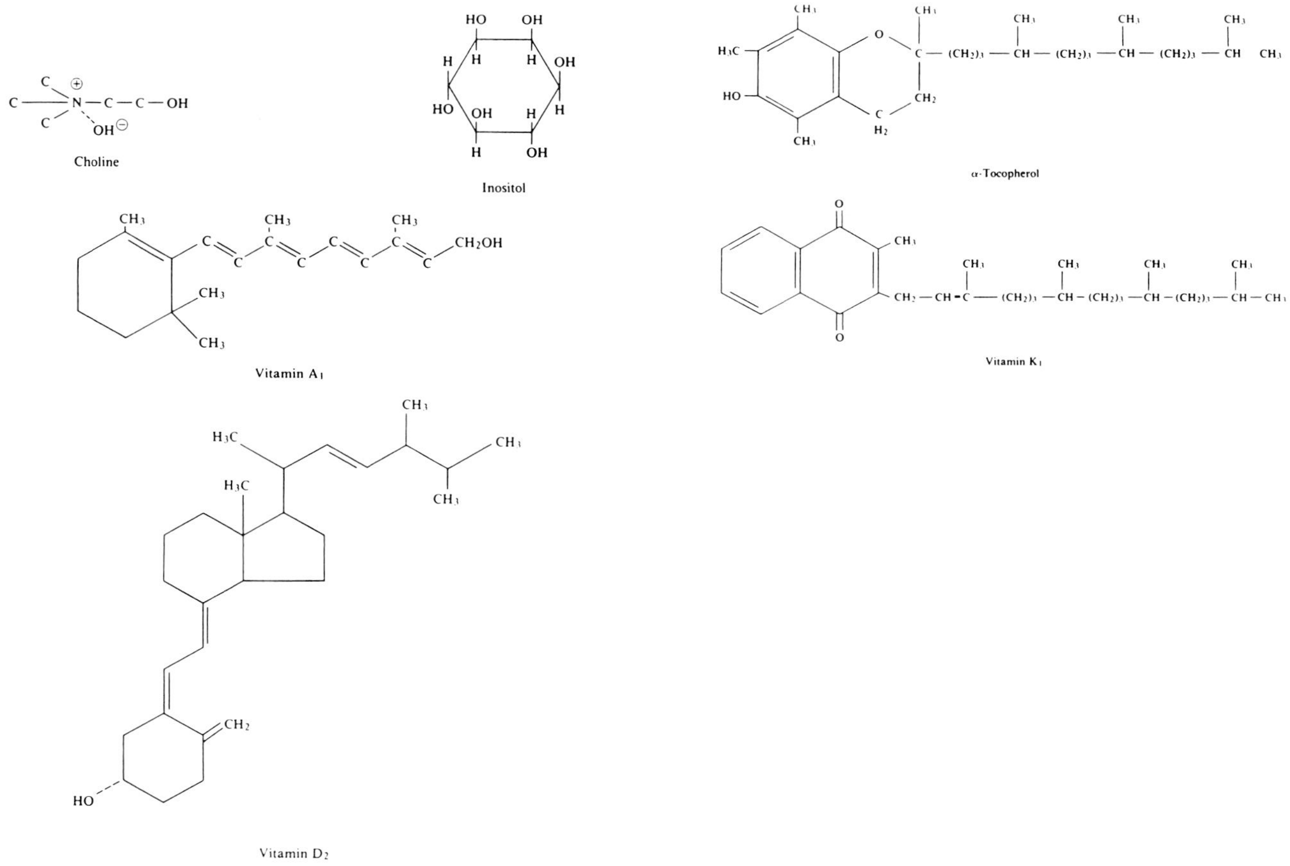

Figure 77. Structural formulas of vitamins required for proper aquatic animal nutrition.

303

TABLE 31. Sources of Vitamins and Signs Associated with Excess or Insufficient Dietary Levels in Fishes (Halver 1972, 1989 Dupree 1977)

Vitamin	Sources	Signs of Hypovitaminosis or Hypervitaminosis[a]
Fat-soluble Vitamins		
A	Fish liver oils, fish meals containing fish oil residues	*Hypovitaminosis:* poor growth, poor vision, night blindness, hemorrhagic areas at base of fins, abnormal bone formation, exophthalmia, edema, retinal degeneration
		Hypervitaminosis: enlargment of liver and spleen, abnormal growth and bone formation, epithelial keritinization
D	Fish oils (some synthesis may occur in fish skin exposed to light)	*Hypovitaminosis:* poor growth, of white skeletal muscle, impaired calcium homeostasis
		Hypervitaminosis: impaired growth, lethargy, dark coloration
E	Wheat germ, soybeans,corn	*Hypovitaminosis:* exophthalmia, reduced growth and survival, anemia, malformation of erythrocytes, elevated body water
		Hypervitaminosis: poor growth, toxic liver reaction, death
K	Green leafy vegetables; soybeans; animal livers	*Hypovitaminosis:* prolonged blood clotting time, anemia, hemorrhagic gills and eyes, lipid peroxidation, reduced hematocrit
Water-soluble Vitamins		
Thiamin	Cereal bran, beans, peas, yeast, fresh organ meats	Poor appetite, loss of equilibirium, lethargy, poor growth, muscle atrophy, convulsions, edema
Riboflavin	Milk, liver, yeast, cereal grains, fresh meats	Opaque eye lens, hemorrhagic eyes, photophobia, dark coloration, poor appetite, poor growth
Pyridoxine	Cereal brans, yeast, egg yolk, liver	Nervous disorders, anemia, loss of appetite, poor growth, flexing of opercles
Panthothenic acid	Cereal bran, yeast, organ meats, fish flesh	Clubbed gill filaments, gill exudate, lethargy, loss of appetite, poor growth

TABLE 31. (*continued*)

Vitamin	Sources	Signs of Hypovitaminosis or Hypervitaminosis[a]
Nicotinic acid	Yeast, legumes, organ meats	Loss of appetite, lesions in colon, tetany, weakness, edema of stomach and colon, poor growth
Folic acid	Yeast, green vegetables, organ meats	Lethargy, dark coloration, poor growth
Biotin	Liver, kidney, egg yolk, yeast, milk products	Loss of appetite, lesions in colon, muscle atrophy, skin lesions, convulsions, poor growth
Cyanocobalamin	Meats and meat byproducts, fish meal	Poor growth, poor appetite, low hemoglobin, anemia
Ascorbic acid	Citrus fruit, beef liver and kidney, fresh fish tissues	Lordosis, scoliosis, impaired collagen formation, hemorrhaging, reduced growth
Inositol	Animal and plant tissue	Poor growth, distended stomach, skin lesions
Choline	Wheat germ, soybean and vegetable meals, organ meats	Hemorrhagic kidneys, enlarged liver, poor food conversion, poor growth

[a] Only signs associated with hypovitaminosis are given for the water-soluble vitamins.

in terms of IU or USP units per gram of carrier; thus, they can still be added on the basis of weight.

The vitamin requirements of a few fish species have been determined in some detail, and limited information exists for others (Table 32). Since some vitamins are degraded by heat, moisture, or light or can be oxidized when exposed to the atmosphere, the apparent requirement for a vitamin can be affected by how a feed is manufactured and stored prior to being offered in a feeding trial. Feed should be stored, as previously indicated, in a cool, dry place. Antioxidants are commonly added to manufactured feeds to retard oxidation of both lipids and vitamins. Frequently used antioxidants include lecithin (phosphatidylcholine), ethoxyquin, BHA, BHT,[135] and vitamin E. Gatlin et al. (1992) found that increasing dietary vitamin E from 60 to 240 mg of dl-α-tocopheryl acetate per kilogram of fish provided increased protection of fillets frozen for 6 months at $-18°C$.

Vitamin C in the form of L-ascorbic acid that has often been supplemented at several hundred percent of the requirement in formulations that are to be extruded. Losses of 80% in vitamin C activity are not uncommon because of the heating that occurs in conjunction with manufacture of extruded feeds. Oxidation of vitamin C

[135] BHA is butylated hydroxyanisol and BHT is butylated hydroxytoluene.

TABLE 32. Vitamin Requirements of Some Aquaculture Species

Vitamin and Units	Amount in Diet					
	Trout[a]	Salmon[a]	Carp[a]	Channel catfish[a]	Tilapia	Shrimp[b]
A (IU)	2000–2500	2000–2500	1000–2000	1000–2000	NA[c]	NA
D (IU)	2400	2400	NA	500–1000	NA	NA
E (mg/kg)	30	30	80–100	30	50–100[d] 10–25[e]	NA
K (mg/kg)	10	10	R[f]	R	NA	NA
Thiamin (mg/kg)	10–12	10–15	2–3	1–3	NA	120
Riboflavin (mg/kg)	20–30 3[h]	20–25	7–10	9	NA	22.3[g]
Pyridoxine (mg/kg)	10–15	15–20	5–10	3	NA	120
Panthothenic acid (mg/kg)	40–50	40–50	30–40	25–50	10[i]	NA
Nicotinic acid (mg/kg)	120–150	150–200	30–50	14	26–121[j]	NA
Folic acid (mg/kg)	6–10	6–10	NA	R	NA	NA
Biotin (mg/kg)	1–1.2	1–1.5	1–1.5	R	NA	NA
Cyanocobalamin (mg/kg)	R	0.015–0.02	NA	R	NA	NA
Ascorbic acid (mg/kg)	100–150	100–150	30–50	60	50[k]	≈10000
Inositol (mg/kg)	200–300	300–400	200–300	R	NA	2000
Choline (mg/kg)	≈700–800[l] >800[o]	6000–8000	1500–2000	400[m]	ND[n]	600

[a] Values in column were compiled by Halver (1989) unless otherwise footnoted.
[b] Values in column were compiled by National Research Council (1983) unless otherwise footnoted.
[c] NA = data not available.
[d] Satoh et al. (1987).
[e] Requirement varies with level of dietary lipid; 10 mg/kg at 3% lipid and 25 mg/kg at 6% lipid (Roem et al. 1990b).
[f] R = required, but precise level not determined.
[g] Chen and Hwang (1992) for *Penaeus monodon*.
[h] For fingerling rainbow trout according to Hughes et al. (1981).
[i] Roem et al. (1991).
[j] 26 mg/kg in diets containing 38% glucose and 121 mg/kg in diets containing 38% dextrin (Shiau and Suen 1992).
[k] Stickney et al. (1984).
[l] Requirement for fingerling rainbow trout decreases with increasing fish size (Rumsey 1991).
[m] Wilson and Poe (1988).
[n] Research has indicated that there is no dietary requirement (Roem et al. 1990a).
[o] Requirement for fry rainbow trout (Poston 1991).

306

after the feed is manufactured is another significant source of loss, which occurs over time. That is one of the reasons that feed should not be stored for long periods.

In recent years, new forms of vitamin C have been developed, two of which are ascorbic acid sulfate and ascorbic acid phosphate. The phosphate form is actually more available, at least to catfish, than the L-ascorbic acid form of the vitamin (Lovell and El Naggar 1988), and it is heat-stable. That form of the vitamin is being increasingly used in fish feeds because of its heat stability and because the amount of vitamin required in the formulation to provide the same level in the final feed can be reduced.

As shown in Table 32, vitamin C is required by all the fish studied to date (Rosenlund et al. 1990, Chavez de Martinez 1990). The same may not be true for all aquaculture species, however. Kean et al. (1985) could not determine an ascorbic acid requirement for juvenile lobsters *(Homarus americanus)* and concluded that the animals may be able to manufacture vitamin C *de novo.*

It has been proposed that large doses of vitamin C provide some protection against disease in animals, including humans. Some studies of the theory with respect to cultured fish have been conducted. For example, Nararre and Halver (1989) found that the resistance of rainbow trout to the bacterium *Vibrio anguillarum* was enhanced if the fish received five to 10 times the dietary vitamin C required for good growth. Megalevels of vitamin C are not routinely fed to aquaculture animals.

Choline is often used in fish diets at a level of about 0.5% of diet (500 mg/kg), yet the requirement has been difficult to demonstrate with respect to some species. Roem et al. (1990a) could not develop a choline deficiency in *Tilapia aurea,* for example. The reason may have been related to the fact that methionine seems to have a sparing effect on choline in at least some fishes (Wilson and Poe 1988), so diets high in methionine may satisfy the choline requirement.

Culture conditions can influence apparent vitamin requirements. Some species, notably penaeid shrimp and tilapia, are able to graze on microflora that grows on the walls of culture chambers. That microflora may contain nutrients that are missing in an experimental feed to the degree that a deficiency cannot be produced. Roem et al. (1990c) were unable to obtain deficiency signs in *Tilapia aurea* fed diets deficient in pantothenic acid or choline in a recirculating system (Roem et al. 1990c). When the study was repeated in a flowthrough system in which ambient sunlight promoted the growth of algae in the culture tanks, the results were the same. In the third repetition of the work, a flowthrough system was used in which extraneous microflora appeared to be eliminated and a requirement for pantothenic acid was established (Roem et al. 1991).

MINERALS

Minerals are required by all animals for various life processes, including the formation of skeletal tissue, respiration, digestion, and osmoregulation. Marine animals live in a medium that contains minerals in concentrations at or above those necessary to meet their requirements, whereas freshwater fish live in a mineral-deficient medium and obtain most of their required minerals from the diet. The concept of salt balance and osmoregulation was considered in Chapter 5.

Minerals are chemical elements that are not produced or destroyed as a result of their functions in life processes. Seven elements, known as the major minerals, are required by animals for proper nutrition. Those minerals account for up to 80% of the inorganic components in the dry weight of an animal. The seven are:

- Calcium
- Phosphorus
- Sulfur
- Sodium
- Chlorine (as chloride ion)
- Potassium
- Magnesium

Additional minerals are required in smaller amounts. Those are known are trace elements and include the following:

- Cobalt
- Copper
- Fluorine
- Iodine
- Iron
- Manganese
- Molybdenum
- Selenium
- Zinc

Not all of the mineral requirements have been determined for many species of aquaculture interest. Some of the available information is summarized in Table 33.

Since the concentrations of minerals in the water is highly variable, it is not surprising that the apparent requirement has varied from one study to another. It has, for example, been difficult to precisely determine the calcium and magnesium requirement for freshwater fish because most fresh waters contain at least some hardness, which, as we have seen, is attributable to those two elements. Studies on the calcium requirements of tilapia and channel catfish conducted by Robinson et al. (1984c, 1986, 1987) were among the first conducted in calcium-free water. In the case of those studies the water source was a well that produced water having a hardness that was virtually zero. The studies on calcium by Robinson and his colleagues, as well as a study on magnesium by Shearer and Aasgaard (1992) clearly demonstrate that the dietary requirement for those elements increases as their concentrations in the water is reduced.

TABLE 33. Mineral Requirements of Representative Aquaculture Species

Mineral and amount in diet	Rainbow trout[a]	Salmon[ab]	Carp[a]	Channel catfish[a]	Tilapia	Eel[a]	Red Sea Bream[a]	Red Drum
Calcium (%)	0.02	0.03	0.03	0.03[a]–0.46[c]	0.17–0.70[de]	0.27	0.34	N[f]
Phosphorus (%)	0.7	0.6–0.7[g]	0.7	0.4[e]–0.8	0.9[hi]	0.3	0.6	0.86[j]
Potassium (%)	NA	0.8[k]	NA	NA	NA	NA	NR[l]	NA
Magnesium (%)	0.05	R[m]	0.05	0.04	NA	0.04	NA	NA
Iron (mg)	NA	60	NA	30	NA	170	NA	NA
Copper (mg)	3	5	3	5	NA	NA	NA	NA
Manganese (mg)	13	20	13	2.4	NA	NA	NA	NA
Zinc (mg)	15–30	R	15–30	20	20[n]	NA	NA	20–25[o]
Iodine (mg)	R	0.6–11[p]	NA	NA	NA	NA	NA	NA
Selenium (mg)	0.15–0.38	R	R	0.25	NA	R	R	NA

[a] Values compiled by Lal (1989) unless otherwise footnoted.
[b] Values for Atlantic salmon unless otherwise footnoted.
[c] Higher value is reported from calcium-free water by Robinson et al. (1986).
[d] Robinson et al. (1984).
[e] Robinson et al. (1987).
[f] Data do not appear to be available.
[g] Values include compilations by Lal (1989) and National Research Council (1983).
[h] Watanabe et al. (1980b) found that the requirement was less than 0.9%.
[i] A dietary level of 0.46% available phosphate was recommended by Haylor et al. (1987).
[j] Davis and Robinson (1987).
[k] Shearer (1988).
[l] Not required.
[m] Required, but precise level not available.
[n] McClain and Gatlin (1988).
[o] Gatlin et al. (1991).
[p] Data from chinook salmon as compiled by Lal (1989).

TABLE 34. Percentage availability of Various Sources of Dietary Phosphorus to Selected Aquaculture Species (National Research Council 1983)

Phosphous Source	Rainbow trout (%)	Carp (%)	Channel catfish (%)
Fish meals			
White fish meal	66	0–18	—
Brown fish meal	74	24	—
Anchovy meal	—	—	40
Menhaden meal	—	—	39
Casein	90	91	90
Plant sources			
Corn, ground	—	—	25
Rice bran	19	25	—
Soybean meal	—	—	29–54
Wheat germ	58	57	—
Wheat middlings	—	—	28
Phytate	$\approx$0–19	8–38	$\approx$0
Phosphates			
Sodium phosphate	$\approx$98	$\approx$94	90
Potassium phosphate	$\approx$98	$\approx$94	—
Calcium phosphate			
Monobasic	94	94	94
Dibasic	71	46	65
Tribasic	64	13	—

The availability of minerals varies from one dietary source to another. Examples of that variability become apparent when Table 34 is examined.

A chemical substance called phytate (a form of phytic acid) is present in various grains. Phytate contains phosphorus but does not seem to be a good source of that element because of poor digestibility. High levels of phytate have been shown to impair growth in various species of fish (Spinelli et al. 1983, Richardson et al. 1986, Satoh et al. 1989a) but do not seem to have a similar effect in shrimp (Civera and Guillaume 1989). In fish, phytate can affect the utilization of such minerals as calcium and zinc (McClain and Gatlin 1988, Gatlin and Phillips 1989, Satoh et al. 1989), and may lead to the development of cataracts (Richardson et al. 1985, 1986).

Most practical aquaculture diets contain suitable levels of trace elements to fulfill the requirements of the target species, though freshwater feeds are generally supplemented with some form of calcium phosphate and, commonly, sodium chloride. Mineral premixes containing trace minerals may also be added, particularly when the level of animal protein in the diet is low (animal proteins such as fish meal and meat and bone meal contain high levels of minerals, particularly in the bone fraction).

OTHER FACTORS

Binders

Practical-feed ingredients usually stick together well and form firm, water-stable pellets when passed through an extruder. Feed ingredients passed through a pellet mill may not bind nearly as well, though the use of steam is often helpful. Wheat is often added to diets at a level of about 2% to aid in binding. Experimental diets are commonly cold-pelleted by running them through a meat grinder and then breaking up the resulting strands of feed into smaller sizes after drying. Chemicals that cause the ingredients to stick together are often added to cold-pelleted feed formulations.

Chemicals that act as binders include agar and alginate (extracts from seaweed), guar gum, carboxymethylcellulose, and gelatin. Most binders are primarily or solely comprised of starch, though gelatin is a protein. Starch binders are often not digestible and therefore add no metabolizable energy to the feed. Storebakken (1985) found that feed intake and the digestibility of protein and fat were reduced in rainbow trout fed diets containing alginate or guar gum as binders. Thus, binders should not be added at levels higher than those necessary to achieve the required pellet stability.

The amount of binding agent required in a diet formulation depends on the composition of the diet and the type of binder that is employed. In most cases 1 or 2% of the diet is comprised of binder; however, added water stability can be obtained by increasing that percentage. Animals that feed slowly by nibbling pieces off feed pellets (e.g., shrimp) may be well served by having pellets that remain intact for several hours. Meyers et al. (1972) found that crustacean feeds with water stability of up to 48 hr could be manufactured from formulations containing between 0.75 and 3% sodium alginate.

Behavior

How a feed particle behaves when introduced into a culture chamber can have a bearing on whether the feed is consumed or ignored. Fish that feed by site may not recognize an inert food particle that is floating at the surface or has landed on the bottom. However, that same fish might strike at a pellet that is slowly sinking. Getting pellets to behave like natural food organisms is not a simple matter, nor is it something to which aquaculturists have devoted much time or attention. In most cases, training finicky fish to accept feed or putting something in the feed formulation that will attract the fish and stimulate feeding are the preferred methods of dealing with the problem.

Attractants

Color, texture, odor, and flavor[136] of a prepared feed may each serve as an attractant to an aquaculture animal. Conversely, each attribute may repulse the animals. There has been very little research conducted on the role of feed color. Most prepared feeds are brown, while semipurified diets are white or nearly white. Animals that have become accustomed to feeding on prepared pellets seem to accept feed of either color equally well as long as the different rations have the same basic nutritional characteristics. Studies in which aquaculture species have been allowed to select from among feeds of various colors do not appear to have been undertaken.

We have already seen how texture can affect the acceptability of a prepared feed (as discussed in the section on weaning). In most cases aquatic animals that would not ordinarily accept a hard pellet can be trained to do so, though the process can be time-consuming.

Various chemicals have been shown to entice a positive response from aquatic animals when those substances are incorporated into feed pellets. Natural feed ingredients, extracts from natural foods, and synthetic amino acids have all been used as attractants. Feedstuffs such as shrimp meal, krill meal, *Spirulina*,[137] fish oil, and fish meal will attract various species. Extracts from various marine organisms have shown attractive properties. Among them are extracts from such organisms as abalone (Zimmer-Faust et al. 1984), silkworm pupae (Murofushi and Ina 1981), polychaete worms (Fuke et al. 1981), fish muscle (Kohbara et al. 1989), fish eggs (Heinsbroek and Kreuger 1992), and bovine spleen (Heinsbroek and Kreuger 1992). Analysis of such extracts has demonstrated that the chemicals responsible for the attractiveness are often amino acids (Takeda et al. 1984, Takii et al. 1984, Mackie and Mitchell 1985, Johnsen and Adams 1986). Glycine, inosine, glutamic acid, aspartic acid, serine, lysine, proline, histidine, and alanine all have been shown to attract certain aquatic species. The L-amino acids appear to be the only ones that attract sea bream *(Chrysophrys major)*, and combinations of various amino acids work synergistically at attractants for that species (Murofushi et al. 1982). Dimethyl-β-propiothetin, dimethylthetin, dipropyl disulfide, dimethylsulfoxide, and dimethylsulfone have various levels of attractiveness for goldfish and carp (Nakajima et al. 1988). Nucleotides extracted from krill have higher attractiveness than amino acids for marbled rockfish *(Sebasticus marmoratus)* according to Takaoka et al. (1990).

Attractants may continue to be required once aquatic animals begin actively feeding on pellets, particularly with respect to species that locate their food primarily or exclusively by sense of smell. In cases where the animals locate food visually, attractants may be of little use, or they may work in concert with vision to help the animal locate feed. When the two senses do work together, a particle (such as a

---

[136] Odor and flavor are, in reality, the same sense. Attractants depend on the animal's sense of smell to entice that animal to consume a feed pellet that might not otherwise be recognized as food.

[137] *Spirulina* is an alga often added to tropical fish foods as both an attractant and a source of pigments. Because of high cost, *Spirulina* is not used in feeds designed for aquatic animals that will be used as human food or for enhancement stocking programs.

prepared feed pellet) that might not be recognized as food might be consumed based on its odor. Once the animal learns that the pellet is a food source, elimination of the odor component may not have an adverse effect on acceptance of the diet.

Pigments

Colors associated with the exterior of an animal or its flesh may be important to the aquaculturist. External color is desirable in such organisms as crawfish and red snapper that are sold for human consumption. Tropical fishes are known for their bright colors. Development and maintenance of those colors are important to the home aquarist. In the case of koi, coloration and its pattern on a particular fish may have a value of thousands or even hundreds of thousands of dollars. While many consumers prefer fish with white flesh, color may be an important consideration. This is particularly true for salmon where pink flesh is often demanded by the consumer.

A group of lipid-soluble chemicals known as carotenoids impart pigmentation to aquatic animals. The carotenoids are responsible for such colors as yellow, orange, and red. The chemicals are not produced by fish but must be obtained from the food.

Examples of carotenoids are astaxanthin, canthaxanthin, zeaxanthin, xanthophyll, and astacene. Common sources of carotenoids are from crustaceans, yeast, and plants (including algae). In addition, synthetic carotenoids are available. Chemical structures of selected carotenoids, sources of the pigments, and the use of them in salmonid aquaculture have been reviewed by Torrissen et al. (1989).

The levels of pure carotenoids that have been used as feed additives tend to range from a few to a few hundred milligrams per kilogram of diet. The effective level of inclusion various among pigments. For example, Torrissen (1989) and Choubert and Storebkken (1989) found that astaxanthin is deposited more efficiently in the flesh of rainbow trout than canthaxanthin. Torrissen (1989) determined that there was differential absorption of the two pigments in the digestive tract. He also indicated that higher total carotenoid concentrations in trout can be achieved from diets containing a combination of the two pigments.

When natural ingredients such as shrimp meal or algae are used as sources of carotenoids, there can be considerable variability both within and between pigment sources.

Currently, the use of synthetic carotenoids in feed is not approved in conjunction with fish being reared for human consumption in the United States. Synthetic carotenoids are widely used as feed supplements in other countries and are available to humans in the United States in tablet form.

Harmful Substances

Various noxious and toxic [138] substances can be present in both natural foods and prepared diets. A variety of trace metals and biocides can lead to toxicity as dis-

---

[138] The term "noxious" tends to be used in conjunction with nontoxic phytoplankton blooms, while the term toxic is confined to substances that are actually poisonous to an organism. The term "harmful" encompasses both terms.

cussed in Chapter 5, but those can be avoided through proper screening of feedstuffs used in the manufacture of prepared rations. Some toxins that are produced through biological activity can also cause problems. Gossypol is one that has been previously discussed in this chapter. That toxin and others, as discussed below, have been reviewed by Roberts and Bullock (1989).

Aflatoxin, a toxic metabolite of a mutant blue-green mold, *Aspergillus flavus,* sometimes occurs as a contaminant of oilseed meals. This substance is known to cause hepatocarcinoma or hepatoma[139] in trout. Both acute toxicity and subchronic toxicity effects of aflatoxin on channel catfish have been studied (Jantrarotai and Lovell 1990, Jantrarotai et al. 1990). The toxicity of aflatoxin to fish varies from species to species (Lovell 1991).

Mold growth with concomitant production of aflatoxin may be induced through improper feed storage. Again, moldy feed should never be offered to aquaculture animals. Winfree and Allred (1992) reported that the addition of 10% bentonite[140] to moistened trout feed led to an average decline in aflatoxin level of 70% within 1 hr. The ability of bentonite to detoxify aflatoxin within fish was not demonstrated in that study, but other authors have reported that the use of bentonite in fish feed has a beneficial effect (Smith 1980), perhaps related to reduced mold toxicity.

Toxic algae, primarily blue-green species, have been reported from freshwater fish ponds (Gorham 1964). Shellfish culturists have experienced problems with toxic algal blooms for many years, and in some regions the problem is increasing in geographic distribution and intensity (Nishitani and Chew 1988). Harmful phytoplankton species have caused the loss of net-pen salmon and other species cultured in net-pens around the world (ICES 1991). In Puget Sound, Washington, fish losses exceeded $10 million during the period from 1987 to 1990 (Horner et al. 1990), primarily due to certain diatoms of the genus *Chaetoceros* that causes hypersecretion of gill mucus leading to suffocation (Rensel in press), and due to massive blooms of the microflagellate *Heterostoma akashiwo,* in which the mode of adverse physiological action on fish remains to be determined (Black et al. 1990).

So-called red tides are caused by marine algae blooms associated with such species as *Alexandrium tamarensis* or *Gymnodinium brevis* and have been recognized for a number of years. As the number of studies concerning red tides and the responsible organisms have expanded, so have the numbers of causative species. For example, *Gymnodinium catenatum* has been recognized as being toxic in recent years, as has *Pyrodinium bahamense* (Canahui et al. 1989, Gaines 1989, La Barbera et al. 1989, LeDoux et al. 1989). These and other harmful algae have been directly responsible for the deaths of millions of fish, been implicated in mass mortality of marine mammals, led to the accumulation of toxins in molluscs, and caused paralytic shellfish poisoning (PSP) in humans who have eaten contaminated shellfish.

Shellfish beds in the United States are routinely monitored for PSP and other

---

[139] Cancer of the liver.

[140] Bentonite is a naturally occurring clay mineral.

toxins. Seasonal closures of contaminated beds are common along most of the U.S. coasts. Cultured molluscs are just as susceptible to pollution-induced PSP as any other shellfish. Harvest of shellfish from contaminated beds is prohibited until the source of the toxin is eliminated[141] and the molluscs have metabolized the toxin. Depuration of contaminated shellfish can also be accomplished by placing the affected animals in seawater that does not contain the toxin.

Devassy and Bhat (1991) reviewed the etiology of red tides and indicated that the phenomenon generally occurs when environmental conditions such as calm, clear weather are present or in areas where two dissimilar water masses meet. Factors associated with human activity that appear to promote red tides include the release to the sea of industrial effluents, domestic sewage, and agricultural fertilizers.

As mariculture expands, problems with toxic red tides, noxious algae blooms, and other sources of toxins can be expected to increase. Experts in the field generally agree that there is a worldwide increase in the occurrence of harmful marine algae in coastal waters (Smayda and White 1990).

In recent years a new toxicity problem associated with shellfish has been identified. First seen in 1987 in people who ate mussels from an area of Prince Edward Island, Canada, the malady causes vomiting, seizures, disorientation, and memory loss. The neurotoxin domoic acid was isolated from the mussels (Wright, et al. 1989) and subsequently caused illness and death of humans and seabirds on the U.S. West Coast. One particular form of the diatom *Nitzschia pungens* appears to be the source of the toxin (Wright 1989, Silvert and Subba Rao 1992). Contaminated mussels can be depurated within about 72 hr, with depuration rate increasing with decreasing mussel size (Novaczek et al. 1992).

To date, ciguatera poisoning has not been a problem in aquaculture, but the potential exists and should not be ignored. Ciguatoxin is produced by tropical dinoflagellates, in particular, *Gambierdiscus toxicus* (Russell and Egen 1991). Herbivores feeding on the algae concentrate the toxin and pass it up the food chain, where it eventually reaches the top carnivores. Fish reared in cages or net pens in tropical regions could conceivably become ciguatoxic by consuming wild fish or invertebrates that are carrying the toxin. The problem is probably not occurring largely because fishes reared in net pens obtain most, if not all, of their feed in the form of prepared rations.

Barracuda are notorious for concentrating the toxin, but other reef-dwelling fishes eaten by humans can also be ciguatoxic. Ingestion of such fishes causes abdominal pain, vomiting, and diarrhea followed by a neurological phase that includes muscle pain, weakness, a burning sensation on the tongue, blurred vision, dizziness, and headache (Geller and Olson 1991). Death can occur. The toxin accumulates in humans as it does in fish; that is, a person may be able to eat fish containing ciguatoxin over a period of time without exhibiting symptoms of poisoning. At

---

[141] Toxicity may last for months or even years after algae cell numbers have been reduced. The source of the toxicity may be algal cysts or other nonvegetative cells.

some threshold accumulation level, however, the toxin will exert an effect. A person who has demonstrated symptoms and recovered should not eat fish that might harbor the toxin as additional accumulation could easily prove fatal.

Off-flavors

A much more common problem in aquaculture products than toxicity is off-flavor. Typically reported as an earthy, musty flavor or the taste of mud, off-flavors have been reported from a number of fish species. Off-flavor in channel catfish has been a particularly serious problem. Tilapia imported to the United States have also gained a reputation for off-flavor though the problem has not occurred to any extent in conjunction with fish grown domestically.

The problem is associated with a chemical called geosmin that is produced by actinomycetes and blue-green algae. Off-flavor commonly occurs during autumn when stocking densities are high, organic loading is also high, and the water is still warm. Feeding rate can also be a factor. Brown and Boyd (1982) found a higher incidence of off-flavor in ponds that received high daily feeding levels (defined as being greater than 50 kg/ha) than in ponds that were fed at rates of below 40 kg/ha/day.

As is true for some of the toxins produced by algae, geosmin can be metabolized by fish. Placing catfish in uncontaminated water for a few days will resolve the problem. Typically, when off-flavor is present in a pond, all the fish will exhibit the problem to one degree or another. Not every pond on a facility will exhibit off-flavor simultaneously, however.

Since most of the catfish marketed in the United States are processed in large processing plants that receive numerous truckloads of fish daily from various farms, a method for detecting off-flavor before the fish are processed is needed. Even a small percentage of off-flavored fish is deemed unacceptable because any customer who purchases an off-flavor catfish will be unlikely to come back again.

A relatively simple method of evaluating catfish for off-flavor was put into place by the catfish-processing industry in the 1980s. Typically, the procedure involves taste-testing fish on three occasions prior to actual processing. A fish selected at random from a pond scheduled for harvest is taken to the processing plant 2 wk in advance of harvest. The tail is cut off the fish and placed in a microwave oven for a couple of minutes. The person responsible for the taste-test checks for odors that might indicate off-flavor and then samples the fish. If the pond passes the 2 wk inspection, another fish is taken to the processing plant 3 days before harvesting. If that test is also negative for off-flavor, the fish are harvested on the appointed day and a randomly selected fish from the truck is sampled before the truck is unloaded. If off-flavor is detected during any of the three tests, the fish will not be accepted at the processing plant. In cases where the person conducting the taste-test is uncertain about the presence or absence of off-flavor, a second opinion will be solicited.

Fish farmers typically do not have facilities for depurating large numbers of fish, so they will usually defer harvesting of a pond containing off-flavor fish until the algae bloom crashes and the fish metabolize the geosmin that has accumulated in

Figure 78. Feed particle size requirements depend on the species under culture and its size. The smallest granules shown are starter feeds for fry such as catfish, tilapia, or trout; the largest may be fed to the same species from fingerling to adult size.

their tissues. The three tests will be repeated once the farmer feels that the problem has been resolved.

Taste-testing of channel catfish has virtually eliminated the problem at the large processing plants. The development of the catfish industry can be attributed, at least in part, to the quality control that has been imposed with respect to solving what was a very serious problem.

STRATEGIES FOR OFFERING PREPARED FEED

Various methods can be applied once fish that initially require live or fresh food have been weaned to prepared feeds. Those same methods can also be used with respect to species where first-feeding fry will accept prepared feed. Examples of the latter are salmonids, tilapia, and channel catfish. In this section the discussion is restricted to the use of prepared feeds.

Strategies for feeding aquatic animals involve selection of feed of the proper particle size (Figure 78), feeding them at the proper rate, and providing food at the proper frequency. Changes in feeding strategy may be required as the animals grow and as water quality conditions change. Providing food of the proper size typically involves selecting a crumble or pellet small enough for the animal to ingest whole.

The particle size is typically increased as the animals grow until the largest available pellet size is reached.

Food Conversion Ratio and Food Conversion Efficiency

Nutritionists often look at food conversion as an indication of performance in a feeding trial. Food conversion is a measurement of how the feed offered is converted into new tissue; that is, how effectively the feed is being used for growth. Generally, rapid growth is accompanied by good food conversion efficiency (FCE), while poor growth and low FCE also go hand in hand.

A food conversion ratio (FCR) is calculated by dividing the dry weight of feed offered in a given period of time by the wet weight gain. FCE is the reciprocal of FCR converted to a percentage. As an example, let us assume that a group of fish are fed for 2 wk. The total amount of feed offered on a dry-weight basis is 100 g/fish over the period, and the fish increase in average weight from 100 to 150 g. Food conversion would be calculated as follows:

$$
\begin{aligned}
FCR &= \text{weight of feed offered/weight gain} \\
&= 100 \text{ g}/(150 - 100) \text{ g} \\
&= 100/50 \\
&= 2.0 \\
FCE &= 1/FCR \times 100 \\
&= 1/2.0 \times 100 \\
&= 0.5 \times 100 \\
&= 50\%
\end{aligned}
$$

As a second example, let us look at another group of fish that are fed for the same period of time and grow from an average weight of 2.0 g to 4.0 g when fed 2 g of feed per fish.

$$
\begin{aligned}
FCR &= 2 \text{ g}/(4 - 2) \text{ g} \\
&= 2/2 \\
&= 1.0 \\
FCR &= 1/1 \times 100 \\
&= 100\%
\end{aligned}
$$

It should be obvious from these examples that low FCR and high FCE are desirable. Compared with terrestrial animals, aquatic species tend to be more efficient at converting food into new tissue. A typical FCR for fingerling fish such as catfish, tilapia, and trout is 1.5. Poultry are also good converters and typically show FCR of about 2.0. Hogs are higher, and cattle are very high, often around 8.0.[142]

---

[142] The values shown are all based on dry weight of feed. If wet weights are used, the values for fish and poultry would increase slightly since prepared feeds contain only a few percent moisture, with the exception of semimoist fish feeds and Oregon Moist Pellet. Cattle feeding on grass might, on the other hand, show FCR values of 15 or so because of the high water content in the feed.

FCR in young, rapidly growing animals tend to be lower (and thus, the FCE is higher) than in fingerling animals. Adults grow very slowly and the FCE is often very poor, though feeding rate is usually low and the object of maintaining the animals in most cases is for the provision of broodstock, not rapid growth and good feed efficiency. For many fishes, FCR values can actually be less than 1.0. Superficially, that would mean that the fish are producing new tissue at a rate higher than their intake of food. One explanation might be that there is natural food in the water system that is making up for the difference. However, FCR less than 1.0 are also common in water systems where there is no natural food. The explanation is actually quite simple. The body of a fish is comprised of over 70% water and the FCR compares the *dry* weight of feed and the *wet* weight of fish. So, a large amount of the weight gain can be attributed to increase in the water volume contained in the tissues of the animal.

There has been a good deal of heated debate as to whether food conversion should be expressed exclusively as FCR or FCE. Individuals in both camps have strong feelings about the subject. Since both methods of reporting food conversion are calculated from the same data, the position can be taken that either is acceptable. At present, there is no consensus among aquaculture scientists, and both FCR and FCE values are commonly seen in the published literature.

Feeding Rate

The first rule of fish feeding is *do not overfeed!* Most aquatic animals can survive for long periods of time without food. They will not grow in the absence of food and may eventually begin to digest their internal organisms, thereby making it impossible for them to digest and assimilate food if it is once again available. In most instances, lack of food for a few days or even a few weeks will not be catastrophic. Overfeeding, on the other hand, can lead to fouling of the water. Impaired water quality [low DO, high ammonia, high biochemical oxygen demand (BOD)] and the stimulation of potentially harmful bacterial and fungal growth can occur as a result of overfeeding.

Now that you have the point about overfeeding has been driven home, it is necessary to point out that there is an important exception to the first rule of fish feeding. Young, rapidly growing animals, such as fry catfish, tilapia, and trout, are typically fed well in excess of what they will consume; some people feed 50% or more of body weight daily to fish fry. The feed is evenly dispersed across the water surface of the culture chamber, with the intent being to ensure that each fish has an opportunity to find food without having to swim long distances in the process. Water quality problems can usually be avoided even though the fish are being fed to extreme excess because the total weight of food being offered is quite small relative to the volume of water.

For some species, very high feeding rates may be required and can be offered without being excessive. Santiago et al. (1987) found that growth and survival of *Tilapia nilotica* fry was better over a 5 wk period when the fish were provided a 35% crude protein crumble diet at 60% of body weight daily as compared with 15,

30, and 45%.[143] Poston and Williams (1991) found, on the other hand, that Atlantic salmon fry grew well when fed between 5.5 and 8.4% of body weight daily. Atlantic salmon fry fed from 2.5 to 5.4% of body weight daily were considered to be underfed.

In ponds, overfeeding of fish fry leads to fertilization and production of plankton blooms that provide additional food. Recall that fry ponds are often fertilized with organic or inorganic compounds in any case. In raceways, excess feed will accumulate and begin to deteriorate. Thus, excess feed should be siphoned from fry raceways on a daily basis if water quality is to be maintained. Otherwise, fry should have food available to them throughout the day.

Once fish reach the fingerling size or crustaceans become postlarvae, feeding rates are usually reduced. Once the fish become sufficiently motile to search out feed, the feeding rate can be reduced. Murai and Andrews (1976) recommended feeding channel catfish 10% of body weight until they reach 0.25 g. The feeding rate should be gradually reduced to about 5% of body weight daily when the fish reach 4 g. Larger fish are usually fed 3%.

The above feeding percentages apply to fish being cultured within their optimum temperature range. The percentage of body weight that an aquatic animal will consume varies as a function of temperature because metabolic rate is under the control of environmental temperature. In general, if the water temperature is above or below the optimum range for the animal, feeding rate will be reduced. Recommendations for channel catfish have been described by Bardach et al. (1972) as follows:

>32°C	No more than 1% of body weight daily
21 to 32°C	3% of body weight daily
16 to 21°C	2% of body weight daily
7 to 16°C	1% of body weight daily
<7°C	Feed may not be accepted.

Many catfish culturists do not feed at all during winter, though in the southern United States some growth can be obtained if the fish are fed on warm days or on alternate days (Lovell and Sirikul 1974).

Juveniles of various fish species grow well when offered pelleted feed at a few percent of body weight daily. For some species, extensive feeding tables have been developed that provide rates at various temperatures and for fish of various sizes (Piper et al. 1982). The maximum rate, as in fingerling catfish, is 3%. The same rate seems optimum for many other species, including pacu, *Colossoma mitrei* (Merola and De Souza 1988). The optimum feeding rate for young white sturgeon, *Acipenser transmontanus,* appears to be 2% at 20°C (Hung and Lutes 1987). Piper et al. (1982) provided information on feeding rates for largemouth bass, smallmouth bass, and striped bass.

By the end of a typical growing season, even a feeding rate of 3% of fish biomass

---

[143] A lower feeding rate might have been optimum if a higher protein percentage had been fed since young, rapidly growing fish typically require higher protein levels than the level used in the cited study.

daily can mean that very large amounts of feed are required within a given culture chamber. Conventional wisdom among catfish farmers once was that ponds should receive no more than about 16 kg/ha/day. Today, rates in excess of 100 kg/ha/day are not uncommon. Cole and Boyd (1986) fed catfish in ponds over a series of rates of up to 224 kg/ha/day. Fish performance was not impacted until the feeding rate exceeded 112 kg/ha/day.

Once the growth rate of fish begins to slow, which it will do even under optimum environmental conditions, and which often corresponds to the onset of maturity, feeding rate should be reduced. Broodfish are often fed at a rate of about 1% of body weight daily.

Feeding Frequency

In addition to reducing the percentage of fish body weight fed daily as animals grow, feeding frequency is often reduced. For example, channel catfish fry weighing less than 1.5 g should be fed every 3 hr (eight times daily) when they are being maintained in an intensive culture system without natural food. The feeding frequency can be reduced to four times daily after the fish exceed 1.5 g (Murai and Andrews 1976). Larger fingerlings (e.g., 5 to 10 g) are usually fed once or twice daily. Warmwater fish are usually fed before midmorning,[144] in the late afternoon, or at both times. Fish should not be fed during the warmest part of the day in systems like ponds where water temperature is susceptible to significant daily fluctuations.

Twice daily feeding is common practice in aquaculture. Andrews and Page (1975) found that feeding channel catfish twice daily to satiation led to good performance. They found no advantage with respect to growth rate or FCE in feeding only once daily or more frequently than twice daily. Andrews and Page conducted their research in flowthrough fiberglass tanks. More recently, Webster et al. (1992) fed channel catfish diets containing 34 or 38% protein in ponds, applying feeding rates of once or twice daily. The saw no significant differences in growth as a function of either dietary protein level or feeding frequency.

Feeding twice a day has been shown to be more effective than once a day feeding for the silurid catfish, *Heteropneustes fossilis* (Singh and Srivastawa 1984). In a study that provided more variables, Kerdchuen and Legendre (1991) showed that feeding the African catfish, *Heterobranchus longifilis,* continuously led to better growth and FCE than providing discrete meals (one to four feedings a day). Also of interest was the fact that growth rate was better in fish fed during the night than in those fed during daylight.

Teshima and Kanazawa (1983) indicated that larval Japanese shrimp, *Penaeus japonicus,* performed better when fed twice a day than once a day; however, more frequent feeding might be in order as demonstrated for the freshwater shrimp, *Mac-*

---

[144] Feed should not be introduced into culture systems until the culturist is certain that the DO level is acceptable.

robrachium lamarrei, which, when provided with a range of frequencies, performed best on three to six daily feedings (Marian et al. 1986).

Weight gain and FCE in *Tilapia nilotica* were not influenced by feeding frequency in a study conducted by Teshima et al. (1986) who fed a series of semipurified diets. Silva et al. (1986) fed *T. nilotica* of four size classes (smallest 0.9 to 1.1 g, largest 16 to 17 g) at various rates and frequencies and found no effect of feeding frequency on performance. Red tilapia hybrids *(T. mossambica × T. nilotica)* did show better growth when fed two times a day, three times a day, or *ad libitum* as compared with feeding frequencies of once a day or every other day in a study conducted by Siraj et al. (1988). *T. nilotica* fry were also found to grow and convert feed better when fed four times a day as compared with twice daily feeding in a study by Bocek et al. (1992).

The snakehead, *Channa striatus,* appears to perform most efficiently when fed once daily (Sampath 1984). Chiu et al. (1987) found that milkfish, *Chanos chanos,* on the other hand, showed significantly improved growth and FCE when feeding frequency was increased from four to eight times daily, while feeding rate (5 or 9% of body weight daily) had no effect.

Culture conditions have been shown to affect optimum feeding frequency. For example, Seymour (1989) indicated that feeding frequency should be adjusted for the European eel, *Anguilla anguilla,* as a function of temperature. Feeding three or four times a day was recommended when the temperature was optimum at 26°C. Holm et al. (1990) recommended making feed more readily available, i.e., increasing feeding frequency for rainbow trout, *Oncorhynchus mykiss,* when densities are extremely high (their final densities were 240 to 450 kg/m^3).

It is obvious that the best feeding frequency to be used in aquaculture will vary with respect to the species under culture; the life history stage; and several culture conditions including water quality, fish density, and even the type of culture system being employed. The availability of natural food will also undoubtedly have a bearing on how often prepared feed should be provided. Studies conducted to date, as is true of most feeding and nutrition studies with aquaculture species, have not been standardized in any manner. Many studies that have attempted to ascertain optimum feeding frequency have not included sufficient variables to be definitive. For example, if a fish is offered feed once or twice daily and performance is improved at the higher feeding frequency, it is not possible to conclude that twice daily feeding is optimum because higher feeding frequencies were not evaluated.

Aquaculture animals should be fed every day of the week, though many commercial operators, particularly those with small operations and few personnel, feed only 6 days a week, or even 5. Most researchers feed 6 or 7 days a week. Some farmers feel fish perform better if given a day off of feed each week. The rationale for that opinion is not supported by any known research data and probably reflects the fact that the fish farmer wants a day off from having to feed the fish.

Feeding Rate and Frequency Interactions

In species that grow best when provided with high daily feeding rates, it may not be possible for them to accept an entire day's ration in one or even two feedings,

Figure 79. Hand feeding channel catfish from a pond bank.

so feeding frequency must be taken into consideration along with feeding rate. Somewhat surprisingly, some species may grow best when provided with a high feeding rate only once daily. Carlos (1988) conducted a study with bighead carp fry in which the fish were fed at 10, 20, or 30% of body weight daily at frequencies of one, three, or five times a day. While a high feeding rate and frequent feeding might be assumed to lead to best performance, the carp receiving 30% of body weight daily in a single feeding grew best.

Methods of Presenting Feed

Hand feeding (Figure 79) is often the technique of choice for small ponds, tanks, raceways, and cages. In large ponds, feed may be distributed by blowers that run off the power takeoff of a tractor. By measuring the amount of feed expelled from the blower per unit of time, the culturist can fairly accurately control the weight of feed offered to each pond. Feed can also be distributed in ponds from boats, and boats are often required in conjunction with cage culture operations (Figure 80). Airplanes have even been used to feed fish in very large ponds. Alternatives to the aforementioned feeding methods include the use of demand feeders and automatic feeders.

Demand feeders are devices that are activated by the fish. A typical demand feeder is shown in Figure 81. Various triggering mechanisms have been developed, but the basic concept involves a container that holds a quantity of feed (usually at least a one-day supply). The container may be as small as a liter or so or as large as a 210 l oil drum. Fish activate the feeder by bumping into a rod suspended into the water. Usually the end of the rod is within a few centimeters of the water surface. When the rod is bumped, a few feed pellets are released.

In many cases, the amount of feed consumed in a day from demand feeders is similar to what the fish would eat if fed *ad libitum* by hand. However, growth in some species can be significantly affected when demand feeders are employed. Meriwether (1986) reported that *Tilapia aurea* gained an average of 72% more weight when fed in cages with a demand feeder than with hand feeding. Rainbow trout, on the other hand, consumed 163% more food and grew significantly better when hand-fed to satiation as compared with fish in ponds equipped with demand feeders (Tidwell et al. 1991).

Demand feeders are suitable for a variety of fishes but are not used in conjunction with invertebrate culture. Many species of fish will quickly learn to bump the activating rod. In most cases the fish learn after one or more individuals accidentally trigger the feeder though some fish will not adapt well to demand feeders. Some species may not even adapt well to demand feeders even when provided with extensive training (Takahashi et al. 1984). Even relatively small fish are typically able to

Figure 80. Boats are commonly used in conjunction with providing feed in cage culture operations.

Figure 81. A demand feeder over a salmonid raceway. Note the activating rod extending from the bottom of the feeder into the water.

activate properly designed feeding mechanisms, so demand feeders are often suitable for use in feeding fingerlings held in raceways or ponds.

A wide array of automatic feeders has been developed. They are designed to provide known amounts of feed either continuously or intermittently. They range in size from small ones designed for feeding fry troughs (Figure 82), to large computerized systems that distribute feed from large bins to several ponds through distribution pipes. Most demand feeders are powered with electricity, though windup models that utilize mechanical clock mechanisms and even solar-powered feeders (McIntyre 1981) have been developed. Commercial automatic feeders designed for various applications are widely available. Unlike demand feeders, automatic feeders are often suitable for use with both fishes and invertebrates.

The major advantage of both demand and automatic feeders is that they can save a considerable amount of labor, particularly in cases where frequent daily feeding is necessary. A primary disadvantage is that if there is a water quality problem or onset of a disease epizootic, the culturist may not recognize the problem as quickly as when hand feeding is practiced. Stress is more readily detectable in conjunction with demand feeders than when automatic feeders are used because if the intake of

Figure 82. An automatic feeder consisting of a motorized continuous belt on which the feed is placed. As the belt slowly moves (perhaps only 1 to 2 cm/hr), the feed drops into the fry tank. Such feeders provide feed on a virtually continuous basis so long as they are filled at appropriate intervals.

feed is reduced or the fish quit feeding altogether, this will be apparent when the culturist goes around to refill the feeders. Automatic feeders will continue to operate even if the fish refuse feed. The waste feed will become an additional stressor once it begins to decay.

To reduce the chance of missing a problem of the type described, demand and automatic feeders should not contain more food than will be used within a few days unless someone routinely ensures that the fish are actively feeding. Good management of an aquaculture facility involves frequent observation of all culture chambers and the animals within.

Automatic feeders may not be appropriate in all situations. Delivery of feed in specific locations rather than spreading it more evenly over the surface of a culture chamber can be a problem with respect to sedentary species and in net pens or cages. For species that are not highly motile, those animals not in the immediate vicinity of the feeder may not have much opportunity to obtain food, particularly if the animals closer to the food source establish territories. Thorpe et al. (1990) found that 1-yr-old net-pen-cultured Atlantic salmon wasted 40.5% of the feed delivered by an automatic feeder but only 1.4% of food distributed evenly over the surface of the water.

Adjusting Feeding Rates

When floating feed is used, the fish can be fed *ad libitum*. The proper amount of feed is that which is completely consumed within 15 to 30 min. In cases where fish are fed on the basis of a percentage of their body weight, it is necessary to know the number of fish in the system and their average weight. It is also necessary to adjust feeding rate to accommodate fish growth. When intermittent harvesting and partial restocking is practiced, keeping track of the biomass and fish numbers in a pond becomes difficult to impossible, and *ad libitum* feeding or the use of demand feeders makes the most sense.

Some aquaculturists feed on the basis of a percentage of the biomass of animals in each culture chamber. Determination of biomass may be based on actually measurement, whereby all of the animals are captured and weighed, or a random sample may be obtained from which total biomass is calculated based on the number of animals thought to be present in the population. Standing-crop biomass can also be calculated from growth equations developed for a given species and size of fish.

Weighing and counting all of the fish in a culture chamber is often undertaken by researchers working in small cages, tanks, or raceways. Intermittent capture of fish from ponds or from large culture systems of any type is difficult and can be highly stressful, so estimates for those types of situations are usually based on a subsample. Typically 30 randomly captured animals will provide a good estimate of average weight.[145] Numbers of animals present in the population are estimated from known numbers of individuals present at stocking less either observed mortality or some average mortality figure (such as 5% a month).

Adjustments to feeding rate can be made on a daily basis if a growth equation is applied (as discussed below). When adjustments are based upon discrete subsampling, they are usually made at intervals of no less than a week, and no more than a month, and most frequently, every 2 wk. Finfish fingerlings tend to double in weight in something less than a month under optimum conditions. Let us assume a fish culturist feeds at 4% of body weight daily and adjusts the feeding rate every 2 wk. Since the animals should be expected to increase in weight each day if feed is provided and the environmental conditions are conducive to growth, each day that passes will mean that the fish are being fed increasingly below the desired rate of 4% since they are constantly gaining weight. By the time the fish are weighed at the next adjustment period, the actual rate of feeding may actually be closer to 2% a day than 4%.

For purposes of illustrating how feeding rates are calculated, let us assume that an aquaculturist wishes to feed a group of shrimp at 3% of their biomass daily and plans to adjust feeding rate at 2 wk intervals. If the shrimp were stocked at 1000/ha at an average weight of 5 g, there would initially be 50,000 g, or 50 kg, of shrimp in the pond. The first day of feeding would require $0.03 \times 50 - 1.5$ kg of feed. The next day the shrimp would be slightly larger, and 1.5 kg of feed would represent slightly less than 3% of the biomass present (unless some mortality had occurred to change the situation). If it was determined that the FCR on the first day of

---

[145] Subsamples as small as 30 individuals are predictive of the average weight of even very large populations of aquatic animals if the subsample is truly random. When the population of fish in a pond ranges widely in size, obtaining a random sample may be difficult and the mean that is arrived at can be highly biased as a result.

feeding was 2.0,[146] the biomass of the population would have increased by 0.5 kg (1.5 kg$/2.0$ = 0.75 kg); thus the total weight of shrimp on the second day would be $50 + 0.75 = 50.75$ kg. To feed at 3% of biomass on the second day would require $0.03 \times 50.75 = 1.52$ kg of feed. The longer the period between weight adjustments, the larger the adjustment that will be required to bring the feeding rate back up to 3%. Again, adjustments are often made at 2 wk intervals so as to reduce the frequency of handling stress without underfeeding at the end of an adjustment period to the extent that growth is significantly impaired.

FCR or FCE should be calculated at the time each feeding adjustment is made. For rapidly growing aquaculture animals, FCR should be fairly constant over the growing season once the animals reach fingerling (fish) or postlarval (invertebrate) size. If there is a considerable change in FCR from one weighing period to the next, it may be a sign that the population subsample was biased, unaccounted-for mortality has occurred, water quality has been impaired, or a disease epizootic is present or imminent. As we have seen earlier, *ad libitum* feeding is one way in which some of the problems can be avoided or perhaps detected before they become serious. With some species, such as shrimp, *ad libitum* feeding is not typically practiced because of the way in which the animals feed (on the bottom and very slowly).

If the FCR for a species is fairly constant, it is possible to calculate the change in feeding rate on a daily basis by assuming a specific growth rate. Two ways of accomplishing daily feeding rate adjustments are presented below. Since FCR values are not necessarily constant, verification of the information used in calculating daily feeding rates should be made by periodically (e.g., once a month or every other month if the animals seem to be growing normally) taking subsamples and calculating actual growth and FCR.

Method 1. $W_t = W_0 + \text{F}/\text{C}$, where

$$W_0 = \text{weight of animals at time 0}$$
$$W_{0+1} = \text{weight of animals at } W_0 + 1 \text{ day}$$
$$\text{F} = \text{feeding rate percentage}$$
$$\text{C} = \text{food conversion ratio}$$

Example. Calculate the feeding rate at W_t if the initial weight of the animals at time W_0 is 1000 kg, feeding rate is 4% daily, and FCR is 1.3.

$$W_t = 1000 \text{ kg} + (1000 \times 0.04/1.3)$$
$$= 1000 + 40/1.3$$
$$= 1000 + 30.8$$
$$= \underline{1030.8 \text{ kg}}$$

---

[146] The reader should be aware that calculating FCR on the basis of 1 day of feeding is not practical. FCR and FCE calculations are usually made from data on weight gain and amount of feed offered over time intervals measured in weeks.

The feeding rate at W_0 was 40 kg/day (1000×0.04). It increased at time W_{t+1} to $1030.8 \times 0.04 = 41.2$ kg, which represents an increase of 1.2 kg after the first day. Similar increases in biomass and amount of feed required can be calculated for subsequent days (see Table 35).

Method 1 assumes that the growth of aquatic animals in linear, although in fact it is generally logarithmic, following a sigmoidal growth curve. Swingle (1967) adapted Beer's law to the growth of fish under aquaculture conditions and developed the second method of determining feeding rates.

Method 2. $W_t = W_0 e^{kt}$, where

$$W_t = \text{weight at time } t$$
$$W_0 = \text{weight at time } 0$$
$$e = \text{the natural logarithm}$$
$$k = \text{ratio of feed percentage to food conversion}$$
$$t = \text{time in days of adjustment period}$$

Example. Calculate the feeding rate at W_t utilizing the same data provided in the previous example ($W_0 = 1000$ kg, FCR $= 1.3$, feeding rate $= 4\%$ daily, $t = 1$ day):

$$W_t = 1000 \; e^{(0.04/1.3)(1)}$$
$$W_t = 1000 e^{0.0308}$$
$$\ln W_t = \ln 1000 + \ln e^{0.0308}$$
$$\ln W_t = 6.9078 + 0.0308$$
$$\ln W_t = 6.9386$$
$$W_t = \underline{1031.3} \text{ kg}$$

Calculate the weight of feed required at $\ln W_t$:

$$1031.3 \times 0.04 = \underline{41.25} \text{ kg.}$$

Thus, the two methods of calculation compare reasonably well during a 1 day period for which growth is predicted. Table 35 presents the results of applying the two calculations over a 30 day period utilizing the initial data from the examples above. One advantage of method 2 is that any number of days can be used in making the calculation. Method 1, on the other hand must be calculated 1 day at a time, with the information generated from 1 day used as input for the next.

Since both methods of calculation assume a constant FCR, neither can be relied on for an extended period. The best practice is to obtain subsamples from each culture unit periodically to update estimates of both biomass and FCR. These can then be put into either formula to form the basis of a new set of daily adjustment

TABLE 35. Comparison of Total Biomass and Feeding Level Values for 30 Days Utilizing Methods 1 and 2: Initial Weight of the Population = 1000 kg; FCR = 1.3; Feeding Rate Is Adjusted Daily; Feed Is Provided at 4% of Biomass Daily

Day	Method 1 ($W_t = W_0 + F/C$)		Method 2 ($W_t = W_0 e^{kt}$)	
	Biomass (kg)	Feeding Rate (kg)	Biomass (kg)	Feeding Rate (kg)
0	1,000.0	40.0	1,000.0	40.0
1	1,030.8	41.2	1,031.3	41.3
2	1,062.5	42.5	1,063.6	42.5
3	1,095.2	43.8	1,096.9	43.9
4	1,128.9	45.2	1,131.2	45.3
5	1,163.6	46.6	1,166.6	46.7
6	1,199.4	48.0	1,203.1	48.1
7	1,236.3	49.5	1,240.7	49.6
8	1,274.3	51.0	1,279.5	51.2
9	1,313.5	52.5	1,319.5	52.8
10	1,353.9	54.2	1,360.8	54.4
11	1,395.6	55.8	1,403.4	56.1
12	1,438.5	57.5	1,447.3	57.9
13	1,482.8	59.3	1,492.6	59.7
14	1,528.4	61.1	1,539.3	61.6
15	1,575.4	63.0	1,587.4	63.5
16	1,623.9	65.0	1,637.1	65.5
17	1,673.9	67.0	1,688.3	67.5
18	1,725.4	69.0	1,741.1	69.6
19	1,778.5	71.1	1,795.6	71.8
20	1,833.2	73.3	1,851.8	74.1
21	1,889.6	75.6	1,909.7	76.4
22	1,947.7	77.9	1,969.4	78.8
23	2,007.6	80.3	2,031.0	81.2
24	2,069.4	82.8	2,094.5	83.8
25	2,133.1	85.3	2,160.0	86.4
26	2,198.7	88.0	2,227.6	89.1
27	2,266.3	90.7	2,297.3	91.9
28	2,336.0	93.4	2,369.2	94.8
29	2,407.9	96.3	2,443.3	97.7
30	2,482.0	99.3	2,519.7	100.8

calculations. Because animals being reared in different tanks, linear raceways, net pens, or ponds will experience somewhat different environmental conditions, it is best to subsample all culture units and adjust each individually.

It should now be clear why *ad libitum* feeding with floating feed or installing demand feeders are two commonly practiced feeding methods. They do not work for all species, however, as we have seen. Finally, reliable feeding tables have been developed for some species, most notably trout. Examples of such tables can be found in Piper et al. (1982).

LITERATURE CITED

Abel, H. J., K. Becker, C. Meske, and W. Friedrich. 1984. Possibilities of using heat-treated full-fat soybeans in carp feeding. Aquaculture, 42: 97–108.

Akand, A. M., M. I. Miah, and M. M. Hague. 1989. Effect of dietary protein level on growth, feed conversion and body composition of shingi (*Heteropneustes fossilis* Bloch). Aquaculture, 77: 175–180.

Akiyama, T., I. Yagisawa, and T. Nose. 1981. Optimum levels of dietary crude protein and fat for fingerling chum salmon. Bull. Natl. Res. Inst. Aquacult., 2: 35–42.

Alava, V. R., and C. Lim. 1983. The quantitative dietary protein requirements of *Penaeus monodon* juveniles in a controlled environment. Aquaculture, 30: 63–62.

Alava, V. R., and F. P. Pascual. 1987. Carbohydrate requirements of *Penaeus monodon* (Fabricius) juveniles. Aquaculture, 61: 211–217.

Anderson, R. J., E. W. Kienholz, and S. A. Flickinger. 1981. Protein requirements of smallmouth bass and largemouth bass. J. Nutr., 111: 1085–1097.

Andrews, J. W., and Y. Matsuda. 1975. The influence of various culture conditions on the oxygen consumption of channel catfish. Trans. Am. Fish. Soc., 104: 322–327.

Andrews, J. W., and J. W. Page. 1975. The effects of frequency of feeding on culture of catfish. Trans. Am. Fish. Soc., 104: 317–321.

Andrews, J. W., and R. R. Stickney. 1972. Interactions of feeding rates and environmental temperature on growth, food conversion, and body composition of channel catfish. Trans. Am. Fish. Soc., 101: 94–99.

Andrews, J. W., L. V. Sick, and G. J. Baptist. 1972. The influence of dietary protein and energy levels on growth and survival of penaeid shrimp. Aquaculture, 1: 341–347.

Aniello, M. S., and T. Singh. 1982. Some studies on the larviculture of the giant prawn *(Macrobrachium rosenbergii)*. pp. 225–232, In: M. B. New (Ed.). Giant prawn farming. Developments in Aquatic Fisheries Science, Vol. 10, Bangkok, Thailand.

Appelbaum, S. 1985. Rearing of the Dover sole, *Solea solea* (L.), through its larval stages using artificial diets. Aquaculture, 49: 209–221.

Arai, S., and T. Nose. 1983. Dietary protein requirement of young ayu, *Plecoglossus altivelis*. Bull. Natl. Res. Inst. Aquacult., 4: 99–105.

Aujero, E. J., E. Tech, and S. Javellana. 1985. Nutritional value of marine yeast fed to larvae of *Penaeus monodon* in combination with algae. p. 168, In: Y. Taki, J. H. Primavera, and J. A. Llobrera (Eds.). Proceedings of the First International Conference on the Culture of Penaeid Prawns/Shrimps, December 4–7, 1984, Iloilo City, Philippines. Southeast Asian Fisheries Development Center, Iloilo City, Philippines.

Balazs, G. H., E. Ross, and C. C. Brooks. 1973. Preliminary studies on the preparation and feeding of crustacean diets. Aquaculture, 2: 369–377.

Baldwin, B. S., R. I. E. Newell, and T. W. Jones. 1989. Omnivorous feeding by *Crassostrea virginica* larvae: a consumption of naturally occurring phytoplankton, protozoa and bacteria. J. Shellfish Res., 8: 473.

Barahona-Fernandes, M. H. 1982a. Problemes lies au sevrage du bar *(Dicentrarchus labrax)*. 1. Influence de la charge dans le bac de sevrage. Piscic. Fr., 65: 5–7.

Barahona-Fernandes, M. H. 1982b. Problemes lies au sevrage du bar *(Dicentrarchus labrax)*. 2. Taille minimale du sevrage. Piscic. Fr., 65: 8–9.

Barahona-Fernandes, M. H. 1982c. Problemes lies au sevrage du bar *(Dicentrarchus labrax)*. 3. Influencce de l'incorporation d'attractants dans le granule. Piscic. Fr., 65: 10–11.

Bardach, J. E., J. H. Ryther, and W. O. McLarney. 1972. Aquaculture. Wiley-Interscience, New York. 868 p.

Bautista, M. N. 1986. The response of *Penaeus monodon* juveniles to varying protein/energy ratios in test diets. Aquaculture, 53: 228–242.

Bayne, D. R., D. Dunseth, and C. G. Ramirios. 1976. Supplemental feeds containing coffee pulp for rearing *Tilapia* in Central America. Aquaculture, 7: 133–146.

Bender, J. A. Y. Vatcharapijarn, and A. Russell. 1989. Fish feeds from grass clippings. Aquacult. Eng., 8: 407–419.

Berger, A., and J. E. Halver. 1987. Effect of dietary protein, lipid and carbohydrate content on the growth, feed efficiency and carcass composition of striped bass, *Morone saxatilis* (Walbaum), fingerlings. Aquacult. Fish. Manage., 18: 345–356.

Black, E. A., J. N. C. Whyte, J. W. Bagshaw, and N. G. Ginter. 1990. The effects of *Heterosigma akashiwo* on juvenile *Oncorhyncus tshawytscha* and its implications for fish culture. J. Appl. Ichthyol., 7: 168–175.

Bocek, A., R. P. Phelps, and T. J. Popma. 1992. Effect of feeding frequency on sex-reversal and on growth of Nile tilapia, *Oreochromis niloticus*. J. Appl. Aquacult., 1: 97–103.

Bonar, S. A., H. S. Sehgal, G. B. Pauley, and G. L. Thomas. 1990. Relationship between the chemical composition of aquatic macrophytes and their consumption by grass carp, *Ctenopharyngodon idella*. J. Fish Biol., 36: 149–157.

Bordner, C. E., L. R. D'Abramo, D. E. Conklin, and N. A. Baum. 1986. Development and evaluation of diets for crustacean aquaculture. J. World Aquacult. Soc., 17: 44–51.

Borlongan, I. G., and R. M. Coloso. 1993. Requirements of juvenile milkfish *(Chanos chanos)* for essential amino acids. J. Nutr., 123: 125–132.

Borlongan, I. G., and M. M. Parazo. 1991. Effect of dietary lipid sources on growth, survival and fatty acid composition of sea bass *(Lates calcarifer,* Bloch) fry. Bamidgeh, 43: 95–102.

Briggs, M. R. P., K. Jauncey, and J. H. Brown. 1988. The cholesterol and lecithin requirements of juvenile prawn *(Macrobrachium rosenbergii)* fed semi-purified diets. Aquaculture, 70: 121–129.

Bromley, P. J., and P. A. Sykes. 1985. Weaning diets for turbot *(Scophthalmus maximus* L.), sole *(Solea solea* L.) and cod *(Gadus morhua* L.). pp. 191–211, In: C. B. Cowey, A. M. Mackie, and J. G. Bell (Eds.). Nutrition and feeding in fish. International Symposium on Nutrition and Feeding in Fish. Aberdeen, Scotland.

Brown, P. B., and E. H. Robinson. 1989. Comparison of practical catfish feeds containing 26 and 30% protein. Prog. Fish-Cult., 51: 149–151.

Brown, M. R., S. W. Jeffery, and C. G. Garland. 1989. Nutritional aspects of microalgae used in mariculture, a literature review. Rep. CSIRO Mar. Lab., 205: 1–44.

Brown, S. W., and C. E. Boyd. 1982. Off-flavor in channel catfish from commercial ponds. Trans. Am. Fish. Soc., 111: 379–383.

Bruggeman, E., P. Sorgeloos, and P. Vanhaecke. 1980. Improvements in the decapsulation technique of *Artemia* cysts. pp. 261–269, In: Persoone, G., P. Sorgeloos, O. Roels, and E. Jaspers (Eds.). Ecology, culturing, use in aquaculture. The Brine Shrimp *Artemia*, Vol. 3. Universa Press, Wetteren, Belgium.

Cai, Z., and L. R. Curtis. 1989. Effects of diet on consumption, growth and fatty acid composition in young grass carp. Aquaculture, 81: 47–60.

Canahui, E., F. Rosales Loessener, L. Miller, E. W. King, and S. Hall. 1989. *Pyrodinium bahamense* and PSP along the Pacific coast of Central America. J. Shellfish Res., 8: 440.

Carlos, M. H. 1988. Growth and survival of bighead carp *(Aristichthys nobilis)* fry fed at different intake levels and feeding frequencies. Aquaculture, 68: 267–276.

Castagna, M., R. S. Bisker, H. Dymsza, and J. N. Kraeuter. 1984. Assessment of supplemental diets for growing seed of *Mercenaria mercenaria* (Linne). J. Shellfish Res., 4: 84–85.

Castell, J. D. 1979. Review of lipid requirements of finfish. pp. 59–84, In: J. E. Halver and K. Tiews (Eds.). Proceedings of the World Symposium on Finfish Nutrition and Fishfeed Technology, June 20–23, 1978. Henneman, Berlin.

Castell, J. D., D. J. Lee, and R. O. Sinhuber. 1972a. Essential fatty acids in the diet of rainbow trout *Salmo gairdneri*): lipid metabolism and fatty acid composition. J. Nutr., 102: 93–99.

Castell, J. D., R. O. Sinhuber, J. H. Wales, and D. J. Lee. 1972b. Essential fatty acids in the diet of rainbow trout *(Salmo gairdneri):* growth, feed conversion and some gross deficiency symptoms. J. Nutr., 102: 77–86.

Catacutan, M. R. 1991. Apparent digestibility of diets with various carbohydrate levels and the growth response of *Penaeus monodon*. Aquaculture, 95: 89–96.

Chatiyanwongse, A., and V. Chuapoehuk. 1982. Protein requirement of walking catfish. Thai Fish. Gaz., 35: 251–260.

Chavez de Martinez, M. C. 1990. Vitamin C requirement of the Mexican native cichlid *Cichlasoma urophthalmus* (Gunther). Aquaculture, 95: 400–416.

Chen, H.-Y., and G. Hwang. 1992. Estimation of the dietary riboflavin required to maximize tissue riboflavin concentration in juvenile shrimp *(Penaeus monodon)*. J. Nutr., 122: 2474–2478.

Chen, H.-Y., and J.-S. Jenn. 1991. Combined effects of dietary phosphatidylcholine and cholesterol on the growth, survival and body lipid composition of marine shrimp, *Penaeus penicillatus*. Aquaculture, 96: 167–178.

Chiang, Y.-M., and J.-I. Young. 1978. The effect of different foods on the growth of a brackish water rotifer, *Brachionus plicatilis*. China Fish. Monogr,. 303: 2–6.

Chiu, Y. N., N. S. Sumagaysay, and M. A. S. Sastrillo. 1987. Effect of feeding frequency and feeding rate on the growth and feed efficiency of milkfish, *Chanos chanos* Forsskal, juveniles. Asian Fish. Sci., 1: 27–31.

Cho, C. Y., S. Kaushik, and B. Woodward. 1992. Dietary arginine requirement of young rainbow trout *(Oncorhynchus mykiss)*. Comp. Biochem. Physiol., 102**A:** 211–216.

Choo, P. S. 1986. Observations on the use of some non-algal feeds in the culture of protozoeae of penaeid prawn. Fish. Bull. Dept. Fish. (Malaysia), 42: 1–6.

Choubert, G., and T. Storebkken. 1989. Dose response to astaxanthin and canthaxanthin pigmentation of rainbow trout fed various deitary carotenoid concentrations. Aquaculture, 81: 69–77.

Chu, F.-L. E., K. L. Webb, D. A. Hepworth, and B. B. Casey. 1987. Metamorphosis of larvae of *Crassostrea virginica* fed microencapsulated diets. Aquaculture, 64: 185–197.

Civera, R., and J. Guillaume. 1989. Effect of sodium phytate on growth and tissue mineralization of *Penaeus japonicus* and *Penaeus vannamei* juveniles. Aquaculture, 77: 145–156.

Cole, B. A., and C. E. Boyd. 1986. Feeding rate, water quality, and channel catfish production in ponds. Prog. Fish-Cult., 48: 25–29.

Colura, R. L. 1987. Saltwater pond fertilization. pp. 51–53, In: G. W. Chamberlain, R. J. Miget, and M. G. Haby (Eds.). Manual on Red Drum Aquaculture, Vol. 3. Texas Agricultural Extension Service, Corpus Christi.

Colura, R. L., B. T. Hysmith, and R. E. Stevens. 1976. Fingerling production of striped bass *(Morone saxatilis)*, spotted seatrout (*Cynoscion nebulosus*), and red drum *(Sciaenops ocellatus)* in saltwater ponds. Proc. World Maricult. Soc., 7: 79–92.

Colvin, L. B., and C. W. Brand. 1977. The protein requirement of penaeid shrimp at various life-cycle stages in controlled environment systems. Proc. World Maricult. Soc., 8: 821–840.

Colvin, P. M. 1976. Nutritional studies on penaeid prawns: protein requirements in compounded diets for juvenile *Penaeus indicus* (Milne Edwards). Aquaculture, 7: 315–326.

Coutteau, P., P. Lavens, and P. Sorgeloos. 1990. Baker's yeast as a potential substitute for live algae in aquaculture diets: *Artemia* as a case study. J. World Aquacult. Soc., 21: 1–9.

Cowey, C. B., and J. R. Sargent. 1977. Lipid nutrition in fishes. Comp. Biochem. Physiol., 57**B**: 269–273.

Craig, S. R., and D. M. Gatlin III. 1992. Dietary lysine requirement of juvenile red drum *Sciaenops ocellatus*. J. World Aquacult. Soc., 23: 133–137.

Crosby, M. P. 1988. Utilization of detrital complexes by the oyster, *Crassostrea virginica* (Gmelin). Ph.D. Dissertation, University of Maryland, College Park. 181 p.

D'Abramo, L. R., C. E. Bordner, D. E. Conklin, and N. A. Baum. 1984. Sterol requirement of juvenile lobsters, *Homarus* sp. Aquaculture, 42: 13–25.

Dabrowski, K. 1977. Protein requirements of grass carp *(Ctenopharyngodon idella)*. Aquaculture, 12: 63–73.

Dabrowski, K., R. Bardega, and R. Przedwojski. 1983. Dry diet formulation study with common carp (*Cyprinus carpio* L.). Z. Tierphysiol. Tierernaehr. Futtermittelkd., 50: 40–52.

Danakusumah, E., A. Basyarie, and M. Mean. 1985. Larval rearing of kuruma prawn (*Penaeus japonicus* Bate) fed with yeast, diatom and rotifers. J. Mar. Fish. Res., 33: 77–82.

Das, K. M., S. N. Mohanty, and S. Sarkar. 1991. Optimum dietary protein to energy ratio for *Labeo rohito* fingerlings. pp. 69–73, In: S. S. DeSilva (Ed.). Fish nutrition research in Asia. Asian Fisheries Society Special Publlication No. 5. Asian Fisheries Society, Manila.

Davies, S. J., and H. Wareham. 1988. A preliminary evaluation of an industrial single cell protein in practical diets for tilapia (*Oreochromis mossamibicus* Peters). Aquaculture, 73: 189–199.

Davies, S. J., S. McCommell, and R. I. Bateson. 1990. Potential of rapeseed meal as an alternative protein source in complete diets for tilapia (*Oreochromis mossambicus* Peters). Aquaculture, 87: 145–154.

Davis, D. A., and E. H. Robinson. 1986. Estimation of the dietary requirement level of the white crayfish *Procambarus acutus acutus*. J. World Aquacult. Soc., 17: 37–43.

Davis, D. A., and E. H. Robinson. 1987. Dietary requirement of juvenile red drum *Sciaenops ocellatus*. J. World Aquacult. Soc., 18: 129–136.

Davis, E. M., G. L. Rumsey, and J. G. Nickum. 1976. Egg-processing wastes as a replacement protein source in salmonid diets. Prog. Fish-Cult., 38: 20–22.

Degani, G., and M. L. Gallagher. 1988. Effects of temperature and dietary protein: energy ratio on growth, body composition and feed utilization of juvenile eels (*Anguilla anguilla*). Bol. Fisiol. Anim. São Paulo, 12: 71–79.

Degani, G., and D. Levanon. 1987. Effects of dietary carbohydrates and temperatures on slow growing juvenile eels *Anguilla anguilla*. Environ. Biol. Fishes, 18: 149–154.

Degani, G., and S. Viola. 1987. The protein sparing effect of carbohydrates in the diet of eels *(Anguilla anguilla)*. Aquaculture, 64: 283–291.

Degani, G., S. Viola, and C. Levanon. 1986. Effects of dietary carbohydrate source on growth and body composition of the European eel (*Anguilla anguilla* L.). Aquaculture, 52: 97–104.

Degani, G., M. L. Gallagher, and D. Levanon. 1987. Effect of dietary protein and energy levels on growth of European eels (*Anguilla anguilla*). Bol. Fisiol. Anim. São Paulo, 11: 95–100.

Dendrinos, P., and J. P. Thorpe. 1987. Experiments on the artificial regulation of the amino acid and fatty acid contents of food organisms to meet the assessed nutritional requirements of larval, post-larval and juvenile Dover sole (*Solea solea* (L.)). Aquaculture, 61: 121–154.

Deshimaru, O., and K. Kuroki. 1974. Studies on a purified diet for prawn. 1. Basal composition of diet. Bull. Jpn. Soc. Sci. Fish., 40: 413–419.

Deshimaru, O., and K. Shigeno. 1972. Introduction to the artificial diet for prawn *Penaeus japonicus*. Aquaculture, 1: 115–133.

Deshimaru, O., and Y. Yone. 1978. Studies on a purified diet for prawn—XII: Optimum level of dietary protein for prawn. Bull. Jpn. Soc. Sci. Fish., 44: 1395–1397.

Desjardins, I. M., B. D. Hicks, and J. W. Hilton. 1987. Iron catalyzed oxidation of trout diets and its effect on the growth and physiological response of rainbow trout. Fish Physiol. Biochem., 3: 173–182.

Devassy, V. P., and S. R. Bhat. 1991. The killer tides. Sci. Rep. (New Delhi), 28: 16–19.

Dhert, P., P. Lavens, M. Duray, and P. Sorgeloos. 1990. Improved larval survival at metamorphosis of Asian seabass *(Lates calcarifer)* using ω3-HUFA-enriched live food. Aquaculture, 90: 63–74.

Dorsa, W. J., H. R. Robinette, E. H. Robinson, and W. E. Poe. 1982. Effects of dietary cottonseed meal and gossypol on growth of young channel catfish. Trans. Am. Fish. Soc., 111: 651–655.

Dupree, H. K. 1977. Vitamins essential for growth of channel catfish. Technical Paper No. 7, U.S. Bureau of Sport Fisheries and Wildlife, Washington, D.C. 12 p.

Duray, M., and T. Bagarinao. 1984. Weaning of hatchery-bred milkfish larvae from live food to artificial diets. Aquaculture, 41: 325–332.

El-Dakour, S. 1986. Effects of different dietary protein:energy ratios on the growth and survival of *Penaeus semisulcatus* de Haan (Decapoda: Penaeidae). Kuwait Bull. Mar. Sci., 213–222.

Ellis, S. C., and R. C. Reigh. 1991. Effects of dietary lipid and carbohydrate levels on growth and body composition of juvenile red drum, *Sciaenops ocellatus*. Aquaculture, 97: 383–394.

Everhart, W. H., and A. E. Eipper, and W. D. Youngs. 1975. Principles of fishery science. Cornell University Press, Ithaca, New York. 288 p.

Fagbenro, D. A. 1990. Food composition and digestive enzymes in the gut of pond-cultured *Clarias ishariensis* (Sydenham 1880), (Siluriformes: Clariidae). J. Appl. Ichthyol., 6: 91–99.

Faranda, F., A. Salleo, G. L. Paro, and A. Manganaro. 1984. Quantitative requirement of *Penaeus kerathurus* for a natural unprocessed diet. Aquaculture, 37: 125–131.

Fermin, A. C., and R. D. Recometa. 1988. Larval rearing of bighead carp, *Aristichthys nobilis* Richardson, using different types of feed and their combinations. Aquacult. Fish. Manage., 19: 283–290.

Forster, J. R. M., and T. W. Beard. 1973. Growth experiments with the prawn *Palaemon serratus* Pennant fed with fresh and compounded foods. Fish. Invest., London, Ser. II Mar. Fish. G.B. Minst. Agric. Fish. Food, 27(7): 1–16.

Fowler, L. G. 1980. Substitution of soybean and cottonseed products for fish meal in diets fed to chinook and coho salmon. Prog. Fish-Cult., 42: 87–91.

Fuchs, J. 1982. Production de juveniles de sole (*Solea solea*) en conditions intensives. 2. Techniques de sevrage entre 1 et 3 months. Aquaculture, 26: 339–358.

Fuke, S., S. Konosu, and K. Ina. 1981. Identification of feeding stimulants for red sea bream in the extract of marine worm *Perinereis brevicirrus*. Bull. Jpn. Sco. Sci. Fish., 47: 1631–1636.

Fukusho, K. 1985. Status of marine larval culture in Japan. pp. 127–139, In: C.-S. Lee and I.-C. Liao (Eds.). Reproduction and culture of milkfish. Tungkang Marine Laboratory, Taiwan.

Fukusho, K., M. Okauchi, H. Tanaka, S. I. Wahyuni, P. Kraisingdecha, and T. Watanabe. 1985. Food value of a rotifer *Brachionus plicatilis,* cultured with *Tetraselmis tetrathele* for larvae of a flounder *Paralichthys olivaceus*. Bull. Natl. Res. Inst. Aquacult., 7: 29–36.

Furuichi, M., H. Taira, and Y. Yone. 1986. Availability of carbohydrate in nutrition of yellowtail. Bull. Jpn. Soc. Sci. Fish., 52: 99–102.

Gaines, G. 1989. *Gymnodinium catenatum:* a recently discovered cause of paralytic shellfish poisoning. J. Shellfish Res., 8: 440–441.

Gallagher, M. L., and A. M. Matthews. 1987. Oxygen consumption and ammonia excretion of the American eel *Anguilla rostrata* fed diets with varying protein energy ratios and protein levels. J. World Aquacult. Soc., 18: 107–112.

Garling, D. L., Jr., and R. P. Wilson. 1976. Optimum dietary protein to energy ratio of channel catfish fingerlings, *Ictalurus punctatus*. J. Nutr., 106: 1368–1375.

Gatesoupe, F.-J. 1982. Nutritional and antibacterial treatments of live food organisms: the influence on survival, growth rate and weaning success of turbot (*Scophthalmus maximus*). Ann. Zootech. (Paris), 31: 353–368.

Gatesoupe, F.-J. 1983. Weaning of sole, *Solea solea,* before metamorphosis achieved with high growth and survival rates. Aquaculture, 32: 401–404.

Gatesoupe, F.-J., and P. Luquet. 1982. Weaning of the sole (*Solea solea*) before metamorphosis. Aquaculture, 26: 359–368.

Gatesoupe, F.-J., T. Arakawa, and T. Watanabe. 1989. The effect of bacterial additives on the production rate and dietary value of rotifers as food for Japanese flounder, *Paralichthys olivaceus*. Aquaculture, 83: 39–44.

Gatlin, D. M., III, and H. F. Phillips. 1989. Dietary calcium, phytate and zinc interactions in channel catfish. Aquaculture, 79: 259–266.

Gatlin, D. M., III, J. P. O'Connell, and J. Scarpa. 1991. Dietary zinc requirement of the red drum, *Sciaenops ocellatus*. Aquaculture, 92: 259–265.

Gatlin, D. M., III, C. B. Sungchul, and M. C. Erickson. 1992. Effects of dietary vitamin E and synthetic antioxidants on composition and storage quality of channel catfish, *Ictalurus punctatus*. Aquaculture, 106: 323–332.

Geller, R. J., and K. R. Olson. 1991. Ciguatera poisoning in San Francisco, California, caused by imported barracuda. West. J. Med., 155: 639–642.

Gomez, G., H. Nakagawa, and S. Kasahara. 1988. Effect of dietary protein/starch ratio and energy level on growth of the giant freshwater prawn *Macrobrachium rosenbergii*. Bull. Jpn. Soc. Sci. Fish., 54: 1401–1407.

Gomez Diaz, G., and H. Nakagawa. 1990. Effects of dietary carbohydrates on growth and body components of the giant freshwater prawn *Macrobrachium rosenbergii*. Aquat. Living Resour., 3: 99–105.

Gorham, P. R. 1964. Toxic algae. pp. 307–336, In: D. F. Jackson (Ed.). Algae and man. Plenum Press, New York.

Greene, D. H. S., and D. P. Selivonchick. 1987. Lipid metabolism in fish. Prog. Lipid Res., 26: 53–83.

Hajra, A., A. Ghosh, and S. K. Mandal. 1988. Biochemical studies on the determination of optimum dietary protein to energy ratio for tiger prawn, *Penaeus monodon* (Fab.), juveniles. Aquaculture, 71: 71–79.

Halver, J. E. (Ed.). 1972. Fish nutrition. Academic Press, New York. 713 p.

Halver, J. E. 1989. The vitamins. pp. 31–109, In: J. E. Halver (Ed.). Fish nutrition. Academic Press, New York.

Hanson, R. H. 1990. *Spirulina* algae improves Japanese fish feeds. Aquacult. Mag., 16(6): 38–43.

Hardy, R. W. 1989. Diet preparation. pp. 475–548, In: J. E. Halver (Ed.). Fish nutrition. Academic Press, New York.

Haylor, G. S. 1991. Controlled hatchery production of *Clarias gariepinus* (Burchell 1822): growth and survival of fry at high stocking density. Aquacult. Fish. Manage., 22: 405–422.

Haylor, G. S., M. C. M. Beveridge, and K. Jauncey. 1988. Phosphorus nutrtion of juvenile *Oreochromis niloticus*. pp. 341–345, In: R. S. V. Pullin, T. Bhukaswan, K. Tonguthai, and J. L. Maclean (Eds.). Second International Symposium on Tilapia in Aquaculture, Bangkok, Thailand, March 16–20, 1987. ICLARM Conf. Proc. No. 15. International Center for Living Aquatic Resources Management, Manila.

Hecht, T. 1981. Rearing of sharptooth catfish larvae (*Clarias gariepinus* Burchell, 1822: Clariidae) under controlled conditions. Aquaculture, 24: 301–308.

Heinsbroek, L. T. N., and J. G. Kreuger. 1992. Feeding and growth of glass eels, *Anguilla*

anguilla L.: The effect of feeding stimulants on feed intake, energy metabolism and growth. Aquacult. Fish. Manage., 23: 327–336.

Hewitt, D. R., and M. G. Iving. 1990. Oxygen consumption and ammonia excretion of the brown tiger prawn *Penaeus esculentus* fed diets of varying protein content. Comp. Biochem. Physiol., 96A: 373–378.

Hidalgo, F., and E. Alliot. 1988. Influence of water temperature on protein requirement and protein utilization in juvenile sea bass, *Dicentrarchus labrax*. Aquaculture, 72: 115–129.

Higgs, D. A., J. R. McBride, B. S. Dosanjh, U. H. M. Fagerlund, C. Archdekin, A.-M. Hammons, and W. C. Clarke. 1988. Nutritive value of plant protein sources for fish with special emphasis on canola products. p. 29, In: Proceedings, Aquaculture International Congress and Exposition, Vancouver, B. C., Canada, Sept. 6–9, 1988. International Congress, Vancouver.

Hilton, J. W., J. L. Atkinson, and S. J. Slinger. 1983. Effect of increased dietary fiber on the growth of rainbow trout *(Salmo gairdneri)*. Can. J. Fish. Aquat. Sci., 40: 81–85.

Holm, J. C., T. Refstie, and S. Boe. 1990. The effect of fish density and feeding regimes on individual growth rate and mortality in rainbow trout *(Oncorhynchus mykiss)*. Aquaculture, 89: 225–232.

Horner, R. A., J. R. Postel, and J. E. Rensel. 1990. Noxious phytoplankton blooms in western Washington waters. A review. pp. 171–176, In: E. Graneli, B. Sundstroem, L. Edler, and D. M. Anderson (Eds.). Fourth International Conference on Toxic Marine Phytoplankton, Lund (Sweden), June 26–30, 1989.

Hubbard, D. M., E. H. Robinson, P. B. Brown, and W. H. Daniels. 1986. Optimum ratio of dietary protein to energy for red crayfish *(Procambarus clarkii)*. Prog. Fish-Cult., 48: 233–237.

Hublou, W. F. 1963. Oregon pellets. Prog. Fish-Cult., 23: 175–180.

Hughes, S. G., G. L. Rumsey, and J. G. Nickum. 1981. Riboflavin requirement of fingerling rainbow trout. Prog. Fish-Cult., 32: 167–172.

Hung, S. S. O., and P. B. Lutes. 1987. Optimum feeding rate of hatchery-produced juvenile white sturgeon (*Acipenser transmontanus*) at 20°C. Aquaculture, 65: 307–317.

Ipinjolu, J. K., A. E. Falaye, and O. O. Tewe. 1989. Assessment of plam kernal meal in the diets of juvenile carp (*Cyprinus carpio* L.). J. West Afr. Fish., 4: 221–230.

James, C. M., A. M. Al-Khara, M. B. Abbas, and A.-A. Al-Ameeri. 1983. Nutritional studies on rotifers and *Artemia* for feeding fish larvae. Annu. Res. Rep. Kuwait Inst. Sci. Res., 8: 61–63.

Jantrarotai, W., and R. T. Lovell. 1990. Subchronic toxicity of dietary aflatoxin B_1 to channel catfish. J. Aquat. Anim. Health, 2: 248–254.

Jantrarotai, W., R. T. Lovell, and J. M. Grizzle. 1990. Acute toxicity of aflatoxin B_1 to channel catfish. J. Aquat. Anim. Health, 2: 237–247.

Jauncey, K. 1982. The effects of varying dietary protein level on the growth, food conversion, protein utilization, and body composition of juvenile tilapias (*Sarotherodon mossambicus*). Aquaculture, 27: 43–54.

Johnsen, P. B., and M. A. Adams. 1986. Chemical feeding stimulants for the herbivorous fish, *Tilapia zillii*. Comp. Biochem. Physiol., 83A: 109–112.

Jones, D. A. 1985. Penaeid larval culture using microencapsulated diets. p. 171, In: Y. Taki, J. H. Primavera, and J. A. Llobrera (Eds.). Proceedings of the First International Conference on the Culture of Penaeid Prawns/Shrimps, December 4–7, 1984, Iloilo City,

Philippines. Southeast Asian Fisheries Development Center, Iloilo City., Iloilo City, Philippines.

Jones, D. A., K. Kurmaly, and A. Arshard. 1987. Penaeid shrimp hatchery trials using microencapsulated dits. Aquaculture, 64: 133–146.

Kanazawa, A., S. Teshima, M. Sakamoto, and A. Shinomiya. 1980a. Nutritional requirements of the puffer fish: purified test diet and the optimum protein level. Bull. Jpn. Soc. Sci. Fish., 46: 1357–1361.

Kanazawa, A., S-I. Teshima, M. Sakamoto, and M. A. Awal. 1980b. Requirements of *Tilapia zillii* for essential fatty acids. Bull. Jpn. Soc. Sci. Fish., 46: 1353–1356.

Kastelein, P. 1983. Survival and growth of elvers (*Anguilla anguilla* L.) reared on an expanded granulated diet. Aquaculture, 30: 155–172.

Katavic, I., J. Jug-Dujakovic, and B. Glamuzina. 1989. Cannibalism as a factor affecting the survival of intensively cultured sea bass *(Dicentrarchus labrax)* fingerlings. Aquaculture, 77: 135–143.

Kaushik, S. J., and E. F. Gomes. 1988. Effect of frequency of feeding on nitrogen and energy balance in rainbow trout under maintenance conditions. Aquaculture, 73: 207–216.

Kaushik, S. J., P. Luquet, D. Blanc, and A. Paba. 1989. Studies on the nutrition of Siberian sturgeon, *Acipenser baeri*. 1. Utilization of digestible carbohydrates by sturgeon. Aquaculture, 76: 97–107.

Kean, J. C., J. D. Castell, and D. J. Trider. 1985. Juvenile lobster *(Homarus americanus)* do not require dietary ascorbic acid. Can. J. Fish. Aquat. Sci., 42: 368–370.

Kean-Howie, J. C., M. A. Silva, and R. K. O'Dor. 1989. The use of microparticulate diet for feeding studies on veliger larvae of *Placopecten magellanicus:* 1. Evidence of acceptability. Bull. Aquacult. Assoc. Can., 89(3): 80.

Kerdchuen, N., and M. Legendre. 1991. Influence de la frequence et de la periode de nourrissage sur la croissance et l'efficacite alimentaire d'un silure africain *Heterobranchus longifilis* (Teleostei, Clariidae). Ressour. Vivantes Aquat., 4: 241–248.

Ketola, H. G., C. E. Smith, and G. A. Kindschi. 1989. Influence of diet and oxidative rancidity on fry of Atlantic and coho salmon. Aquaculture, 79: 417–423.

Kim, I.-B., and S.-H. Lee. 1981. Fish growth experiment in a green water recirculating system. Bull. Korean Fish. Soc., 14: 233–238.

Kim, K.-I., T. B. Kayes, and C. H. Amundson. 1991. Purified diet development and reevaluation of the dietary protein requirement of fingerling rainbow trout *(Oncorhynchus mykiss)*. Aquaculture, 96: 57–67.

Kim, K.-I., T. B. Kayes, and C. H. Amundson. 1992. Requirements for lysine and arginine by rainbow trout *(Oncorhynchus mykiss)*. Aquaculture, 106: 333–344.

Kitajima, C., M. Yoshida, and T. Watanabe. 1980a. Dietary value for ayu *(Plecoglossus altivelis)* for rotifer *Brachionus plicatilis* cultured with baker's yeast *Saccharomyces cerevisiae* supplemented with cuttlefish liver oil. Bull. Jpn. Soc. Sci. Fish., 46: 47–50.

Kitajima, C., T. Arakawa, F. Ooma, S. Fujita, O. Imada, T. Watanabe, and Y. Yone. 1980b. Dietary value for red sea bream larvae of rotifer *Brachionus plicatilis* cultured with a new type of yeast. Bull. Jpn. Soc. Sci. Fish., 46: 43–46.

Kohbara, J., K. Fukuda, and I. Hidaka. 1989. The feeding-stimulatory effects of jack mackerel muscle extracts on the young yellowtail *Seriola quinqueradiata*. Bull. Jpn. Soc. Sci. Fish., 55: 1343–1347.

Koshio, S., R. K. O'Dor, and J. D. Castell. 1990. The effect of different dietary energy levels on growth and survival of eyestalk ablated and intact juvenile lobsters *Homarus americanus*. J. World Aquacult. Soc., 21: 160–169.

Kungvankij, P., L. B. Tiro, B. P. Pudadera, and I. O. Potestas. 1986. Induced spawning and larval rearing of grouper (*Epinephelus salmoides* Maxwell). pp. 26–31, In: J. L. Maclean, L. B. Dizon, and L. V. Hosilios (Eds.). Proceedings of the First Asian Fisheries Forum, Manila, Philippines, May 26–31. Asian Fisheries Society, Manila.

Kurmaly, K., D. A. Jones, A. B. Yule, and J. East. 1989. Comparative analysis of the growth and survival of *Penaeus monodon* (Fabricius) larvae, from protozoea 1 to post-larva 1, on live feeds, artificial diets and on combinations of both. Aquaculture, 81: 27–45.

La Barbera, A., G. Estrella, L. Miller, E. W. King, and S. Hall. 1989. *Alexandrium* sp., *Gymnodinium catenatum,* and PSP in Venezuela. J. Shellfish Res., 8: 442.

Laing, I., and C. G. Verdugo. 1991. Nutritional value of spray-dried *Tetraselmis suecica* for juvenile bivalves. Aquaculture, 92: 207–218.

Laing, I., A. R. Child, and A. Janke. 1990. Nutritional value of dried algae diets for larvae of Manila clam *(Tapes philippinarum)*. J. Mar. Biol. Assoc. U.K., 70: 1–12.

Lal, S. P. 1989. The minerals. pp. 219–257, In: J. E. Halver (Ed.). Fish nutrition. Academic Press, New York.

Langdon, C. J. 1989. Comparison of two capsule types for the delivery of dietary protein to the Pacific oyster, *Crassostrea gigas*. J. Shellfish Res., 8: 414.

Langdon, C. J., and R. I. E. Newell. 1990. Utilization of detritus and bacteria as food sources by two bivalve suspension-feeders, the oyster *Crassostrea virginica* and the mussel *Gaukensia demissa*. Mar. Ecol. Prog. Ser., 58: 299–310.

Leber, K. M., and G. D. Pruder. 1988. Using experimental microcosms in shrimp research: the growth-enhancing effect of shrimp pond water. J. World Aquacult. Soc., 19: 197–203.

LeDoux, M., J. M. Fremy, E. Nezan, and E. Erard. 1989. Recent occurrence of paralytic shellfish poisoning (PSP) toxins from the northwestern coasts of France. J. Shellfish Res., 8: 486.

Lee, D. L. 1971. Studies on the protein utilization related to growth in *Penaeus monodon*. Aquaculture, 1: 1–13.

Liang, Y., and Wenjuan, J. 1986. Protein requirement in formulated diets for penaeid shrimp (*Penaeus orientalis* (Kishenouye)) in different growth stage. Mar. Fish. Res. Shandong, 7: 79–87.

Lim, C., and W. Dominy. 1992. Substitution of full-fat soybeans for commercial soybean meal in diets for shrimp, *Penaeus vannamei*. J. Appl. Aquacult., 1(3): 35–46.

Lim, C., S. Sukhawongs, and F. P. Pascual. 1979. A preliminary study on the protein requirements of *Chanos chanos* (Forskal) fry in a controlled environment. Aquaculture, 17: 195–201.

Lovell, R. T. 1991. Mycotoxins in fish feeds. Feed Manage., 42(11): 42–44.

Lovell, R. T., and G. El-Naggar. 1988. New source of vitamin C for fish feeds. Highlights Agric. Res., 38(4): 15.

Lovell, R. T., and B. Sirikul. 1974. Winter feeding of channel catfish. Proc. Southeast. Assoc. Game Fish Comm., 28: 208–216.

Ludwig, G. M., and D. L. Tackett. 1991. Effects of using rice bran and cottonseed meal as organic fertilizers on water quality, plankton, and growth and yield of striped bass, *Morone saxatilis,* fingerlings in ponds. J. Appl. Aquacult., 1: 79–94.

Machiels, M. A. M. 1987. A dynamic simulation model for growth of the African catfish, *Clarias gariepinus* (Burchell 1822). 4. The effect of feed formulation on growth and feed utilization. Aquaculture, 64: 305–323.

Mackie, A. M., and A. I. Mitchell. 1985. Identification of gustatory feeding stimulants for fish-applications in aquaculture. pp. 177–189, In: C. B. Cowey, A. M. Mackie, and J. G. Bell (Eds.). Proceedings, Symposium on Feeding and Nutrition of Fish, Aberdeen, Scotland.

Manzi, J. J., M. B. Maddox, and P. Sorgeloos. 1980. Requirements for *Artemia* nauplii in *Macrobrachium rosenbergii* (de Man) larviculture. pp. 313–329, In: G. Persoone, P. Sorgeloos, O. Roels, and E. Jaspers (Eds.). Ecology, culturing, use in aquaculture. The Brine Shrimp *Artemia.* Vol. 3. Universa Press, Wetteren, Belgium.

Mao, Y.-Q., F.-S. Cai, and D. Lin. 1985. Studies on the daily requirements of protein, carbohydrate, fat, minerals and fiber of juvenile grass carp (*Ctenopharyngodon idellus* C. and V.). Trans. Chin. Ichthyol. Soc., 4: 81–92.

Marian, M. P., T. J. Pandian, S. Mathavan, S. Muragadass, and D. R. D. Premkumar. 1986. Suitable diet and optimum feeding frequency in the eyestalk ablated prawn, *Macrobrachium lamarrei.* pp. 589–592, In: J. L. Maclean, L. B. Dizon, and L. V. Hosillos (Eds.). Proceedings, First Asian Fisheries Forum, Manila, Philippines, May 26–31. Asian Fisheries Society, Manila.

Mazid, M. A., Y. Tanaka, T. Katayama, K. L. Simpson, and C. O. Chichester. 1978. Metabolism of amino acids in aquatic animals. III. Indispensable amino acids for *Tilapia zillii.* Bull. Jpn. Soc. Sci. Fish., 44: 739–742.

Mazid, M. A., M. A. Rahman, S. Gheyasuddin, M. A. Hussin, and M. B. Rashid. 1987. Nutritional requirements of major carp. 1. Optimum level of dietary protein for *Labeo rohita.* Bangladesh J. Fish., 10: 75–82.

McClain, W. R., and D. M. Gatlin III. 1988. Dietary zinc requirement of *Oreochromis aureus* and effects of dietary calcium and phytate on zinc bioavailabiilty. J. World Aquacult. Soc., 19: 103–108.

McGeachin, R. B. 1977. Algae fed *Artemia salina* as a food source for larval *Cynoscion nebulosus.* M.S. Thesis, Texas A&M University, College Station. 39 p.

McIntyre, D. C. 1981. Cage rearing of rainbow trout with a solar-charged automatic feeder. Prog. Fish-Cult., 43: 12–15.

Medland, T. E., and F. W. H. Beamish. 1985. The influence of diet and fish density on apparent heat increment in rainbow trout, *Salmo gairdneri.* Aquaculture, 47: 1–10.

Meriwether, F. H. 1986. An inexpensive demand feeder for cage-reared tilapia. Prog. Fish-Cult., 48: 226–228.

Merola, N., and J. H. De Souza. 1988. Preliminary studies on the culture of the pacu, *Colossoma mitrai,* in floating cages: effect of stocking density and feeding rate on growth performance. Aquaculture, 68: 243–248.

Metailler, R., B. Menu, and P. Moriniere. 1981. Weaning of Dover sole *(Solea vulgaris)* using artificial diets. J. World Maricult. Soc., 12: 111–116.

Meyers, S. P., D. P. Butler, and W. H. Hastings. 1972. Alginates as binders for crustacean rations. Prog. Fish-Cult., 34: 9–12.

Millikin, M. R. 1982. Qualitative and quantitative nutrient requirements of fishes: a review. Fish. Bull., 80: 655–686.

Moon, H.-Y., and D. M. Gatlin III. 1991. Total sulfur amino acid requirement of juvenile red drum, *Sciaenops ocellatus*. Aquaculture, 95: 97–106.

Mossmann, R. L., B. Moraes, D. R. Azevedo, G. F. Rey, I. P. Silveira, M. Frozza, P. K. Serdiuk, and T. M. Azevedo Viana. 1990. Crescimento de *Macrobrachium rosenbergii* De Man (Crustacea/Decapoda/Palaemonidae) mantidos em "nursery" e tratados com racao alternativa. Estud. Leopold., 26: 118: 13–26.

Murai, T., and J. W. Andrews. 1976. Effects of frequency of feeding on growth and food conversion of channel catfish fry. Bull. Jpn. Soc. Sci. Fish., 42: 159–161.

Murai, T., A. Sumalangcay, Jr., and F. Piedad-Pascual. 1985. Supplement of various attractants to a practical diet for juvenile *Penaeus monodon* Fabricius. Fish. Res. J. Philipp., 8: 61–67.

Murofushi, S., and K. Ina. 1981. Survey of feeding stimulants for the sea bream present in the dried pupae of silkworms. Agric. Biol. Chem., 45: 1501–1504.

Murofushi, S., A. Sano, and K. Ina. 1982. Structure-activity relationships of neutral amino acids as feeding stimulants for sea bream *Chrysophrys major*. 9. Feeding stimulants for fishes. J. Agric. Chem. Soc. Jpn., 56: 255–259.

Naas, K. E. 1989. Extensive startfeeding of marine fry. Can. Tech. Rep. Fish. Aquat. Sci., 176: 137–141.

Naess, T., O. Bergh, T. Harboe, K. E. Naas, H. Rabben, and L. H. Skjolddal. 1990. Green water in larviculture—an experiment with natural phytoplankton in tanks for first feeding of halibut larvae (*Hippoglossus hippoglossus* L.). International Council for the Exploration of the Sea, Copenhagen, Denmark. 22 p.

Nakajima, K., A. Uchida, and Y. Ishida. 1988. A new feeding attractant, dimethyl-beta-propiothetin, for freshwater fish. Bull. Jpn. Soc. Sci. Fish., 55: 689–695.

National Research Council. 1973. Nutrient requirements of trout, salmon and catfish. National Academy of Sciences Press, Washington, D. C. 57 p.

National Research Council. 1981. Nutrient requirements of coldwater fishes. National Academy of Sciences Press, Washington, D. C. 63 p.

National Research Council. 1983. Nutrient requirements of warmwater fishes and shellfishes. National Academy Press, Washington, D.C. 102 p.

Navarre, D., and J. E. Halver. 1989. Disease resistance and humoral antibody production in rainbow trout fed high levels of vitamin C. Aquaculture, 79: 207–221.

New, M. B. 1976. A review of dietary studies with shrimp and prawns. Aquaculture, 9: 101–144.

New, M. B. 1980. A bibliography of shrimp and prawn nutrition. Aquaculture, 21: 101–128.

Nishitani, L., and K. K. Chew. 1988. PSP toxins in the Pacific coast states: monitoring programs and effects on bivalve industries. J. Shellfish Res., 7: 653–669.

Norman-Boudreau, K. E., and D. E. Conklin. 1984. Protein requirement of juvenile lobster *Homarus* sp. J. Shellfish Res., 4: 96.

Nose, T., and S. Arai. 1972. Optimum level of protein in purified test diet for eel, *Anguilla japonica*. Bull. Freshwater Fish. Res. Lab. Tokyo, 22: 145–155.

Novaczek, I., M. S. Madhyastha, R. F. Ablett, A. Donald, G. Johnson, M. S. Nijjar, and

D. E. Sims. 1992. Depuration of domoic acid from live blue mussels *(Mytilus edulis)*. Can. J. Fish. Aquat. Sci., 49: 312–318.

Oduro-Boateng, F., and A. Bart-Plange. 1988. Pito brewery waste as an alternative protein source to fishmeal in feeds for *Tilapia busumana*. pp. 357–360, In: R. S. V. Pullin, T. Bhukaswan, K. Tonguthai, and J. L. Maclean (Eds.). Second International Syposium on Tilapia in Aquaculture, Bangkok, Thailand, March 16–20, 1987. ICLARM Conference Proceedings No. 15. International Center for Living Aquatic Resources Management, Manila.

Ogino, C. 1980. Protein requirements of carp and rainbow trout. Bul. Jpn. Soc. Sci. Fish., 46: 385–388.

Okauchi, M. 1988. Studies on the mass culture of *Tetraselmis tetrathele* (West, G.S.) Butcher as a food organism. Bull. Natl. Res. Inst. Aquacult., 14: 1–123.

Orme, L. E., and C. A. Lemm. 1973. Use of dried sludge from paper processing wastes in trout diets. Feedstuffs, 45(51): 28–30.

Owen, J. M., J. W. Adron, C. Middleton, and D. B. Cowey. 1975. Elongation and desaturation of dietary fatty acids in turbot, *Scophthalmus maximus* L., and rainbow trout, *Salmo gairdneri*. Lipids, 10: 528–531.

Person-Le Ruyet, J., B. Menu, M. Cadena-Roa, and R. Metailler. 1983. Use of expanded pellets supplemented with attractive chemical substances for the weaning of turbot *(Scophthalmus maximus)*. J. World Maricult. Soc., 14: 676–678.

Persoone, G., P. Sorgeloos, O. Roels, and E. Jaspers (Eds.). 1980a. Morphology, genetics, radiobiology, toxicology. The Brine Shrimp *Artemia,* Vol. 1. Universa Press, Wetteren, Belgium. 318 p.

Persoone, G., P. Sorgeloos, O. Roels, and E. Jaspers (Eds.). 1980b. Physiology, biochemistry, molecular biology. The Brine Shrimp *Artemia,* Vol. 2. Universa Press, Wetteren, Belgium. 636 p.

Pezzato, L. E., N. Castagnolli, N. Viega, J. J. Souze, and A. C. Pezzato. 1981. Farinha de visceras de aves como alternativa proteica na alimentacao de carpa (*Cyprinus carpio* L.). pp. 74–75, In: Annals, 2nd Brazilian Symposium on Aquaculture, Jaboticabal, Brazil.

Phillips, A. M., Jr. 1972. Calorie and energy requirements. pp. 1–28, In: J. E. Halver (Ed.). Fish nutrition. Academic Press, New York.

Piedat-Pascual, F., R. M. Coloso, and C. T. Tamse. 1983. Survival and some hisstological changes in *Penaeus monodon* Fabricius juveniles fed various carbohydrates. Aquaculture, 31: 168–180.

Piper, R. G., I. B. McElwain, L. E. Orme, J. P. McCraren, L. G. Fowler, and J. R. Leonard. 1982. Fish hatchery management. U.S. Fish and Wildlife Service, Washington, D.C. 517 p.

Planas, M., and A. Estevez. 1989. Effects of diet on population development of the rotifer *Brachionus plicatilis*. Helgol. Meeresunters., 43: 171–181.

Poston, H. A. 1991. Choline requirement of swim-up rainbow trout fry. Prog. Fish-Cult., 53: 220–223.

Poston, H. A., and R. C. Williams. 1991. Influence of feeding rate on performance of Atlantic salmon fry in an ozonated water reuse system. Prog. Fish-Cult., 53: 111–113.

Pyen, C.-K., and J.-Y. Jo. 1982. Seed production of red-sea-bream, *Chrysophrys major*. Bull. Korean Fish. Soc., 15: 161–170.

Rainuzzo, J. R., Y. Olsen, and G. Rosenlund. 1989. The effect of enrichment diets on the fatty acid composition of the rotifer *Brachionus plicatilis*. Aquaculture, 79: 157–161.

Raj, S. P. 1989. Evaluation of *Clitoria* leaf as a protein supplement in the feed of *Cyprinus carpio* var. *communis*. J. Ecobiol., 3: 195–202.

Rajadevan, P., and M. Schramm. 1989. Nutritional value of cabbage and kikuyu grass as food for grass carp, *Ctenopharyngodon idella* Val. S. Afr. J. Anim. Sci. 19: 67–70.

Ramos Elorduy, J. 1990. Nota sobre el uso de los insectos como una alternativa en la aquicultura. pp. 313–316, In: G. de la Lanza Espino and J. Arredondo Figueroa (Eds.). Aquaculture in Mexico: from concepts to production.

Reece, D. L., D. E. Wesley, G. A. Jackson, and H. K. Dupree. 1975. A blood meal-rumen contents blend as a partial or complete substitute for fish meal in channel catfish diets. Prog. Fish-Cult., 37: 15–19.

Refstie, T., and E. Austreng. 1981. Carbohydrate in rainbow trout diets. 3. Growth and chemical composition of fish from different families fed four levels of carbohydrate in the diet. Aquaculture, 25: 35–49.

Reigh, R. C., and R. R. Stickney. 1989. Effects of purified dietary fatty acids on the fatty acid composition of freshwater shrimp, *Macrobrachium rosenbergii*. Aquaculture, 77: 157–174.

Rensel, J. E. 1993. Severe blood hypoxia of Atlantic coast salmon *(Salmo salar)* exposed to the marine diatom *Chaetoceros concavicornis*. pp. 625–630. In: T. J. Smayda and Y. Shimizu (Eds.) Toxic phytoplankton blooms in the sea. Elsevier, New York.

Richardson, N. L., D. A. Higgs, R. M. Beames, and J. R. McBride. 1985. Influence of dietary calcium, phosphorus, zinc and sodium phytate level on cataract incidence, growth and histopathology in juvenile chinook salmon *(Oncorhynchus tshawytscha)*. J. Nutr., 115: 553–567.

Richardson, N. L., D. A. Higgs, and R. M. Beames. 1986. The susceptibility of juvenile chinook salmon (*Oncorhynchus tshawytscha*) to cataract formation in relation to dietary changes in early life. Aquaculture, 52: 237–243.

Roberts, R. J., and A. M. Bullock. 1989. Nutritional pathology. pp. 423–473, In: J. E. Halver (Ed.). Fish nutrition. Academic Press, New York.

Robinson, E. H., and R. P. Wilson. 1985. Nutrition and feeding. pp. 323–404, In: C. S. Tucker (Ed.). Channel catfish culture. Elsevier, New York.

Robinson, E. H., S. D. Rawles, and R. R. Stickney. 1984a. Evaluation of glanded and glandless cottonseed products. Prog. Fish-Cult., 46:92–97.

Robinson, E. H. S. D. Rawles, P. W. Oldenburg, and R. R. Stickney. 1984b. Effects of feeding glandless or glanded cottonseed products to *Tilapia aurea*. Aquaculture, 38: 145–154.

Robinson, E. H., S. D. Rawles, H. E. Yette, and L. W. Greene. 1984c. An estimate of the dietary calcium requirement of fingerling *Tilapia aurea* reared in calcium-free water. Aquaculture, 41: 389–393.

Robinson, E. H., S. D. Rawles, P. B. Brown, H. E. Yette, and L. W. Greene. 1986. Dietary calcium requirement of channel catfish *Ictalurus punctatus*, reared in calcium-free water. Aquaculture, 53: 263–270.

Robinson, E. H., D. LaBomascus, P. B. Brown, and T. L. Linton. 1987. Dietary calcium and phosphorus requirements of *Oreochromis aureus* reared in calcium-free water. Aquaculture, 64: 267–276.

Roem, A. J., C. C. Kohler, and R. R. Stickney. 1990a. Inability to detect a choline requirement for the blue tilapia *Oreochromis aureus*. J. World Aquacult. Soc., 21: 238–240.

Roem, A. J., C. C. Kohler, and R. R. Stickney. 1990b. Vitamin E requirement for the blue tilapia *Oreochromis aureus* (Steindachner), in relation to dietary lipid level. Aquaculture, 87: 155–164.

Roem, A. J., R. R. Stickney, and C. C. Kohler. 1990c. Vitamin requirements of blue tilapia in a recirculating water system. Prog. Fish-Cult., 52: 15–18.

Roem, A. J., R. R. Stickney, and C. C. Kohler. 1991. Dietary pantothenic acid requirement of the blue tilapia. Prog. Fish-Cult., 53: 216–219.

Rosenlund, G., L. Joergensen, R. Waagboe, and K. Sandnes. 1990. Effects of different levels of ascorbic acid in plaice (*Pleuronectes platessa* L.). Comp. Biochem. Physiol., 96A: 395–398.

Rumsey, G. L. 1991. Choline-betaine requirements of rainbow trout (*Oncorhynchus mykiss*). Aquaculture, 95: 107–116.

Russell, F. E., and N. B. Egen. 1991. Ciguateric fishes, ciguatoxin (CTX) and ciguatera poisoning. J. Toxicol.: Toxin Rev., 10: 37–62.

Sambasivam, S., P. Subramanian, and K. Krishnamurthy. 1982. Observations on growth and conversion efficiency in the prawn *Penaeus indicus* (H. Milne Edwards) fed on different protein levels. Symp. Ser. Mar. Biol. Assoc. India, 6: 406–409.

Sampath, K. 1984. Preliminary report on the effects of feeding frequency in *Channa striatus*. Aquaculture, 40: 301–306.

Santiago, C. B., and O. S. Reyes. 1991. Optimum dietary protein level for growth of bighead carp *(Aristichthys nobilis)* fry in a static water system. Aquaculture, 93: 155–165.

Santiago, C. B., M. Banes-Aldaba, and M. A. Laron. 1982. Dietary crude protein requirement of *Tilapia nilotica* fry. Kalikasan, 11: 255–265.

Santiago, C. B., M. B. Aldaba, and O. S. Reyes. 1987. Influence of feeding rate and diet form on growth and survival of Nile tilapia *(Oreochromis niloticus)* fry. Aquaculture, 64: 277–282.

Satoh, S., T. Takeuchi, and T. Watanabe. 1987. Requirement of *Tilapia* for alpha-tocopherol. Bull. Jpn. Soc. Sci. Fish., 53: 119–124.

Satoh, S., W. E. Poe, and R. P. Wilson. 1989a. Effect of supplemental phytate and/or tricalcium phosphate on weight gain, feed efficiency and zinc content in vertebrae of channel catfish. Aquaculture, 80: 155–161.

Satoh, S., W. E. Poe, and R. P. Wilson. 1989b. Studies on the essential fatty acid requirement of channel catfish. Aquaculture, 79: 121–128.

Scura, E. D., J. Fischer, and M. P. Yunker. 1985. The use of microencapsulated feeds to replace live food organisms in shrimp hatcheries. p. 171, In: Y. Taki, J. H. Primavera, and J. A. Llobrera (Eds.). Proceedings of the First International Conference on the Culture of Penaeid Prawns/Shrimps, December 4–7, 1984, Iloilo City, Philippines. Southeast Asian Fisheries Development Center, Iloilo City.

Sedgwick, R. W. 1979. Influence of dietary protein and energy on growth, food consumption and food conversion efficiency in *Penaeus merguiensis* de Man. Aquaculture, 16: 7–30.

Seikai, T. 1985. Reduction in occurrence frequency of albinism in juvenile flounder *Paralichthys olivaceus* hatchery-reared on wild zooplankton. Bull. Jpn. Soc. Sci. Fish., 8: 1261–1267.

Seymour, E. A. 1989. Devising optimum feeding regimes and temperatures for the warm-water culture of eel, *Anguilla anguilla* L. Aquacult. Fish. Manage., 20: 311–323.

Shcherbina, M. A., L. N. Trofimova, I. A. Salkova, and A. V. Grin. 1987. Availability of amino acids in yeast raised on hydrocarbons for carp, *Cyprinus carpio*. J. Ichthyol., 27: 23–28.

Shearer, K. D. 1988. Dietary potassium requirement of juvenile chinook salamon. Aquaculture, 73: 119–129.

Shearer, K. D., and T. Aasgaard. 1992. The effect of waterborne magnesium on the dietary magnesium requirement of the rainbow trout *(Oncorhynchus mykiss)*. Fish Physiol. Biochem., 9: 387–392.

Shewbart, K. L., W. L. Mies, and P. D. Ludwig. 1973. Nutritional requirements of the brown shrimp, *Penaeus aztecus*. U.S. Deptartment of Commerce, NOAA Sea Grant Report No. COM–73–11794. National Oceanic and Atmospheric Administration, Rockville, Maryland. 52 p.

Shiau, S.-Y., and C. C. Kwok. 1989. Effects of cellulose, agar, carrageenan, guar gum and carboxymethylcellulose on tilapia growth. World Aquacult., 20: 60.

Shiau, S.-Y., and G.-S. Suen. 1992. Estimation of the niacin requirements for tilapia fed diets containing glucose and dextrin. J. Nutr., 122: 2030–2036.

Shiloh, S., and S. Viola. 1973. Experiments in the nutrition of carp growing in cages. Bamidgeh, 25: 17–31.

Shimeno, S., H. Hosokawa, M. Takeda, H. Kajiyama, and T. Kaisho. 1985. Effect of dietary lipid and carbohydrate on growth, feed conversion and body composition of young yellowtail. Bull. Jpn. Soc. Sci. Fish., 51: 1893–1898.

Shimma, Y., H. Shimma, and K. Ikeda. 1980. Effects of supplemental oils to single cell protein feeds on the growth and fatty acid composition of auy, *Plecoglossus altivelis*. Bull. Natl. Res. Inst. Aquacult., 1: 47–60.

Silva, S. S. de, R. M. Gunasekera, and C. Keembiyahetty. 1986. Optimum ration and feeding frequency in *Oreochrimis niloticus* young. pp. 559–564, In: J. L. Maclean, L. B. Dizon, and L. V. Hosillos (Eds.). Proceedings, First Asian Fisheries Forum, Manila, Philippines, May 26–31. Asian Fisheries Society, Manila.

Silvert, W., and D. V. Subba Rao. 1992. Dynamic model of the flux of domoic acid, a neurotoxin, through a *Mytilus edulis* population. Can. J. Fish. Aquat. Sci., 49: 400–405.

Singh, B. N. 1990. Protein requirement of young silver carp, *Hypophthalmichthys molitrix* (Val.). J. Freshwater Biol., 2: 89–95.

Singh, B. N., and K. K. Bhanot. 1988. Protein requirement of the fry of *Catla catla* (Ham.). Proceedings, First Indian Fisheries Forum, Mangalore, Karnataka, India.

Singh, R. P., and A. K. Srivastawa. 1984. Effect of feeding frequency on the growth, consumption and gross conversion efficiency in the siluroid catfish, *Heteropneustes fossilis* (Bloch). Bamidgeh, 36: 80–89.

Siraj, S. S., Z. Kamaruddin, M. K. A. Satar, and M. S. Kamarudin. 1988. Effects of feeding frequency on growth, food conversion and survival of red tilapia *(Oreochromis mossambicus/O. niloticus)* hybrid fry. pp. 383–386, In: R. S. V. PUllin, T. Bhukaswan, K. Tonguthai, and J. L. Maclean (Eds.). Second International Symposium on Tilapia in Aquaculture, Bangkok, Thailand, Mar. 16–20, 1987. ICLARM Conference Proceeding No. 15. International Center for Living Aquatic Resources Management, Manila.

Smayda, T. J., and A. W. White. 1990. Has there been a global expansion of algal blooms?

If so, is there a connection with human activities? pp. 516–517, In: E. Graneli, B. Sundstrom, L. Elder, and D. M. Anderson (Eds.). Toxic marine phytoplankton. Elsevier, New York.

Smith, R. R. 1980. Recent advances in nutrition: clay in trout diets. Salmonid, Nov./Dec.: 16–18.

Smith, R. R. 1989. Nutritional energetics. pp. 1–29, In: J. E. Halver (Ed.). Fish nutrition, Academic Press, New York.

Soliman, A. K., R. J. Roberts, and K. Jauncey. 1983. The pathological effects of feeding rancid lipid in diets for *Oreochromis niloticus* (Trewavas). pp. 193–199, In: L. Fishelson, and Z. Yaron (compilers). Proceedings, International Symposium on Tilapia in Aquaculture, Nazareth, Israel, May 8–13. Tel Aviv University Press, Tel Aviv.

Sorgeloos, P. 1980. The use of the brine shrimp *Artemia* in aquaculture. pp. 25–46, In: G. Persoone, P. Sorgeloos, O. Roels, and E. Jaspers (Eds.). Ecology, culturing, use in aquaculture. The Brine Shrimp *Artemia,* Vol. 3. Universa Press, Wetteren, Belgium.

Sorgeloos, P. 1986. Live animal food for larval rearing in aquaculture: the brine shrimp *Artemia*. pp. 199–214, In: M. Bilio, H. Rosenthal, and C. J. Sindermann (Eds.). Realism in aquaculture: achievements, constraints, perspectives. World Conference on Aquaculture, Venice, Italy. European Aquaculture Society, Ghent.

Sorgeloos, P., P. Lavens, P. Leger, W. Tackaert, and D. Versichele. 1986. Manual para el cultivo y uso de *Artemia* en aquacultura. Food and Agriculture Organization of the United Nations, Brasilia, Brazil. 301 p.

Sorgeloos, P., P. Leger, and P. Lavens. 1988. Improved larval rearing of European and Asian seabass, seabream, mahi-mahi, siganid and milkfish using enrichment diets for *Brachionus* and *Artemia*. World Aquacult., 19(4): 78–79.

Spannhof, L., and H. Plantikow. 1983. Studies on carbohydrate digestion in rainbow trout. Aquaculture, 30: 95–108.

Spinelli, J., C. R. Houle, and J. C. Wekell. 1983. The effect of phytates on the growth of rainbow trout *(Salmo gairdneri)* purified diets containing varying quantities of calcium and magnesium. Aquaculture, 30: 71–84.

Stickney, R. R. 1971. Effects of dietary lipids and lipid-temperature interactions on growth, food conversion, percentage lipid and fatty acid composition of channel catfish. Ph.D. Dissertation, Florida State University, Tallahassee. 96 p.

Stickney, R. R. 1977. Lipids in channel catfish nutrition. pp. 14–18, In: R. R. Stickney and R. T. Lovell (Eds.). Nutrition and feeding of channel catfish. Southern Cooperative Series Bulletin No. 218. Auburn University, Auburn, Alabama.

Stickney, R.R., and R.W. Hardy. 1989. Lipid requirements of some warmwater species. Aquaculture, 79: 145–156.

Stickney, R. R., and R. B. McGeachin. 1983. Responses of *Tilapia aurea* to semipurified diets of differing fatty acid composition. pp. 346–355, In: L. Fishelson, and Z. Yaron (compilers). Proceedings, International Symposium on Tilapia in Aquaculture, Nazareth, Israel, May 8–13. Tel Aviv University Press, Tel Aviv.

Stickney, R. R., and S. E. Shumway. 1974. Occurrence of cellulase activity in the stomachs of fishes. J. Fish Biol., 6: 779–790.

Stickney, R. R., H. B. Simmons, and L. O. Rowland. 1977. Growth responses of *Tilapia aurea* to feed supplemented with dried poultry waste. Tex. J. Sci., 29: 93–99.

Stickney, R. R., R. B. McGeachin, D. H. Lewis, J. Marks, A. Riggs, R. F. Sis, E. H.

Robinson, and W. Wurts. 1984. Response of *Tilapia aurea* to dietary vitamin C. J. World Maricult. Soc., 14: 179–185.

Storebakken, T. 1985. Binders in fish feeds. 1. Effect of alginate and guar gum on growth, digestibility, feed intake and passage through the gastrointestinal tract of rainbow trout. Aquaculture, 47: 11–16.

Sumagaysay, N. G., and Y. N. Chiu-Charn. 1991. Effects of fiber in supplemental feeds on milkfish (*Chanos chanos* Forsskal) production in brackishwater ponds. Asian Fish. Sci., 4: 189–199.

Swingle, H. S. 1967. Estimation of standing crops and rates of feeding fish in ponds. FAO Fish. Rep., 44: 416–423.

Tacon, A. G. J., J. V. Haaster, P. B. Featherstone, K. Kerr, and A. J. Jackson. 1983. Studies on the utilization of full-fat soybean and solvent extracted soybean meal in a complete diet for rainbow trout. Bull. Jpn. Soc. Sci. Fish., 49: 1437–1443.

Takahashi, M., S. Murachi, S. Moriwaki, and S. Ogawa. 1984. The response level of blue-gill sunfish *Lepomis macrochirus*, to a demand feeder. Bull. Jpn. Soc. Sci. Fish., 50: 1475–1480.

Takaoka, O., K. Takii, M. Nakamura, H. Kumai, and M. Takeda. 1990. Identification of feeding stimulants for marbled rockfish. Bull. Jpn. Soc. Sci. Fish., 56: 345–351.

Takeda, M., K. Takii, and K. Matsui. 1989. Identification of feeding stimulatns for juvenile eel. Bull. Jpn. Soc. Sci. Fish., 50: 645–651.

Takeuchi, T., S. Satoh, and T. Watanabe. 1983. Requirement of *Tilapia nilotica* for essential fatty acids. Bull. Jpn. Soc. Sci. Fish., 49: 1127–1134.

Takii, K., M. Takeda, and Y. Nakao. 1984. Effects of supplement of feeding stimulants to formulated feeds on feeding activity and growth of juvenile eel. Bull. Jpn. Soc. Sci. Fish., 50: 1039–1043.

Tandler, A., and F. W. H. Beamish. 1981. Apparent specific dynamic action (SDA), fish weight and level of caloric intake in largemouth bass, *Micropterus salmoides* Lacepede. Aquaculture, 23: 231–242.

Teng, S., T. Chua, and P. Lim. 1978. Preliminary observations on the dietary protein requirement of estuary grouper, *Epinephelus salmoides* Maxwell, cultured in floating net-cages. Aquaculture, 15: 257–271.

Teshima, S., and A. Kanazawa. 1983. Effects of several factors on growth and survival of the prawn larvae reared with micro-particulate diets. Bull. Jpn. Soc. Sci. Fish., 49: 1893–1896.

Teshima, S., and A. Kanazawa. 1986. Nutritive value of sterols for the juvenile prawn. Bull. Jpn. Soc. Sci. Fish., 52: 1417–1422.

Teshima, S., A. Kanazawa, and M. Sakamoto. 1981. Attempt to culture the rotifers with microencapsulated diets. Bull. Jpn. Soc. Sci. Fish., 47: 1575–1578.

Teshima, S., A. Kanazawa, and Y. Uchiyama. 1986. Effect of several protein sources and other factors on the growth of *Tilapia nilotica*. Bull. Jpn. Soc. Sci. Fish., 52: 525–530.

Teshima, S., A. Kanazawa, S. Koshio, and N. Kondo. 1989. Nutritive value of sitosterol for the prawn *Penaeus japonicus*. Bull. Jpn. Soc. Sci. Fish., 55: 153–157.

Thorpe, J. E., C. Talbot, M. S. Miles, C. Rawlings, and D. S. Keay. 1990. Food consumption in 24 hours by Atlantic salmon (*Salmo salar* L.) in a sea cage. Aquaculture, 90: 41–47.

Tidwell, J. H., C. D. Webster, and R. S. Knaub. 1991. Seasonal production of rainbow trout, *Oncorhynchus mykiss* (Walbaum), in ponds using different feeding practices. Aquacult. Fish. Manage., 22: 335–341.

Torrissen, O. J. 1989. Pigmentation of salmonids: interactions of astaxanthin and canthaxanthin on pigment deposition in rainbow trout. Aquaculture, 79: 363–374.

Torrissen, O. J., R. W. Hardy, and K. D. Shearer. 1989. Pigmentation of salmonids—carotenoid deposition and metabolism. Rev. Aquat. Sci., 1: 209–225.

Tucker, J. W., Jr. 1987. Snook and tarpon snook culture and preliminary evaluation for commercial farming. Prog. Fish-Cult., 49: 49–57.

Tuncer, H., R. M. Harrell, and E. D. Houde. 1990. Acceptance and consumption of food by striped bass and hybrid larvae. J. World Aquacult. Soc., 21: 225–234.

Tung, P.-H., and S.-Y. Shiau. 1991. Effects of meal frequency on growth performance of hybrid tilapia, *Oreochromis niloticus* × *O. aurea*. Aquaculture, 92: 343–350.

Ufodike, E. B. C., and A. J. Matty. 1986. Nutrient digestibility and growth response of rainbow trout (*Salmo gairdneri*) fed different carbohydrate types. Proc. Ann. Conf. Fish. Soc. Nigeria, 3: 76–83.

Van Ballaer, E., F. Amat, F. Hontoria, P. Leger, and P. Sorgeloos. 1985. Preliminary results on the nutritional evaluation of ω3-HUFA-enriched *Artemia nauplii* for larvae of the sea bass, *Dicentrarchus labrax*. Aquaculture, 49: 223–229.

Venkataramiah, A., G. J. Lakshmi, and G. Gunter. 1975. Effect of protein level and vegetable matter on growth and food conversion efficiency of brown shrimp. Aquaculture, 6: 115–125.

Verreth, J., and M. van Tongeren. 1989. Weaning time in *Clarias gariepinus* (Burchell) larvae. Aquaculture, 83: 81–88.

Viola, S., S. Mokady, and Y. Arieli. 1983. Effects of soybean processing methods on the growth of carp *(Cyprinus carpio)*. Aquaculture, 32: 27–38.

Wang, K.-W., T. Takeuchi, and T. Watanabe. 1985. Optimum protein and digestible energy levels in diets for *Tilapia nilotica*. Bull. Jpn. Soc. Sci. Fish., 51: 141–146.

Watanabe, T. 1982. Lipid nutrition in fish. Comp. Biochem. Physiol., 73**B**: 3–15.

Watanabe, T., F. Oowa, C. Kitajima, and S. Fujita. 1980a. Relationship between dietary value of brine shrimp *Artemia salina* and their content ω3 highly unsaturated fatty acids. Bull. Jpn. Soc. Sci. Fish., 46: 35–41.

Watanabe, T., T. Takeuchi, A. Murakami, and C. Ogino. 1980b. The availability to *Tilapia nilotica* of phosphorus in white fish meal. Bull. Jpn. Soc. Sci. Fish., 46: 897–899.

Watanabe, T., C. Kitajima, and S. Fujita. 1983. Nutritional values of live organisms used in Japan for mass propagation of fish: a review. Aquaculture, 34: 115–143.

Webster, C. D., J. H. Tidwell, and D. H. Yancey. 1992. Effect of protein level and feeding frequency on growth and body compositon of cage-reared channel catfish. Prog. Fish-Cult., 54: 92–96.

Webster, C. D., J. H. Tidwell, J. A. Clark, and D. H. Yancey. 1992. Effects of feeding diets containing 34 or 38% protein at two feeding frequencies on growth and body composition of channel catfish. J. Appl. Aquacult., 1: 67–80.

Wee, K. L. 1991. Use of non-conventional feedstuff of plant origin as fish feeds—is it practical and economically feasible? pp. 13–32, In: S. S. De Silva (Ed.). Fish nutrition research in Asia. Asian Fisheries Society Special Publication No. 5. Asian Fisheries Society, Manila, Philippines.

Wee, K. L., and S.-W. Shu. 1989. The nutritive value of boiled full-fat soybean in pelleted feed for Nile tilapia. Aquaculture, 81: 303–314.

Wee, K. L., and A. G. J. Tacon. 1982. A preliminary study on the dietary protein requirement of juvenile snakehead. Bull. Jpn. Soc. Sci. Fish., 48: 1463–1468.

White, A., P. Handler, and E. L. Smith. 1964. Principles of biochemistry. McGraw-Hill, New York. 1106 p.

Wilson, R. P. 1977a. Carbohydrates in channel catfish nutrition. pp 19–20, In: R. R. Stickney and R. T. Lovell (Eds.). Nutrition and feeding of channel catfish. Southern Cooperative Series Bulletin No. 218. Auburn University, Auburn, Alabama.

Wilson, R. P. 1977b. Energy relationships in catfish diets. pp. 21–25, In: R. R. Stickney and R. T. Lovell (Eds.). Nutrition and feeding of channel catfish. Southern Cooperative Series Bulletin No. 218. Auburn University, Auburn, Alabama.

Wilson, R. P. 1989. Amino acids and proteins. pp. 111–151, In: J. E. Halver (Ed.). Fish nutrition. Academic Press, New York.

Wilson, R. P., and W. E. Poe. 1985a. Apparent digestible protein and energy coefficients of common feed ingredients for channel catfish. Prog. Fish-Cult., 47: 154–158.

Wilson, R. P., and W. E. Poe. 1985b. Effects of feeding soybean meal with varying trypsin inhibitor activities on growth of fingerling channel catfish. Aquaculture, 46: 19–25.

Wilson, R. P., and W. E. Poe. 1987. Apparent inability of channel catfish to utilize dietary mono- and disaccharides as energy sources. J. Nutr., 117: 280–285.

Wilson, R. P., and W. E. Poe. 1988. Choline nutrition of fingerling channel catfish. Aquaculture, 88: 65–71.

Windell, J. T., R. Armstrong, and J. R. Clinebell. 1974. Substitution of brewer's single cell protein into pelleted fish feed. Feedstuffs, 46(20): 22–23.

Winfree, R. A., and A. Allred. 1992. Bentonite reduces measurale aflatoxin B_1 in fish feed. Prog. Fish-Cult., 54: 157–162.

Winfree, R., A., and R. R. Stickney. 1981. Effects of dietary protein and energy on growth, feed conversion efficiency and body composition of *Tilapia aurea*. J. Nutr., 111: 1001–1012.

Winfree, R. A., and R. R. Stickney. 1984. Starter diets for channel catfish: effects of dietary protein on growth and carcass composition. Prog. Fish-Cult., 46: 79–86.

Wolf, L. E. 1952. Some pathological symptoms in brown trout on an all-meal diet. Prog. Fish-Cult., 14:110–112.

Wright, J. L. C. 1989. Domoic acid, a new shellfish toxin: the Canadian experience. J. Shellfish Res., 8: 444.

Wright, J. L. C., R. K. Boyd, A. S. W. de Freitas, M. Falk, R. A. Foxall, W. D. Jamieson, M. V. Laycock, A. W. McCulloch, and A. G. McInnes. 1989. Identification of domoic acid, a neuroexcitatory amino acid, in toxic mussels from eastern Prince Edward Island. Can. J. Chem., 67: 481–490.

Xu, X., and A. Li. 1988. Studies on the daily requirements and optimum contents of protein, carbohydrate, fiber and fat in the compound diet of *Penaeus orientalis*. Mar. Sci., 6: 1–6.

Xu, X., W. Ji, Y. Li, and C. Gao. 1991. A preliminary study on protein requirement of juvenile black sea bream *(Sparus macrocephalus)*. pp.63–67, In: S. S. De Silva (Ed.).

Fish nutrition research in Asia. Asian Fisheries Society Special Publication No. 5. Asian Fisheries Society, Manila, Philippines.

Yang, G. J. Li, L. Guo, and D. Gu. 1981. Optimum level of protein in diet for black carp fingerlings. J. Fish. China, 5: 49–55.

Yone, Y. 1976. Nutritional studies of red sea bream. pp. 39–64, In: K. S. Price, W. N. Shaw, and K. S. Danberg (Eds.). Proceedings, First International Conference on Aquaculture Nutrition, Lewes/Rehoboth, Delaware, Oct. 14–15, 1975. University of Delaware, Newark.

Yone, Y. 1982. Essential fatty acids and nutritive value of dietary lipids for marine fish. pp. 251–259, In: B. R. Melteff and R. A. Neve (Eds.). Proceedings, North Pacific Aquaculture Symposium. Alaska Seat Grant Program, University of Alaska, Fairbanks.

Yong, W-Y., T. Takeuchi, and T. Watanabe. 1989. Relationship between digestible energy contents and optimum energy to protein ratio in *Oreochromis niloticus* diet. Bull. Jpn. Soc. Sci. Fish., 55: 869–873.

Zein-Eldin, Z. P., and J. Corliss. 1976. The effect of protein levels and sources on growth of *Penaeus aztecus*. pp. 592–595, In: T. V. R. Pillay and W. A. Dill (Eds.). Advances in aquaculture. Fishing News Books, Surrey, England.

Zhou, H., and Y. Wang. 1991. Nutritional requirements of cholesterone and phospholipid for Chinese prawn, *Penaeus orientalis*. J. Fish. China, 15: 148–154.

Zimmer-Faust, R. K., W. C. Michel, J. E. Tyre, and J. F. Case. 1984. Chemical induction of feeding in California spiny lobster, *Panulirus interruptus* (Randall): responses to molecular weight fractions of abalone. J. Chem. Ecol., 10: 957–965.

7 Reproduction, Selective Breeding, and Genetics

REPRODUCTIVE STRATEGIES

Successful aquaculture is often dependent, at least in part, on the ease with which culture animals can be reproduced in captivity. Unless captive spawning and rearing are achieved, the culturist has little or no control on the genetic makeup of the stock, and is thus unable to attempt to improve the characteristics of the animals through selective breeding. While virtually all current aquaculture stocks must still be considered essentially wild animals,[147] there have been some significant improvements in some species through selective breeding. Breeding studies are often aimed at improving growth rate, disease resistance, dress-out percentage, or fecundity.

Success in captive spawning and larval rearing has been achieved for the majority of the species under culture today. For a few species, captive spawning has been achieved by researchers, but the technology has not been sufficiently developed whereby commercial aquaculturists can routinely apply it on a large scale. Examples are milkfish *(Chanos chanos)*, dolphin *(Coryphaena hippurus)*, white sturgeon *(Acipenser transmontanus)*, Pacific halibut *(Hippoglossus stenolepis)*,[148] and some species of penaeid shrimp.

Aquatic animals exhibit a wide variety of reproductive strategies; however, virtually all the important aquaculture species rely on some form of sexual reproduction. Most species are dioecious; that is, the sexes are separate and distinct. There are some hermaphroditic species of aquaculture interest, but most rarely fertilize themselves. Instead, they exchange gametes with other individuals within their species.

Viviparity and ovoviviparity are not common in species of aquaculture interest, though both strategies are found in aquatic animals and, once again, are employed by some species of interest to one or another branch of aquaculture. Most of the species commonly cultured today are oviparous. The types and sizes of eggs that are produced, the environment in which the eggs are incubated, and the amount of parental care that is provided varies considerably among aquaculture species.

---

[147] An argument might be made that koi and goldfish have been domesticated.

[148] Atlantic halibut *(Hippoglossus hippoglossus)* culture has developed in Norway and Scotland to the point where commercial culture is being attempted, but economic success is still tenuous as of this writing.

352

In this chapter, examples are presented that provide some insight into the diversity of reproductive strategies. No attempt has been made to cover all species or variations, however, so the reader should delve into specific literature on species that are not covered and should probe more deeply into the literature on those discussed here to obtain more detail.

CONTROL OF SPAWNING

The reproductive physiology of various species of vertebrates and invertebrates has been extensively studied, and despite the apparent role of various endocrine glands in sexual activity, the pituitary gland and the gonads seem to be the most important endocrine gland. In fishes, hypophysectomy [149] results in blockage of both ovulation and spermatogenesis in all fish except agnathans (Hoar 1969). Interest in hormones by aquaculturists has long involved investigation of their use in the induction of spawning and in sexreversal. In recent years there has been increased interest in transferring growth hormone from mammals to fish through genetic engineering (discussed below). Salmonid culturists have also investigated the role of endocrine glands in smoltification.

While hormone injection has been used to induce spawning in various species, and eyestalk ablation is commonly employed in penaeid shrimp hatcheries,[150] both fish and invertebrates can often be induced to spawn through manipulation of the environment. Temperature modification will induce spawning in some instances, or temperature and photoperiod control may be used in others as described by Arnold et al. (1977). Fores et al. (1990) found that a sudden change in photoperiod can be used to induce gametogenesis in turbot *(Scophthalmus maximus)*. Photoperiod modification has been used to induce spawning in a number of other fishes including ayu, *Plecoglossus altivelis* (Yamaguchi and Okubo 1984); gilthead sea bream, *Sparus aurata* (Kadmon et al. 1985, Micale and Perdichizzi 1988); and milkfish, *Chanos chanos* (C. S. Lee et al. 1987a). Some species even spawn in conjunction with a particular phase of the moon, as is the case with certain rabbitfishes, for example. Other instances of photoperiod or photoperiod and temperature control to induce spawning are mentioned in the sections that follow.

Invertebrate spawning is also often controlled by temperature, photoperiod, or interactions between those two environmental variables. Spawning in the American lobster *(Homarus americanus)* is controlled by temperature and not photoperiod in nature (Aiken and Waddy 1988, 1989, 1990, Waddy and Aiken 1989). However, in the laboratory, photoperiod can influence spawning if the temperature is elevated above normal during the winter. Exposure of lobsters to water temperatures above 13 to 14°C and short daylengths (8 hr of light to 16 hr of dark) after mid-January

---

[149] Removal of the pituitary gland.

[150] Hormones associated with spawning in penaeid shrimp are located in the eye region. But cutting off one or both eyestalks, a hormone that inhibits spawning is blocked, allowing gametogenesis to proceed. Eyestalk ablation is discussed in greater detail later in this chapter.

will induce spawning in advance of the normal July spawning period in Canadian waters (Waddy and Aiken 1990).

Another environmental variable that has been thought to play a role in spawning for some species is pressure. However, even fish and invertebrates that spawn in relatively deep water can usually be successfully spawned in hatcheries at ambient water pressures. Examples are Atlantic and Pacific halibut and various species of shrimp.

The aquaculturist may take an active role in preparing animals for spawning and ensuring that the proper conditions are established, or the animals may be left to their own devices. Another option is to collect ripe adults from the wild for spawning. In some cases (e.g., *Tilapia* spp.) it is possible to stock adults in ponds and capture schools of fry that are subsequently produced. In others (e.g., *Ictalurus punctatus* and most marine species), special conditions to encourage spawning must be provided for the adults. If the appropriate conditions are not provided, the females may resorb their eggs, spawn nonviable eggs, or produce viable eggs that have little chance of normal development because they lack some required environmental condition. In some species (e.g., Pacific halibut) death of females may occur or they are unable to spawn. Random mating is allowed in many instances, though pairing of selected individuals has become increasingly popular for several species. Milt collected from an individual male may even be used to fertilize the eggs from two or more females (common practice in spawning salmonids).

DIFFERENTIATING SEX

The mere recognition of sex among aquaculture animals can be difficult. Sex determination is particularly difficult in conjunction with many species of fish of culture interest. While the sexes may become readily separable immediately before and during the spawning season, it may be difficult to differentiate between them on the basis of external characteristics during the bulk of the year. As the spawning season approaches, many adults (usually males) develop distinctive coloration patterns. There may also be differences in body shape. The female abdomen in many species becomes distended with eggs as the time of ovulation approaches. In Pacific salmon, the lower jaw of the male becomes extended and misshapen, forming what is known as a kipe.

The appearance of the genital openings may also change as the time of spawning approaches. There is often reddening and sometimes swelling around the female vent at that time. Significant size differences may also be exhibited between the sexes of adult fish. Adult female flounders (e.g., *Paralichthys* spp.) are often considerably larger than males. Adult males may be only a few hundred grams, whereas the females may reach several kilograms. This is also true of another group of flatfishes, the halibut (*Hippoglossus* spp.). Males over 100 cm in length are uncommon, whereas the majority of the females are larger than 100 cm.

When external sex differentiation cannot be reliably accomplished, analysis of the steroid hormones present in the blood plasma can often be used. Not only is the presence or absence of certain hormones indicative of sex, but the cycles of hormone levels can provide an indication of when spawning will occur. For example, Liu (1988) was able to sex and predict the time of spawning in Pacific halibut *(Hippoglossus stenolepis)* by looking at circulating hormone levels in the blood.

Most crustaceans of interest to aquaculturists are readily separable by sex. Some species of current or potential aquaculture interest, including the blue crab *(Callinectes sapidus),* Florida lobster *(Panulirus argus),* and freshwater shrimp *(Macrobrachium rosenbergii)* carry their eggs externally; thus the females can be readily identified once the eggs have been extruded. In most of those cases, other external characteristics exist that allow culturists to quickly and accurately determine invertebrate sex throughout the year. For example, in the blue crab and other crab species, the abdomen of juvenile females is triangular, with the apex of the triangle located anterior to the base. The apex of the abdominal triangle becomes rounded in adult females. In males the abdomen is narrow in the lateral aspect and long in the anterior-posterior aspect (Figure 83). From the dorsal aspect the sexes appear identical.

Male marine and freshwater shrimp have an organ called the petasma which is located on the first pair of pleopods and can be seen as an accessory organ protruding from those appendages. The petasma is used to transfer a packet of sperm, called a spermatophore, to the female. The spermatophore is attached to an organ called the thelycum, located between and just anterior to the most posterior pair of walking legs.

EGG SIZE AND RELATIVE FECUNDITY

The amount of time required to rear an animal from birth to market is related to its initial size, although the ultimate size at slaughter is also an important factor. For example the absolute weight increase required to take a calf from birth to market is much greater than that of a channel catfish or rainbow trout; however, the number of times the weight of the newborn calf must be doubled between birth and market size (approximately 225 to 275 kg) is considerably lower than that of the catfish or trout marketed at 0.5 kg.

Chickens are often used as an example in demonstrating the accomplishments that are possible with domestication and selective breeding. Broilers can be produced from chicks in as little as 6 weeks, which is a fraction of the time required for wild chickens to reach the same size. It has been suggested that similar improvements in growth rate could be achieved with fish. The problem lies in the fact that fish begin life at a very small size compared with chickens. Significant improvements in fish growth have, in fact, been achieved with respect to a number of species, but the probability that fish culturists will be able to develop a 6 week catfish or trout is extremely low as long as the size of eggs and fry remain small. Various species of fish have extremely small eggs and can be expected to grow even less rapidly to the same market size because of the added time required for their larvae to reach the size of catfish or trout fry.

Dorsal

Ventral

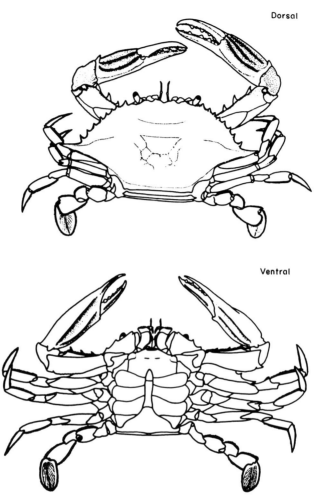

Figure 83. Dorsal and ventral views of a male blue crab, *Callinectes sapidus.* (Original drawings by Daniel Perlmutter.)

Channel catfish females lay 6,600 to 8,800 eggs per kilogram of body weight (Clemens and Sneed 1957). Egg number will vary depending on the age and general condition of the adult female. When hatched from the approximately 5-mm-diameter eggs, the 20 mg fry can be reared to 0.5 kg in from 6 to 10 months under the proper environmental conditions and when provided with the proper feed (longer periods of rearing are often required depending on the rearing strategy used).[151]

---

[151] Catfish may be reared at high density during the first growing season and overwintered as fingerlings of about 10 cm that are grown out to market size the second year, for a total rearing period of about 18 months.

Since few adult catfish can supply the number of fingerlings required by the typical farmer, there is not typically a great deal of expense involved in maintaining broodstock.

The eggs of tilapia are similar in size to those of channel catfish, but the most commonly cultured species produce no more than a few hundred eggs per spawn (Bardach et al. 1972). Since tilapia are multiple spawners, the annual production rate per female in terms of eggs per kilogram may be similar to the channel catfish.

Trout and salmon are other species with relatively large eggs. Spawns of several thousand eggs per female are typical, and egg size exceeds 5 mm. Growout time varies as a function of the size desired at harvesting. Unlike catfish and tilapia, which are usually harvested at 0.5 to 1.0 kg, salmonid harvest sizes range from somewhat less than 0.5 kg (pan-sized trout) to at least a few kilograms (Atlantic salmon).

It has been suggested that fish with much larger eggs could be reared more rapidly than any of the fishes previously mentioned, since the initial size of the fry would be larger. In light of what is known about chickens and other terrestrial livestock species, this appears to be a reasonable hypothesis. The problem involves finding species of fish, or invertebrates for that matter, that not only produce large eggs but would also enjoy consumer acceptance. One such species is the gafftopsail catfish, *Bagre marinus,* a marine fish in the family Ariidae. Female *B. marinus* lay eggs that are more than 1.0 cm in diameter. After fertilization the eggs are picked up and incubated in the mouth of the male. The fry are maintained in the mouth of the male until yolk sac absorption is complete, after which the fry begin foraging for food on their own (Gudger 1918). Because of the protection afforded the developing eggs and fry, and because the newly hatched fish weigh several grams, survival is generally excellent.

While the prospect for culturing *B. marinus* might appear bright from the information presented above, there is a major problem. Because of egg size, only a few eggs can be accommodated in the mouth of the male. The number of eggs produced by females is generally less than 50. Thus, a very large population of broodfish would have to be maintained to provide sufficient numbers of fry for growout. The economics of maintaining the broodfish dictate against culture of the species.

B. marinus is an exception among marine fishes in that it lays very large eggs. Most marine species broadcast large numbers of very small eggs (typically ranging from 1 to no more than 3 mm in diameter)[152] into the water. For many species the numbers of eggs produced runs into the hundreds of thousands to millions per kilogram of female body weight. Little or no parental care is provided, and contrary to the high survival rate exhibited by the marine catfish, the mortality rate for most marine fishes (and invertebrates) is well in excess of 99.99%. In nature, replacement of a fish population requires that two of the millions of eggs that are laid over a female's lifetime survive and grow to adulthood. Aquaculturists who work with species with high fecundity attempt to provide conditions that will support good survival of eggs and fry, though with the large numbers produced, survivals of 25

---

[152] Halibut have eggs of about 3 mm diameter, which is very large for marine pelagic spawners.

to 50% are often considered to be incredibly good. High fecundity also means that small numbers of broodfish are required. The costs of maintaining those fish may be quite low.

SPAWNING AND LARVAL REARING TECHNIQUES

This section provides details associated with the spawning of several species of aquaculture importance, including marine and freshwater animals, and both vertebrates and invertebrates. Many species are not covered, and the intent is primarily to show the diversity of techniques that are currently being applied. Included for some species is information on selective breeding and genetics. Additional information on those topics is presented in later sections of this chapter.

Channel Catfish

In nature, channel catfish can reach sizes in excess of 20 kg, but broodfish maintained by catfish farmers typically range from 0.9 to 4.5 kg (Martin 1967). Relatively small animals are not only easier to handle than larger fish but also produce more eggs per kilogram of body weight. Females of 0.5 to 1.8 kg produce an average of about 8,800 eggs/kg, whereas fish larger than 1.8 kg produce about 6,600 eggs/kg (Clemens and Sneed 1957).

Channel catfish spawn in the late spring and summer, depending on the strain of fish and the geographic region in which they are being reared. Spawning in the continental United States does not usually occur much before May and is usually concluded by early August, although exceptions can occur. Spawning typically occurs when water temperature is between 21 and 29°C (Clemens and Sneed 1957). The optimum temperature for spawning is reportedly 25.6°C (Wellborn and Schwedler 1981). Bondari et al. (1985) evaluated 5-yr-old catfish of various sizes with respect to spawning characteristics and found that larger males were less successful spawners than smaller ones and that female size influenced the time required for eggs to hatch.

Catfish can be induced to spawn by manipulation of temperature. Exposing broodfish to an artificial winter by lowering the water temperature for a period of time and then allowing it to increase slowly to that required for spawning may induce gamete development. The minimum refractory period required between spawns (approximately 1 yr in nature) has yet to be determined.

Following spawning, broodfish are often maintained in ponds at densities of about 375 fish/ha (Nelson 1960), but higher densities have been reported (Martin 1967). During the early years of catfish culture, when good-quality prepared feeds were not available, the stocking of minnows and crawfish in broodstock ponds as forage was recommended (Canfield 1947). During late winter and early spring, it has been common practice to provide organ meats such as beef liver and heart for additional nutrition. Now that nutritionally complete feeds have been developed,[153] good spawning results can be obtained without augmenting the pelleted ration with live food or organ meats.

During most of the year it is difficult to separate the sexes; however, during the spawning season certain secondary sexual characteristics become apparent. Females show a characteristically well-rounded abdomen as the eggs develop. The ovaries become soft and palpable, and the genital pore becomes raised and inflamed (Clemens and Sneed 1957). Male channel catfish generally have heads somewhat wider than the rest of the body (which is not true of females), dark pigment develops under the lower jaw and on the abdomen, and the genital papilla becomes well formed and tubular (Clemens and Sneed 1957).

Prior to spawning, the broodstock may be seined from holding ponds for selection and stocking into ponds, with or without spawning pens, or they can be placed in tanks or aquaria. Fish should be handled gently during the spawning season and should be returned to the water as quickly as possible.

Pond Spawning. Captive spawning of channel catfish may date back to the 1890s (Martin 1967), when it was discovered that the species readily spawn in ponds if some type of nest is made available. In nature, channel catfish typically spawn under logs or in depressions along stream banks. Providing some type of artificial nest in a pond is generally a requirement for spawning since suitable sites are not available in the typical aquaculture pond.

Artificial spawning nests of various types have been used, including but not restricted to milk or cream cans (Figure 84), nail kegs, beer kegs, metal drums and cans of various sizes (e.g., grease cans and jerry cans), wooden boxes (held down with weights), and ceramic and concrete drain tiles. Commercial spawning nests have also been developed. Transformer cases should not be used as spawning containers because they may contain PCB (polychlorinated biphenyl) residues (Anonymous 1981).

The number of nests that should be placed in a pond relative to the number of broodfish present varies depending on whether the eggs will be allowed to hatch in the nests or removed to a hatchery. In the former case each nest will be occupied for a week or more between spawns, so more nests will be required than if the eggs are removed. Usually there are fewer nests than pairs of catfish.

When pond spawning is practiced, broodfish are stocked at densities of 60 to 375 fish/ha (Martin 1967). Some culturists prefer to stock a slightly higher number of females than males (e.g., four females for every three males), but other culturists stock equal numbers of each sex. Bondari (1983a) found no difference in number of spawns produced, weight of the eggs, or hatching success when ratios of male to

---

[153] It can be argued that the feeds used for channel catfish have been tested only on fingerlings, and feeds specifically designed for broodfish remain to be developed. While that is true, excellent results in terms of egg quantity and quality are being produced when those feeds are the sole source of nourishment provided by the culturist.

Figure 84. Milk cans make excellent catfish spawning containers, but they are becoming increasingly hard to find.

female were 1:1, 1:2, 1:3, or 1:4. Since each male can spawn with two or more females, there is no reason to stock more males than females.

Spawning nests are generally placed in 15 to 150 cm of water with the open end of each nest facing the middle of the pond. While fish farmers who have ponds of the types described in Chapter 3 place their spawning nests on the bottom, it is possible to obtain spawns in nests suspended about 1 m beneath the water surface of deep ponds and lakes (Moy and Stickney 1987).

Nests that are open at both ends (e.g., drain tiles) should have one end covered or placed against the pond bank in a manner that effectively closes it off. Nest placement at between 60 cm and 1.2 m has been shown to have no effect on spawning success when dissolved oxygen (DO) level remains above 4 mg/l (Steeby 1987). That study also showed some preference of a metal spawning can with a 16 cm opening to a similar can with a larger opening and two types of plastic containers.

Following broodfish stocking the culturist may wish to check the nests daily; if the fish are disturbed in the act of spawning, however, they may fail to resume spawning activity. Thus, many aquaculturists prefer to examine the nests only two or three times weekly. Nest inspection is a good idea even if the eggs are going to be allowed to hatch in the pond, since it gives the culturist an indication of spawning success and at least a rough estimate of fry production. If the eggs are to be hatched indoors, inspection at 3 day intervals is sufficient for egg collection because hatching requires at least 5 days under normal pond temperatures.

Some caution should be exercised when the nests are inspected. Following spawning the male catfish expels the female and guards the eggs during incubation, fanning them with his fins to keep a flow of well-oxygenated water moving through the egg mass. Catfish lay up to a few tens of thousands of eggs in an adhesive mass. During the spawning and incubation period the male may become highly aggressive and can inflict a painful bite to the careless fish culturist who sticks a hand into the spawning container. It is usually good practice to lift the spawning container to the water surface and slowly empty it until the interior can be visualized. This reduces the possibility of exciting the male (or both adults if they are in the act of spawning). Excited adults may quit spawning and can destroy the eggs that have been laid by thrashing around in the spawning container.

If left alone to hatch in the pond, the male will guard the sac fry until they disperse from the nest about a week after hatching. Good survival can be expected because of parental attention, but it is difficult for culturists to determine how many fry have been produced.

Pen Spawning. In open pond spawning the aquaculturist has no control over the pairings that will occur unless only two fish are released into each spawning pond or a single male is placed in a pond with two or more females. Because this is impractical from the standpoint of economical use of facilities, other means of segregating paired fish have been developed. The most widely utilized technique is pen spawning.

Spawning pens may be located in the middle of a pond (Figure 85) or along the edges (Figure 86). Pens along the pond bank provide easy access though construction may be complicated by the presence of the sloped levee.

The number of pens that can be placed in a given pond is variable, but in no case

Figure 85. Channel catfish spawning pens in a 0.1 ha pond.

Figure 86. Spawning pens constructed along the pond edge. Note the wooden spawning containers with concrete block weights.

should the number be so large that water quality is impaired by the overstocking of broodfish. Maximum fish density should be in line with the figures presented above for open pond spawning (60 to 375 fish/ha).

Pens are generally constructed by stapling steel wire mesh or plastic-covered wire mesh of suitable size over a wooden frame. The mesh must be fine enough to prevent escapement of the broodfish. The bottom of the mesh should be sunk below the substrate surface to prevent the fish from digging under and escaping. Pens of any convenient size can be used, but most are no larger than 2 × 3 m. They should be sufficiently large to allow entry by the culturist since it will be necessary for people to enter the pens to check the spawning containers.

Each pen must be provided with a suitable nest and should be stocked with a selected pair of broodfish. Care should be taken to ensure that the sexes are correctly determined, for if two males or two females are placed together, not only will spawning not occur, but in the former instance there may be fighting that results in the death of one or both males. The female should be slightly smaller than the male since the male is expected to chase the female from the spawning container after the eggs are laid. Prolonged fighting within the nest will cause disruption of the egg mass. If the female is successful in driving the male away from the nest, she may either ignore the eggs (allowing them to die) or eat them. The female should be removed from the pen as soon as possible following spawning. If the eggs are transferred to a hatchery, as is usually the case, the pen may be restocked with a second female, or the male may also be removed and a new pair of broodfish introduced. In most cases a single male is allowed to spawn with no more than two females during a given spawning season.

Aquarium and Tank Spawning. The aquarium spawning method was developed by Clemens and Sneed (1962). It provides the culturist with the greatest degree of control over spawning; however, it is not particularly desirable in terms of producing large numbers of fry. The method, which can also be used to spawn catfish in circular tanks, is best applied when direct observation of spawning activity is desired, as is the case in a teaching laboratory or where behavioral studies are being conducted. Hormone injection of the females is required for the induction of spawning when this technique is utilized. Although not generally a requirement for successful spawning, hormone injections may also be employed in conjunction with the pen spawning technique. It is not necessary to inject males with hormones in any catfish spawning technique.

Hormones are not used to induce gonadal development but only to initiate ovulation. Once resorption of eggs has begun (as in the case of females that are not provided with an opportunity to spawn at the time their ovaries become ripe), it is too late to reverse the process, and hormone injections will not alter the situation.

The most popular hormones for use with channel catfish are fish pituitary extracts (carp pituitary being among the most popular) and human chorionic gonadotropin (HCG). Both are available commercially, though fresh fish pituitary can be surgically obtained.

Hormones are injected intraperitoneally or intramuscularly (Figure 87). Culturists have often incorporated 10,000 IU of penicillin with the hormone injection to prevent secondary infections. Spawning generally follows the injection of about 13

Figure 87. Administration of an intraperitoneal hormone injection to an adult female channel catfish.

Figure 88. Channel catfish eggs are deposited in an adhesive mass.

mg/kg of body weight using pituitary extract (Clemens and Sneed 1962) or an average of 1760 mg/kg of HCG (Sneed and Clemens 1959).

In the aquarium spawning method, fish are paired in a running-water aquarium large enough to permit the animals to maneuver. The female is injected with hormones, after which the pair is checked frequently so they can be removed as soon as spawning is completed. If the fish do not spawn within 24 hr of the first injection, subsequent injections may be required. If spawning does not occur after three to five hormone injections, the female should be replaced since she may be unable to spawn. In this technique the eggs are virtually always removed to a hatchery.

Egg Development and Hatching. Channel catfish eggs are deposited in a yellow, adhesive mass (Figure 88). Optimum temperature for hatching is 25.6°C (Wellborn and Schwedler 1981), though good hatching success occurs over a range of 18 to 29°C (Martin 1967). Temperatures below 18°C and above 29°C may lead to reduced hatchability and fry survival (Wellborn and Schweler 1981).

When eggs are incubated within the optimum temperature range, a period of 5 to 10 days is required for hatching (Toole 1951). The development of catfish eggs has been discussed by Murphree (1940), Clemens and Sneed (1957), and Saksena et al. (1961).

If the eggs are hatched by the male in a pond, the fry will remain in the nest for several days following hatching and can be removed by driving off the male and pouring the offspring into a suitable container for transfer to a fry rearing pond. Alternatively, the broodfish can be seined from the pond at the end of the spawning season, leaving the fry to develop into fingerlings. A third alternative is to remove the eggs from the nests and place them in a hatchery.

Figure 89. A channel catfish hatching trough with a series of paddles that rotate to move water through the egg masses, which are placed in baskets suspended in the trough; the motor that moves the paddles is located in the foreground.

One advantage of removing the eggs to a hatchery is that the spawning containers can almost immediately be occupied by other adult catfish. This is particularly desirable, as previously indicated, when the pen spawning technique is employed. Also, the culturist can keep a fairly accurate account of the number of eggs hatched when artificial hatching is practiced. In addition, the culturist can be certain that the fry are well established on artificial feed before they are stocked into rearing ponds.

Channel catfish eggs can be hatched in jars receiving running water, as was once common practice in salmonid hatcheries (Canfield 1947); however, in most cases catfish eggs are hatched in troughs. The first catfish hatching trough was described by Clapp (1929). It consisted of a small raceway fitted with paddles that rotated through the water to simulate the fanning action of the adult male. The paddles, fitted on an axle, were turned by a water wheel outside the laboratory. Modern hatching trough paddles are operated by electric motors (Figure 89).

When hatching troughs are employed, egg masses are put into hardware cloth baskets, which are placed between the paddles in the hatching troughs. Large egg masses may be broken into two or more pieces, each of which is placed in a separate basket.

If hatching jars are going to be used for incubation of channel catfish eggs, it is desirable to separate the eggs, which means that the matrix surrounding them needs to be dissolved. Fungal problems and the amount of labor associated with hatching channel catfish can be reduced when hatching jars are used (Ringle et al. 1992). Those authors indicated that 1.5% Na_2SO_3, 1.5% Na_2SO_3 plus 0.2% papain, 1.5% L-cystine-HCl plus 0.2% papain, and 1.0% Na_2SO_3 plus 0.5% L-cystine and 0.2% papain are all effective in egg separation.

The percentage of eggs that hatch when the trough method is employed is generally very high if care is taken to prevent the establishment of fungus on the developing egg masses (see Chapter 8). In addition, fry survival is usually good, especially if inbreeding is not great and good management practices are employed. Survival rates from egg to fry often range between 80 and 90%.

Fry Stocking. Ponds should be treated for predaceous insects and unwanted vegetation before fry are stocked. In addition, a plankton bloom should be established. Insects can be controlled by spreading a thin film of diesel fuel on the pond surface. When the insect larvae surface to breathe, they will be killed by the fuel. Herbicides and fertilizers can be used to control nuisance vegetation and establish plankton blooms, respectively, as discussed in Chapter 4. Draining ponds during winter will reduce problems with both predators and nuisance vegetation and will also allow organic matter in the pond bottoms to oxidize.

If fry are allowed to remain in the broodfish pond, that pond should be properly prepared before the adults are stocked. The number of broodfish should be limited so that no more than 625,000 fry are produced per hectare (Martin 1967). Fry can easily be maintained at high density in fertilized ponds and provided supplemental feed until they become fingerlings 2 to 4 cm long. They should then be stocked into growout ponds or restocked at lower densities for growout to larger fingerling sizes.

When channel catfish eggs are hatched in troughs or jars, the fry may be transferred to small raceways until they are established on feed. The yolk sac of fry will generally be absorbed within about a week following hatching (Figure 90), at which time the fry will surface in search of food. Yolk sac absorption is accompanied by a change in fry color from pink or orange to black. When the fry surface for feed, a nutritionally complete ration, high in protein and energy, should be provided (see Chapter 6).

After the fry have become accustomed to eating prepared feed, they may be stocked into ponds as indicated. In instances where more intensive culture is practiced, the fry may be maintained in raceways or tanks throughout the period until harvest.

Genetic Studies. By the 1970s, channel catfish culture was developing rapidly, and interest in improving growth rate and enhancing other desirable characteristics was beginning to be aroused in geneticists. Early studies involved crossbreeding fish collected from various fish farms and public hatcheries. Few improvements were obtained as a result of selective breeding with those fish, possibly because most (or perhaps all) of them probably could be traced back to one or a few federal government hatcheries as the points of origin. By the end of the same decade, attempts were being made to find genetically distinct groups of catfish. For the first time, researchers searched for wild strains that they could use to mate with other wild strains and with fish that had been in culture for a number of generations.

Broussard and Stickney (1981) found that the time of spawning varied in catfish from various locations. Fish from northern Minnesota, for example, spawned in Texas in early May, well in advance of fish obtained from the southern United

Figure 90. Channel catfish fry. The yolk sacs of these fish are nearly absorbed, so exogenous feeding will begin within a day or two.

States. It was not possible, at least under natural conditions, to cross the Minnesota fish with those from other available strains. Bondari (1983b) attempted to spawn wild and domestic strains of channel catfish and was unable to spawn wild fish with one another but could obtain spawns from wild strains crossed with domesticated ones. Growth trials showed that domestic × domestic crosses outperformed domestic × wild crosses.

Dunham et al. (1983) compared timing of spawning between fish paired within and between strains and found that between-strain (crossbred) fish tended to spawn earlier. Differences in spawning timing between domesticated catfish strains from Alabama and Kansas were observed by Smitherman et al. (1984).

Experience has generally shown that the fish produced from a given spawn of channel catfish will grow at variable rates. Fast, slow, and intermediate growth-rate fish will be produced. If the largest fish produced from a number of spawns are retained and used as broodstock, their offspring will also produce fish that grow at fast, slow, and intermediate rates. While there is at least one study indicating that selection of large fish will enhance the growth rate of offspring (Dunham and Smitherman 1982), in general, selecting fast-growing broodfish has not seemed to improve performance in the next generation. One theory, based on the author's experience, is that selection for rapid growth may actually in some cases be selection for aggression (since the most aggressive fish will be those that obtain the most feed and, therefore, grow most rapidly).

Tilapia

The species of *Tilapia* that are of interest to aquaculturists are nearly all mouth-brooders. Included are *T. aurea* (Figure 91), *T. nilotica, T. mossambica,* and red hybrids that have been produced by crossing those and other species. Culture of tilapia has recently been reviewed by Stickney (1992).

Males of mouth-brooding species construct nests (Figure 92). They then attract females that deposit eggs in the nests. Following fertilization, the female picks up the eggs (usually only a few hundred in number) in her mouth and retreats from the nest site. The male may then mate with other females.

The eggs hatch in the female's mouth after a few days, and the fry remain there until their yolk sacs have absorbed (typically within a week after hatching). The fry will then venture out into the water seeking food but will remain close to the mother and retreat into her mouth when danger threatens. Because of the degree of parental attention afforded the eggs and fry of mouth-brooding tilapia, survival to the fingerling stage is generally very high (commonly approaching 100%).

While the fecundity of tilapia is low compared with other successful aquaculture species, that problem is largely overcome through multiple spawning. Females of most species of aquaculture interest will spawn when the water temperature is above 20 to 23°C (Uchida and King 1962). As a consequence, spawning can occur year round in the tropics and takes place over variable periods in temperate climates. While a given female may not spawn every month, individual females of *T. mossambica* can spawn from six to 11 times within 1 yr (Chimits 1955). Other species may not spawn that many times.

While multiple spawns can be considered an advantage, there are some problems that must be considered. First, female tilapia can begin spawning not only when

Figure 91. *Tilapia aurea* fingerling and adult.

Figure 92. Tilapia nests along the edge of a pond. Their small size, as shown in comparison with the human hand at the top of the photograph, is an indication of the size of the males that constructed the nests.

they are only a few months old, but also when they are quite small, perhaps only about 10 cm in length for some species (Atz 1954). Once tilapia reach maturity, growth of females can slow dramatically, so males grow to market size significantly more rapidly (Chimits 1955, Lowe-McConnell 1958, Avault and Shell 1968). Much of the food energy that the female obtains will go to the support of egg development. Once the eggs are laid, the female picks them up in her mouth and may retain them for two weeks or longer, during which time she cannot feed. Once the fry are released, food energy goes once again to egg development, so growth is minimized. Since tilapia mature at sizes much below those at which they become marketable in most countries, the presence of females in growout ponds is generally not desirable.

In ponds where tilapia are allowed to spawn freely, stunting has widely been reported. Once the submarketable fish begin to spawn, the density and biomass of fish in a pond will be greatly augmented by the presence of fry and fingerlings. The overall effect is, at least in theory, to cause stunting of the originally stocked fish due to overcrowding and competition for food. In reality, there is little scientific documentation of stunting in tilapia ponds, though anecdotal evidence abounds. The author's observation has been that while total fish numbers increase considerably when the originally stocked fish begin to spawn, subsequent spawning does not add appreciably to the population. Cannibalism, particularly by fingerlings on fry (which would account for the good survival of initially spawned fish and poor survival of those spawned later), appears to occur. Little research on tilapia cannibalism has been conducted, though one study by Pantastico et al. (1988) demonstrated the phenomenon in a laboratory study.

A variety of methods have been developed to overcome the presumed stunting problem caused by enhancing fish numbers in growout ponds due to uncontrolled reproduction, and to prevent the disparate growth between the sexes that occurs when the females mature. One method that helps in the removal of unwanted fry but has no effect on their production involves the stocking of predators. The predator should, of course, be sufficiently small when stocked that it cannot eat the fish that are being grown out. As reviewed by Stickney (1992), among the species that have been stocked as predators in tilapia ponds are snakeheads, peacock bass, Oscars, sea bass, Nile perch, tarpon, largemouth bass, and mudfish.

Other methods of maintaining stable tilapia populations include the use of cage culture, monosex hybridization, and sex reversal through exposure to hormones. Cage culture tends to restrict or eliminate spawning because no nest can be constructed, and if the female does lay eggs, they generally fall through the cage bottom and are lost. This method is generally successful, but instances of successful reproduction of caged tilapia have been reported (Bardach et al. 1972, Pagan-Font 1975).

The rearing of all-male populations has been widely touted as the best means of maintaining rapid and relatively uniform growth rates in tilapia. By hybridizing various species of tilapia, offspring sex ratios of higher than 50% males, including 100% males in some instances, can be produced. Crosses that have been shown to produce all-male offspring include *T. mossambica* female × *T. hornorum* male (Hickling 1960, Chen 1969), *T. nilotica* female × *T. hornorum* male (Pruginin 1967), *T. nilotica* × *T. macrochir* (Jalabert et al. 1971), and *T. nilotica* female × *T. aurea* male (Fishelson 1962, Pruginin 1967).[154] Since tilapia freely hybridize and the hybrids are usually fertile, finding pure strains of fish is becoming increasingly difficult. If a fish that appears to be a nonhybrid is crossed with a species that the literature says will produce all-male fry, the offspring may not be 100% males because one or both of the adults may actually have been a hybrid that morphologically appeared normal. Some culturists believe that virtually all so-called species of tilapia being maintained by culturists are, in reality, hybrids.

Even if unhybridized tilapia can be found, it is difficult to maintain two or more species entirely separate, mix them for spawning, and then get them properly separated once again. Errors inevitably occur in replacing adults into the proper holding ponds, and there will eventually be hybrids that become mixed with the unhybridized fish. The result is that producing all-male offspring is difficult.

A technique has been devised by which all-male populations of tilapia can be produced from adults spawned under normal conditions. By providing small amounts of male hormones to first-feeding fry, sex reversal is possible as first demonstrated in *Tilapia aurea* by Guerrero (1975). Fry are collected from spawning tanks or ponds and are fed a standard fish feed treated with 60 μg/ml diet of 17-α-ethynyltestosterone. The hormone-treated feed is offered for about 3 weeks. Treated

---

[154] A more complete review can be found in Stickney (1992).

fry are frequently 95 to 100% male. As reviewed by Stickney (1992) other species of tilapia, including *T. mossambica* and *T. nilotica,* have also been successfully sex-reversed using the technique.

Hand sexing is another means by which all-male populations can be stocked. Before tilapia can be visually sexed, they must be 6 cm long or more so a few months of culture are required before the separation can be achieved. Complicating the process and increasing the expense is the fact that each fish must be examined individually by persons who have been trained to distinguish between the sexes by viewing the genital pore area. Mistakes will inevitably be made. Hand sexing, once popular in some places, such as Israel, is rarely used today.

Largemouth Bass

Captive spawning of largemouth bass is conducted by private, state, and federal hatcheries in the United States for the purpose of producing fingerlings for stocking. Production of largemouth bass for direct marketing as foodfish is prohibited by law in some jurisdictions. Spawning has traditionally been conducted in specially prepared ponds, though at least some hatcheries have been turning to raceway spawning in recent years. Pond culture techniques have recently been reviewed by Williamson et al. (1992), and the following comments include information from that review.

Prior to the introduction of adults to spawning ponds the ponds are drained, disked, smoothed, and allowed to dry during fall and winter. Herbicides may also be applied to control vegetation. Gravel may be placed in selected areas in the pond, or artificial turf mats may be placed at least 2 m apart on the pond bottom as nesting sites. Artificial turf mats are also used as nests in concrete raceway spawning with one mat being placed every 2 or 3 m.

Adults are typically stocked at rates of two or three males for each female in ponds. Ratios of 1:1 are used by some culturists involved in raceway spawning, while others prefer higher numbers of females than males. The total number of fish stocked is dictated by the male/female ratio and the fact that no more females should be stocked than there are available nests. Forage is normally not stocked into spawning ponds or raceways. The adults are allowed to remain therein only a few days until they spawn and while those trained to accept pellets may be provided with feed, forage fish should not be provided since they could prey on eggs and fry and might introduce diseases.

Spawning tends to occur during late afternoon or early morning. The male constructs a shallow nest. The nest will be about the same diameter as the length of the male's body. Males leave their nests in search of a ripe female and to chase away intruding fish. When a ripe female is lured to the nest, the pair spawns over the nest. Multiple spawning can occur in conjunction with both sexes. Females should be removed after spawning. They can be transferred to reconditioning ponds and later reintroduced to the spawning ponds or raceways. If spawning mats are collected and removed to a hatchery, males may also be removed after spawning takes place. The male will guard the nest and protect the eggs from predators. If spawning mats are used and each nest is removed to a hatchery once eggs are deposited, a new spawning mat can be placed in the pond. The male will defend the new nest and entice another female to spawn.

Spawning typically occurs from spring through summer when water temperatures are in the range of 15 to 24°C. Spawning time varies as a function of latitude. Largemouth bass have been observed spawning as early as January in Florida, but spawning may not take place until late spring at the northern extreme of the range.

Fecundity rates reportedly range from 2,000 to 176,000 eggs per kilogram of female body weight, with the number of eggs spawned in a given nest ranging from about 5,000 to 43,000. As previously indicated, both sexes are capable of multiple spawning, so the entire number of eggs carried by a given female may be deposited in a number of nests over the spawning period and will, in many cases, be fertilized by a different male each time.

Egg hatching time depends on temperature, ranging from 4 days upward. Yolk sac absorption typically requires 5 to 7 days. Eggs may be transferred on spawning mats from spawning raceways to ponds or indoor raceways for hatching. If allowed to hatch in spawning ponds, fry may be captured with nets when they school at the time of swim-up. Stocking densities of fry in fingerling ponds vary greatly, with recommendations ranging from 120,000 to 240,000 fry/ha if 25 to 50 mm fingerlings are desired. Lower densities are employed when larger fry are desired.

Striped Bass and Hybrid Striped Bass

Striped bass *(Morone saxatilis)* represent a species that has almost exclusively been produced from wild broodstock. In nature the species has declined dramatically in recent years, so striped bass production is being approached in a two-pronged manner. Some culturists are interested in restoring natural populations through augmentation stocking, while others are involved in producing striped bass in captivity for direct sale as human food. Foodfish culturists are also involved in the production of hybrid striped bass. Hybrids are produced by crossing striped bass with white bass *(M. chrysops),* white perch *(M. americana),* or yellow bass *(M. missippiensis).* The culture of these fishes has recently been reviewed by Kerby (1992) and was also reviewed by Piper et al. (1982). Unless otherwise noted, the information presented in this section comes from those reviews.

Striped bass are anadromous, although landlocked strains have been found. They move into freshwater spawning areas in the spring. The traditional means of spawning striped bass and its hybrids is to collect broodstock when they move into freshwater and transport them to hatcheries. HCG is generally used to induce the fish to spawn. Females are most commonly injected with hormones, but males may also be injected.

Woods et al. (1990) reported a case of noninduced spawning of a female striped bass that was reared in captivity. Henderson-Arzapalo and Colura (1987) were able to induce striped bass maintained for 13 months in captivity to spawn without hormone injection by manipulating temperature and photoperiod. Hybrid striped bass (striped bass female × white bass male) have also been matured in captivity with temperature and photoperiod control and spawned to produce a second generation of hybrids (Harrell 1984, Smith and Jenkins 1984). Hormone implants have been used to induce natural production of hybrids, but the fertilization rate has been low. Attempts to spawn hybrid males with pure-strain females have not been successful (Harrell 1984).

From 20 to 28 hr after initial hormone injections a small sample of eggs is taken with a small-diameter glass or plastic catheter inserted through the urogenital pore into the ovary. The approximate time to ovulation can be determined by microscopic examination of the eggs. Details of how eggs are staged in advance of hatching were first presented by Bayless (1972).

When the females ovulate, eggs are expressed into a pan from fish that have been either sacrificed or anesthetized. If anesthetized, a common method has been to spray quinaldine on the gills, though other anesthetics are becoming more commonly used since quinaldine can be toxic to people handling the fish. Milt is then stripped from running-ripe males, and the eggs and sperm are mixed for 2 or 3 min. The fertilized eggs are then placed in McDonald hatching jars where hatching will occur after 40 to 56 hr, depending on temperature. Larvae swim over the lip of the hatching jars and are collected in aquaria.

Broodfish have also been induced to spawn in circular tanks. Both males and females are injected with HCG and placed in tanks of 1.2 to 2.4 m diameter at rates of one or two females to two to four males. The fish should spawn in from 36 to 62 hr after injection.

Swim-bladder inflation has been a problem in striped bass culture, particularly when fry are reared in tanks rather than being released into ponds soon after hatching. Survival of fish that do not properly inflate their swim bladders can be poor. Injecting the thyroid hormone triiodothyronine (T_3) into females prior to spawning has been shown to significantly increase the percentage of fry that undergo proper swimbladder inflation (Brown et al. 1988). A method by which striped bass fry with inflated swim bladders can be separated from those with uninflated swim bladders was developed by Henderson-Arzapalo et al. (1992). The researchers anesthetized the young fish with tricaine methanesulfonate (MS–222) and found that those with inflated swim bladders floated while those without remained on the bottom.

Trout and Salmon

The spawning of trout and salmon in nature is quite similar in many respects, though there are considerable differences as well. Both types of fish spawn in gravel nests, called redds. Most species spawn in streams, but lake-spawning species and strains can also be found. It was once convenient to distinguish between fish that died after spawning (genus *Oncorhynchus*) from those that can recover and spawn in subsequent years (*Salmo, Salvelinus,* and others). Now that the taxonomy of the rainbow trout, which can spawn annually for a number of years after reaching adulthood, has been changed, even that simple distinction has been blurred. The rainbow trout has been changed from *Salmo* to *Oncorhynchus* (Robins 1991), so now the rainbow trout is a salmon (at least if members of the genus *Oncorhynchus* are considered to be salmon). Of course, the Atlantic salmon has posed a dilemma of sorts to fishery biologists, since it has always been classified in the genus *Salmo,* so it may technically be a trout. The life cycles of various salmonids of aquaculture interest have been recently reviewed by Stickney (1991).

Pacific salmon die after spawning (except for rainbow trout, of course, which has an anadromous form called the steelhead trout), while trout do not.[155] This distinction has some importance in the hatchery, where salmon that will die after spawning can be sacrificed and the eggs removed by opening the abdomen, whereas eggs from trout are expressed by carefully squeezing the abdomen. Eggs are typically removed from females and placed in a pan to which milt is added (Figures 93 and 94). The mixture is then stirred and allowed to stand for a few minutes, and then the milt is washed from the eggs.

Salmon and trout eggs are large, so only a few thousand eggs can be obtained from even relatively large adults. Various levels of selection have been used in conjunction with spawning salmonids, depending on the strategy being used and the number of spawners available. Fertilized egg batches are commonly mixed and incubated in Heath incubators (Figure 95). The eggs rest on screens, and the larvae fall through the screens into a pan underneath after hatching. Eggs are subject to attack by fungus, so they are inspected frequently during incubation. Dead and dying eggs become opaque and can be removed either manually or with specialized egg-picking machines.

Most hatcheries place swim-up fry in raceways and adapt them to prepared feeds. Fish released from enhancement hatcheries and those grown out in marine netpens are retained in fresh water until smolting occurs. The time from hatching to smoltification varies from a few months to a year depending on species. Fish that are to be reared to market size in fresh water (primarily trout, but also a technique applicable to Pacific salmon)[156] can be stocked in growout raceways at any size considered suitable by the culturist.

Trout and salmon spawning have become standardized over the past several decades, but research continues to determine if improvements can be made. For example, broodstock diet has been evaluated to determine if reproductive performance can be improved. Washburn et al. (1990) found that feeding rainbow trout broodfish a low-protein, high-carbohydrate diet or one with intermediate protein and carbohydrate led to significantly higher fry survival to the eyed stage as compared with broodfish fed a high-protein, low-carbohydrate diet for the 9 months preceding spawning.

An area that has received a considerable amount of attention by researchers involves the effect of light on salmonid reproduction. Some culture strategies may require extending the spawning season, and that can be accomplished by manipulating photoperiod. Exposure of rainbow trout to long days early in the year followed

---

[155] Many Atlantic salmon survive spawning to spawn in subsequent years.

[156] The entire life cycle of the Pacific salmon that were introduced some years ago into the Great Lakes occurs in fresh water.

Figure 93. Salmon that will die after spawning are often sacrificed for egg collection. In this photo the eggs are being put into a bucket with the eggs from a number of other females.

by short days later has been shown to advance the time of spawning (Bromage et al. 1984, Elliott et al. 1984). Various photoperiod strategies will extend the normal spawning date of rainbow trout by 2 to 4 months (Bromage et al. 1984, Boulier and Billard 1985). Delayed spawning has also been induced in chinook salmon through photoperiod manipulation (Johnson 1984). Scott et al. (1984) placed rainbow trout under a constant 18:6 hr (light/dark) photoperiod immediately after spawning and were able to respawn the fish again 6 months earlier than normal.

Nakari et al. (1988) experimented with the effects of photoperiod on rainbow trout by exposing the fish to advanced or delayed photoperiod cycles before the normal time of spawning. They found that fish exposed to the advanced photoperiod matured early but did not spawn until the normal time of year, while those exposed to delayed photoperiod produced eggs that were smaller than normal.

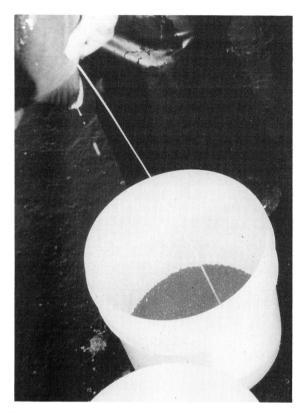

Figure 94. Milt from a male rainbow trout being expressed into a bucket containing eggs from a single female.

Red Drum

As reviewed by Henderson-Arzapalo (1992), adult female red drum *(Sciaenops ocellatus)* repeatedly produce large batches of eggs during the spawning season, which occurs when the light/dark photoperiod is 12:12 and temperature is between 20 and 25°C. Those conditions (which naturally occur from August to November depending upon location) will allow a given female to spawn seven or more times within a month.

Techniques by which red drum females can be matured with hormone injections and then stripped have been developed, but the most widely used method involves manipulation of temperature and photoperiod as first described by Arnold et al. (1977). The technique (reviewed by Arnold 1988, McCarty 1990, and Henderson-Arzapalo 1992) was later refined to compress the annual temperature and photoperiod cycle into 3 months, after which the environmental conditions on the spawning grounds are maintained and the adults will spawn repeatedly. Eggs can be expected from a given female every few days, and it has been possible to maintain spawning

Figure 95. Stacks of Heath trays used to incubate salmon and trout eggs. Water flows in the top and trickles down through each tray stack.

activity year-round. While a particular group of adults are capable of spawning for periods of many months, egg numbers and spawning frequency decrease over time. Therefore, the adults should be exposed to winter conditions periodically to allow them to recover. This can be accomplished by lowering the water temperature to below 20°C.

Red drum eggs, which are about 1 mm in diameter, will hatch in about 30 hr at 21 to 23°C. Larvae are from 1.7 to 1.8 mm long when hatched and develop functional mouth parts in about 3 days at 25°C. The swim bladder develops, and the fish begin seeking food 3 to 4 days after hatching (reviewed by Henderson-Arzapalo 1992).

The Texas Parks and Wildlife Department red drum hatcheries incubate batches of about 1 million eggs in conical-bottomed tanks of 1,890 l capacity. The water is maintained at about 30 ppt salinity and between 20 and 30°C. Gentle aeration is provided to keep the eggs and larvae suspended and to maintain the DO concentration above 5.0 mg/l, but there is little or no water exchange until the fish hatch (Henderson-Arzapalo 1992). After hatching, a slow water exchange may be used to help maintain water quality without impinging the larvae on the screen-covered drains.

When the larvae reach the time of first feeding, they can be transferred to production ponds that have been properly fertilized to initiate a plankton bloom. (The process has been reviewed by Henderson-Arzapalo 1992). Ponds are dried and disked to allow the oxidation of organic matter. They are smoothed, and filling is initiated 10 to 14 days before stocking. Organic fertilizer such as chicken manure, cottonseed meal, or peanut meal is added at 284 kg/ha to stimulate a plankton bloom. Salt water of 15 to 45 ppt is used, and the pond is only filled to about 0.5 m and then left for several days. Inorganic fertilizer may be added when the water level is brought up.

Stocking densities of larvae range widely, with 750,000 larvae/ha being considered optimum for the production of 25 mm fingerlings (Colura et al. 1976). When the proper zooplankton level is present, red drum fry will grow from 3 to 6 mm/wk (depending on temperature), with 25 mm fingerlings being produced in 21 to 40 days. Commercial trout or salmon starter feeds can be offered after about 2 wk (Henderson-Arzapalo 1992).

Much of the current production of red drum is for enhancement stocking, so the 25 mm fingerlings are released into coastal waters when they reach that size. However, the fish can also be moved to ponds or more intensive culture systems, stocked at the appropriate densities, and grown to foodfish size once they are harvested from the ponds. While it is possible to rear the larvae in tanks under more intensive culture conditions, the costs involved and poorer survival that has been observed (Arnold et al. 1977) support the conclusion that pond rearing is the method of choice.

Red drum broodstock have historically been obtained from nature; however, if domesticated red drum are to be developed for the commercial foodfish industry, maintenance of several generations in captivity will be required. The first step in the process was achieved by Arnold (1991), who was able to spawn red drum produced in captivity at the age of 19.5 months after hatch.

Penaeid Shrimp

A variety of penaeid shrimp species are commercially fished around the world, and many of them have been considered for aquaculture or are being reared by aquaculturists. Much of the early emphasis on penaeid shrimp aquaculture in the United States was directed toward the three most commonly occurring species of the southeastern Atlantic and Gulf of Mexico coasts: the white shrimp *(Penaeus setiferus)*, pink shrimp *(P. duorarum)*, and brown shrimp *(P. aztecus)*. While a great deal of information was developed on those species, particularly during the 1970s, successful aquaculture remained elusive, largely because of difficulties associated with the induction of spawning. Attention in the United States turned to exotic species, with two from Latin America, *P. stylirostris* and *P. vannamei* receiving the most attention and leading to the few commercial operations that have been developed.[157] Some interest in an Asian species, *P. monodon*, has also developed, but that species does not seem to be receiving active consideration at the present time in the United States. It is the most widely cultured species in Asia.

Many of the initial successes in spawning penaeid shrimp involved sourcing females to which spermatophores had been attached, a technique that is still practiced. (Sourcing involves dragging otter trawls in short tows behind vessels and

[157] It has been reported that there are nine shrimp farms in South Carolina (Rhodes 1991), fewer than 10 in Texas (Chamberlain 1991), and two major farms in Hawaii (Pruder 1991).

sorting through the catch to find the subject females. They are placed in buckets on deck and will usually spawn within 24 hr.) Inducing gonadal development and copulation in the laboratory remained largely unsuccessful for a number of years after sourcing became common practice.[158]

While sourcing remained the primary method used for obtaining spawners, it was demonstrated that ovarian development can be induced by eyestalk ablation (Caillouet 1972), in which one (or sometimes both) of the eyestalks of adult females are surgically removed or crushed (enucleated). Ablation prevents the production of a hormone that inhibits gonadal maturation and, when accompanied by placement of adult females in the proper environmental conditions, can induce maturation and spawning.

Eyestalk ablation is a rather extreme measure and is not reversible, so once the eyestalks have been ablated, further induction of spawning in a given female is not possible. In any case, the technique continues to be widely used. Wyban et al. (1987) concluded that unilateral enucleation is the best method of maturing female *P. vannamei*. The authors also indicated that shrimp size is an important factor.

Since shrimp spawn in nature without apparent difficulty, and certainly without sacrificing their eyestalks, it is clear that some factor is missing from the equation. Examination of the areas in which shrimp naturally spawn has provided information on temperature and salinity requirements, and those are being maintained within the appropriate ranges by shrimp culturists. Dim lights are often used in shrimp hatcheries, but the quantity of light that is routinely used may still be too high, as theorized by Wurts and Stickney (1981), who found that most hatcheries use light levels much higher than those that occur on natural spawning grounds of *P. setiferus*.

Chamberlain and Lawrence (1981) attempted to mature both ablated and unablated *P. vannamei* and *P. stylirostris* under various light levels, including total darkness (in a maturation tank covered with black plastic). Dim light and darkness did not enhance maturation success to any great degree, and ablated adults spawned more often than those that were not ablated. Dim blue or green light (80 to 120 lux) was found to be effective in inducing spawning in ablated and unablated *P. monodon* (Lin et al. 1990). Among the conclusions that can be drawn is that different species of penaeid shrimp may have different responses to ambient light level during maturation.

A number of studies have been conducted comparing such things as the efficacy of the ablation technique on maturation rate, differences in maturation rate of ablated and unablated shrimp, differences in the maturation rate of wild and pond-raised ablated shrimp of various species, the rematuration of ablated shrimp, and the effects of unilateral as opposed to bilateral ablation. Examples of such studies include those of Vincente et al. (1990), Tan-Fermin (1991), and Chu and Chow (1992).

---

[158] The other option, still practiced in some regions, is to collect wild postlarvae from nature and stock them or, in low-intensity operations, to accept whatever stocking density occurs when wild postlarvae enter as the ponds are flooded.

Photoperiod has been hypothesized as having an influence on maturation in penaeid shrimp. Nakamura (1988) reared young *P. japonicus* from April to July under a variety of photoperiods and concluded that molting and maturation were not influenced by that environmental variable. However, Chamberlain and Gervais (1984) had earlier determined that the manipulation of temperature and photoperiod produced comparable results in terms of survival, maturation rate, fecundity, and egg hatching rate in *P. stylirostris*. Similarly, Crocos and Kerr (1986) were able to spawn both ablated and unablated *P. esculentus* by providing proper temperature (26°C) and extended photoperiods (14.5 hr of light).

Other ways in which induction of maturation has been approached in penaeid shrimp include modifications in diet. For example, D'Croz et al. (1988) indicated that the polychaete *Americonuphis reesei* is widely used in Panama as a supplemental food to induce maturation of penaeid shrimp. Those authors found that the polychaete contains high concentrations of prostaglandins (a type of lipid) and related compounds that influence gonadal maturation. Millamena et al. (1986) also examined the relationship between diet and reproductive performance in penaeids. They found a positive response to feeds supplemented with cod liver oil. Bloodworms have been used as a food that will stimulate maturation in *P. vannamei* (Ogle 1991a, b). Lytle et al. (1990) suggested that the ratio of n–3 to n–6 fatty acids in bloodworms may be important in the effectiveness of that animal as a maturation promoter. Ascorbic acid (vitamin C) concentration in broodstock feed has been shown to affect embryonic development in *P. indicus* (Cahu et al. 1991).

As demonstrated by the white, pink, and brown shrimp of the United States, penaeids go through a series of larval stages following hatching. These include five naupliar (nonfeeding) stages, three stages of protozoea, and three mysis stages. The animals then metamorphose into postlarvae that resemble the adult (Pearson 1939, Heegaard 1953, Dobkin 1961, Cook and Murphy 1971).

Larval penaeids begin feeding at the second protozoeal stage (Cook and Murphy 1969). A variety of species of phytoplankton have been successfully utilized in conjunction with rearing larvae using the green water technique. When the larvae reach the mysis stage, they may be fed a mixture of phytoplankton and brine shrimp nauplii or other types of zooplankton. Postlarvae can be stocked into fertilized ponds and will accept prepared feeds.

Freshwater Shrimp

Macrobrachium rosenbergii is the most popular species of freshwater shrimp now being reared by aquaculturists. Through the 1970s a great deal of interest was expressed in the culture of freshwater shrimp, but storage and marketing problems, cannibalism, and breakthroughs in the marine shrimp industry led to collapse of the industry in many nations. Only modest production levels exist today in countries where the animal continues to be grown.

Hanson and Goodwin (1977) reviewed the cultural practices developed for *M. rosenbergii*.[159] Most of the information, while many years old, remains valid today.

---

[159] Unless otherwise indicated, the information presented in this section is from Hanson and Goodwin (1977).

M. rosenbergii is a tropical species native to Southeast Asia. It has been introduced widely into other parts of the world where either suitable temperatures naturally occur or environmental temperature can be controlled by the culturist. Attempts at freshwater shrimp culture in the United States have been made in South Carolina, Texas, and Hawaii, with Hawaii being the only state that provides year-round temperatures high enough to support the growth and reproduction of *M. rosenbergii*. The lower temperature tolerance of the species is from 10 to 12°C. Other states have conducted trials with *M. rosenbergii* and in at least one of them, Mississippi, there continues to be some interest.

Given the proper environment, *M. rosenbergii* will spawn throughout the year. Males do not molt once they have attained sexual maturity. Females, on the other hand, undergo a premating molt in advance of spawning. Broodstock of both sexes should be maintained together since the males protect the females from cannibalism during the prespawning molt.

Freshwater shrimp, like their marine cousins, produce hormones associated with maturation in their eyestalks, the histology of which has been described by Dietz (1983). However, it has not been necessary to ablate freshwater shrimp to induce spawning.

A typical adult female will produce up to 30,000 eggs per spawn and may spawn twice within a period of 5 months. The eggs are extruded and attached to the pleopods of the female from 6 to 20 hr following mating. Hatching requires about 19 days.

In nature, *M. rosenbergii* females move into brackish water to spawn, thus laboratory culture methods incorporate low-salinity water as a part of the larval rearing process. The eggs will develop properly and hatch in freshwater, but larvae produced in that medium will survive only a few days. For survival to occur the larvae should be hatched in or moved to brackish water. Most culturists employ water of 12 ppt salinity for hatching and larval rearing. The planktonic larvae undergo 11 molts before metamorphosing into postlarvae. Feeding may begin as early as 24 to 48 hr after hatching. *Artemia* nauplii or other zooplanktonic species are provided as first food. The animals can be converted to prepared feeds in later stages.

Larval development varies as a function of temperature and salinity, with 29°C and 12 ppt appearing to be near optimum. Under those conditions postlarvae can be produced in 35 to 40 days. Postlarval shrimp become benthic and, in nature, will begin to migrate upstream. Therefore, in culture, salinity is gradually reduced to that of freshwater after the postlarval stage has been reached. The remainder of the life cycle is spent in fresh water, though gravid females being utilized as broodstock are intermittently placed in brackish water to spawn.

Crawfish

The red swamp crawfish [160] *(Procambarus clarkii)* and the white river crawfish *(P. acutus)* are widely cultured, with the greatest amount of production being in Louisi-

---

[160] While "crayfish" has long been the accepted spelling for this group of decapod crustaceans, the American Fisheries Society now recognizes the word "crawfish" in conjunction with some North American species (American Fisheries Society 1989).

ana, Texas, and South Carolina. There is also some culture of *Orconectes immunis* in the north and central United States (Huner 1989). The culture of the *Procambarus* species has been widely reviewed (Avault 1973, LaCaze 1976, de la Bretonne 1988, de la Bretonne and Romaire 1989, Huner 1988, 1989). In Louisiana, about 60% of the approximately 40 million pounds of crawfish harvested annually (with a total value about $65 million) are from the capture fishery in the Atchafalaya Basin, with the remainder from aquaculture (de la Bretonne 1988). The general information that follows is taken from the reviews cited above. Specific data and other information is attributed to those and other authors as warranted.

Approximately 80% of current U.S. crawfish production occurs in Louisiana, where some 65,000 ha of culture ponds are located (de la Bretonne and Romaire 1989). The Texas industry probably accounts for the bulk of the remainder, and South Carolina comes in third with about 50 farms averaging less than 8 ha each (Eversole and Pomeroy 1989). Production levels in South Carolina average between 500 and 800 kg/ha/yr (Eversole and Pomeroy 1989), while production in excess of 2,000 kg/ha/yr appears common in Louisiana (Lawson and Wheaton 1983, Garces and Avault 1985). Estimated U.S. production in 1988 was at least 60,000 tons, with *P. clarkii* accounting for 90% of all crawfish harvested in the United States, Europe, Asia, and Australia[161] (Huner 1988).

Both *P. clarkii* and *P. acutus* broodstock are stocked in the spring. Stocking rates of 20 to 65 kg/ha of adults are used (Landau 1992). The ponds are drained within a few weeks after stocking, which forces the animals to burrow into the pond bottom where reproduction takes place.[162] Crawfish burrows may be very complex structures, 1 m or more deep. At the surface they may be merely plugged with a mud cap, or a mud chimney may protrude several centimeters above the pond bottom. Burrows can also extend into and weaken pond levees.

During mating the male deposits sperm in a receptacle organ on the female. The eggs are not extruded and fertilized until September, during which time they are carried on the swimmerets of the female and held in place by a sticky substance called glair (LaCaze 1976). Hatching takes place within about 2 or 3 wk following extrusion and fertilization for the red swamp crawfish. White river crawfish eggs require from 3 to 8 days longer for hatching. The average number of young produced (not all the eggs develop and hatch) is approximately 400 for the red swamp crawfish, with maximum production being about 700 juveniles. The white river crawfish produces somewhat fewer young per female on average. Production of red swamp crawfish in an experimental hatchery averaged 223 juveniles per female (Trimble and Gaude 1988).

Vegetation has traditionally been planted in the summer to serve as forage for the crawfish when they emerge from their burrows. Emergence occurs after pond flooding in the September-October period, which coincides with the time of hatching. Such aquatic plants as alligatorweed, smartweed, and water primrose were the

---

[161] The species is native to North America but has been widely introduced into other parts of the world.
[162] Crawfish are also reproductively active in the spring, but there may be no recruitment of young produced during that period (Romaire and Lutz 1989).

traditional plant crops often used in conjunction with crawfish culture, as was rice (which went unharvested). More recently, double cropping systems have been developed wherein marketable grains such as rice or sorghum or a combination of rice and soybeans is harvested but the stubble is left for crawfish forage (Brunson 1988, 1989, Brunson and Griffin 1988, de la Bretonne 1988). Research on the best cultivars of sorghum for use in double cropping systems has been conducted by Brunson and Taylor (1987).

Once sufficient water is present, the young crawfish that have been clinging to the abdomen of the female become free swimming and leave the burrow to enter the pond. The young are approximately 1 cm long at that time. If sufficient water is not present, some of the young crawfish may be released inside the burrows, where crowding and lack of food prevent their growth. In some cases the females leave the burrows and attempt to move overland to find water. Desiccation and predation often take a toll during such migrations, and it is not in the interest of the crawfish culturist to have the broodstock and their offspring leave the ponds in search of more desirable environments.

Harvesting is done with baited traps from November through May or June. Both the adults and their offspring are captured and marketed.

Crawfish farming requires only shallow ponds; water depths of over 0.5 m are not necessary. Hatcheries are not necessary since the animals will reproduce efficiently in the ponds if proper management is employed. The entire culture process is one of taking the proper actions at the proper times to enhance production. Major problems that occur during the growout stage include oxygen depletions and overcrowding. Both can be solved by proper culture system management (Avault et al. 1974). Studies on the responses of *P. acutus* and *P. clarkii* to low oxygen and high temperature stress demonstrated that neither species was more tolerant than the other (Huner 1987). Aerated water can be circulated through crawfish ponds to prevent hypoxia (Lawson and Wheaton 1982, 1983).

Oysters

Oyster culture in the United States was, for many years, largely limited to spreading shell (cultch material) on the bottom of bays with unsuitable substrates so that oyster spat would settle and grow. In recent years, interest in culture involving the production of spat in a hatchery has been developed to the point of commercialization. The spat are allowed to settle on bags of cultch material in culture tanks, after which the bags are placed into an appropriate environment (Figure 96) until the young oysters become a few millimeters long. Shells might then be inspected and excessive oysters removed. The shell material with attached oysters are then placed in growout beds. This technique is most extensively practiced by commercial producers in Washington State with Pacific oysters *(Crassostrea gigas)*. Some culturists of oysters along the Atlantic and Gulf of Mexico coasts use similar techniques with the American oyster *(C. virginica)*. Disease problems with the American oysters have severely affected wild stocks in such important harvest grounds as Chesapeake Bay (mid-Atlantic coast) and Apalachicola Bay (Florida Gulf coast). To meet the

Figure 96. Sacks of oyster shell onto which spat has been allowed to settle. One settlement has been completed in fiberglass tanks, the bags of shell are placed in the intertidal zone (shown here at low tide) for a period of time after which the shell is distributed on oyster beds. The photograph was taken in Puget Sound, Washington, and the shell is from *Crassostrea gigas*.

demand in the eastern United States, much of the oyster production in the Pacific Northwest is shipped to that region.

The following information is a summary concerning the reproduction of the American oyster (discussed in detail by Galtsoff 1964), which is illustrative of oyster reproduction in general. *C. virginica* spawns when the temperature is in the range of 21 to 27°C. Oysters can be conditioned to spawn during winter within a period of about six weeks if exposed to temperatures of 23 to 24°C (Hidu et al. 1969), though spring spawning is normal. The most common method of inducing spawning involves temperature and chemical stimulation (Loosanoff and Davis 1963) in which ripe adults are exposed to a rapid rise in water temperature to 30°C and to a sperm suspension from a sacrificed male. Each female may release several million eggs following this treatment.

Fertilized eggs can be removed from the water with a fine-mesh sieve and transferred to well-aerated, filtered seawater. At 30°C the eggs will hatch into veliger larvae within 48 hr (Landers 1968). The larvae are provided with phytoplankton that can be easily produced with mass culture techniques (as described below and in Chapter 6). Natural phytoplankton pumped into the laboratory can also be used to feed oyster larvae, with growth comparable to that of oysters reared on cultured algae (Ogle 1982).

Metamorphosis of American oysters can occur in as few as 10 days at 30°C if proper levels of phytoplankton are maintained (Landers 1968). Extremely dense or dilute phytoplankton levels will cause an increase in the time for larval setting. The setting spat are generally placed in shallow tanks containing cultch material, which is most commonly empty oyster or clam shells, though a variety of other materials can be used. Depending on expected level of mortality, the larvae may be placed in the setting tanks at densities of 10 to 50 spat per oyster shell cultch (Landers, 1968).

Spat-laden cultch can be spread over a pond bottom, distributed in a leased area in an estuary, reared in culture tanks, or suspended in the water column from rafts or other structures. The last method, often called string culture, provides protection of the oysters from various benthic predators such as oyster drills, and it has received at least some attention from U.S. aquaculturists beginning in the 1960s (Shaw 1960, 1962, 1968, Linton 1968, Marshall 1968, 1969, May 1968, 1969). The spreading of oysters in leased estuarine areas is the least intensive form of culture and is closely related to traditional commercial oyster fishing in which oyster shell is returned to the sea following shucking so the shell can provide natural cultch. The difference is that the culturist establishes a hatchery and plants cultch that has already been seeded with spat.

Cultchless oyster production has also been developed by culturists. The technique involves settling spat on sheets of plastic or other flexible material in the hatchery. They are removed after attachment and cannot reattach to another substrate. The small oysters are ultimately transferred to trays in which they are allowed to grow. When the oysters become large enough, they can be moved into natural areas in an estuary, or the whole system may be maintained in a shoreside facility. A large supply of phytoplankton-rich water must be available in either case.

Of importance to aquaculture is the common practice of collecting oysters from waters polluted with potentially pathogenic bacteria and moving them to clean areas for depuration. Increasing numbers of natural and aquaculture oyster beds are becoming contaminated, largely with human sewage (and also by livestock and even marine mammals in some instances). After several days in clean water, the contaminating bacteria will be removed and the oysters will be safe for human consumption. Public health agencies have become increasingly active in inspecting shellfish for contamination from bacteria and also from other toxins such as PSP (paralytic shellfish poisoning) and domoic acid.

Commercial hatcheries in the state of Washington produce *C. gigas* in a manner similar to that described above for *C. virginica,* with the major exception that *C. gigas* grows well at lower temperatures. Spat produced in the hatchery can be wrapped in fine plankton netting and shipped to growers. A ball of spat about the size of a baseball will contain approximately 8 million individuals.[163] One commercial hatchery in Washington has cultured unispecific algae in fiberglass tanks of about 10,000 l capacity under high intensity lamps of the type used to light athletic stadiums. At harvest, the algae can be centrifuged down to a volume of about 1 l. The resulting algae paste can be kept frozen until needed. Feeding involves scooping some of the paste into spat settling tanks to provide the spat and newly settled

---

[163] Kenneth K. Chew, personal communication.

oysters with food until the cultch is distributed into the natural environment for further growout.

Some attempts at selective breeding, a first step toward domestication of oysters, have begun, but in many cases no selective breeding is practiced. Mass spawning, the mixing of spawns, and "haphazard" selecting of oyster broodstock by the industry will have to be curtailed if progress toward oyster domestication is to be achieved (Hedgecock et al. 1988). Several interesting and positive results of selective oyster breeding have been reported by researchers. Much of the work has been conduced with *C. gigas* in the Pacific Northwest, where selective breeding studies at the University of Washington began in 1976 (Perdue et al. 1984). Stocks developed by researchers are resistant to summer mortality; those same stocks also show the desirable trait of having increased levels of glycogen (Perdue et al. 1984, Hershberger et al. 1988). Selected Pacific oysters have also demonstrated increased survival (Beattie 1984). Pongthana (1990) indicated that selective breeding of *C. gigas* could lead to increased carbohydrate content in the meat. Selective breeding studies have also been reported in conjunction with the European oyster, *Ostrea edulis* (Newkirk 1988).

HORMONE SPAWNING

As we have seen in some of the above examples, hormones can be used to induce spawning. A variety of hormones have been used to achieve that end. Fish pituitary, particularly from carp, was a early source of hormones to induce spawning. HCG, which is prepared from the urine of pregnant humans, has also been widely used since it is readily available in known concentration. A number of other hormones have also been used successfully; one of those that has recently seen broad use is luteinizing hormone-releasing hormone (LHRH) and its analogs. Some of the species that have been induced to spawn through hormone injections and the types of hormones that have been used are presented in Table 36. In most cases it is only necessary to inject females, though male injections are used in conjunction with some species. While injections are common, hormone implants have also been developed.

SEX CONTROL

Control of sex in the rearing of tilapia is often quite important to prevent overpopulation and, perhaps more importantly, to avoid the differential growth rates of males and females. Other species also show differential growth rates between the sexes, and their culture may benefit from sex control. For exotic species, controlling the sex will reduce the chance of establishing wild populations through escapement. An excellent example is the grass carp, *Ctenopharyngodon idella* (Figure 97).

Fear that release of grass carp in the natural waters of the United States as a biological weed control agent will result in competitive exclusion of native, more

TABLE 36. Examples of Aquaculture Species That Have Been Induced to Spawn with Hormones, and Examples of Hormones That Have Been Used in Conjunction with Those Species

Species	Type(s) of Hormone(s) Used[a]	Reference(s)
Aristichthys nobilis (bighead carp)	LHRH[a]	Ngamvongchon et al. (1987)
Catla catla (catla)	Ovaprim-C	Shetty (1990)
Chanos chanos (milkfish)	LHRH GnRH[b] HCG,[c] GnRH LHRH, testosterone	Lee, et al. (1986a,b) Marte et al. (1987) Marte et al. (1988) Tamaru et al. (1988)
Cirrhina mrigala (mrigal)	LHRH, pimozide Ovaprim-C GnRH	Kaul and Rishi (1986) Shetty (1990) Halder et al. (1991)
Clarias batrachus (Asian catfish)	pimozide, LHRH	Manickam and Joy (1989)
Clarias gariepinus (sharptooth catfish)	HCG, FPE,[d] FSH[e] luteocytic hormone oxytocin	Hecht et al. (1982)
Clarias macrocephalus (catfish)	LHRH	Ngamvongchon et al. (1987)
Ctenopharyngodon idella (grass carp)	LHRH	Rottmann and Shireman (1985)
Cynoscion nebulosus (spotted seatrout)	LHRH HCG, FPE	Thomas and Boyd (1988, 1989) Colura et al. (1990)
Cynoscion xanthulus (orangemouth corvina)	LHRH	Thomas and Boyd (1988)
Dicenrarchus labrax (seabass)	LHRH	Barnabe and Barnabe-Quet (1985)
Hypophthalmichthys molitrix (silver carp)	HCG HCG, FPE LHRH	Banerjee et al. (1984) Dwivedi et al. (1986) Ngamvongchon et al. (1987)
Ictalurus punctatus (channel catfish)	LHRH	Busch and Steeby (1990)
Labeo rohita (rohu)	Ovaprim-C GnRH	Shetty (1990) Halder et al. (1991)

TABLE 36. (*Continued*)

Species	Type(s) of Hormone(s) Used[a]	Reference(s)
Lates calcarifer (sea bass)	LHRH	Harvey et al. (1985)
	LHRH	Garcia (1989, 1990)
Lutjanus campechanus (red snapper)	HCG	Minton et al. (1983)
Mugil cephalus (grey mullet)	FPE	Yashouv (1969)
	FPE	Shehadeh and Ellis (1970)
	FPE	Shehadeh et al. (1973)
	HCG	Kuo et al. (1973)
	LHRH	Lee et al. (1987b)
	FPE, HCG, LHRH	Lee et al. (1988)
Mylio berda (white sea bream)	Synahorin, HCG	Mok (1985)
Morone saxatilis (and hybrid striped bass)	HCG	Reviewed by Kerby (1992)
Oncorhynchus mykiss (rainbow trout)	LHRH	Crim et al. (1983)
Salmo salar (Atlantic salmon)	LHRH	Crim et al. (1986)
	Testosterone	Crim et al. (1989)
Sciaenops ocellatus (red drum)	LHRH	Thomas and Boyd (1988)
	HCG	Henderson-Arzapalo (1992)
Siganus guttatus (rabbitfish)	LHRH	Harvey et al. (1985)
	HCG	Ayson (1991)
Sparus aurata (gilthead sea bream)	HCG, LHRH	Colombo et al. (1989)
Hybrid carp (grass carp × bighead carp and grass carp × silver carp)	LHRH	Rottmann and Shireman (1985)

[a] LHRH = luteinizing hormone-releasing hormone
[b] GnRH = gonadotropin-releasing hormone
[c] HCG = human chorionic gonadotropin
[d] FPE = fish pituitary extract
[e] FSH = follicle stimulating hormone

desirable fishes has led to the banning of grass carp in many states. At one time more than 30 states had outlawed the species. Many have now approved the stocking of fish that have been certified sterile. Sterility is assured by one of three methods: e.g., using hybrids (grass carp × bighead carp hybrids), triploids (having an extra set of chromosomes which renders the fish unable to successfully reproduce), or gynogenetic fish (developed from an ovum following sperm penetration but with-

Figure 97. The grass carp, *Ctenopharyngodon idella*, is widely utilized in aquatic vegetation control and aquaculture. Hybridization of grass carp with bighead carp can lead to polyploid individuals that are sterile.

out fusion of the gametes). These methods are discussed in the following three sections.

Some consideration has also been given to the stocking of monosex grass carp, though the rationale for that approach seems somewhat flawed unless every contiguous jurisdiction agrees.

Hybridization

While the classic definition of a species includes the requirement that intraspecific mating will produce fertile offspring while presumably interspecific mating will not, many hybrids between closely related fish species will be fertile, while others do not. Hybridization may be used to produce fish that perform better than either parental species in aquaculture (hybrid striped bass), or to ensure that fish released into the environment will not reproduce (hybrid grass carp).

Reproductively successful hybrids of some species of aquaculture interest can be readily produced. Examples are tilapia (various species) and crosses among blue, white, and channel catfish. Hybrid production success can vary to some extent depending on which sex is selected from which species. For example, Tave and Smitherman (1982) found that hybrids between male blue catfish and female channel catfish are easier to obtain than when female blue and male channel catfish are used.

Triploidy

The production of animals with three sets of chromosomes (triploidy) can be used as a means of producing sexually dysfunctional aquaculture species. Normally, eggs and sperm are haploid (containing 1n chromosomes). When the two combine, diploid (2n) zygotes are produced. By using various chemical and physical methods of interfering with meiosis in one of the parents, it is possible to produce offspring with three sets of chromosomes (3n). Triploid individuals are sterile since they cannot produce haploid gametes. Triploid (and tetraploid animals—those with 4n chromosomes) may grow faster than normal individuals, presumably because their individual cells are larger to accommodate the extra intracellular material, though the fact that they are sterile is often of more interest. Triploidy has been induced through exposure of eggs to temperature and hydrostatic shock (discussed by Brydges and Benfey 1991), and through exposure to such chemicals as cholchicine and formalin (Chernenko 1985). Subjecting eggs to hydrostatic pressure seems to be the method that most consistently produces triploidy in fish (Rottmann et al. 1991).

Johnstone et al. (1991) reported on the performance of triploid Atlantic salmon *(Salmo salar)* in culture. Triploids performed as well as diploids in fresh water. In saltwater, triploids did not grow as well as maturing diploids, though growth of nonmaturing diploids and triploids was similar. The advantage of triploid Atlantic salmon in net-pen culture is that if farmed fish were to escape they could not inter-breed with wild fish.

Sugama et al. (1992) studied the growth and development of triploid red sea bream, *Pagrus major*. They reported that survival of triploids, produced with tem-perature shock, was lower than that of diploids but that growth was comparable. The gonads of triploid fish were reportedly smaller than those of diploids.

Triploidy in oysters has received a great deal of attention. Normal oysters con-vert their glycogen stores, which are the basis of the excellent flavor of the meat, into substances that are required for the development of gametes during the spawn-ing season. The oysters become milky and undesirable in quality as a result. This has led to the closure of commercial oyster harvesting in months without an 'R' (May, June, July, and August). While many believe that oysters are unsafe during the closure period, the actual reason is related to the poor quality of the meat. If oysters do not become sexually mature, the loss of glycogen does not occur and the quality of oysters can be maintained during the summer. Triploid, or sexless, oysters have been produced by exposing newly fertilized eggs of the Pacific oyster, *Crassostrea gigas,* to cytochalasin B (Allen and Downing 1986). Triploids of the European flat oyster, *Ostrea edulis,* and Manila clam *(Tapes semidecussatus)* have been produced through thermal shock (Gendreau et al. 1989, Gosling and Nolan 1990).

Gynogenesis

Gynogenesis is the development of an ovum following sperm penetration but with-out fusion of the gametes. In other words, a ripe egg is penetrated by a spermato-zoan, but the genetic material contained in the sperm cell is not incorporated into the nucleus of the egg, even though the egg is stimulated to develop. In most cases the larvae formed from this activity are haploid (having 1n chromosomes) and die within a few days after fertilization. However, in some instances diploid (2n) eggs are produced that contain only the chromosomes from the mother. Those eggs develop into females that should be genetically identical with the mother. Some question about the latter statement was raised by Carter et al. (1991), who found evidence, through DNA fingerprinting, that in gynogenetically produced *Tilapia aurea* there was some paternal DNA present. Gynogenetic *T. nilotica,* on the other hand, contained exclusively maternal DNA.

Aquaculturists have been interested in developing gynogenetic individuals in large numbers for stocking. The production and rearing of only females of such species as grass carp ensures that no fish that escape will spawn (assuming males from other facilities have not already been introduced into the same receiving water body).

Two ways in which haploid ova can become diploid without receiving chromosomes from a male's sperm were proposed by Stanley and Sneed (1974). In the first, the polar body that is normally released during the second meiotic, or reduction division in gamete formation recombines with the oocyte to double the haploid number of chromosomes. Successive mitotic cell divisions produce the diploid embryo. The second mechanism involves the normal release of the first polar body from the meiotic division to form a haploid egg. Upon proper stimulation, the chromosomes in the ovum are induced to replicate as in normal mitosis, but incomplete cell division occurs. The two sets of chromosomes are incorporated into a single nucleus, and subsequent cell divisions produce a diploid embryo.

The presence of male sperm is not always required for the induction of gynogenesis, particularly when humans become involved. Stanley and Sneed (1974) discussed how gynogenesis can be induced by dipping a needle in the serum or whole blood of a fish and then pricking the eggs to stimulate cell division. An alternative approach is to expose fish eggs to weak electrical currents. For gynogenetic development that follows the definition more explicitly, sperm from a distantly related species may be utilized or irradiated sperm from the same species may be used to stimulate cell division in eggs. Sufficient exposure to radiation (e.g., x-ray or UV light) will denature the DNA in sperm without destroying motility. Methods for inducing gynogenetic development in common carp *(Cyprinus carpio)* have been described by Hollebecq et al. (1986) and Linhart et al. (1986). John et al. (1984) described methods used to induce gynogenetic development in the eggs of two species of Indian carp *(Labeo rohita* and *Catla catla).*

No matter what technique is used, the percentage of viable gynogenetic individuals remains low. Temperature shock used in conjunction with one of the above techniques can increase the production of diploid ova, though the total will remain small.

The use of hybridization to produce sterile animals and the development of technology to produce triploids have reduced the amount of interest in gynogenesis as a means of sex control in aquaculture, though there have been some reports of its use in conjunction with other sex control techniques. Pandian and Varadaraj (1990) used a combination of selective breeding, hormone sex reversal, and gynogenesis in conjunction with tilapia production. The authors selected fish for use in gynogenetic development and then subjected offspring to hormones to create males with female chromosomes (XX). Earlier, the same authors used similar techniques to produce what they referred to as "supermale tilapia" which carried only male (YY) chromosomes (Varadaraj and Pandian 1989).

CRYOPRESERVATION

Cryopreservation is the freezing of gametes in viable condition so they can be used at some date after they were initially available from the mature animal. In fish, cryopreservation has worked well for milt but not eggs. Sperm from oysters, including *C. gigas* (Yankson and Moyse 1991), and shrimp, *Sicyonia ingentis* (Anchordoguy et al. 1988), have also been successfully cryopreserved, and blue mussel *(Mytilus edulis)* embryos have survived freezing (Toledo et al. 1989). Cryopreservation techniques include freezing sperm in cryotubes (Rana and McAndrew 1989), as pellets (Yamano et al. 1990), or in straws (Wheeler and Thorgaard 1991).

Some type of cryoprotectant chemical, called an extender, is used in conjunction with freezing. Liquid nitrogen ($-196°C$) is often used for freezing and subsequent storage, though various other temperatures have been used. The standard cryoprotectant has been dimethylsulfoxide (DMSO), though Gallant and McNiven (1991) found that dimethylacetamide (DMA) was a better chemical to use in conjunction with rainbow trout sperm. Thorogood and Blackshaw (1992) evaluated glycerol and DMSO as extenders for yellowfin bream *(Acanthopagrus australis)* sperm and determined that glycerol was the better of the two. Milt from tilapia (species not given) protected with 12.5% methanol in fish Ringers solution and frozen in liquid nitrogen remained viable for at least 13 months according to Rana and McAndrew (1989). Baynes and Scott (1987) used a sucrose-based extender. They found that the addition of 5 to 20% hen's egg yolk to the extender improved fertility of rainbow trout sperm as compared with the extender alone. Similarly, Scheerer and Thorgaard (1989) found an improved fertilization rate for rainbow trout when the eggs were fertilized in a buffered saline activator solution containing 5 nannomoles of theophylline (a compound related to caffeine).

There are practical reasons for utilizing cryopreservation in conjunction with fish culture. Some species, for example, may have populations that do not spawn at the same time, or even individuals within the population that cannot be mated because gametogenesis does not overlap. That has been observed in Pacific halibut *(Hippoglossus stenolepis),* for example, by Liu (1991). Male halibut in the group being held for spawning generally became running ripe in December and continued to produce good-quality milt through January. Females, on the other hand, did not mature until late January, and most ovulation occurred in February and into March. While it was possible to obtain viable gametes from both sexes for brief periods, by cryopreserving milt it was possible to fertilize eggs obtained after the males had ceased producing sperm. The technique for cryopreserving halibut sperm was described by Bolla et al. (1987).

Of significant current interest is the restoration of species [164] that have been listed as threatened or endangered. Cryopreservation can be used to help keep animals from becoming extinct. Efforts that have recently been made to restore the Redfish Lake sockeye salmon *(Oncorhynchus nerka)* are illustrative. The fish must travel hundreds of kilometers up the Columbia River system to reach Redfish Lake in Idaho. A series of dams on the rivers make the upward migration extremely difficult and also take a high toll on outmigrating smolts. As a result, the population has declined to the point of near extinction. In 1991, a total of four adult fish arrived at Redfish Lake. They were captured and the milt from the three males was used to fertilize the eggs from the lone female. Some milt was also cryopreserved for use in subsequent spawning seasons. The offspring were separated into two groups; one was sent to Seattle, Washington, where the fish were placed in a hatchery operated by the National Marine Fisheries Service; the second group was maintained by the Idaho Department of Fish and Game at their Eagle Hatchery near Boise. The fish are to be used as broodstock for the production of subsequent generations that will ultimately be released into the natural environment.

In 1992, a single male sockeye returned to Redfish Lake. His milt was also cryopreserved and will be used if a female appears in the future. Alternatively, the milt from that fish can be used to fertilize eggs from the fish of the 1991 brood year, which are being reared in captivity.

SELECTIVE BREEDING AND GENETICS

Attempts to improve aquaculture stocks by commercial producers usually involve selective breeding with little knowledge of genetic consequences or relationships. The possible selection for aggressiveness rather than rapid growth in channel catfish has already been mentioned. There has been a considerable amount of selective breeding research conducted on some species. Catfish and oysters are discussed above, and other species are mentioned in this section. It can still be said, however, that no aquaculture species has been truly domesticated, though significant strides have been made in certain instances. The development of the Donaldson strain of rainbow trout is one excellent example of an early success story that can be attributed to selective breeding (Hines 1976).

Inbreeding has long been a subject of discussion among aquaculturists. Breeding closely related animals can lead to inbreeding depression (reduced performance) and an increased incidence of deformities in animals. Warnings of inbreeding depression in aquaculture animals have been issued, but there remains some question as to the severity of the problem. Some species, such as *Tilapia aurea, T. nilotica,*

---

[164] The Endangered Species Act that was enacted in the United States defines a species as being any reproductively isolated group of organisms. As a result, many species that are not threatened or endangered as a group have been found to have isolated populations that can be listed under the act. The Redfish Lake sockeye salmon discussed in this section is one example.

and *Macrobrachium rosenbergii* were established in aquaculture facilities in the United States on the basis of small numbers of individuals, so a high level of inbreeding was inevitable, though there appear to have been few signs of performance depression as a result.

Reduced shell size and meat weights have been demonstrated in inbred Pacific oysters (Beattie et al. 1987). An increased incidence of albinism has occurred as a result of inbreeding, and has even been promoted through selective inbreeding in such species as channel catfish and rainbow trout.[165] On the other hand, Dunham et al. (1991) reported that some abnormalities that have been observed in channel catfish (stumpbody and taillessness) are not inherited but are more likely related to the environment in which the fish are produced.

A new concept called outbreeding depression, has been proposed and is currently being debated with respect to whether it actually can occur. The concept was developed by Gharrett and Smoker (1991) after a study they conducted in which pink salmon *(Oncorhynchus gorbuscha)* that return in odd and even years were hybridized and the F_1 and F_2 generations evaluated. Low F_2 returns and an increase in bilateral asymmetry were seen as signs of outbreeding depression. The application of that theory to fish other than salmon that would not normally find themselves in a position to interbreed remains to be determined.

Rather than concern themselves with the problems associated with inbreeding, many commercial aquaculturists commonly select on the basis of some trait that they consider valuable (e.g., fast growth, high food conversion efficiency, high dress-out percentage, good body configuration) without regard to the kinships that might be associated with the fish or shellfish that are selected as broodstock. It is generally assumed that each desirable characteristic is controlled by a single dominant gene, when that may not be the case. Not only is control by multiple genes ignored, but the influence of environment on the characteristics of interest may be discounted. For example, in a carefully controlled experiment, a group of fish or invertebrates might be reared at 30°C from egg to adulthood. If the most rapidly growing individuals were mated, their offspring might also be expected to grow rapidly under the same experimental conditions. However, if animals of the F_1 generation were to be placed into a commercial outdoor facility where temperature fluctuates daily and seasonally, the animals might grow even less rapidly than those produced from random matings. Superficially, the development of laboratory strains of fish or invertebrates for aquaculture may seem desirable, but aquacultured species, unlike laboratory white mice, are not and cannot be cultured under identical conditions at every facility (whether that facility is a commercial fish farm, public hatchery, or research laboratory). No two facilities can be expected to have identical culture systems or water quality, and an aquatic species developed under highly controlled conditions in a given laboratory cannot be expected to perform in the same manner at other facilities. Maintenance of genetic diversity is important to commercial aquaculture. It is also important to maintain genetic diversity in order

[165] Albinos can sometimes be sold at premium prices if the proper marketing techniques are employed.

to minimize the potential impacts of cultured animals on wild populations, as discussed by Doyle et al. (1991).

For selective breeding to be successful, there must be a degree of variance in one or more characteristics that are considered beneficial to the culturist. In many cases such variance is difficult to ascertain, let alone quantify. For example, dressout percentage or percentage of fat deposited in the body cavity of a fish may vary by several percent within a given population, though the fish may be morphologically indistinguishable from external characteristics. If the animals are sacrificed and dressed to ascertain differences, it will be difficult to conduct selective breeding experiments with the species.[166] Noninvasive means of assessing differences in body composition with respect to lipid content have been developed (M. H. Lee et al. 1992), so it should be possible to measure improvements in dressout percentage without sacrificing animals. As similar techniques are developed for other variables, it may be possible to evaluate improvement in other characteristics without sacrificing animals and subjecting their tissues to chemical analyses.

The portion of total variance within a population of animals attributable to certain kinds of genetic effects is known as heritability. The concept was developed in detail by such authors as Dobzhansky (1970) and Strickberger (1976). When heritability is low, large-scale breeding experiments involving many pairings and probably extending over several generations may be required to improve the stock.

Heritability estimates have been developed for at least a few traits in some aquaculture species. Some of those evaluated in recent years with respect to one or more traits are the hard clam *Mercenaria mercenaria* (Hadley et al. 1991); blue mussel, *Mytilus edulis* (Stroemgren and Nielsen 1989); the crawfish *Procambarus clarkii* (Lutz and Wolters 1989); the shrimp *Penaeus vannamei* (Lester and Lawson 1990); *Tilapia* spp. (Jarimopas 1986, Kronert et al. 1989, Huang and Chiu 1990, Jarimopas 1990); channel catfish, *Ictalurus punctatus* (Bondari 1983c, Dunham and Smitherman 1983); rainbow trout, *Oncorhynchus mykiss* (Gall and Huang 1988a, b, Gjerde and Schaeffer 1989, Siitonen and Gall 1989, Rye and Lillevik 199, Rye et al. 1990, Tipping 1991, Yamamoto et al. 1991); Atlantic salmon, *Salmo salar* (Bailey and Friars 1990, Rye and Lillevik 1990, Rye et al. 1990); chinook salmon, *Oncorhynchus tshawytscha* (Withler 1987, Beacham and Withler 1991); and coho salmon, *O. kisutch* (Hershberger et al. 1990, Iwamoto et al. 1990, Swift et al. 1991). Those studies examined such traits as age at maturity, disease resistance, growth, pigmentation, reproductive performance, survival during early life stages, timing of spawning, and age at which smolting occurs (in salmonids). Heritability estimates ranged widely. One study reported a lack of response to mass selection in *Tilapia nilotica* in terms of heritability for growth early in the life of the fish (Teichert-Coddington and Smitherman 1988).

Heterosis, or hybrid vigor, is the response shown by organisms that are mated to distantly related individuals of the same species. This is known as outcrossing or

---

[166] Salmon and other species could be sacrificed and their gametes collected and used for production of F_1 fish, but the process of gametogenesis might change the body-fat content significantly from what would normally have been present had the fish reached market size.

outbreeding and can lead to an increase in heterozygosity, with resulting improvements in performance, increased egg survival, and so on. Thus, it is the opposite of inbreeding. (However, recall that there is currently a theory that outbreeding can also cause depression). Hybrid vigor usually disappears in the second generation because heterozygosity is decreased. Thus, it is necessary to maintain two or more inbred lines of brood animals that when crossed, will produce vigorous offspring for stocking in growout culture chambers. Ideally, the culturist would prefer to have separate inbred lines that are homozygous for certain traits, such that on parent is homozygous dominant and the other homozygous recessive for each trait. The parents might have genotypes as follows:

$$male = aa \quad BB \quad cc \quad DD \quad ee$$
$$female = AA \quad bb \quad CC \quad dd \quad EE$$

When the adults depicted above are mated, the offspring will be heterozygous for each gene:

$$F_1 = aA \quad bB \quad cC \quad dD \quad eE$$

Strictly random mating, another way in which inbreeding can be reduced, calls for a fairly large population of brood animals, and even then a detectable level of inbreeding occurs. Eventually, inbreeding can become so severe that new broodstock strains will have to be introduced.

A third scheme for reducing inbreeding involves the maintenance of three lines of brood animals and the use of rotational line mating (Kincaid 1976, 1977). The initial mating lines generally are established by separating a strain that will produce progeny having the desired traits into three groups, identified as lines A, B, and C. During each breeding season the males from line A are mated with females from line B, males from line B with females from line C, and males from line C with females from line A (Figure 98). The offspring from the matings are maintained together if the parents are to be utilized in subsequent matings, or offspring are maintained separately in sufficient numbers to replace the broodstock during years when that is desired. The extent of inbreeding in the rotational-line mating scheme is somewhat less than that which might occur from random matings, and this scheme solves the broodstock replacement problem that exists when only two breeding lines are maintained.

One problem associated with the rotational-line mating scheme is maintenance of separate populations of brood animals. To ensure that only the desired crosses occur, it is almost essential that each line be maintained separately (i.e., in different culture chambers). Needless to say, one of the most important considerations in the employment of this mating scheme is the maintenance of adequate records to prevent mistakes in crosses.

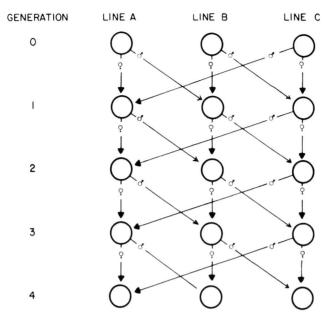

GENERATION   LINE A   LINE B   LINE C

Figure 98. Rotational line mating scheme for minimizing the extent of inbreeding within a population of animals. Adapted from Kincaid (1976, 1977).

GENETIC ENGINEERING

The molecular biology revolution has made it possible to actually create new species of organisms in the laboratory, and perhaps of more immediate importance, to transplant genes from one species to another to produce transgenic organisms. Aquaculturists have begun learning molecular biological techniques in recent years and have viewed the technology as a means by which rapid improvement in the desirable characteristics of farmed fish and shellfish can be achieved. Gene transfer could, for example, result in greatly improved growth rate, the development of tilapia that tolerate low water temperatures, trout that can tolerate and grow more rapidly in warm water, or even salmonids that can grow well at unusually cold temperatures.

While transgenic aquatic animals could revolutionize aquaculture production, their potential impact on wild populations, should they escape, is unknown, which has prompted governmental agencies, at least in the United States, to proceed very cautiously with the approval of such animals for use in aquaculture. In the United States, the Department of Agriculture's Animal and Plant Health Inspection Service (APHIS) is responsible for regulation of transgenic fish. That role has recently been summarized by Medley and Brown (1992) in a paper that includes a description of how the agency worked with Auburn University to develop appropriate facilities to handle transgenic fish.

Much of the work to date has been conducted merely to determine if foreign

DNA could be transferred into fish embryos and be expressed in the genome of the receiving species. Fishes in which foreign DNA has been transferred and expressed include trout, salmon, carp, tilapia, medaka, goldfish, zebrafish, loach, walleye, northern pike, and catfish (Hallerman et al. 1990, Houdebine and Chourrout 1991). As discussed by Hallerman et al. (1990), gene transfer in fish began about 1985, but by 1990 at least 14 species of transgenic fishes had been produced in 15 or more countries.

Instances of the incorporation of foreign genes into fishes are becoming increasingly more frequent, and in many instances the research is being conducted with species of aquaculture interest. Examples of confirmed or suspected production of transgenic fishes have involved the transfer of the growth hormone gene (from humans and fish) into channel catfish (Dunham et al. 1987) and *Tilapia nilotica* (Brem et al. 1988), transfer of a bacterial galactosidase gene and an antifreeze gene into Atlantic salmon (McEvoy et al. 1988), and transfer of an antifreeze gene into Atlantic salmon (Fletcher et al. 1988, 1990, Shears et al. 1991).

The maintenance of transgenic fish in U.S. research facilities is carefully controlled to prevent escapement and the mixing of the genetically altered animals with wild stocks. The American Fisheries Society policy statement on the development of transgenic fishes also cautions against uncontrolled release because the potential ecological impacts on natural ecosystems have not been determined (Kapuscinski and Hallerman 1990).

Aquaculture, not unlike other agricultural disciplines, is at a crossroads with respect to genetic engineering. The potential exists to greatly increase productivity through the development of transgenic organisms, but consideration of potential environmental consequences cannot be ignored. Extremists would prohibit all production of transgenic animals, but the realistic approach seems to be that such animals should be developed under fully controlled conditions where the chances of escapement are minimal, and that they should only be released for commercial production once it has been determined that they pose no undue threat to natural communities. Assurance of the latter requirement may involve the coupling of genetic engineering with the production of sterile individuals.

Gene-transfer candidates of aquaculture interest are not limited to fishes, of course, and interest in applying the technique to invertebrates has developed. Cadored et al. (1991), for example, discussed using the technique to impart disease resistance to invertebrates of mariculture interest.

Molecular biology techniques can also be used to develop vaccines to combat diseases. For example, viruses such as the infectious hematopoietic necrosis (IHN) virus of salmon (see Chapter 8) can be sampled from an infected fish, and amplified through a technique known as the polymerase chain reaction, and the strain of the virus can then be identified. That may allow the development of a vaccine to combat the disease.

Another way in which genetic engineering can be used in aquaculture involves the production of genetically engineered substances that can enhance the performance of fish. One example is the injection of recombinant bovine growth hormone into channel catfish fingerlings (Wilson et al. 1988). Increased growth and food consumption occurred in fish that received the growth hormone, though at one dosage the controls performed as well as injected fish. The authors attributed that result to handling stress (the controls had not been handled or sham-injected).

HATCHERY BASHING

Declining returns of Pacific salmon in some regions have led to the establishment of unprecidented expenditures of funds to determine causes and develop recovery plans. The most visible and well-funded example is in the Columbia River system, where research on salmon is currently funded at over $60 million annually. Hydroelectric dams, the first of which was constructed in the 1930s have impeded, and in some cases prohibited the upstream migration of salmon to their traditional spawning grounds. The reservoirs behind the dams have changed the environment significantly, changing riverine environments into lakes. When the first dams were constructed, there was acceptance by the government that spawning habitat and access by anadromous salmonids to some of the remaining spawning grounds would be lost. To mitigate against fish losses, a large number of fish hatcheries were established in the lower Columbia River. Years later, the Endangered Species Act (ESA) was passed by the United States Congress, and national policy required that efforts be made to protect species that are threatened or endangered.[167] Recently, the Act has been used to protect some groups of salmon in the Columbia River system.

Dam construction in the Columbia River Basin has provided decades of low cost electricity from hydroelectric power generating stations. In addition, the reservoirs that were created behind the dams have provided sources of water for irrigation of croplands. Taming the rivers, in conjunction with the construction of locks around the dams, provided an aquatic highway for barges carrying grain and other goods up and down the river system. Returns to many of the hatcheries have been poor and opponents of the hatchery system claim that the mixing of hatchery fish with their wild counterparts has had a negative impact on the wild fish.

In reality, hatchery practices have not produced fish that are fully competitive when they are released. Hatchery fish are maintained at high densities in raceways where they are fed prepared feeds. They have not experienced the need to forage for live foods, are not afraid of predators (a shadow passing over a raceway can be associated with a person coming to provide feed rather than with a bird interested in an easy meal), and may lack the stamina of their wild counterparts.

Some scientists have indicated that hatcheries should be eliminated. Wild salmon production could be augmented, according to them, by providing appropriate spawning habitat. While that might work in some river systems, it would only be possible in the Columbia River Basin if the dams were removed, and that would have enormous impacts of the economy of the West Coast.

---

[167] Under ESA the definition is extremely broad. Any genetically distinct stock of fish is considered as a species under ESA.

Realistically, the problem can only be resolved by both protecting and upgrading natural spawning habitat and altering hatchery procedures in a manner than will lead to the production of cultured fish that behave and survive similarly to their wild counterparts. Some effort toward that end is currently underway, and significant progress can be expected if fisheries scientists work together in a cooperative effort to bring the salmon back to historical levels. Geneticists have postulated that selective breeding in hatcheries has led to a change in the genotype of hatchery fish as compared with their wild counterparts, but the evidence to support that claim has yet to be provided. If the genetic diversity associated with hatchery fish is not different from that of wild salmon it should be possible to produce hatchery fish identical both phenotypically and genotypically to wild fish. The pressure is on aquaculturists to respond to the challenge.

LITERATURE CITED

Aiken, D. E., and S. L. Waddy. 1988. Temperature-photoperiod control of spawning by American lobsters: a facultative regulatory system. J. Shellfish Res., 7: 196.

Aiken, D. E., and S. L. Waddy. 1989. Interaction of temperature and photoperiod in the regulation of spawning by American lobsters *(Homarus americanus).* Can. J. Fish. Aquat. Sci., 46: 145–148.

Aiken, D. E., and S. L. Waddy. 1990. Winter temperature and spring photoperiod requirements for spawning in the American lobster, *Homarus americanus,* H. Milne Edwards, 1837. J. Shellfish Res., 9: 41–43.

Allen, S. K., Jr., and S. L. Downing. 1986. Performance of triploid Pacific oysters, *Crassostrea gigas* (Thunberg). 1. Survival, growth, glycogen content, and sexual maturation in yearlings. J. Exp. Mar. Biol. Ecol., 102: 197–208.

American Fisheries Society. 1989. Common and scientific names of aquatic invertebrates from the United States and Canada: decapod crustaceans. American Fisheries Society Special Publication No. 17. American Fisheries Society, Bethesda, Maryland. 77 p.

Anchordoguy, T., J. H. Crowe, F. J. Griffin, and W. H. Clark, Jr. 1988. Cryopreservation of sperm from the marine shrimp *Sicyonia ingentis.* Cryobiology, 25: 238–243.

Anonymous. 1981. Use of transformer cases for spawning catfish dangerous. pp. 1–2, In: For fish farmers. Mississippi State University, Mississippi State.

Arnold, C. R. 1988. Controlled year-round spawning of red drum *Sciaenops ocellatus* in captivity. Contrib. Mar. Sci., 30: 65–70.

Arnold, C. R. 1991. Precocious spawning of red drum. Prog. Fish-Cult., 53: 50–51.

Arnold, C. R., W. H. Bailey, T. D. Williams, A. Johnson, and J. L. Lasswell. 1977. Laboratory spawning and larval rearing of red drum and southern flounder. Proc. Southeast. Assoc. Fish Wildl. Agencies, 31: 437–440.

Atz, J. W. 1954. The peregrinating *Tilapia.* Anim. Kingdom, 57: 148–155.

Avault, J. W., Jr. 1973. Crawfish farming in the United States. pp. 239–250, In: S. Abrahamsson (Ed.). Freshwater crayfish; Papers from the First International Symposium on Freshwater Crayfish, Hinterthal, Austria, September, 1972. Studenlitt, Lund, Sweden.

Avault, J. W., Jr., and E. W. Shell. 1968. Preliminary studies with the hybrid *Tilapia nilotica* × *Tilapia mossambica*. FAO Fish. Rep., 44: 237–242.

Avault, J. W., L. W. de la Bretonne, and J. V. Huner. 1974. Two major problems in culturing crayfish in ponds: oxygen depletion and overcrowding. pp. 139–144, In: J. W. Avault, Jr. (Ed.). Freshwater crayfish, Papers from the Second International Symposium on Freshwater Crayfish, Louisiana State University, April 7–11. Louisiana State University, Baton Rouge.

Ayson, F. G. 1991. Induced spawning of rabbitfish, *Siganus guttatus* (Bloch) using human chorionic gonadotropin (HCG). Aquaculture, 95: 133–137.

Bailey, J. K., and G. W. Friars. 1990. Inheritance of age at smolting in hatchery-reared Atlantic salmon *(Salmo salar)*. Aquaculture 85: 317.

Banerjee, S., D. Saha, and S. Podder. 1984. Induced spawning of silver carp *Hypopthhalmichthys molitrix* by human chorionic gonadotrophic hormone (HCG) administration. Environ. Ecol., 2: 153–156.

Bardach, J. E., J. H. Ryther, and W. O. McLarney. 1972. Aquaculture. Wiley-Interscience, New York. 868 p.

Barnabe, G., and R. Barnabe-Quet. 1985. Advancement and improvement of induced spawning in the sea-bass *Dicentrarchus labrax* (L.) using an LHRH analogue injection. Aquaculture 49: 125–132.

Bayless, J. D. 1972. Artificial propagation and hybridization of striped bass, *Roccus saxatilis*. South Carolina Wildlife and Marine Resources Department, Columbia. 135 p.

Baynes, S. M., and A. P. Scott. 1987. Cryopreservation of rainbow trout spermatozoa: the influence of sperm quality, egg quality and extender composition on post-thaw fertility. Aquaculture, 66: 53–67.

Beacham, T. D., and R. E. Withler. 1991. Genetic variation in mortality of chinook salmon, *Oncorhynchus tshawytscha* (Walbaum), challenged with high water temperatures. Aquacult. Fish. Manage., 22: 125–133.

Beattie, J. H. 1984. Increased survival among Pacific oysters: the result of selective breeding. J. Shellfish Res., 4: 109.

Beattie, J. H., J. Perdue, W. Hershberger, and K. Chew. 1987. Effects of inbreeding in the Pacific oyster *(Crassostrea gigas)*. J. Shellfish Res., 6: 25–28.

Bolla, S., I. Holmefjord, and T. Refstie. 1987. Cryogenic preservation of Atlantic halibut sperm. Aquaculture 65: 371–374.

Bondari, K. 1983a. Efficiency of male reproduction in channel catfish. Aquaculture, 35: 79–82.

Bondari, K. 1983b. Performance of domestic and domestic × wild crosses of channel catfish in two water temperatures. J. World Maricult. Soc., 14: 661–667.

Bondari, K. 1983c. Response to bidirectional selection for body weight in channel catfish. Aquaculture, 33: 73–81.

Bondari, K., G. O. Ware, B. G. Mullinix, Jr., and J. A. Joyce. 1985. Influence of brood fish size on the breeding performance of channel catfish. Prog. Fish-Cult., 47: 21–33.

Boulier, A., and R. Billard. 1985. Delayed gametogenesis and spawning in rainbow trout *(Salmo gairdneri)* kept under permanent light during the first and second reproductive cycles. Aquaculture, 43: 259–268.

Brem, G., B. Brenig, G. Hoerstgen-Schwark, and E-L. Winnacker. 1988. Gene transfer in tilapia *(Oreochromis niloticus)*. Aquaculture, 68: 209–219.

Bromage, N. R., J. A. K. Elliott, J. R. C. Springate, and C. Whitehead. 1984. The effects of constant photoperiods on the timing of spawning in the rainbow trout. Aquaculture, 43: 213–223.

Broussard, M. C., Jr., and R. R. Stickney. 1981. Evaluation of reproduction characters for four strains of channel catfish. Trans. Am. Fish. Soc., 110: 502–506.

Brown, C. L., S. I. Doroshov, J. M. Nunez, C. Hadley, J. Vaneenennaam, R. S. Nishioka, and H. A. Bern. 1988. Maternal triiodothyronine injections cause increases in swimbladder inflation and survival rates in larval striped bass, *Morone saxatilis.* J. Exp. Zool., 248: 168–176.

Brunson, M. W. 1988. Forage and feeding systems for commercial crawfish culture. J. Shellfish Res., 7: 210.

Brunson, M. W. 1989. Forage and feeding systems for commercial crawfish culture. J. Shellfish Res., 8: 277–280.

Brunson, M. W., and J. L. Griffin. 1988. Comparison of rice-crayfish and grain sorghum-crayfish double cropping systems. Aquaculture, 72: 265–272.

Brunson, M. W., and R. W. Taylor. 1987. Evaluation of three sorghums as potential alternative forages for crawfish culture. J. World Aquacult. Soc., 18: 247–252.

Brydges, K., and T. J. Benfey. 1991. Triploid brown trout *(Salmo trutta)* produced by hydrostatic pressure shock. Bull. Aquacult. Assoc. Can., 91(3): 31–33.

Busch, R. L., and J. A. Seeby. 1990. An evaluation of a leuteinizing hormone-releasing hormone analog to induce spawning of channel catfish *Ictalurus punctatus.* J. World Aquacult., Soc., 21: 10–15.

Cadored, J. P., E. Mialhe, S. Gendreau, M. Ohresser, R. M. LeDeuff. 1991. Gene transfer: potential application to pathology of farmed marine invertebrates. ICES Council Meeting Papers, 1991. International Council for the Exploration of the Sea, Copenhagen. 5 p.

Cahu, C., M.-F. Gouillou-Coustans, M. Fakhfakh, and P. Quazuguel. 1991. The effect of ascorbic acid concentration in broodstock feed on reproduction of *P. indicus.* ICES Council Meeting Papers. International Council for the Exploration of the Sea, Copenhagen. 10 p.

Calliouet, C. W. Jr. 1972. Ovarian maturation induced by eyestalk ablation in pink shrimp, *Penaeus duorarum* Burkenroad. Proc. World Maricult. Soc. 3: 205–225.

Canfield, H. L. 1947. Artificial propagation of those channel cats. Prog. Fish-Cult., 9: 27–30.

Carter, R. E., G. C. Mair, D. O. F. Skibinski, D. T. Parkin, and J. A. Beardmore. 1991. The application of DNA fingerprinting in the analysis of gynogenesis in tilapia. Aquaculture, 95: 41–52.

Chamberlain, G. W. 1991. Status of shrimp farming in Texas. pp. 36–57, In: P. A. Sandifer (Ed.). Shrimp culture in North America and the Caribbean. World Aquaculture Society, Baton Rouge, Louisiana.

Chamberlain, G. W., and N. F. Gervais. 1984. Comparison of unilateral eyestalk ablation with environmental control for ovarian maturation of *Penaeus stylirostris.* J. World Maricult. Soc., 14: 29–30.

Chamberlain, G. W., and A. L. Lawrence. 1981. Effect of light intensity and male and female eyestalk ablation on reproduction of *Penaeus stylirostris* and *P. vannamei.* J. World Maricult. Soc., 12: 357–372.

Chen, F. Y. 1969. Preliminary studies on the sex-determining mechanism of *Tilapia mossambica* Peters and *T. hornorum* Trewavas. Verh. Int. Ver. Limnol., 17: 719–724.

Chernenko, E. V. 1985. Induction of triploidy in Pacific salmons (Salmonidae). J. Ichthyol., 25: 124–130.

Chimits, P. 1955. Tilapia and its culture. A preliminary bibliography. FAO Fish. Bull., 8: 1–23.

Chu, K. H., and W. K. Chow. 1992. Effects of unilateral versus bilateral eyestalk ablation on moulting and growth of the shrimp, *Penaeus chinensis* (Osbeck, 1765) (Decapoda, Penaeidea). Crustaceana, 62: 225–233.

Clapp, A. 1929. Some experiments in rearing channel catfish. Trans. Am. Fish. Soc., 59: 114–117.

Clemens, H. P., and K. E. Sneed. 1957. The spawning behavior of the channel catfish, *Ictalurus punctatus*. U. S. Department of the Interior, Special Scientific Report—Fisheries, No. 219. 11 p.

Clemens, H. P., and K. E. Sneed. 1962. Bioassay and the use of pituitary materials to spawn warmwater fishes. U.S. Fish and Wildlife Service Resources Report 61. U.S. Dept. of the Interior, Washington, D.C. 30 p.

Colombo, I. A. Francescon, A. Barharo, P. Belvadere, and P. Melotti. 1989. Induction of spawning in the gilthead sea bream, *Sparus aurata* L., by elevation of water temperature and salinity and by HCG and LHRH analogue treatments. Riv. Ital. Aquacult., 24: 187–196.

Colura, R. L., B. T. Hysmith, and R. E. Stevens. 1976. Fingerling production of striped bass *(Morone saxatilis)*, spotted seatrout *(Cynoscion nebulosis)*, and red drum *(Sciaenops ocellatus)* in saltwater ponds. Proc. World Maricult. Soc., 7: 79–92.

Colura, R. L., A. F. Maciorowski, and A. Henderson-Arzapalo. 1990. Induced spawning of spotted seatrout with selected hormone preparations. Prog. Fish-Cult., 52: 205–207.

Cook, H. L., and M. A. Murphy. 1969. The culture of larval penaeid shrimp. Trans. Am. Fish. Soc., 98: 751–754.

Cook, H. L., and M. A. Murphy. 1971. Early development stages of the brown shrimp, *Penaeus aztecus* Ives, reared in the laboratory. Fish. Bull., 69: 223–239.

Crim, L. W., A. M. Sutterlin, D. M. Evans, and C. Weil. 1983. Accelerated ovulation by pelleted LHRH analogue treatment of spring-spawning rainbow trout *(Salmo gairdneri)* held at low temperature. Aquaculture, 35: 299–307.

Crim, L. W., B. D. Glebe, and A. P. Scott. 1986. The influence of LHRH analog on oocyte development and spawning in female Atlantic salmon, *Salmo salar*. Aquaculture, 56: 139–149.

Crim, L., C. Wilson, Y. So, D. Idler, and F. Johnston. 1989. Influence of testosterone treatment on the rematuration/spawning performance of female Atlantic salmon kelt reared in captivity. Bull. Aquacult. Assoc. Can., 89(3): 31.

Crocos, P. J., and J. D. Kerr. 1986. Factors affecting induction of maturation and spawning of the tiger prawn, *Penaeus esculentus* (Haswell), under laboratory conditions. Aquaculture, 58: 203–214.

D'Croz, L., V. L. Wong, G. Justine, and M. Gupta. 1988. Prostaglandins and related compounds from the polychaete worm *Americonuphis reesei* Fauchald (Onuphidae) as possible inducers of gonad maturation in penaeid shrimps. Rev. Biol. Trop., 36: 331–332.

de la Bretonne, L. W. 1988. Commercial crawfish cultivation practices. J. Shellfish Res., 7: 210.

de la Bretonne, L. W., and R. P. Romaire. 1989. Commercial crawfish cultivation practices: a review. J. Shellfish Res., 8: 267–275.

Dietz, R. A. 1983. Eyestalk histology and the effects of eyestalk ablation on the gonads of the shrimp, *Macrobrachium rosenbergii* (DeMan). Ph.D. Dissertation, Texas A&M University, College Station. 110 p.

Dobkin, S. 1961. Early developmental stages of the pink shrimp, *Penaeus duorarum,* from Florida waters. Fish. Bull., 61: 321–349.

Dobzhansky, T. 1970. Genetics of the evolutionary process. Columbia University Press, New York. 505 p.

Doyle, R. W., N. L. Shackell, Z. Basiao, S. Uraiwan, T. Matricia, and A. J. Talbot. 1991. Selective diversification of aquaculture stocks: a proposal for economically sustainable genetic conservation. Ecological and genetic implications of fish introductions symposium. Can. J. Fish. Aquat. Sci., 48 (Suppl. 1): 148–154.

Dunham, R. A., and R. O. Smitherman. 1982. Effects of selecting for growth on reproductive performance in channel catfish. Proc. Southeast. Assoc. Fish and Wildl. Agencies, 36: 182–189.

Dunham, R. A., and R. O. Smitherman. 1983. Response to selection and realized heritability for body weight in three strains of channel catfish, *Ictalurus punctatus,* grown in earthen ponds. Aquaculture, 33: 89–96.

Dunham, R. A., J. Eash, J. Askins, and T. M. Townes. 1987. Transfer of metallothionein-human growth hormone fusion gene into channel catfish. Trans. Am. Fish. Soc., 116: 87–91.

Dunham, R. A., R. O. Smitherman, J. L. Horn, and T. O. Bice. 1983. Reproductive performances of crossbred and pure-strain channel catfish brood stocks. Trans. Am. Fish. Soc., 112: 436–440.

Dunham, R. A., R. O. Smitherman, and K. Bondari. 1991. Lack of inheritance of stump-body and taillessness in channel catfish. Prog. Fish-Cult., 53: 101–105.

Dwivedi, S. N., C. S. Chaturvedi, and P. K. Varshney. 1986. Breeding of silver carp in combination of human chorionic gonadotropin and constant dosage of pituitary hormone. Agric. Biol. Res., 2: 8–11.

Elliott, J. A. K., N. R. Bromage, and J. R. C. Springate. 1984. Changes in reproductive function of three strains of rainbow trout exposed to constant and seasonally-changing light cycles. Aquaculture, 43: 23–34.

Eversole, A. G., and R. S. Pomeroy. 1989. Crawfish culture in South Carolina: an emerging aquaculture industry. J. Shellfish Res., 8: 309–313.

Fishelson, L. 1962. Hybrids of two species of the genus *Tilapia* (Cichlidae, Teleostei). Fishermen's Bull. Haifa, 4: 14–19.

Fletcher, G. L., M. A. Shears, M. J. King, P. L. Davies, and C. L. Hew. 1988. Evidence for antifreeze protein gene transfer in Atlantic salmon *(Salmo salar).* Can. J. Fish. Aquat. Sci., 45: 352–357.

Fletcher, G. L., S.-J. Du, M. A. Shears, C. L. Hew, and P. L. Davies. 1990. Antifreeze and growth hormone gene transfer in Atlantic salmon. Bull. Aquacult. Assoc. Can., 90(4): 70–71.

Fores, J., J. Iglesias, M. Olmedo, F. J. Sanchez, and J. B. Peleteiro. 1990. Induction of spawning in turbot (*Scophthalmus maximus* L.) by a sudden change in the photoperiod. Aquacult. Eng., 9: 357–366.

Gall, G. A. E., and N. Huang. 1988a. Heritability and selection schemes for rainbow trout: body weight. Aquaculture, 73: 43–56.

Gall, G. A. E., and N. Huang. 1988a. Heritability and selection schemes for rainbow trout: female reproductive performance. Aquaculture, 73: 57–66.

Gallant, R. K., and M. A. McNiven. 1991. Cryopreservation of rainbow trout spermatozoa. Bull. Aquacult. Assoc. Can., 91(3): 25–28.

Galtsoff, P. S. 1964. The American oyster *Crassostrea virginica* Gmelin. Fish. Bull., 64: 1–480.

Garces, C. A., and J. W. Avault, Jr. 1985. Evaluation of rice *(Oryza sativa)*, volunteer vegetation, and alligatorweed *(Alternanthera phyloxeroides)* in various combinations as crawfish *(Procambarus clarkii)* forages. Aquaculture, 44: 177–186.

Garcia, I. M. 1989. Spawning response of mature female sea bass, *Lates calcarifer* (Bloch), to a single injection of luteinizing hormone-releasing hormone analogue: effect of dose and initial oocyte size. J. Appl. Ichthyol., 5: 177–184.

Garcia, I. M. 1990. Spawning response latency and egg production capacity of LRHRa-injected mature female sea bass, *Lates calcarifer* Bloch. J. Appl. Ichthyol., 6: 167–172.

Gendreau, S., J. P. Cadoret, and H. Grizel. 1989. Triploidy in the larviparous European flat oyster, *Ostrea edulis* L. pp. 111–112, In: R. Billard, and N. de Pauw (Comps.). Aquaculture Europe '89. Special Publication of the European Aquaculture Society No. 10. Ghent, Belgium.

Gharrett, A. J., and W. W. Smoker. 1991. Two generations of hybrids between even- and odd-year pink salmon *(Oncorhynchus gorbuscha):* a test for outbreeding depression? Can. J. Fish. Aquat. Sci., 48: 1744–1749.

Gjerde, B., and L. R. Schaeffer. 1989. Body traits in rainbow trout. 2. Estimates of heritabilities and of phenotypic and genetic correlations. Aquaculture, 80: 25–44.

Gosling, E. M., and A. Nolan. 1990. Triploidy induction by thermal shock in the Manila clam, *Tapes semidecussatus*. Aquaculture, 85: 324.

Gudger, E. W. 1918. Oral gestation in the gaff-topsail catfish *Felichthys felis*. pp. 25–52, In: Papers from the Department of Marine Biology of the Carnegie Institute of Washington, Vol. 12.

Guerrero, R. D. III. 1975. Use of androgens for the production of all-male *Tilapia aurea* (Steindachner). Trans. Am Fish. Soc., 104: 342–348.

Hadley, N. H., R. T. Dillon, Jr., and J. J. Manzi. 1991. Realized heritability of growth rate in the hard clam *Mercenaria mercenaria*. Aquaculture, 93: 109–119.

Halder, S., S. Sen, S. Bhattacharya, A. K. Ray, A. Ghosh, and A. G. Jhingran. 1991. Induced spawning of Indian major carps and maturation of a perch and a catfish by murrel gonadotropin releasing hormone, pimozide and calcium. Aquaculture, 97: 373–382.

Hallerman, E. M., A. R. Kapuscinski, P. B. Hachett, Jr., A. J. Faras, and K. S. Guise. 1990. Gene transfer in fish. pp. 35–49, In: M. N. Voigt, and J. R. Botta (Eds.). Advances in fisheries technology and biotechnology for increased profitability. Atlantic Fisheries Technological Conference and Seafood Biotechnology Workshop, St. John's, Newfoundland, August 27 – September 1, 1989. University of Alaska, Fairbanks.

Hanson, J. A., and H. L. Goodwin. 1977. Shrimp and prawn farming in the western hemisphere. Dowden, Hutchinson & Ross, Stroudsburg, Pennsylvania. 439 p.

Harrell, R. M. 1984. Tank spawning of first generation striped bass × white bass hybrids. Prog. Fish-Cult., 46: 75–78.

Harvey, B., J. Nacario, L. W. Crim, J. V. Juario, and C. L. Marte. 1985. Induced spawning of sea bass, *Lates calcarifer,* and rabbitfish, *Siganus guttatus,* after implantation of pelleted LHRH analogue. Aquaculture, 47: 53–59.

Hecht, T., J. E. Saayman, and L. Polling. 1982. Further observations on the induced spawning of the sharptooth catfish, *Clarias gariepinus* (Clariidae: Pisces). Water S. A. (Pretoria), 2: 101–107.

Hedgecock, D., F. Sly, K. Cooper, B. Hershberger, and X. Guo. 1988. Pedigreed broodstocks for culture and breeding of Pacific oysters. J. Shellfish Res., 7: 550.

Heegaard, P. E. 1953. Observation on spawning and larval history of the shrimp, *Penaeus setiferus* (L.). Publ. Inst. Mar. Sci. Univ. Tex., 3: 73–105.

Henderson-Arzapalo, A. 1992. Red drum aquaculture. Rev. Aquat. Sci., 6: 479–491.

Henderson-Arzapalo, A., and R. L. Colura. 1987. Laboratory maturation and induced spawning of striped bass. Prog. Fish-Cult., 49: 60–63.

Henderson-Arzapalo, A., C. Lemm, J. Hawkinson, and P. Keyes. 1992. Tricaine used to separate phase-I striped bass with uninflated gas bladders from normal fish. Prog. Fish-Cult., 54: 133–135.

Hershberger, W. K., J. H. Beattie, N. Pongthana, and K. K. Chew. 1988. Genetic improvement of the Pacific oyster *(Crassostrea gigas)* for commercial production. J. Shellfish Res., 7: 163.

Hershberger, W. K., J. M. Myers, R. N. Iwamoto, W. C. Macauley, and A. M. Saxton. 1990. Genetic changes in the growth of coho salmon *(Oncorhynchus kisutch)* in marine net-pens, produced by ten years of selection. Aquaculture, 85: 187–197.

Hickling, C. F. 1960. The Malacca *Tilapia* hybrids. J. Genet. 57: 1–10.

Hidu, H., K. G. Drobeck, E. A. Dunnington, Jr., W. Roosenburg, and R. L. Beckett. 1969. Oyster hatcheries for the Chesapeake Bay region. University of Maryland Natural Resources Institute Special Report No. 2. University of Maryland, College Park. 18 p.

Hines, N. O. 1976. Fish of rare breeding. Salmon and trout of the Donaldson strains. Smithsonian Institution Press, Washington, D.C. 167 p.

Hoar, W. S. 1969. Reproduction. pp. 1–72, In: W. S. Hoar and D. J. Randall (Eds.). Fish physiology, Vol. 3. Academic Press, New York.

Hollebecq, M. G., D. Chourrout, G. Wohlfarth, and R. Billard. 1986. Diploid gynogenesis induced by heat shocks after activation with UV-irradiated sperm in common carp. Aquaculture, 54: 69–76.

Houdebine, L. M., and D. Chourrout. 1991. Transgenesis in fish. Experientia, 47: 891–897.

Huang, C.-M., and L. I. Chiu. 1990. Response to mass selection for growth rate in *Oreochromis niloticus.* Aquaculture, 85: 199–205.

Huner, J. V. 1987. Tolerance of the crawfishes *Procambarus acutus acutus* and *Procambarus clarkii* (Decapoda, Cambaridae) to acute hypoxia and elevated thermal stress. J. World Aquacult. Soc., 18: 113–114.

Huner, J. V. 1988. Overview of international and domestic freshwater crawfish production. J. Shellfish Res., 7: 209–210.

Huner, J. 1989. Crayfish culture in North America. pp. 126–135, In: J. Skurdal, K. Westman, and P. I. Bergan (Eds.). Crayfish culture in Europe. Report from the Workshop on Crayfish Culture, Trondheim, Norway, Nov. 16–19, 1987.

Iwamoto, R. N., J. M. Myers, and W. K. Hershberger. 1990. Heritability and genetic correlations for flesh coloration in pen-reared coho salmon. Aquaculture, 86: 181–190.

Jalabert, B., P. Kammacher, and P. Lessent. 1971. Sex determination in *Tilapia macrochir* × *Tilapia nilotica* hybrids. Investigations on sex ratios in first generation × parent crossings. Ann. Biol. Anim. Biochim. Biophys., 11: 155–165.

Jarimopas, P. 1986. Realized response of Thai red tilapia to weight-specific selection for growth. pp. 109–111, In: J. L. Maclean, L. B. Dizon, and L. V. Hosillos (Eds.). Proceedings of the First Asian Fisheries Forum, Manila, Philippines, May 26–31. Asian Fisheries Society, Manila.

Jarimopas, P. 1990. Realized response of Thai red tilapia to 5 generations of size-specific selection for growth. pp. 519–522, In: R. Hirano and I. Hanyu (Eds.). Proceedings of the Second Asian Fisheries Forum, Tokyo, Japan, April 17–22, 1989. Asian Fisheries Society, Manila.

John, G., P. V. G. K. Reddy, and S. D. Gupta. 1984. Artificial gynogenesis in two Indian major carps, *Labeo rohita* (Ham.) and *Catla catla* (Ham.). Aquaculture, 42: 161–168.

Johnson, W. S. 1984. Photoperiod induced delayed maturation of freshwater reared chinook salmon. Aquaculture, 43: 279–287.

Johnstone, R., H. A. McLay, and M. V. Walsingham. 1991. Production and performance of triploid Atlantic salmon in Scotland. pp. 15–36, In: V. A. Pepper (Ed.). Proceedings of the Atlantic Canada workshop on methods for the production of non-maturing salmonids, Dartmouth, Nova Scotia, Feb. 19–21. Canada Techical Reports in Fisheries and Aquatic Sciences No. 1789. Department of Fisheries and Oceans, Ottawa.

Kadmon, G., H. Gordin, and Z. Yaron. 1985. Breeding-related growth of captive *Sparus aurata* (Teleostei, Perciformes). Aquaculture, 46: 299–305.

Kapuscinski, A. R., and E. M. Hallerman. 1990. Transgenic fishes (AFS position statement). Fisheries, 15(4): 2–5.

Kaul, M., and K. K. Rishi. 1986. Induced spawning of the Indian major carp, *Cirrhina mrigala* (Ham.) , with LH-RH analogue or pimozide. Aquaculture, 54: 45–48.

Kerby, J. H. 1992. The striped bass and its hybrids. pp. 251–306, In: R. R. Stickney (Ed.). Culture of nonsalmonid freshwater fishes. CRC Press, Boca Raton, Florida.

Kincaid, H. L. 1976. Inbreeding in salmonids. pp. 33–37, In: T. Y. Nosho and W. K. Hershberger (Eds.). Salmonid genetics: status and role in mariculture. University of Washington Sea Grant Report WSG WO 76–2. Seattle, Washington.

Kincaid, H. L. 1977. Rotational line crossing: an approach to the reduction of inbreeding accumulation in trout brood stocks. Prog. Fish-Cult., 39: 179–181.

Kronert, U., G. Hoerstgen-Schwark, and H.-J. Langholz. 1989. Prospects of selecting for late maturity in tilapia *(Oreochromis niloticus)* 1. Family studies under laboratory conditions. Aquaculture, 77: 113–121.

Kuo, C.-M., Z. H. Shehadeh, and C. E. Nash. 1973. Induced spawning of captive grey mullet (*Mugil cephalus* L.) females by injection of human chorionic gonadotropic (HCG). Aquaculture, 1: 429–432.

LaCaze, C. 1976. Crawfish farming, revised edition. Louisiana Wildlife and Fishery Commission Fishery Bulletin No. 7. Baton Rouge, Louisiana. 27 p.

Landau, M. 1992. Introduction to aquaculture. John Wiley & Sons, New York. 440 p.

Landers, W. S. 1968. Oyster hatcheries in the northeast. pp. 35–40, In: T. L. Linton (Ed.). Proceedings of the oyster culture workshop, July 11–13. Marine Fisheries Division, Georgia Game and Fish Commission, Brunswick.

Lawson, T. B., and F. W. Wheaton. 1982. Pond culturing of crawfish in the southern United States. Aquacult. Eng., 1: 311–317.

Lawson, T. B., and F. W. Wheaton. 1983. Crawfish culture systems and their management. J. World Maricult. Soc., 14: 325–335.

Lee, C-S., C. S. Tamaru, C. D. Kelley, and J. E. Banno. 1986a. Induced spawning of milkfish, *Chanos chanos,* by a single application of LHRH-analogue. Aquaculture, 58: 87–98.

Lee, C.-S., C. S. Tamaru, J. E. Banno, C. D. Kelley, A. Bocek, and J. A. Wyban. 1986b. Induced maturation and spawning of milkfish, *Chanos chanos* Forsskal, by hormone implantation. Aquaculture, 52: 199–205.

Lee, C.-S., C. S. Tamaru, and G. M. Weber. 1987a. Studies on the maturation and spawning of milkfish *Chanos chanos* Forsskal in a photoperiod-controlled room. J. World Aquacult. Soc., 18: 253–259.

Lee, C.-S., C. S. Tamaru, G. T. Miyamoto, and C. D. Kelley. 1987b. Induced spawning of grey mullet *(Mugil cephalus)* by LHRH-a. Aquaculture, 62: 327–336.

Lee, C.-S., C. S. Tamaru, and D. D. Kelley. 1988. The cost and effectiveness of CPH, HCG and LHRH-a on the induced spawning of grey mullet, *Mugil cephalus.* Aquaculture, 73: 341–347.

Lee, M. H., A. G. Cavinato, D. M. Mayes, and B. A. Rasco. 1992. Noninvasive short-wavelength near-infrared spectroscopic method to estimate the crude lipid content in the muscle of intact rainbow trout. J. Agric. Food Chem., 40: 2176–2181.

Lester, L. J., and K. S. Lawson. 1990. Inheritance of size as estimated by principal component analysis at two temperatures in *Penaeus vannamei.* Aquaculture, 85: 323.

Lin, R., J. He, and H. Qiu. 1990. Inducement on ovarian development, maturation and spawning in *Penaeus monodon* Fabricius grow-out from earth ponds. J. Fish. China, 14: 277–285.

Linhart, O., P. Kvasnicka, V. Slechtova, and J. Pokorny. 1986. Induced gynogenesis by retention of the second polar body in the common carp, *Cyprinus carpio* L., and heterozygosity of gynogenetic progeny in transferrin and Ldh-B[1] loci. Aquaculture, 54: 63–67.

Linton, T. L. 1968. Feasibility studies of raft-culturing oysters in Georgia. pp. 69–73, In: T. L. Linton (Ed.). Proceedings of the Oyster Culture Workshop, Brunswick, Georgia, July 11–13. Marine Fisheries Division, Georgia Game and Fish Commission, Brunswick.

Liu, H. W. 1988. Seasonal changes in sex steroids of Pacific halibut *Hippoglossus stenolepis.* M.S. Thesis. University of Washington, Seattle. 33 p.

Liu, H. W. 1991. Laboratory studies on the spawning and the early life history of Pacific halibut *(Hippoglossus stenolepis).* Ph.D. Dissertation, University of Washington, Seattle. 205 p.

Loosanoff, V. L., and H. C. Davis. 1963. Rearing of bivalve mollusks. pp. 1–136, In: F. S. Russel (Ed.). Advances in marine biology, Vol. 1. Academic Press, London.

Lowe-McConnell, R. H. 1958. Observations on the biology of *Tilapia nilotica* Linné in East Africa waters. Rev. Zool. Bot. Afr. 57: 131–170.

Lutz, C. G., and W. R. Wolters. 1989. Estimation of heritabilities for growth, body size, and processing traits in red swamp crawfish, *Procambarus clarkii* (Girard). Aquaculture, 78: 21–23.

Lytle, J. S., T. F. Lytle, and J. T. Ogle. 1990. Polyunsaturated fatty profiles as a comparative tool in assessing maturation diets of *Penaeus vannamai*. Aquaculture, 89: 287–299.

Manickam, P., and K. P. Joy. 1989. Induction of maturation and ovulation by pimozide-LHRH analogue treatment and resulting high quality egg production in the Asian catfish, *Clarias batrachus* (L.). Aquaculture, 83: 193–199.

Marshall, H. L. 1968. Three-dimensional oyster culture research in North Carolina. pp. 62–66, In: T. L. Linton (Ed.). Proceedings of the Oyster Culture Workshop, Brunswick, Georgia, July 11–13. Marine Fisheries Division, Georgia Game and Fish Commission, Brunswick.

Marshall, H. L. 1969. Development an evaluation of new cultch materials and techniques for three-dimensional oyster culture. Division of the Commission on Sports Fisheries, North Carolina Department of Conservation and Development, Special Report No. 17. 34 p.

Marte, C. L., N. M. Sherwood, L. W. Crim, and B. Harvey. 1987. Induced spawning of maturing milkfish (*Chanos chanos* Forsskal) with gonadotropin-releasing hormone (GnRH) analogues administered in various ways. Aquaculture, 60: 303–310.

Marte, C. L., N. Sherwood, L. Crim, and J. Tan. 1988. Induced spawning of maturing milkfish *(Chanos chanos)* using human chorionic gonadotropin and mammalian and salmon gonadotropin releasing hormone analogues. Aquaculture, 73: 333–340.

Martin, M. 1967. Techniques of catfish fingerling production. pp. 13–22, In: Proceedings of the Commercial Fish Farming Conference, Texas A&M University, Feb. 1–2. Texas A&M University, College Station.

May, E. B. 1968. Raft culture of oysters in Alabama. pp. 76–77, In: T. L. Linton (Ed.). Proceedings of the Oyster Culture Workshop, Brunswick, Georgia, July 11–13. Marine Fisheries Division, Georgia Game and Fish Commission, Brunswick.

May, E. B. 1969. Feasibility of off bottom oyster culture in Alabama. Ala. Mar. Res. Bull., 3: 1–14.

McCarty, C. E. 1990. Design and operation of a photoperiod/temperature spawning system for red drum. pp. 44–45, In: G. W. Chamberlain, R. J. Miget, and M. G. Haby (Eds.). Red drum aquaculture. Texas A&M University Sea Grant Program, College Station.

McEvoy, T., M. Stack, B. Keane, T. Barry, J. Sreenan, and F. Gannon. 1988. The expression of a foreign gene in salmon embryos. Aquaculture, 68: 27–37.

Medley, T. L., and C. L. Brown. 1992. Procedures for research involving the planned introduction into the environment of organisms with deliberately modified hereditary traits. pp. 47–50, In: R. DeVoe (Ed.). Proceedings of the Conference & Workshop Introductions and Transfers of Marine Species, Hilton Head Island, South Carolina, Oct. 30–Nov. 2, 1991. South Carolina Sea Grant Program, Charleston.

Micale, V., and F. Perdichizzi. 1988. Photoperiod effects on gonadal maturation in captivity-born gilthead bream, *Sparus aurata* (L.): early findings. J. Fish Biol., 32: 793–794.

Millamena, O. M., J. H. Primavera, R. A. Pudadera, and R. V. Caballero. 1986. The effect of diet on the reproductive performance of pond-reared *Penaeus monodon* Fabricius broodstock. pp. 593–596, In: J. L. Maclean, L. B. Dizon, and L. V. Hosillos (Eds.). Proceedings of the First Asian Fisheries Forum, Manila, Philippines, May 26–31. Asian Fisheries Society, Manila.

Minton, R. V., J. P. Hawke, and W. M. Tatum. 1983. Hormone induced spawning of red snapper, *Lutjanus campechanus*. Aquaculture, 30: 363–368.

Mok, T. K. 1985. Induced spawning and larval rearing of the white seabream, *Mylio berda*. Aquaculture, 44: 41–49.

Moy, P. B., and R. R. Stickney. 1987. Suspended spawning cans for channel catfish in a surface-mine lake. Prog. Fish-Cult., 49: 76–77.

Murphree, J. M. 1940. Channel catfish propagation. Privately printed by T. J. Rennick. 24 p.

Nakamura, K. 1988. Photoperiod influences on molting cycle and maturation of the prawn *Penaeus japonicus*. Mem. Fac. Fish. Kagoshima Univ. 37: 135–139.

Nakari, T. A. Soivio, and S. Pesonen. 1988. The ovarian development and spawning time of *Salmo gairdneri* reared in advanced and delayed annual photoperiod cycles at naturally fluctuating water temperature in Finland. Ann. Zool. Fenn., 25: 335–340.

Nelson, B. 1960. Spawning of channel catfish by use of hormone. Proc. Southeast. Assoc. Game Fish Comm., 14: 145–148.

Newkirk, G. F. 1988. Response to selection for growth in *Ostrea edulis:* second generation. J. Shellfish Res., 7: 172.

Ngamvongchon, S., O. Pawaputanon, W. Leelapatra, and W. E. Jonnson. 1987. Effectiveness of an LHRH analogue for the induced spawning of carp and catfish in northeast Thailand. Aquaculture, 74: 35–40.

Ogle, J. T. 1982. Operation of an oyster hatchery utilizing a brown water culture technique. J. Shellfish Res., 2: 153–156.

Ogle, J. T. 1991a. Design and operation of a small tank system for ovarian maturation and spawning of *Penaeus vannamei*. Gulf Res. Rep., 8: 285–289.

Ogle, J. T. 1991b. Maturation of *Penaeus vannamei* based upon a survey. Gulf Res. Rep., 8: 296–297.

Pagan-Font, F. A. 1975. Cage culture as a mechanical method for controlling reproduction in *Tilapia aurea*. Aquaculture, 6: 243–247.

Pandian, T. J., and K. Varadaraj. 1990. Development of monosex female *Oreochromis mossambicus* broodstock by integrating gynogenetic technique with endocrine sex reversal. J. Exp. Zool., 225: 88–96.

Pantastico, J. B., M. M. A. Dangilan, and R. V. Eguia. 1988. Cannibalism among different sizes of tilapia *(Oreochromis niloticus)* fry/fingerlings and the effect of natural food. pp. 465–468, In: R. S. V. Pullin, T. Bhukaswan, K. Tonguthai, and J. L. Maclean (Eds.). Second International Symposium on Tilapia in Aquaculture, Bangkok, Thailand, March 16–20, 1987. ICLARM Conference Proceedings No. 15. International Center for Living Aquatic Resources Management, Manila.

Pearson, J. C. 1939. The early life histories of some American Penaeidae, chiefly the commercial shrimp, *Penaeus setiferus* (Linn.). Bull. U.S. Bur. Fish., 49: 1–73.

Perdue, J. A., H. Beattie, W. K. Hershberger, and K. Chew. 1984. Selective breeding for improved meat quality in the Pacific oyster *Crassostrea gigas* (Thunberg) in Washington state. J. Shellfish Res., 4: 98.

Piper, R. G., I. B. McElwain, L. E. Orme, J. P. McCraren, L. G. Fowler, and J. R. Leonard. 1982. Fish hatchery management. U.S. Fish and Wildlife Service, Washington, D.C. 517 p.

Pongthana, N. 1990. The use of selective breeding to increase the carbohydrate content of the Pacific oyster, *Crassostrea gigas*. pp. 527–530, In: R. Hirano, and I. Hanyu (Eds.). Proceedings of the Second Asian Fisheries Forum, Tokyo, Japan, April 17–22, 1989. Asian Fisheries Society, Manila.

Pruder, G. D. 1991. Shrimp culture in North America and the Caribbean: Hawaii 1988. pp. 58–69, In: P. A. Sandifer (Ed.). Shrimp culture in North America and the Caribbean. World Aquaculture Society, Baton Rouge, Louisiana.

Pruginin, Y. 1967. Report to the Government of Uganda on the experimental fish culture project in Uganda, 1965–66. FAO/UNDP (Technical Assistance). Reports on Fisheries. TA Report 2446. Food and Agriculture Organization of the United Nations, Rome. 19 p.

Rana, K. J., and B. J. McAndrew. 1989. The viability of cryopreserved tilapia spermatozoa. Aquaculture, 76: 335–345.

Rhodes, R. J. 1991. Will US shrimp farms survive? The South Carolina experience. pp. 202–215, In: P. A. Sandifer (Ed.). Shrimp culture in North America and the Caribbean. World Aquaculture Society, Baton Rouge, Louisiana.

Ringle, J. P., J. G. Nickum, and A. Moore. 1992. Chemical separation of channel catfish egg masses. Prog. Fish-Cult., 54:73–80.

Robins, R. 1991. A list of common and scientific names of fishes from the United States and Canada. American Fisheries Society, Bethesda, Maryland. 1991. 183 p.

Romaire, R. P., and C. G. Lutz. 1989. Population dynamics of *Procambarus clarkii* (Girard) and *Procambarus acutus acutus* (Girard) (Decapoda: Cambaridae) in commercial ponds. Aquaculture, 81: 253–274.

Rottmann, R. W., and J. V. Shireman. 1985. The use of synthetic LH-RH analogue to spawn Chinese carps. Aquacult. Fish. Manage., 16: 1–6.

Rottmann, R. W., J. V. Shireman, and F. A. Chapman. 1991. Induction and verification of triploid in fish. Southern Regional Aquaculture Center Publication 427. Delta Branch Experiment Station, Stoneville, Mississippi. 2 p.

Rye, M., and K. M. Lillevik. 1990. Survival in the early fresh-water period in Atlantic salmon *(Salmo salar)* and rainbow trout *(Salmo gairdneri)*: heritabilities for survival and genetic correlation between survival and growth. Aquaculture, 85: 328–329.

Rye, M., K. M. Lillevik, and B. Gjerde. 1990. Survival in early life of Atlantic salmon and rainbow trout: estimates of heritabilities and genetic correlations. Aquaculture, 89: 209–216.

Saidin, T. 1986. Induced spawning of *Clarias macrocephalus* (Guncher). pp. 26–31, In: J. L. Maclean, L. B. Dizon, and L. V. Hosillos (Eds.). Proceedings of the First Asian Fisheries Forum, Manila, Philippines, May 26–31. Asian Fisheries Society, Manila.

Saksena, V. P., K. Yamamoto, and C. D. Riggs. 1961. Early development of the channel catfish. Prog. Fish-Cult., 23: 156–161.

Scheerer, P. D., and G. H. Thorgaard. 1989. Improved fertilization by cryopreserved rainbow trout semen treated with theophylline. Prog. Fish-Cult., 51: 179–182.

Scott, A. P., S. M. Baynes, O. Skarphedinsson, and V. J. Bye. 1984. Control of spawning time in rainbow trout, *Salmo gairdneri,* using constant long daylengths. Aquaculture, 43: 225–233.

Shaw, W. N. 1960. A fiberglass raft for growing oysters off the bottom. Prog. Fish-Cult., 22: 154.

Shaw, W. N. 1962. Raft culture of oysters in Massachusetts. Fish Bull., 61: 481–495.

Shaw, W. N. 1968. Raft culture of oysters in the united States. pp. 5–31, In: T. L. Linton (Ed.). Proceedings of the Oyster Culture Workshop, Brunswick, Georgia, July 11–13. Marine Fisheries Division, Georgia Game and Fish Commission, Brunswick.

Shears, M. A., G. L. Fletcher, C. L. Hew, S. Gauthier, and P. L. Davies. 1991. Transfer, expression, and stable inheritance of antifreeze protein genes in Atlantic salmon *(Salmo salar)*. Mol. Mar. Biol. Biotechnol., 1: 58–63.

Shehadeh, Z. H., and J. N. Ellis. 1970. Induced spawning of the striped mullet *Mugil cephalus* L. J. Fish Biol., 2: 355–360.

Shehadeh, Z. H., C.-M. Kuo, and K. K. Milisen. 1973. Induced spawning of grey mullet *Mugil cephalus* L. with fractionated salmon pituitary extract. J. Fish Biol., 5: 471–478.

Shetty, H. P. C. 1990. Induced spawning of Indian major carps through single application of Ovaprim-C. pp. 581–585, In: B. Hirano and I. Hanyu (Eds.). Proceedings of the Second Asian Fisheries Forum, Tokyo, Japan, April 17–22. Asian Fisheries Society, Manila.

Siitonen, L., and G. A. E. Gall. 1989. Response to selection for early spawn date in rainbow trout, *Salmo gairdneri*. Aquaculture, 78: 153–161.

Smith, T. I. J., and W. E. Jenkins. 1984. Controlled spawning of F_1 hybrid striped bass *(Morone saxatilis* × *M. chrysops)* and rearing of F_2 progeny. J. World Aquacult. Soc., 14: 147–161.

Smitherman, R. O., R. A. Dunham, T. O. Bice, and J. L. Horn. 1984. Reproductive efficiency in the reciprocal pairings between two strains of channel catfish. Prog. Fish-Cult., 46: 106–110.

Sneed, K. E., and H. P. Clemens. 1959. The use of human chorionic gonadotropin to spawn warm-water fishes. Prog. Fish-Cult., 21: 117–120.

Stanley, J. G., and K. E. Sneed. 1974. Artificial gynogenesis and its application in genetics and selective breeding in fishes. pp. 527–536, In: J. H. S. Blaxter (Ed.). The early life history of fishes. Springer-Verlag, New York.

Steeby, J. A. 1987. Effects of spawning container type and placement depth on channel catfish spawning success in ponds. Prog. Fish-Cult., 49: 308–310.

Stickney, R. R. 1991. Salmonid life histories. pp. 1–20, In: R. R. Stickney (Ed.). Culture of salmonid fishes. CRC Press, Boca Raton, Florida.

Stickney, R. R. 1992. Tilapia. pp. 81–115, In: R. R. Stickney (Ed.). Culture of nonsalmonid freshwater fishes. Second edition. CRC Press, Boca Raton, Florida.

Strickberger, M. W. 1976. Genetics. Macmillan, New York. 914 p.

Stroemgren, T., and M. V. Nielsen. 1989. Heritability of growth in larvae and juveniles of *Mytilus edulis*. Aquaculture, 80: 1–6.

Sugama, K., N. Taniguchi, S. Seki, and H. Nabeshima. 1992. Survival, growth and gonad development of triploid red sea bream, *Pagrus major* (Temminck & Schlegel): use of allozyme markers for ploidy and family identification. Aquacult. Fish. Manage., 23: 149–159.

Swift, B. D., R. G. Peterson, and A. Winkelman. 1991. Heritability and genetic correlations estimates for weight, length and survival of s_0 and s_1 coho salmon *(Oncorhynchus kisutch)*. Bull. Aquacult. Assoc. Can., 91: 19–21.

Tamaru, C. S., C. S. Lee, C. D. Kelley, and J. F. Banno. 1988. Effectiveness of chronic LHRH-analogue and 17 alpha-methyltestosterone therapy, administered at different times to the spawning season on the maturation of milkfish *(Chanos chanos)*. Aquaculture, 70: 159–167.

Tan-Fermin, J. D. 1991. Effects of unilateral eyestalk ablation on ovarian histology and oocyte size frequency of wild and pond-reared *Penaeus monodon* (Fabricius) broodstock. Aquaculture, 93: 77–86.

Tave, D., and R. O. Smitherman. 1982. Spawning success of reciprocal hybrid pairings between blue and channel catfishes with and without hormone injection. Prog. Fish-Cult., 44: 73–74.

Teichert-Coddington, D. R., and R. O. Smitherman. 1988. Lack of response by *Tilapia nilotica* to mass selection for rapid early growth. Trans. Am. Fish. Soc., 117: 297–300.

Thomas, P., and N. Boyd. 1988. Induced spawning of spotted seatrout, red drum and orangemouth corvina (family: Sciaenidae) with luteinizing hormone-releasing hormone analog injection. Contrib. Mar. Sci., 30: 43–48.

Thomas, P., and N. W. Boyd. 1989. Dietary administration of an LHRH analogue induces spawning of spotted seatrout *(Cynoscion nebulosus)*. Aquaculture, 80: 363–370.

Thorogood, J., and A. Blackshaw. 1992. Factors affecting the activation, motility and cryopreservation of the spermatozoa of the yellowfin bream, *Acanthopagrus australis* (Guenther). Aquacult. Fish. Manage., 23: 337–344.

Tipping, J. M. 1991. Heritability of age at maturity in steelhead. N. Am. J. Fish. Manage., 11: 105–108.

Toledo, J. D., H. Kurokura, and S. Kasahara. 1989. Preliminary studies on the cryopreservation of the blue mussel embryos. Bull. Jpn. Soc. Sci. Fish., 55: 1661.

Toole, M. 1951. Channel catfish culture in Texas. Prog. Fish-Cult., 13: 3–10.

Trimble, W. C., and A. P. Gaude III. 1988. Production of red swamp crawfish in a low-maintenance hatchery. Prog. Fish-Cult., 50: 170–173.

Uchida, R. N., and J. E. King. 1962. Tank culture of tilapia. Fish. Bull., 14: 21–52.

Varadaraj, K., and T. J. Pandian. 1989. First report on production of supermale tilapia by integrating endocrine sex reversal with gynogenetic techniques. Curr. Sci. (Bangalore), 58: 434–441.

Vincente, H. J., A. E. Openiano, P. L. Openiano, Jr., L. S. Valdez, and G. D. Pagalan. 1990. Maturation and spawning in capacity of *Penaeus indicus* H. Milne Edwards and *Penaeus merguiensis* de Man in experimental and commercial scale. pp. 613–616, In: R. Hirano and I. Hanyu (Eds.). Proceedings of the Second Asian Fisheries Forum, Tokyo, Japan, April 17–22, 1989. Asian Fisheries Society, Manila.

Waddy, S. L., and D. E. Aiken. 1989. Control of spawning in the American lobster: winter temperature and spring photoperiod requirements. Bull. Aquacult. Assoc. Can., 89(3): 94–96.

Waddy, S. L., and D. E. Aiken. 1990. Induction of spawning in preovigerous American lobsters, *Homarus americanus*. Bull. Aquacult. Assoc. Can. 90(1): 83–85.

Washburn, B. S., D. J. Frye, S. S. O. Hung, S. I. Doroshov, and F. S. Conte. 1990. Dietary effects on tissue composition, oogenesis and the reproductive performance of female rainbow trout *(Oncorhynchus mykiss)*. Aquaculture, 90: 179–195.

Wellborn, T. L., Jr., and T. E. Schwedler. 1981. Handling and hatching catfish eggs at different temperatures. pp. 4–5, In: For fish farmers. Mississippi State University, Mississippi State.

Wheeler, P. A., and G. H. Thorgaard. 1991. Cryopreservation of rainbow trout semen in large straws. Aquaculture, 93: 95–100.

Williamson, J. H., G. J. Carmichael, K. G. Graves, B. A. Simco, and J. R. Tomasso. 1992. Centrarchids. pp. 145–197, In: R. R. Stickney (Ed.). Culture of nonsalmonid freshwater fishes. Second edition. CRC Press, Boca Raton, Florida.

Wilson, R. P., W. E. Poe, T. G. Nemetz, and J. R. MacMillan. 1988. Effect of recombinant bovine growth hormone administration on growth and body composition of channel catfish. Aquaculture, 73: 229–236.

Withler, R. E. 1987. Genetic variation in flesh pigmentation of chinook salmon *(Oncorhynchus tshawytscha).* pp. 421–429, In: K. Tiews (Ed.). Selection, hybridization and genetic engineering in aquaculture. Vol. 1. Schriften der Bundesforschungsansalt für Fischerei, Hamburg.

Woods, L. C., III, J. G. Woiwode, M. A. McCarthy, D. D. Theisen, and R. D. Bennett. 1990. Noninduced spawning of captive striped bass in tanks. Prog. Fish-Cult., 52: 201–202.

Wurts, W. A., and R. R. Stickney. 1981. An hypothesis on the light requirements for spawning penaeid shrimp, with emphasis on *Penaeus setiferus*. Aquaculture, 41: 93–98.

Wyban, J. A., and C. S. Lee, J. N. Sweeney, and W. K. Richards, Jr. 1987. Observations on development of a maturation system for *Penaeus vannamei*. J. World Aquacult. Soc., 18: 198–200.

Yamaguchi, M., and H. Okubo. 1984. Production of ayu-fish, *Plecoglossus altivelis,* seedlings from spring spawned eggs. 1. Control of sexual maturity and spawning by photoperiodicity. Bull. Tokai Reg. Fish. Res. Lab., 114: 133–140.

Yamamoto, S., I. Sanjyo, R. Sato, M. Kohara, and H. Tahara. 1991. Estimation of the heritability for resistance to infectious hematopoietic necrosis in rainbow trout. Bull. Jpn. Soc. Sci. Fish., 57: 1519–1522.

Yamano, K., N. Kasahara, E. Yamaha, and F. Yamazaki. 1990. Cryopreservation of masu salmon by the pellet method. Bull. Fac. Fish. Hokkaido Univ., 41: 149–154.

Yankson, K., and J. Moyse. 1991. Cryopreservation of the spermatozoa of *Crassostrea tulipa* and three other oysters. Aquaculture, 97: 259–267.

Yashouv, A. 1969. Preliminary report on induced spawning of *M. cephalus* (L.) reared in captivity in freshwater ponds. Bamidgeh, 21: 19–24.

8 Disease, Predation, and Cannibalism

MORTALITY IN AQUACULTURE

Natural mortality as a result of old age, if it happens at all, occurs only in aquacultured animals that are utilized for broodstock, since marketable individuals are often sold prior to, or within a few months after, reaching sexual maturity.[168] Adults capable of multiple spawning are often not maintained throughout their lives because fecundity tends to decline as the animals age, and some species become difficult to handle at larger sizes. Various sources of mortality other than old age have been discussed in other chapters. They include degraded water quality, nutritional imbalance, toxicants in improperly stored feed, poaching, and pollutants. In this chapter we look at the impact of disease, predation, and cannibalism on aquacultured animals.[169]

Disease organisms of interest to aquaculturists include viruses, fungi, bacteria, and parasites (including protozoans, cestodes, nematodes, and copepods). While the species present vary as a function of environment, organisms of each type listed are cosmopolitan in aquatic environments. Yet, only a small percentage of aquaculturists experience severe disease epizootics in any given year, and in most instances, the onset of those epizootics can often be attributed to some type of stress. Physical damage is one form of stress, but more common stressors are crowding, handling, transportation, and incidents of sublethal water quality degradation. Exposure to such parameters as significant changes in temperature, reduced dissolved oxygen (DO), and increased ammonia can be lethal. In many instances the problem can be corrected before mortality occurs, although often not before the animals are stressed. As a result of such stress, the immune response of the culture animals may be suppressed allowing disease organisms to proliferate (Bejerang and Sarig 1991). The result can be development of an epizootic as soon as 24 to 48 hr after the stress event or as long after that event as 2 wk.

Stress can also lead to mortality in the absence of a disease. Tomasso and Carmichael (1988) reported that transporting red drum *(Sciaenops ocellatus)* for 5 hr led to 1% immediate mortality; however the cumulative mortality 10 days after the fish were hauled ranged from 12 to 51%.

---

[168] Salmonid species that die immediately after spawning are an obvious exception.
[169] Parasitism is included in use of the term ''disease'' throughout this chapter.

Corticosteroid levels may increase in stressed aquatic animals. Cortisol levels have been shown to increase in largemouth bass (Carmichael et al. 1984) and red drum (Robertson et al. 1987, 1988) exposed to stress. Cortisol levels have been shown to rise in both blood and bile in rainbow trout exposed to stress (Pottinger et al. 1992). Fevolden et al. (1991) hypothesized that the relationship between cortisol increase and stress may have a genetic component at least in Atlantic salmon and rainbow trout. Other physiological responses to stress include elevated plasma glucose and decreased plasma chloride concentration (Carmichael et al. 1984).

Proximity of diseased animals to those that are not exhibiting signs of disease provides a pathway for spread of the vector. Animals that are moribund or dead should be removed from the culture system. That will not necessarily prevent spread of the disease, but it will reduce the density of the disease organism to some extent and thereby reduce the level of exposure that the healthy animals experience.

The issue of proximity of diseased or potentially diseased animals to healthy ones has also arisen in conjunction with the net-pen farming industry. Environmentalists have raised the possibility that wild fish will become infected with diseases borne by cultured fish in net pens, while net-pen culturists have expressed the fear that their animals can become infected by pathogens carried by wild fish. Documented examples of disease transmission in either direction are rare as discussed by Brackett (1991). Most of the instances that have been documented have involved parasites. There have been documented incidents of transmission of bacterial kidney disease, furunculosis, and the infectious pancreatic necrosis (IPN) virus from wild to farmed salmon.

Disease epizootics are uncommon in water systems that are well managed and in which the animals are not stressed. Animals in aquaculture systems may become stressed even when good management practices are in place. As indicated above, handling is a stressor. The capture of fish for sorting, adjustment of feeding rates, or for other purposes imposes stress on the animals that can lead to a disease epizootic. Even feeding can cause stress. For example, if the water temperature becomes unusually high during a summer day, the animals may be reluctant to eat and the food will then add to the biochemical oxygen demand in the culture system and serve as a substrate for the growth of microorganisms. Aquatic species in outdoor systems commonly experience rapid fluctuations in water temperature that provide a form of stress that is largely unavoidable. The aquaculturist should constantly be on the alert for stressors and be prepared to face an epizootic as a result of stressful conditions.

Each time animals are exposed to stress, they should be carefully watched for at least 2 wk to ensure that an epizootic is not developing. Reluctance to eat, unusual behavior, discoloration of the integument, and lesions are signs of developing disease problems. Proper diagnosis and treatment must be made as soon as a problem is observed. Some culturists employ prophylactic disease treatments either routinely or after each stress event. Because of the cost of chemicals for disease treatment, the paucity of approved drugs (discussed in a later section of this chapter), and the potential for development of disease organisms that are resistant to the available drugs, most fish health experts recommend treatment only after a disease outbreak has been confirmed. In many instances it is possible to rely on good animal husbandry practices to avoid stress and resulting epizootics.

Often disease epizootics are caused by a single organism. Secondary infections can occur, however. As the number of disease organisms expands, so will the treatment protocols and required treatment chemicals. Since drugs used for disease treatment are sources of stress themselves, the problems may become exacerbated to the point that massive mortalities cannot be avoided. The following example is illustrative of that point.

In 1970, a closed culture system at the Skidaway Institute of Oceanography in which channel catfish *(Ictalurus punctatus)* averaging about 1 kg were held at a density of about 600 g/l. The fish were healthy under those conditions for several months, then a power failure caused the biofilter to become anaerobic for a period of time resulting in exposure of the fish to a high concentration of ammonia. Several days later, an epizootic of a protozoan parasite (*Costia* sp.) occurred. Successful treatment of the parasite was effected with formalin, but the stress of the disease coupled with treatment stress led to epizootics by a variety of organisms about a week later. All of the fish succumbed to the secondary infestations.

Many aquaculturists attempt to diagnose disease and initiate treatment protocols themselves. In matters of fish and shellfish diseases one cannot carry the affected animals to the local veterinarian because few vets have received any training in aquatic animal diseases. Specialists in fish diseases can be found on the campuses of universities (primarily the Land Grant universities) in states where aquaculture is an important agricultural activity. Some state and federal laboratories also provide diagnostic services on fish diseases, though the fish farmer may lose a crop while waiting for fish shipped to such a laboratory to return a diagnosis. Each fish farmer should have at least some training in recognizing diseases that commonly occur in the species being reared. Further, each aquaculturist should make an effort to remain current on the status of chemicals that have been approved for use in disease treatment. While some nations have no regulations in this regard, others are quite strict and violations can carry heavy fines and even imprisonment.

Frequent examination of each culture chamber is important if disease epizootics and the potential for their development are to be recognized. Leaving culture chambers untended for several days while relying on automatic feeders and remote monitoring of water quality can be a poor decision if an epizootic develops beyond the stage where effective treatment can be implemented. Part of good management is on-site observation, in person, on a frequent basis.

SANITATION

When an epizootic is detected, every attempt should be made to contain it. The careless transfer of equipment from one pond to another without treating that equipment with antiseptic can lead to the spread of an otherwise localized disease. Such items as dip nets, seines, feed pails, and even the bodies and clothing of personnel who work around the aquaculture animals should be sanitized after exposure to diseased animals, their culture chambers, associated equipment, and the water in which they live. It is also a good idea to sanitize equipment between uses even if no epizootic is occurring. Keeping nets and other such items in disinfectant solutions of chlorine, formalin, merthiolate, or one of various commercial solutions will help prevent the spread of disease. Assigning individual nets, feed pails, and other small items to a specific pond, tank, or raceway will also help eliminate the spread of disease. If strong disinfectant solutions are used, the nets and other gear should be rinsed thoroughly before being used.

Large items such as seines can be soaked in vats of disinfectant, though this may not always be practical. Alternatively, seines can be rinsed with uncontaminated water, then dried in the sun between uses. Caution should be taken with monofilament nylon nets since they are subject to degradation when exposed to direct sunlight. Such nets are not bulky and will often fit in a vat for treatment with a disinfectant. A spray treatment with chlorine solution or some other type of disinfectant may also be effective.

Some laboratories and even commercial facilities, particularly hatcheries, have taken what might appear to be unusual steps to ensure that disease organisms are not inadvertently introduced. In some cases there are restrictions on who can enter a hatchery or research facility. In others, visitors and workers are required to walk through an iodine bath to disinfect their shoes. The next step is to wear clean coveralls over street clothing, put on a hat and exchange street shoes for rubber boots when entering the facility. At the most extreme, different colors of coveralls are worn in different rooms and no one is allowed in a room unless they are wearing the appropriate color. Such steps, along with careful disinfection of all supplies and equipment, prevent the introduction and spread of disease organisms.

TREATMENT AND PREVENTION METHODOLOGY

In the past, with the exception of prophylactic chemical use, control of diseases in aquaculture animals was largely reactive; that is, treatment was effected when a problem was detected. During the 1980s effective fish vaccines were first developed against certain pathogenic organisms including some viruses and bacteria.

Vaccines

Vaccines can be administered in various ways. Each individual animal can be injected with the vaccine, the vaccine can be administered orally, or it can be either absorbed either after immersion of the animals in water containing the vaccine or by spraying it on the body surface. The degree of protection varies from one mode of administration to another as has been demonstrated in Atlantic salmon *(Salmo salar)* by Hjeltnes et al. (1989). In general, injection is the most effective method of adminsitration. Immersion sometimes works, though variable results have been obtained. Administering vaccines orally can be effective so long as the vaccine is not destroyed by digestive enzymes. For example, an enteric-coated vaccine against the bacterium *Vibrio anguillarum* was developed for use in salmonids by Wong et al. (1992).

In addition to efficacy of treatment, cost is an important factor with respect to the administration of vaccines. Individual injection of fish requires a significant amount of labor and is only cost-effective in conjunction with fish that have high individual value. Lillehaug (1989a) indicated that injection of salmonids can be cost-effective if used on fish larger than about 40 g. Significant monetary savings are associated with the relatively small amount of vaccine required for injection as compared with immersion. A review of delivery methods and cost-effectiveness associated with vaccines has been published by Dunn et al. (1990).

A number of vaccines have been developed and tested to date. Among them are those designed to protect rainbow trout *(Oncorhynchus mykiss)* against the furunculosis bacteriim, *Aeromonas salmonicida* (Rodgers 1990), and against *Vibrio* (Lillehaug 1989b). Atlantic salmon *(Salmo salar)* have been vaccinated against the coldwater vibrio, *Vibrio salmonicida* (Lillehaug 1990, 1991). A vibriosis vaccine has also been developed to protect ayu, *Plecoglossus altivelis,* as described by Itami and Kusuda (1980a, b). A vaccine to prevent the channel catfish virus was developed and tested by Awad et al. (1989) for use on *Ictalurus punctatus.*

Invertebrates of aquaculture interest have also been involved in the process. A vaccine has also been developed to protect the penaeid shrimp *Penaeus japonicus* against vibriosis (Itami et al. 1989). The bacterium *Aerococcus viridans,* causes a disease in the American lobster, *Homarus americanus,* called gaffkemia. An effective injectable vaccine against gaffkemia was described by Keith et al. (1988).

A great deal of progress in the development of vaccines for use on aquatic animals has been made, but much has yet to be accomplished. Many of the most devastating viruses have thus far resisted attempts by researchers to develop effective vaccines, and many bacterial diseases continue to be treated with antibiotics, not vaccines. A significant amount of research is underway, and breakthroughs can be expected.

Isolation and Environmental Manipulation

When an epizootic has been detected or is anticipated, the first step that should be taken is to isolate, insofar as is possible, the affected individuals or groups of animals. In instances where the water supply to each culture system is separate, achieving isolation is possible. As indicated above, once a group of animals or culture chambers has been isolated, it is important to avoid contamination through transfer of disease organisms on nets, feed scoops, and other equipment.

Once the affected animals have been quarantined, the physical environment should be adjusted for the treatment method that has been selected. For example, if a static-water bath treatment is to be used in a pond, tank, or raceway, inflow water should be turned off. In tanks and raceways and sometimes in ponds, supplemental aeration should be provided so oxygen depletions can be avoided during the treatment period. In running-water bath treatments, the flow rate should be adjusted to ensure that the flush rate is appropriate to provide sufficient residence time for the chemical to be effective.

In some cases it is possible to effect treatment through manipulation of the culture environment by altering either a physical or a chemical factor. Certain parasites, such as the protozoan *Ichthyophthirius multifiliis* (commonly known as Ich), can be controlled by increasing or reducing temperature or by increasing salinity. That protozoan also responds to various treatments with harsher chemicals, but the two methods mentioned are often effective and place less stress on the culture animals.

Another nonchemical treatment method consists of interrupting some portion of the life cycle of a disease organism. Various trematodes that infect fish have rather complex life cycles, usually involving one or more intermediate hosts, one of which is often a snail. If snails or their larvae can be eradicated from the culture system, the life cycle of the parasite can be broken. Similarly, birds of many kinds, in addition to being predators on aquatic animals, carry diseases that may infect fish. Also, birds that consume wild fishes can subsequently transfer diseases by means fecal droppings to fish in culture ponds, thereby establishing the potential for epizootics.

Treatment techniques vary to some extent depending on the type of culture system being used, particularly when a treatment chemical is to be added directly to the water. Differences among water systems with respect to disease treatment were discussed in Chapter 3. Recall that care must be taken in treating recirculating water systems to avoid destruction of the microflora required for proper functioning of biofiltration devices. Treatment of open raceways or tanks may be accomplished with or without first reducing or stopping the flow of water, although different amounts of chemical are required to effect control depending on which technique is used. In any intensive culture situation, provision for aeration must be made when flow is curtailed for any length of time. Cages may be treated in a variety of ways (Chapter 3), and ponds can be treated by adding chemicals to the water under static conditions. Pond levels may have to be reduced to some extent to conserve on chemical quantity (which can present a major expense to the culturist).

When drugs are added to feed, no special precautions are required to protect water quality. Definitive research on the effects of antibiotics in the feed on biofilter microflora remains to be conducted. Opponents of net-pen culture have expressed the fear that the use of antibiotics in feed will destroy useful bacteria in the natural environment around and under the net pens and lead to antibiotic-resistant strains of bacteria. Those and other concerns associated with net-pen salmon culture have been discussed by Stickney (1988). Since that review, additional information on the rate of degradation of the antibiotic oxytetracycline in seawater and in sediments has become available (Samuelsen 1989).

Treating large areas of water can be difficult. Small ponds (e.g., those 0.5 ha or less) can generally be treated from the bank. Chemicals are usually dissolved or diluted in water and dispersed as evenly as possible over the surface of the pond by broadcasting them in some manner. Buckets, hand-operated sprayers of the type used to apply herbicides and pesticides, and long-handled dippers have been effectively utilized to treat small ponds.

It is often difficult to reach all areas of large ponds from the bank. In those cases chemicals may be diluted with water and poured into the wake of an outboard motor mounted on a boat that covers as much of the surface area as possible while the chemical is being distributed. If the pond has relatively deep water in one or more areas, more chemical should be utilized in those regions to ensure even distribution. Aircraft may be employed to spray treatment chemicals over very large ponds, though the culturist should be certain that the spray tanks and nozzles of crop-dusting planes have been thoroughly cleaned of any residual toxic compounds before being used for the spreading of treatment chemicals.

Topical treatments are sometimes effective but are often impractical in large aquaculture operations because of the logistical problems associated with capturing infected animals from large, basically healthy populations, and applying a chemical to local infections. In most instances a whole culture chamber will receive treatment if a disease has been found to affect any portion of the animals in that chamber.

Bath treatments are of three general types. Pond treatments (described above) can be thought of as long-term or indeterminate baths, since dissipation of the chemical is not by dilution but through degradation, which may require several days or even weeks. Extended baths require several hours and are conducted in static water that must be replaced with new water after treatment is complete. Such baths may be conducted in ponds or smaller culture chambers. Aeration is necessary in linear raceways, tanks, and cages if extended baths are utilized, since water flow must be curtailed for the duration of the bath.

Short-term treatments of large numbers of animals can be accomplished through dips or flushes. Dip treatment usually involves capturing the animals and dipping them, in groups or individually, in one or more treatment chemicals. Dips can last from a few seconds to a few minutes and usually involve fairly strong chemical concentrations. Following dip treatment the culture animals are returned to their culture chambers. In flush treatments the chemical is added directly to the culture chambers but the water is not turned off. The concentration of chemical used is higher than in static baths, but since the water continues to run, the chemical will be rapidly diluted and flushed from the culture chambers.

Oral treatment of diseases involves the incorporation of drugs into feed. This can be an effective means of treatment, and in fact, some diseases cannot be effectively treated in any other manner. The method requires consumption of treated feed by the culture animals. Since cessation of feeding is an early sign of an epizootic, it is important for the culturist to diagnose and begin treatment before the animals refuse to feed.

TABLE 37. Compounds Approved for Use in Disease Treatment and for Disinfection by the U.S. Food and Drug Administration (Meyer and Schnick 1989)

Chemical	Usage	Comments
Sodium chloride	Osmoregulatory enhancer	No withdrawal period
Vinegar (acetic acid)	Parasiticide for fish	No withdrawal period
Formalin	Used to treat parasites in various species of fish and eggs	No withdrawal period
Copper (elemental)	Antibacterial in penaeid shrimp	No withdrawal period
Oxytetracycline (Terramycin)	Antibacterial in fish	21 days withdrawal
Sulfadimethoxine and ormetoprim (Romet-30 and Romet-B)	Antibacterial for use in salmonids and catfish	6 wk withdrawal for salmonids, 3 days for catfish
Calcium hypochlorite	Disinfectant and sanitizer in culture tanks and equipment. Also used in algae control	No withdrawal time established
Povidone-iodine compounds (Betadine, Wescodyne, etc.)	Disinfection of eggs	No withdrawal required
Quaternary ammonium compounds (Hyamine 1622, Hyamine 3500, Roccal)	Disinfecton of water, equipment, and culture chambers	Exempted from registration by FDA; no withdrawal required

Chemical Control of Disease

In some countries there is no control on the use of drugs to treat aquatic animal diseases, though there have been recommendations for banning the use of certain chemicals in conjunction with aquaculture even in developing countries (Anonymous 1993). In the United States the use of drugs, herbicides, pesticides, fish toxicants, and a variety of other chemicals on fish and other aquatic animals is regulated by the U.S. government through the U.S. Food and Drug Administration (FDA) and the U.S. Environmental Protection Agency (EPA). Evaluation of drugs by the FDA is an ongoing process. For a period early in the 1990s, even sodium chloride was banned for use in treating disease. Some chemicals can be used on fish other than foodfish. Sportfish come under the category of nonfood fish even though they may ultimately be caught and consumed by humans. Table 37 presents information on treatment chemicals approved for use in U.S. aquaculture as of 1989.

The costs involved in clearing a chemical for use in the treatment of disease in humans, terrestrial livestock, and aquatic animals (among others) is enormous. Approval can require not only millions of dollars but also a number of years of data collection. In most instances involving drugs that might be of use in aquaculture, the time and expense are unwarranted because the ultimate market for the drug would not be sufficiently great.

Not only should aquaculturists exclusively use approved chemicals in treating diseases; they should be meticulous in following dosage recommendations and not treat for longer than recommended periods. Prophylactic treatment is not recommended, particularly with respect to antibiotics since disease-resistant bacteria may be developed, rendering the drugs useless in future epizootics. Withdrawal times should also be scrupulously followed.[170]

In the sections on specific types of disease that appear below, drugs other than those approved for use in the United States are sometimes mentioned. Those drugs have been used experimentally in the United States, were once used legitimately but are no longer approved for use (and in some cases are no longer available), or are used in nations (other than the United States) that have less rigorous governmental control. Mention of drugs other than those in Table 37 does not represent endorsement of their use by the author. As previously stated, only approved drugs should be used in the treatment of diseases of aquaculture animals.

Calculation of Treatment Levels

Before the amount of chemical to be added to a pond, tank, or raceway can be calculated, the volume of the culture chamber should be determined. In circular tanks this is readily accomplished by multiplying the depth of the chamber by πr^2 (where r is the radius of the circle). Similarly, in square or rectangular tanks or raceways, the volume can be calculated by multiplying length by width by depth. The calculation of pond volumes is somewhat more involved, especially when shape is irregular and depths vary considerably from one part of the pond to another. In large ponds with flat bottoms it is often sufficient to multiply the surface area by depth at the base of the levee to determine volume. However, in small ponds, the volume of water lost as a function of bank slope can be significant. If not taken into consideration, the excess treatment chemical that would be used could be toxic. In small ponds the volume should be calculated more precisely by subtracting the portion lost under the slopes of the levees from the volume assumed from total surface area. This calculation is simplified if all pond banks have the same slope. In most ponds great precision in measurement is not required since slight errors will not appreciably influence the amount of chemical added, and there is usually some latitude between therapeutic and lethal doses of the drugs currently in use.

Most ponds constructed in the United States are laid out in acres rather than hectares; thus volume is often measured in acre-feet. However, treatment levels are usually presented in milligrams per liter (mg/l) or parts per million (ppm). It is often convenient to convert pond volumes to the metric system for calculating treatment levels. The conversion units presented in Appendix A can be helpful in making those calculations.

As an example of a treatment level calculation, assume that you have a 0.5 acre pond with a mean depth of 3 ft. You want to treat the pond with a chemical at the

---

[170] For some chemicals, a minimum period of time must be allowed to pass after the drug treatment has been discontinued and before the animals are harvested and marketed.

rate of 5 mg/l. How many kilograms of the chemical should be added to obtain the proper dosage?

0.5 acre × 3 ft	=	1.5 acre-ft
1.5 acre-ft × 43,560 ft³/acre-ft	=	65,340 ft³
65,340 ft³ × 7.5 gal/ft³	=	490,050 gal
490,050 gal × 3.8 gal/l	=	1,862,190 l
1,862,190 l × 5 mg/l	=	9,310,950 mg
9,310,950 mg / 1,000,000 mg/kg	=	9.3 kg

The example can be simplified to some extent if the transition from the English to metric system is made at an earlier stage:

0.5 acre × 3 ft	=	1.5 acre-feet
1.5 acre-feet × 1233.51 m³/acre-ft	=	1850.27 m³
1850.27 m³ × 1000 l/m³	=	1,850,270 l
1,850,270 l × 5 mg/l	=	9,251,350 mg
9,251,350 mg/1,000,000 mg/kg	=	9.3 kg

The precise number of milligrams derived in the two methods differed considerably (9,310,950 vs. 9,251,350), yet rounding the final number to the nearest 0.1 kg leads to the same treatment dosage (9.3 kg). The difference between the two answers at the milligram level is associated with the imprecision of the conversion factors employed. In the treatment of large volumes of water, minor differences are unimportant in terms of the effectiveness of treatment or potential overdose toxicity. Of more serious consequence are mathematical errors that may result in gross overdosage or undertreatment.

In fairly large bodies of water it is generally sufficient to figure the level of chemical addition to the nearest 0.1 kg, though in aquaria, small culture chambers, or small containers used for dip treatments, chemical dosage should be measured to the nearest 0.1 g or even the nearest milligram. Common sense should be used when chemicals are being measured. For example, if 50 kg of a substance is to be added to a pond, it would be foolish to measure to the nearest milligram. On the other hand, if a small tank is to receive 10 mg of a chemical, an error in measurement of only 1 or 2 mg could result in underdosage (an no response) or overdosage (with perhaps lethal results).

If ponds are constructed on the basis of hectares, or if the culturist converts acreage to hectares for purposes of calculating dosages, the procedure can be even further simplified from that presented above. For example, if an aquaculturist wishes to add 5 mg/l of a chemical to a pond that is 0.25 ha in area and has an average depth of 0.5 m, the following calculation can be made:

0.25 ha × 0.5 m × 10,000 m²/ha	=	1,250 m³
1,250 m³ × 1000 l/m³	=	1,250,000 l
1,250,000 l × 5 mg/l	=	6,250,000 mg
6,250,000 mg/1,000,000 mg/kg	=	6.3 kg

EXAMINATION OF ANIMALS FOR DISEASE

Bacterial characterization in aquatic animals, as in terrestrial organisms, requires isolation of the bacteria from an infected animal, culture of the bacteria in one or more types of nutrient medium, and additional procedures such as staining. Once the bacteria have been identified, colonies can be challenged with various antibiotics to determine sensitivity prior to treatment. Because of the ease with which contamination can occur, good sterile technique is required throughout the necropsy procedure, incubation of cultures, and subsequent examination of culture results.

All materials that come in contact with the cultures must be sterilized. Personnel conducting bacteriological examinations should wear clean laboratory-type clothing, wash their hands before and after each examination, and not smoke or drink in the laboratory. Extreme caution must be taken to ensure that any bacteria that grow on the culture media were obtained during necropsy, not introduced through contamination.

Amlacher (1970) stressed that fishes should be examined and cultured for external bacteria before they are opened or otherwise cut in any way. External cultures can be obtained from swabs of the integument and gills. The swabs are subsequently used to streak culture plates. Care must be taken in collecting the animals so that contamination is avoided. Live animals can be transported in clean ice chests or plastic bags partially filled with water from the culture chamber in which the fish were being reared. Dead or moribund animals can be placed in sterile plastic bags and transported on ice.

When bacterial cultures of internal organs are taken, the animals should be humanely killed and swabbed with alcohol or other disinfectant before incisions are made. Samples from the liver, spleen, heart, and kidney can be obtained with a sterile loop, which is then used to streak culture plates.

Most pathogenic bacteria present on fish stain gram-negative. Identification is often possible through a combination of staining and examination of the shape and color of the colonies produced on the culture plates, along with determination of whether colonies will grow on certain kinds of culture media.

Some aquaculturists undertake their own diagnostic work or have personnel working for them who can conduct such work. Most have to rely on others for microbiological testing. Some state and federal laboratories provide at least limited diagnostic services, and a few universities will also perform those services. Where diagnostic services are available is typically related to regions where aquaculture is an important industry. The aquaculturist who is not located in the center of a major production region may not have the services available.

Even when bacterial diagnostic services are available, whether on site or through some public or private laboratory, the culturist does not typically have the luxury of waiting for the results of microbiological testing before effecting treatment. Bacterial diseases of fish often take the form of fin, gill, or integument erosion; the presence of white patches; or the formation of what appear to be boils on the outer body surface. There are internal bacteria that do not produce external lesions but may be diagnosed by examination of internal organs for discoloration or hemorrhaging, or by changes in behavior. When an epizootic is predicted, as might be the case when a few animals begin to behave in a peculiar manner or develop lesions, treatment should be initiated immediately, even if confirmation of the type of bacteria may not be forthcoming for a day or more. Given the limited number of antibiotics available to aquaculturists, at least in the United States, verification of which drug will be most effective may be irrelevant. Lack of action until the etiology of the disease is known may result in mass mortality that could have been prevented.

When examining fish or invertebrates for external and gill parasites, a fresh specimen should be obtained so that living parasites will be present. Not only do parasites often drop off dead animals, making the chances of finding the cause of an epizootic difficult, but identification and location of parasites is facilitated when living specimens of the aquaculture animals are available. If an animal must be transported for some distance before being examined, it may be necessary to preserve it in formalin. However, since parasites may be distorted in formalin, identification may be difficult.

As a first step in examination of animals for parasites, gills and fins, or parts thereof, should be removed and placed under a cover glass on a microscope slide. A drop of water should be added. Body scrapings obtained with a scalpel can be handled in the same manner. Low-power microscopic examination of the slide will often be sufficient for identification of helminths, crustaceans, and large protozoan parasites. Higher magnification may be required for small protozoans.

When the blood of fishes is examined for protozoans, the method of Strout (1962) has worked well. A few drops of blood are placed in a vial and allowed to clot. Then a drop of the clear serum is pipetted onto a microscope slide and examined under high power. In another method a drop of whole blood is placed on a microscope slide, allowed to dry, and stained with either Giemsa's or Wright's stain (Kudo 1954) prior to examination under the microscope.

Fish blood can be obtained in several ways. For fingerlings it is usually best to sever the tail across the caudal peduncle and allow the blood to drop into a collection vial. In larger fish, sacrifice is not usually necessary. A needle and syringe can be used to obtain blood by either heart puncture or from the caudal vein. Obtaining blood by heart puncture requires more practice than does caudal vein puncture, but both methods are effective and neither permanently damages the fish. In the caudal puncture technique the fish is placed with the ventral side uppermost and the needle is inserted into the midline of the caudal peduncle until it comes in contact with the vertebral column. The needle is then withdrawn slightly, and, if has been properly placed, it is then in the caudal vein.

When the external examination for parasites is complete, the animal may be opened and the internal organs (especially the intestines, mesenteries, liver, gonads, kidneys, gall bladder, and urinary bladder) examined. Small pieces of tissue can be placed on microscope slides and examined in the manner indicated above.

SOME COMMON AQUACULTURE DISEASES

A large variety of diseases occur in aquatic organisms. Disease organisms are cosmopolitan in the aquatic environment, and while many species that have troubled aquaculturists are rare or even unreported in wild populations, they do exist. Often, no one had ever looked for them, and they only became interesting to aquatic health specialists when epizootics began to occur in captive populations.

There are literally thousands of diseases that could impact aquaculture species. It is certainly not within the scope of this book to cover even a fraction of those that have impacted aquaculture to date. Some pathogenic organisms attack a wide variety of species, while others are highly host-specific. New diseases affecting aquaculture animals are being described each year, so any treatment of the subject is bound to be incomplete.

As an introduction to aquatic animal diseases, the approach taken here is to describe some of the common diseases that have posed problems to aquaculturists. The information presented should provide the reader with at least an indication of the array of diseases that have caused problems and how those diseases can be handled.

Nutritional diseases represent a group of disorders that are not caused by invasive organisms. Included are vitamin deficiencies or excesses, amino acid imbalances, and insufficiencies in essential fatty acids. Those topics have been covered in Chapter 6 and will not be reconsidered here.

Viral Diseases

Channel Catfish Virus Disease (CCV). CCV is the only known pathogenic virus disease that affects channel catfish. The disease was first reported in 1968 (Fijan 1968, Fijan et al. 1970). Epizootics may result in losses as high as 95% among fry and fingerlings (Plumb 1971a). The virus attacks fish smaller than about 15 cm (Anonymous 1981). Larger fish are either carriers or immune. Bowser and Munson (1986) examined serum antibody titers of catfish adults that had survived epizootics and concluded that the fish could carry the disease for as long as 4 yr after exposure.

The agent responsible for CCV has been characterized as a herpesvirus (Wolf and Darlington 1971). It has been isolated from fingerling catfish during epizootics (Plumb 1972) and can be maintained in tissue culture using brown bullhead cells. It was first suspected that channel catfish broodstock carried the virus and that those fish transmitted the disease by way of the reproductive cells and/or the fluids associated with reproduction (Wellborn et al. 1969, Plumb 1971b). Until the advent of molecular biological techniques and their application to fish in recent years, it was not possible to verify the presence of carrier adults in catfish populations. Nusbaum and Grizzle (1987a) questioned the vertical transmission of CCV from male channel catfish reproductive fluids when they could not demonstrate adherence of the virus to catfish sperm. Wise et al. (1988) used a nucleic acid probe to find the virus in adult catfish that were subsequently mated. The offspring of those matings tested positive for the virus, thereby verifying vertical transmission of the disease by at least females if not by both sexes.

It has also been demonstrated that infected fish can transmit CCV to other fry and fingerlings through the water (Plumb 1972). The virus enters by way of the gills and becomes concentrated in the gut and liver within 48 hr (Nusbaum and Grizzle 1987b). The virus can also be isolated from other organs in infected fish. Plumb (1971c) found high levels of viral activity in the kidneys.

Mortality from CCV may occur as early as 32 hr following infection (Plumb 1971a). Affected fish may swim erratically or hang vertically in the water column with the head uppermost. Other signs of the disease include distention of the abdomen with the presence of fluid in the peritoneal cavity; exophthalmia; anemia; hemorrhaging of the gills, fin bases, skin, kidneys, and other internal organs; and absence of food in the intestines (Plumb 1971a, 1972). Epizootics may be influenced by poor culture conditions; e.g., low DO, high temperature, crowding, secondary infections, and improper handling (Plumb 1971a, 1973). The disease strikes when the water temperature is above 20°C (Anonymous 1981).

The signs of CCV described above are among those found in a variety of other diseases. Fluid in the peritoneal cavity is a common sign of internal bacterial infections, and empty intestines are not uncharacteristic of any disease that causes fish to refuse food. It is probable in the years immediately following recognition of the disease that misidentification of CCV led to the destruction of fish that were exhibiting a treatable disease. Elimination of CCV for several years called for the destruction of all broodstock associated with an epizootic. The only methods of identifying exposed stocks that were immune was by evaluating their previous history and examining them for the presence of a serum-neutralizing antibody to CCV (Amend and McDowell 1983). Now it is possible to detect the virus through employment of the polymerase chain reaction (Boyle and Blackwell 1991).

Fingerlings that survive a CCV outbreak can be reared to market size but should not be maintained for broodstock. It appears that a high level of immunity exists in fish that do not contract the disease during an epizootic (Heartwell 1975), and immunity has been shown to persist for at least 2 yr (Hedrick et al. 1987).

Injection of serum from catfish that have CCV neutralizing activity into juveniles has been shown to impart a high level of immunity (Hedrick and McDowell 1987). Major drawbacks are that the fish must be large enough to be injected (thus susceptible to the disease for at least several weeks after hatching) and must be handled and injected individually. A vaccine that can be administered through immersion of catfish eggs or fry has been developed (Awad et al. 1989). Vaccinating the eggs or fry at 1 wk and then revaccinating both groups at 2 wk led to survivals to CCV exposure of 81 and 89%.

When CCV occurs, all equipment exposed to diseased fish should be properly disinfected, and all culture chambers, including ponds, should be sterilized after fish with CCV have been maintained in them (Plumb 1972). Chlorine solutions appear to be effective in sterilizing ponds (Fijan et al. 1970).

Infectious Hematopoietic Necrosis (IHN). One of several viral diseases of salmonids, IHN is particularly important because of its virulence and the fact that no reliable treatment has been developed. The species primarily affected are Pacific salmon (Amend et al. 1969). It has caused significant mortality in chinook and sockeye salmon. Coho salmon have also been found to carry the disease. LaPatra et al. (1989a) found that chinook salmon and rainbow trout alevins were highly susceptible to the virus obtained from adult coho and chinook salmon.

The disease is caused by a rhabdovirus and is typically seen in young salmonids being produced in culture. The disease typically affects alevins, fry, and early juveniles. LaPatra et al. (1990) found that rainbow trout as large as 7.2 g could exhibit the disease. Originally confined to the Pacific Northwest of the United States, IHN was spread, probably due to transfer of infected eggs to various other states, as well as to Japan and Europe (reviewed by Plumb 1993).

During epizootics, mortalities in affected populations of fry and fingerling salmonids may approach 100%. Most epizootics occur when the water temperature is at or below 12°C, though a few outbreaks have been reported at higher temperatures (Wolf 1988). Amend (1970) reported that increasing temperature when an epizootic occurs can reduce the development of the disease, though surviving animals will be carriers.

Fish affected with IHN swim lethargically and then display intervals of erratic swimming or swimming in spirals (Amend et al. 1969). The fish have difficulty maintaining their equilibrium and position in flowing water. Other signs of the disease may include exophthalmia, pale gills, hemorrhaging at the base of the fins, bleeding beneath the skin, and abdominal swelling. The body cavity may contain clear, pale yellow fluid, and there may be hemorrhaging in internal organs and tissues (Plumb 1993).

Winton et al. (1988) identified four different isolates of the virus and concluded that they differed by geographic origin. IHN can be detected and reliably identified with a fluorescent antibody test (LaPatra et al. 1989b).

The disease in the hatchery is typically controlled through management practices, since reliable vaccines have yet to be developed. Restrictions on the transport of fish from infected areas to other watersheds have been imposed. Any fish that are moved from an infected area have to be certified as noncarriers.

Infectious Hypodermal and Hematopoietic Necrosis (IHHN). The first outbreak of IHHN disease in the United States occurred in Pacific blue shrimp, *Penaeus stylirostris,* introduced from Costa Rica and Ecuador (Lightner et al. 1983a, b). Those authors found that the virus could also effect *P. vannamei* and *P. monodon.*

Confirmation that IHHN was a viral disease was confirmed by Bell and Lightner (1984). A method for isolating and purifying the virus was outlined by Lu et al. (1989).

Lightner et al. (1986) indicated that IHHN is one of at least six viruses that affect cultured penaeid shrimp. They also added to the list of species that can contract the disease. Included are the commercially important shrimp of the United States Gulf

Figure 99. *Paralichthys* sp. with lymphocystis nodules prominent around the mouth and on the caudal fin.

of Mexico and southeast Atlantic coasts; *P. aztecus, P. duorarum,* and *P. setiferus.*

The disease has occurred widely in North and South America, Asia, and the Pacific Basin. IHHN seems to have greater impact on small shrimp as compared with larger animals (Bell and Lightner 1987). IHHN and other shrimp viruses have led to catastrophic losses in some instances, while in others the effects of the disease have been insignificant to moderate (Lightner et al. 1988). Freshwater shrimp are also susceptible to viral disease (Anderson et al. 1990), though apparently not to IHHN.

Lymphocystis. Perhaps the most common and best-known viral disease affecting a wide variety of marine and estuarine fishes is lymphocystis. This disease also occurs in fresh water but has not been a significant problem to culturists except those working in saline environments. According to Menezes et al. (1987), nearly 100 species of marine and freshwater fish species have been observed with signs of the disease. Lymphocystis generally occurs as whitish nodules on the fins, head, and sometimes the body of fish (Figure 99). The nodules are caused by the increase in size and encapsulation of connective tissue cells (Sindermann 1970c).

The viral origin of lymphocystis was suspected long before the etiology of the disease was provided by Walker (1962) and Walker and Wolf (1962). Lympho- cystis is not generally fatal unless it forms around the mouth of the fish to an extent that greatly interferes with the ingestion of food, but the disease is highly contagious (Sindermann 1970c).

Under culture conditions lymphocystis spreads rapidly and can have an adverse effect on consumer acceptance of some species. For example, flounders sold with the head and fins intact would be hard to market if those areas were infested with lymphocystis nodules. The disease has been reported from tank-cultured flounders of the genus *Paralichthys* (Stickney and White 1974) and appears to have entered the culture facility in the water since the problem did not recur after a UV sterilization device was added to the incoming water treatment system. Fish within the same genus have also been infected with the virus in Japan (Tanaka et al. 1984).

Paperna et al. (1982) observed a low incidence of lymphocystis in sea bream *(Sparus aurata)* reared in sea cages. Heavy infestations occurred when infected fish were transferred to onshore holding tanks, though the infection disappeared within one month. Infection was not promoted during subsequent transfers from cages to tanks. Lymphocystis is the only viral disease reported thus far from tilapia (Roberts and Sommerville 1982).

There is no known cure for lymphocystis. Fish affected with the disease should be destroyed to prevent its spread. Culture chambers and equipment should be sterilized with chlorine solutions following removal of the fish.

Bacterial Diseases

The gram-negative bacteria *Aeromonas hydrophila, A. salmonicida, Edwardsiella ictaluri,* and *Pseudomonas fluorescens* are among the species that produce diseases variously known as hemorrhagic septicemia, infectious abdominal dropsy, red mouth disease (Bullock and McLaughlin 1970), and enteric septicemia (Plumb and Sanchez 1983). *A. salmonicida* is found primarily in salmonids and leads to the development of open sores resembling boils, hence, the common name of the disease, furunculosis. *E. ictaluri* has been most widely observed in channel catfish, though it has also affected other species (Plumb and Sanchez 1983).

While not widely recognized by fish culturists, the transfer of pathogenic bacteria from one region to another in frozen processed fish has been recognized by some people as a possible means of transmission. In New Zealand, for example, where live rainbow trout populations have not been augmented in the 19th century, incoming frozen fish are inspected to ensure that disease organisms are not present. Brady and Vinitnantharat (1990) found that viable *A. hydrophila* could be recovered from catfish after 20 days of frozen storage. *E. ictaluri* survived 30 days, while *P. fluorescens* were recovered after 50 days from frozen fish.

Aeromonas hydrophila. The incubation period for hemorrhagic septicemia varies depending on environmental conditions and the physical condition of affected fish, but it is usually not longer than 10 to 14 days. Ventura and Grizzle (1987) found that the bacteria could enter catfish via the integument or intestinal tract. Fish with abraded skins, cultured under crowded conditions, or exposed to high temperatures showed increased susceptibility to the disease.

External signs of fish infected with *Aeromonas hydrophila* include shallow grayish or red ulcers, inflammation around the mouth, exophthalmia, and distention of the abdomen associated with the presence of bloody or slightly opaque fluid in the peritoneal cavity.[171] The kidneys may be swollen and soft, the liver pale or green,

---

[171] Note that some of the signs of this bacterial disease are the same as those previously discussed with respect to viral diseases.

and there may be blood in the intestine (Snieszko and Bullock 1968). Treatments that have been effective include feeding 50 to 75 mg/kg of feed of chloramaphenicol or oxytetracycline for 10 days (Snieszko and Bullock 1968, Anonymous 1970). Nitrofurazone can also be fed for 10 days at 90 mg/kg of feed. If the disease has progressed to the point where the fish will not feed, oxytetracycline or nitrofurazone may be added to the water (Meyer and Hoffman 1976).

Aeromonas salmonicida. Furunculosis has been known to occur in both fresh and salt water. It is primarily a pathogen of salmonids, but it has infected other freshwater and marine fishes (Fryer and Rohovec 1993). Pathology includes necrosis of the musculature and various internal organs, and hemorrhaging at the base of the pectoral and pelvic fins.

Oxytetracycline, sulfamerazine, oxolinic acid, amoxicillin, and various other drugs, along with immunostimulants, have been used to control furunculosis (Barnes et al. 1991b, Giles et al. 1991, Nikl et al. 1991, 1992, Elston 1992, Fryer and Rohovec 1993). Results have varied to some extent with fish species. Bowser et al. (1990) evaluated oxolinic acid and enrofloxacin against *A. salmonicida* in Atlantic salmon and found a lack of efficacy (which may have been related to the use of low dosages). Barnes et al. (1991a), on the other hand, found that the chemical flumequine was more efficacious than oxolinic acid. Barnes et al. (1990) evaluated a group of chemicals known as 4–quinolones and found many of them to be more effective than oxolinic acid. Tsoumas et al. (1989) cautioned against the misuse of 4-quinolone because of the potential for development of resistance to the drug. Vaccines have also been developed and tested (Adams et al. 1988, Rodgers 1990, Bricknell et al. 1991, Ellis et al. 1992, Turgeon and Elazhary 1992), though according to Fryer and Rohovec (1993) none of them is in common use.

As has been demonstrated in conjunction with other fish diseases, vitamin status may play a role in disease resistance to furunculosis. Hardie et al. (1990) found that Atlantic salmon depleted in vitamin E and subsequently exposed to the disease experienced significantly higher mortality than fish that had received the proper level of vitamin E. Atlantic salmon deficient in vitamin C have also been shown to be more susceptible to furunculosis than those that received normal or high levels of the vitamin (Hardie et al. 1991).

Enzyme-linked immunosorbent assay (ELISA) has been applied by Adams and Thompson (1990) and Bernoth (1990). The assay, which can be conducted in from 90 minutes to 4 hr, reduces the time required to accurately identify *A. salmonicida* in infected fish.

Edwardsiella ictaluri. Identification of *E. ictaluri* as an organism now known to cause enteric septicemia in catfish did not occur until the late 1970s (Plumb and Schwedler 1982), though the disease was quickly determined to be a primary cause of catfish mortality. Confirmation of *E. ictaluri* can now be reliably obtained through an indirect ELISA (Klesius et al. 1991)

From one-quarter to one-third of catfish mortalities in culture can be attributed to enteric septicemia each year (Plumb and Quinlan 1986, Beleau and Plumb 1987).

In 1988, for example, 2,456 cases of the disease were reported to the Mississippi Cooperative Extension Service's two disease laboratories located in the heart of the Mississippi catfish growing region (Durborow et al. 1991). Signs of *E. ictaluri* infections in channel catfish include inflammation of various internal organs, swelling of the gill lamellae, reductions in erythrocytes and leucocytes, and necrosis of the liver, pancreas, spleen, and kidney (Areechon and Plumb 1983, Shotts et al. 1986).

Antibiotic treatment with the potentiated sulfonamide Romet–30 has been effective at greatly reducing mortalities in diseased catfish. A dose rate of 50 mg/kg of fish appears to be effective (Plumb et al. 1987). A vaccine against *E. ictaluri* has also been developed (Plumb 1988). Liu et al. (1989) examined the relationship between high dosages of vitamin C to immunize fish and their subsequent resistance to the disease. Prior to the development of effective vaccines, Li and Lovell (1985) had shown that mortality of fish fed high levels of vitamin C were less susceptible to enteric septicemia than fish that had not received high doses of the vitamin. Sheldon and Blazer (1991) found a relationship between bactericidal activity in immunized fish and the level of n–3 fatty acids in the diet.

Flexibacter columnaris. Columnaris disease is produced in species such as channel catfish by the myxobacterium *F. columnaris* (formerly *Chondrococcus columnaris*). The disease is characterized by discoloration on the body (grayish or yellowish areas), which may develop into shallow ulcerations. The gills may also be affected, with eventual destruction of the gill lamellae and filaments (Snieszko and Ross 1969, Meyer and Hoffman 1976). High mortalities may result from the disease if it is not properly diagnosed and treated (Figure 100).

The bacteria may be transmitted by infected fish, carrier fish that are asymptomatic, and possibly through the water (Snieszko and Ross 1969). Incubation may be as brief as 24 hr, depending on the conditions under which the fish are maintained. All sizes and ages of catfish are equally susceptible (Snieszko and Ross 1969).

As a means of preventing the disease, fish should not be overcrowded, especially during warm weather. Prophylaxis includes the addition of nitrofurazone to hauling and holding tanks at 5 to 10 ppm active ingredient. Copper sulfate may be used at 1 ppm in soft water and 2 ppm in hard water to prevent or possibly control outbreaks of columnaris disease (Snieszko and Ross 1969).

For fish that have developed external columnaris infections, diquat may be used as a 4 day bath, with the best dosage being about 2 to 4 ppm of diquat cation concentration. Antibiotics added to the water may also be effective in controlling the disease. Snieszko and Ross (1969) recommended 10 to 20 ppm of chlortetracycline or 50 to 10 ppm of chloramphenicol for aquarium fishes. For systemic infections, the addition of 11 g/100 kg of dry feed of sulfamerazine or 6.6 g/100 kg of dry feed of oxytetracycline may be effective when the feed is presented over a period of 10 days. The nitrofuran compound Furanase may be effective in the form of 1 hr baths. Attempts have been made to control the disease in channel catfish through vaccination (Moore et al. 1990).

MacFarlane et al. (1986) reported that striped bass could be protected from infec-

Figure 100. Channel catfish fingerling afflicted with bacterial disease.

tion with *F. columnaris* by exposing them to trace metals at levels above normal. A mixture of arsenic, cadmium, copper, lead, and selenium at four to 10 times typical environmental concentrations provided protection from the bacteria. Increased levels of copper alone provided protection against infection and cadmium alone provided marginal protection. Increased levels of arsenic enhanced susceptibility to the disease, while increased levels of lead and selenium had no observed effect.

Vibrio *sp.* Diseases caused by species in the genus *Vibrio* affect many species of fish in the marine environment and the genus has also been implicated in outbreaks of hemorrhagic septicemia in fresh water (Ross et al. 1968, Egusa 1969). At least one species, *V. parahaemolyticus,* can affect not only fish but also humans, leading to a form of food poisoning (Bullock and McLaughlin 1970). The most common species that have been isolated from cultured marine fish are *V. anguillarum, V. ordalii,* and *V. salmonicida* (Fryer and Rohovec 1993). *V. salmoninarum* has been a problem for Atlantic salmon culturists in Norway and Scotland. Mortalities from vibriosis in salmonids and other cultured fishes can reach levels in excess of 90% (Cisar and Fryer 1969).

Control of vibriosis can be obtained through the use of antibiotics such as oxytetracycline (Cox and Rainnie 1991, Giles et al. 1991), Romet–30, oxolinic acid, sarafloxacin, erythromycin, and streptomycin (Giles et al. 1991). In addition, the three most common species of *Vibrio* can be controlled through the use of vaccines

(Hayashi et al. 1964, Fryer et al. 1978, Amend and Johnson 1981, Holm and Jorgensen 1987).

Vibriosis has also been reported from invertebrates of aquaculture interest. The oysters *Crassostrea virginica* and *Ostrea edulis* have been subjected to epizootics (Elston et al. 1982, Sindermann and Lightner 1988). Epizootics have been of significance to shrimp culturists (Stewart 1993). Systemic infections of red swamp crawfish by *V. mimicus* and *V. cholerae* have been reported from fresh water. Mortalities of about 25% have occurred, usually in conjunction with low DO and high environmental temperature (Thune et al. 1991).

Bacterial Kidney Disease (BKD). Caused by *Renibacterium salmoninarum,* BKD has long been a significant problem to anadromous salmonid culturists, infecting the fish during the fish during the freshwater phase of their life histories, but also causing mortality in seawater (reviewed by Fryer and Rohovec 1993). External signs of BKD may include exophthalmia, eye lesions, blisters containing blood on the body surface, and swollen abdomen. Internally, white or gray pustules in the kidney can be observed. High levels of mortality can occur in infected fish populations, with chinook salmon being the most highly susceptible of the commercially cultured species (Fryer and Sanders 1981).

R. salmoninarum is difficult to culture, so diagnosis of the disease is often by methods such as fluorescent antibody techniques and ELISA (reviewed by Fryer and Rohovec 1993). Vertical transmission involves transfer of a disease from parent to offspring, while horizontal transmission involves transfer of a disease through the water. BKD can be transferred in both ways (Elliott et al. 1991). Treatment of the disease is difficult. The most effective antibiotic against the disease is erythromycin (Groman and Klontz 1983, Austin 1985, Moffitt and Bjornn 1989, Moffitt 1992); however, that antibiotic is not approved for use on fish in the United States. Erythromycin may prevent transmission of the disease to the offspring of adult salmon fed a diet containing the antibiotic (Evelyn et al. 1986), or injected erythromycin and other antibiotics may be effective (Brown et al. 1990).

McCarthy et al. (1984) and Sakai et al. (1989) were able to vaccinate rainbow trout against BKD with some success. Bowser et al. (1988) found that the addition of fluoride to the diet of rainbow trout at levels from 4 to 16 mg/kg of ration helped control BKD.

Fungal Infections

Fish eggs are particulãrly susceptible to fungal infections during incubation. Dead eggs will quickly develop a covering of such fungi as *Saprolegnia* sp. which are cosmopolitan in fresh water. Healthy eggs will become infected if dead eggs are not removed. Fungus can strike fish of all sizes after disruption of the integument, but those infestations tend to affect individual fish and not entire populations. *Saprolegnia* sp. occurs as a white cottony growth on eggs and the bodies of fish.

The fungus can also attack fish after hatching. Carballo and Muñoz (1991) found that rainbow trout showed increased susceptibility to *Saprolegnia parasitica* infections when they had been exposed to sublethal concentrations of unionized ammonia and nitrate.

Channel catfish eggs are susceptible to fungal attack when they are hatched indoors in hatching troughs. In nature, the adult male tends the eggs and removes any that die and are attacked by fungus. This type of activity is labor-intensive and generally is avoided by culturists who utilize indoor hatching facilities. The problem may be reduced to some extent by the use of well water rather than surface water, since the former is less likely to become contaminated with fungus. In any case, chemical treatment is frequently required to limit losses of eggs due to fungi. Treatment generally involves dipping egg masses in fungicides two or more times weekly. Various concentrations and times of treatment with malachite green have been utilized, with a 30 sec treatment in a 1:15,000 solution once being popular. (Malachite green is an aniline dye having fungicidal properties but currently banned in the United States as carcinogenic.) Burrows (1949) recommended a 5 ppm malachite green dip treatment for 1 hr. Today, antiseptics such as Betadine are widely used. A concentration of 1% can be used as a 10 min dip. Singhal et al. (1986) reported that *Saprolegnia* could be controlled with both malachite green and sodium chloride.

Treatment of fingerling or adult channel catfish for fungi usually involves dipping the animals in a fungicidal solution. If only one or a few fish are affected with fungus, their destruction to prevent the spread of the problem may be the most expeditious approach.

Protozoan Diseases

A variety of protozoans are known to parasitize freshwater fishes. Some of the more common ones are discussed in the following sections. Information is concentrated on channel catfish and salmonids for convenience, though many additional species can be impacted by the protozoans discussed. A number of parasitic protozoans affecting aquaculture species, such as *Epistylis* sp., *Ambiphrya* sp. (= *Scyphidia*, Figure 101), and *Apisoma* sp. (= *Glossataella*), are not included in this discussion.

Ichthyophthirius multifiliis. The ciliated protozoan *Ichthyophthirius multifiliis*, commonly known as "Ich" or "white spot disease," is the causative agent of the disease ichthyophthiriasis (Figure 100). The parasite attacks the integument of fish in fresh water. The result is thickening of the epithelium and the production of excess mucus in many species (Meyer and Hoffman 1976). The parasites feed on the epithelial tissues of fish and can produce high mortality. Histopathology and variability in susceptibility of various fish species to ichthyophthiriasis has been described by Ventura and Paperna (1985).

Infected fish may be found congregated at the water inflow or drain in ponds or other types of culture chambers and may attempt to rub themselves against the bottom or sides of culture chambers, apparently in an attempt to scratch or dislodge the parasites (Meyer 1966a, Meyer and Hoffman 1976). Infected fish exhibit small

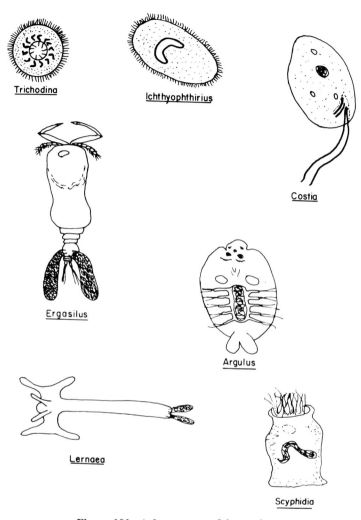

Figure 101. A few common fish parasites.

white nodules or pustules on their body surfaces and gills. In advanced cases the entire surface of the fish may be covered with the nodules (Figure 102).

The life cycle of *I. multifiliis* involves development of the adult parasite as trophonts in the nodules on the body of the fish. The adults eventually leave the fish to become free-swimming for a few hours, after which they form cysts on suitable substrates. Within each cyst the parasite undergoes multiple cell division, forming numerous juveniles, called theronts, that finally emerge from the cyst and swim through the water, seeking a fish on which to attach. Once established on the body of a fish, the parasite grows into the adult form and the cycle is repeated (Meyer

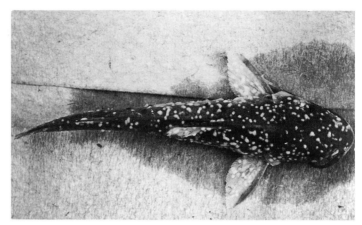

Figure 102. Fingerling channel catfish exhibiting white pustules on the body surface indicative of *Ichthyophthirius multifiliis*. (Photograph by S. K. Johnson.)

and Hoffman 1976). Ewing et al. (1988) found evidence that there may also be reproduction of trophonts within the epithelial tissues of infected fish.

No effective treatment methods have been found for Ich during the cyst stage or when the parasites are attached to the fish; however, they can be destroyed during the free-swimming stage. Since the various life-history stages of the protozoan can exist simultaneously, a single treatment is generally ineffective in controlling an epizootic. Post and Vesely (1983) tested several commonly used protozoacidal chemicals and were unable to kill the trophonts on infected fish. Several treatments or long-term exposures are necessary to ensure that each individual parasite is destroyed during a vulnerable period.

Among the treatments that have been used are malachite green (Allison 1957, Johnson 1961, Beckert and Allison 1964), formalin (Davis 1953, Allison 1957, Meyer and Collar 1964, van As et al. 1984), combinations of formalin and malachite green (Leteux and Meyer 1972), copper sulfate at 0.5 ppm weekly in ponds (Meyer 1966a, Leteux and Meyer 1972), and increased salinity (Allen and Avault 1970). Johnson (1976) tested a variety of chemicals on channel catfish infected with Ich and found that new infestations occurred with all the types of treatment mentioned above except the addition of salinity to 2 ppt.[172] Selosse and Rowland (1990) successfully eliminated the disease in five species of Australian fishes through the addition of 5 g/l of sodium chloride. Theronts and trophonts were eliminated from host fishes within 7 days when temperature was 19 to 26°C and in 14 days at 11 to 18°C. Other effective treatments included chloramine T, quinine sulfate, and quinine bisulfate (Johnson 1976).

Acquired immunity to Ich has been demonstrated in catfish and tilapia exposed to sublethal infections (Clark et al. 1988, Subasinghe and Sommerville 1986). Vac-

---

[172]Mixtures of formalin and malachite green were not tested by S. K. Johnson (1976).

cines against the disease that provide protection for channel catfish and rainbow trout have been developed by Goven et al. (1981) and Wolf and Markiw (1982). Ultraviolet light can be used to control Ich in recirculating water systems (Gratzek et al. 1983).

The optimum temperature range for Ich is from 21 to 24°C, so one effective treatment is to raise the water temperature of culture tanks to 29 to 32°C (Meyer 1966a, Johnson 1976). Because of the rather restricted temperature range for the development of epizootics, the disease is most commonly encountered during the spring and fall.

Trichodina *sp.* Among the most common external parasites to affect fish are various species in the genus *Trichodina* (Figure 101). The parasite may occur on the body, fins, and gills of infected fish. Infestations of parasites within the genus have also been reported from Pacific oysters *(Crassostrea gigas)* by Bower (1988).

Under magnification *Trichodina* sp. appear as circular transparent animals with an internal disc structure that contains hooks. The animal is a protozoan with cilia located completely around the circumference of the cell. The cilia may not be readily apparent under the microscope (Davis 1953, Meyer 1966b). The aboral surface of the organism contains an adhesive disk, although the animals are often found to be in rapid motion on the surface of the host (Davis 1947, 1953).

Epizootics can be treated with dips of 30 ppt salt water, 1:500 solutions of acetic acid, or 1:4,000 solutions of formalin (Davis 1953). In ponds, the addition of sufficient formalin to make a 15 ppm solution may be effective (Amlacher 1970). Eversole and Field (1982) found that potassium permanganate was more effective than sodium chloride for the control of *Trichodina* in elvers of the eel, *Anguilla rostrata*. Chun (1983) was able to control *Trichodina,* as well as *Costia,* in *A. japonica* with 4 ppm of potassium permanganate. Formalin (250 ppm for 30 to 40 min) and potassium permanganate (5 ppm for 10 to 15 min) controlled epizootics in *Tilapia nilotica* (Nguenga 1988).

Costia *spp.* Among the flagellated protozoans that attack fish, those in the genus *Costia* create some of the most frequent problems. *Costia* sp. may attack the gills and external surfaces of fish (Rogers and Gaines 1975). As the result of excessive mucus production, affected fish display a characteristic grayish-white or bluish cast in the film that covers the external surface. Epidermal necrosis may result from epizootics (Amlacher 1970). Costiasis may produce high levels of mortality and is especially serious when it occurs on fingerling fish (Davis 1953).

Depending on the species, *Costia* is essentially an oval or pear-shaped organism (Figure 101) that moves rapidly when viewed under the microscope (Davis 1953). Control of the parasite has been effected with 1:500 acetic acid dips (Davis 1953) and 1:4,000 formalin solutions (Fish 1940), as well as methylene blue, sodium chloride, potassium permanganate, and copper sulfate (Amlacher 1970).

Myxosporidian Parasites. Parasites responsible for such afflictions as whirling disease in salmonids caused by a myxosporidian protozoan have received a great deal of attention by fish disease specialists. Another myxosporidian, *Henneguya* sp., can lead to epizootics in warmwater fishes (McCraren et al. 1975).

Infections by *Henneguya* sp. appear as white cysts (which are larger than those of *I. multifiliis*) on the skin, gills, and various other tissues of fish, including internal organs. Spores of the parasite resemble spermatozoa.

Perkinsus marinus. An epizootic killer of the American oyster, *Crassostrea virginica,* the protozoan *P. marinus* (= *Dermocystidium marinum*) was first identified in the 1950s. Once thought to be a fungal disease, the problem is now known to be caused by a parasitic protozoan. As reviewed by Couch (1993), the parasite leads to mortality mostly during the late summer and early fall when temperature and salinity are high. Distribution of the parasite may also be affected by pollutants associated with industrial and agricultural land use in areas near oyster beds (Wilson et al. 1990). Crosby and Roberts (1990) found no correlation between infection and oyster size in South Carolina.

In recent years the parasite has been found in low-salinity waters that were once free of the disease. Ragone and Burreson (1989) exposed infected oysters to various salinities and found that while the parasite remained present at all salinities, mortality was much higher in oysters maintained at 20 ppt than at 12 ppt or lower.

The disease has been observed in American oysters from New Jersey to Texas (Perkins 1993). *P. marinus,* or perhaps a few closely related species within the genus, are known to parasitize a large number of molluscs throughout the world in temperate, subtropical, and tropical waters (reviewed by Perkins 1993). Wilson et al. (1990) found no relationship between intensity of infection and sex or reproductive stage in American oysters from the Gulf of Mexico.

Protozoans within the genus have also become a common problem for Bay of Fundy Atlantic salmon farmers (Cawthorn et al. 1991). Atlantic salmon smolts can die without exhibiting signs of disease, though subacute signs of infection include lethargy, darkening in color, and changes in behavior. Smolts typically exhibit the disease in the summer and fall, while older fish are impacted during spring, summer, and fall.

Zoospores released by the parasites into the water can infect other oysters. Transmission of the disease may also occur when scavengers eat infected oyster tissue and excrete spores of the parasite in their feces (Meyers and Burreson 1989). The parasites multiply mainly in the connective tissue and between epithelial cells of oysters (Perkins 1993). No effective chemical control for *P. marinus* has been found.

MSX Disease. Another disease of oysters is caused by the sporozoan parasite, *Haplosporidium nelsoni*. The common name for the disease, MSX, stands for Multinucleated Sphere Unknown (Sparks 1993). The disease affects American oysters from Massachusetts to North Carolina, though there is some indication that the disease has spread to Maine and Florida (Kern 1988). Also first reported in the 1950s, MSX has been responsible for high levels of mortality in American oysters.

The origin of the disease and even the life cycle of the parasite remain largely unknown.

Higher than normal salinities have been associated with mortalities attributed to MSX, and there may be a temperature influence. The parasite appears unable to tolerate low salinities (Ford and Haskin 1988) and, in fact, will die *in vitro* within a few minutes at salinities below 10 ppt (Haskin and Ford 1989). Transfer of infected oysters to water of salinities of 10 ppt or less and temperatures above 20°C will eliminate the parasite within 2 wk (Ford 1985). Returning those oysters to high-salinity water does not lead to reinfection. As discussed by Perkins (1993), cold winters often precede summers in which MSX incidence is reduced, indicating that the parasite may not be able to withstand cold.

While mortality is the most significant effect of MSX, sublethal impacts have also been demonstrated. For example, Ford and Figueras (1988) found that gametogenesis in oysters with systemic infections was depressed. Oysters that only had infected gills did not show the same response.

Confirmation of MSX has required histological methods that are time-consuming and expensive. Figueras et al. (1988) reported on development of an ELISA procedure that was moderately successful in identifying the parasite in oysters with advanced infections.

No treatment for MSX has been found, but there has been some success associated with the development of MSX-resistant oysters (Haskin and Ford 1987, Ewart et al. 1988, Chintala and Fisher 1989, Matthiessen et al. 1990). Chintala and Fisher (1989) indicated that oysters that had been developed with MSX resistance were not resistant to infections with *Perkinsus marinus*. Burreson et al. (1989) came to the same conclusion.

Helminth Parasites

Various parasitic worms are commonly found in aquacultured animals, but in most cases infestations are not sufficiently severe to cause significant problems in wild fish populations. Epizootics of significance can occur, however, in aquaculture where crowding occurs. While there are hundreds of thousands, or perhaps millions of helminth species, not all are parasitic. Flatworms (Platyhelminthes) in the families Trematoda and Cestoda are exclusively parasitic. Certain species within the spiny-headed (Acanthocephala) and roundworm (Nematoda) phyla are also parasitic. Many epizootics in aquaculture animals are attributable to monogenetic trematodes.

The distinction between monogenetic and digenetic trematodes varies to some extent depending on which authority is consulted. Chandler and Read (1961) indicated that monogenetic trematodes differ from digenetic species in that the former do not undergo asexual development during any stage of the life cycle, whereas asexual development does occur in digenetic species. Villee et al. (1963) divided the two groups on the basis of number of hosts—monogenetic trematodes having life cycles with a single host and digenetic ones with two or more hosts. In reality, both definitions are valid since species that utilize only one host (Monogenea) reproduce only sexually, and species having two or more hosts (Digenea) employ both sexual and asexual reproduction within their life cycles.

The life cycle of digenetic trematodes often involves the presence of the adult parasite in the body of a fish. Eggs are released from the adult trematode and enter the water, where they hatch into larvae. The larval trematodes enter a gastropod mollusc and undergo further development. The later larval stages of the parasite may then enter a mayfly naiad and become encysted within the body of the host. A fish eating the mayfly becomes infected when the cyst ruptures and releases larval parasites, which then develop into adults (Anonymous 1970). Alternatively, fish may serve as an intermediate, rather than the final host for digenetic trematodes. Typical species of digenetic trematodes that infect channel catfish include *Alloglossidium corti,* which is found in the intestines, and *Clinostomum marginata,* which occurs in cysts in the flesh of fish (Meyer 1966b). The latter is referred to as the yellow grub of fishes and reaches adulthood only when consumed by the proper bird host. This parasite normally is not a problem if fish-eating birds are kept away from culture chambers. The life cycle of various digenetic trematodes can also be interrupted by eliminating snails, which often serve as intermediate hosts.

The most common monogenetic trematodes infecting catfish are *Gyrodactylus* sp. (found primarily on the body and fins, but also occasionally on the gills) and *Cleidodiscus* sp. (found only on the gills). These parasites have been treated effectively with 25 ppm formalin in ponds or 250 ppm formalin as 1 hr bath (Meyer 1966b). Dylox, potassium permanganate at 5 ppm, and potassium dichromate at 20 ppm have also been effectively utilized to control monogenetic trematodes (Meyer 1966b, Anonymous 1970).

Copepods

A variety of parasitic copepods have proved troublesome to aquaculturists. Many species have become highly modified in their physical appearance, with these modifications adapting them for their parasitic existence. Among the species that commonly infect channel catfish, *Ergasilus* sp. is found on the gills, where *Argulus* sp. (known as the fish louse) is found on the surface of the body and may resemble scales (Meyer 1966b). Both species (Figure 101) feed on body fluids.

Ergasilus sp. look similar to the free-cyclopoid copepods, except that one pair of antennae are developed into stout hooks and the mouth parts are designed for biting (Bowen 1966). The parasite has been a problem in freshwater aquaculture and has also been reported from brackish water in conjunction with Atlantic salmon rearing in New Brunswick, Canada (Hogans 1989).

Respiratory impairment, anemia, and poor growth are among the symptoms of serious *Ergasilus* infection. Moreover, the disruption of the tissues associated with the parasite often leads to the outbreak of bacterial and other types of secondary infection. Suggested treatment for *Ergasilus* sp. is 0.25 ppm Dylox weekly in ponds (Anonymous 1970).

Argulus sp. grossly resembles the horseshoe crab *(Limulus polyphemus)* in many respects, although the parasite never attains a size larger than a few millimeters.

Epizootics of *Argulus* sp. are often initiated when parasitized fish are released into culture facilities that were previously free of the parasite (Bowen and Putz 1966). Rushton-Mellor (1992) indicated that *A. japonicus* was originally described from Japan but has now become established worldwide. Other species of *Argulus* also exist and have created problems for aquaculturists. Menezes et al. (1990) described failure in an attempt to develop trout culture in the Azores because of infestations by *A. foliaceus,* a parasite that was apparently introduced accidentally from Europe when the lake was stocked with fish originating there.

Secondary infections associated with *Argulus* sp. frequently occur. Heavily infested fish may swim erratically and will flash.[173] Weight loss is a common sign of epizootics (Bowen and Putz 1966). Shimura et al. (1983) examined the life cycle of *A. coregoni* in Japan in conjunction with salmon and rainbow trout and determined that the parasite produces eggs that overwinter, and that there are one or two generations per year. Larvae hatched from overwintered eggs were most abundant from May through July.

Because of the size of the parasite and its location on the body surface, diagnosis is relatively simple. Various chemicals have been used in the treatment of *Argulus* sp. One of the more effective ones, Dylox, is typically used at 0.25 ppm (reviewed by Bowen and Putz 1966).

Another parasitic copepod, *Lernaea cyprinacae* (Figure 101), often called the anchor parasite or anchorworm, has been known to attack channel catfish when that species is reared in conjunction with scaled fishes (Meyer 1966b). The copepod has also been a problem for various other species of freshwater fishes such as Indian carp, bighead carp, silver carp, koi, goldfish, and rainbow trout (Scott and Fogle 1983, Seenappa et al. 1985, Ali et al. 1989, Shariff and Roberts 1989, Berry et al. 1991). Once attached to a fish, the head of the parasite is modified and literally becomes anchored in the host's flesh. Dylox at 0.25 ppm has been effectively used to control the parasite in catfish (Meyer 1966b, Anonymous 1970). Unden (2-iso-propoxy-phenyl-*N*-methylcarbamate) and Dipterex (0,0-dimethyl-2,2,2-trichloro-1-hydroxyethylphosphate) have also been used to control the parasite (Shariff et al. 1986). Dempster et al. (1988) found that maintaining chloride concentrations of 20 to 40 mg/l would control anchorworms without causing harm. They tested that method of control on a variety of freshwater fishes. Exposure of the parasite to low pH can also be used as a form of treatment. If the parasite is exposed to acid water of pH 4.0 to 4.4 for 5 min, the parasites will be killed. Twenty minute exposures are required when the pH is 4.8, with 30 minutes being required for pH levels of 5.2 to 6.0 (Seenappa et al. 1985).

PREDATION

Predation on aquaculture species can come from a variety of sources. Snakes, turtles, otters, crabs, predaceous fish and insect larvae, and most commonly birds have

---

[173] Flashing involves a fish twisting on its long axis causing one of its sides (which are often white or silvery in color) to briefly reflect light entering the water from above the surface.

posed problems with respect to a variety of freshwater and marine species. The threat of predation by many of the types of animals mentioned is almost intuitive, though many may not be aware that there are a number of predatory insect larvae. Dragonfly larvae are often a significant problem, and even damselfly nymphs can be predatory on postlarval marine shrimp reared in low- to intermittent-salinity water (Huang et al. 1986).

Among the birds that have posed significant problems are herons, kingfishers, cormorants, seagulls, and loons, to name but a few. Schramm et al. (1987) indicated that at least 35 bird species were potential aquaculture predators in Florida.

Filtering incoming water can reduce the presence of predators that enter an aquaculture facility in instances where surface water is being used. Trapping is sometimes effective for certain predators, and lethal means may also be appropriate (and legal) in certain instances. Many farmers keep rifles or shotguns handy to shoot snakes and turtles, for example. Most bird predators of fish are protected, at least in the United States, by state or federal laws and cannot be legally killed.

Operators of state and federal hatcheries that release fish into the environment are not only plagued by bird predation when the fish are in the hatchery but also subject to losses when the fish are released into the environment. This may be particularly true with respect to anadromous fish releases. For example, 20 yr of study on the effects of bird predation on Atlantic salmon released from hatcheries in Norway demonstrated bird predation rates up to 10.8% (Reitan et al. 1987).

Fish farmers that are prohibited from killing predatory birds have developed a variety of approaches for dealing with the problem. Some of them, and something about their effectiveness are as follows:

- Sound cannons and other noisemakers to drive away birds: Many species become accustomed to the noise after a period of time and ignore it.
- Placement of bird netting over culture chambers: Bird netting can be expensive, particularly in conjunction with large facilities, though it has been used effectively with raceway systems. There are some inconveniences to personnel who have to negotiate around the bird netting to conduct their daily activities.
- Continuous human activity provided by people who drive or walk around the facility during all periods when birds are active can be effective, though the cost may be prohibitive.
- Use of dogs to chase birds has been effective in many instances. As is the case with continuous human presence, it can be anticipated that birds that are unable to prey upon fish in a facility will eventually go elsewhere when they are continuously frustrated for a period of about 3 days. They will return eventually, so dog or human activity should not be discontinued in the absence of bird predators.
- Releases of fish from enhancement aquaculture facilities after the time when migratory birds have left the area can be an effective way to reduce predation on salmonids, as discussed by Mace (1983).

• Trapping and relocation of predatory birds has met with some success, as reviewed by Mott (1978).

In addition to direct predation, birds may also transmit diseases to aquaculture
animals by carrying pathogens into culture chambers on body surfaces or in feces.
Birds serve as intermediate hosts for a number of parasites that can impact aquaculture species, for example.

CANNIBALISM

Species that have been considered, or are being reared, by aquaculurists range with
respect to their propensity for cannibalism. Channel catfish, trout, oysters and mussels are examples of fish and invertebrates that are not cannibalistic.[174] At the other
extreme are such animals as crabs and lobsters, which are highly cannibalistic; so
much so, in fact, that they have to be reared as individuals in isolation from one
another. Cannibalism occurs immediately after molting when the animals are de-
fenseless. Neither crabs nor lobsters are currently being cultured commercially,
largely because of the cannibalism problem. Other species that are highly cannibalistic are being effectively produced by aquaculturists for food or enhancement. Examples are northern pike, walleye, largemouth bass, freshwater shrimp, and penaeid shrimp. Problems with cannibalism in fish can sometimes be managed through
frequent feeding to keep the animals satiated. Shrimp culturists can reduce cannibalism by providing hiding places for newly molted individuals. Pieces of PVC pipe
are commonly put in culture chambers for that purpose. Vertical nets hung in culture chambers can also provide protection.

Several fish species are highly cannibalistic during the early stages of their lives
but may become less cannibalistic once they are established on prepared feeds.
Examples are northern pike, muskellunge, and hybrids between the two. Hatchery
managers may feed such species every few minutes. Frequent grading can also be
advantageous since it is difficult for fish of the same size to cannibalize one another.
In sea bass *(Dicentrarchus labrax),* for example, the cannibal must be twice the
length of the prey (Katavic et al. 1989).

Some species are not generally cannibalistic but large individuals in a population
may feed on larvae or fry of their own species. An example is tilapia. Adults do not
seem to be significant predators on fry, but fingerlings produced early in the spawning season do appear to cannibalize fry from subsequent generations, thereby reducing the extent of overpopulation and stunting that occur.

---

[174] Large catfish may eat catfish eggs, and oysters and mussels may consume eggs or larvae of their own
species as a result of filter feeding, but none of the species mentioned has a proclivity for cannibalism
of large individuals on smaller ones.

LITERATURE CITED

Adams, A., and K. Thompson. 1990. Development of an enzyme-linked immunosorbent assay (ELISA) for the detection of *Aeromonas salmonicida* in fish tissue. J. Aquat. Anim. Health, 2: 281–288.

Adams, A., N. Auchinachie, A. Bundy, M. F. Tatner, and M. T. Horne. 1988. The potency of adjuvanted injected vaccines in rainbow trout (*Salmo gairdneri* Richardson) and bath vaccines in Atlantic salmon (*Salmo salar* L.) against furunculosis. Aquaculture, 69: 15–26.

Allen, K. O., and J. W. Avault, Jr. 1970. Effects of brackish water on ichthyophthiriasis of channel catfish. Prog. Fish-Cult., 32: 227–230.

Ali, N. M., F. T. Mhaisen, E. S. Abul-Eis, and L. S. Kadim. 1989. Parasites of the silver carp *Hypophthalmichthys molitrix* from Babylon Fish Farm, Hilla, Iraq. Riv. Idrobiol., 28: 151–154.

Allison, R. 1957. Some new results in the treatment of ponds to control some external parasites of fish. Prog. Fish-Cult., 19: 58–63.

Amend, D. F. 1970. Control of infectious hemaatopoietic necrosis virus disease by elevating the water temperature. J. Fish. Res. Board Can., 27: 265–270.

Amend, D. F., and K. A. Johnson. 1981. Current status and future needs of *Vibrio anguillarum* bacterins. Dev. Biol. Stand., 49: 403–417.

Amend, D. F., and T. McDowell. 1983. Current problems in the control of channel catfish virus. J. World Aquacult. Soc., 14: 261–267.

Amend, D. F., W. T. Yasutake, and R. W. Mead. 1969. A hematopoietic necrosis (IHN) virus in Washington salmon. Prog. Fish-Cult., 34: 143–147.

Amlacher, E. 1970. Textbook of fish diseases. T.F.H. Publications, Jersey City, New Jersey. 302 p.

Anderson, I. G., A. T. Law, M. Shariff, and G. Nash. 1990. A parvo-like virus in the giant freshwater prawn, *Macrobrachium rosenbergii*. J. Invertebr. Pathol., 55: 447–449.

Anonymous. 1970. Report to the fish farmers. U.S. Bureau of Sport Fisheries and Wildlife Resource Publication No. 83, Washington, D.C. 124 p.

Anonymous. 1981. Tis the season to be wary—of catfish disease. For Fish Farmers, 1981: 6–7.

Anonymous. 1993. Scientists urge ban on selected chemicals in aquaculture. Southeast Asian Fisheries Development Center Newsl., 15(3): 8–9.

Areechon, N., and J. A. Plumb. 1983. Pathogenesis of *Edwardsiella ictaluri* in channel catfish, *Ictalurus punctatus*. J. World Maricult. Soc., 14: 249–260.

Austin, B. 1985. Evaluation of antimicrobial compounds for the control of bacterial kidney disease in rainbow trout, *Salmo gairdneri* Richardson. J. Fish Dis., 8: 209–220.

Awad, M. A., K. E. Nusbaum, and Y. J. Brady. 1989. Preliminary studies of a newly developed subunit vaccine for channel catfish virus disease. J. Aquat. Anim. Health, 1: 233–237.

Barnes, A. C., C. S. Lewin, T. S. Hastings, and S. G. B. Amyes. 1990. *In vitro* activities of 4-quinolones against the fish pathogen *Aeromonas salmonicida*. Antimicrob. Agents Chemother., 34: 1819–1820.

Barnes, A. C., C. S. Lewin, T. S. Hastings, and S. G. B. Amyes. 1991a. *In vitro* susceptibility of the fish pathogen *Aeromonas salmonicida* to flumequine. Antimicrob. Agents Chemother., 35: 2634–2635.

Barnes, A. C., C. S. Lewin, S. G. B. Amyes, and T. S. Hastings. 1991b. Susceptibility of Scottish isolates of *Aeromonas salmonicida* to the antibacterial agent amoxycillin. ICES Council Meeting Papers. International Council for the Exploration of the Sea , Copenhagen. 5 p.

Beckert, H., and R. Allison. 1964. Some host responses of white catfish to *Ichthyophthirius multifiliis* Fouquet. Proc. Southeast. Assoc. Game Fish Comm., 18: 438–441.

Bejerang, I., and S. Sarig. 1991. Stress-induced infection of *Sarotherodon aureus* under laboratory conditions. pp. 69–79, In: H. Rosenthal, and O. H. Oron (Eds.). Research on Intensive aquaculture. Special Publication No. 9, European Aquaculture Society, Ghent, Belgium.

Beleau, M. H., and J. A. Plumb. 1987. Channel catfish culture methods used in the United States. Vet. Hum. Toxicol., 29: 52–53.

Bell, T. A., and D. V. Lightner. 1984. IHHN virus: infecteivity and pathogenicity studies in *Penaeus stylirostris* and *P. vannamei*. Aquaculture, 38: 185–194.

Bell, T. A., and D. V. Lightner. 1987. IHHN disease of *Penaeus stylirostris:* effects of shrimp on disease expression. J. Fish Dis., 10: 165–170.

Bernoth, E.-M. 1990. Screening for the fish disease agent *Aeromonas salmonicida* with an enzyme-linked immunosorbent assay (ELISA). J. Aquat. Anim. Health, 2: 99–103.

Berry, C. R., Jr., G. J. Babey, and T. Shrader. 1991. Effect of *Lernaea cyprinacea* (Crustacea: Copepoda) on stocked rainbow trout *(Oncorhynchus mykiss)*. J. Wildl. Dis., 27: 206–213.

Bowen, J. T. 1966. Parasites of freshwater fish. IV. Miscellaneous. 4. Parasitic copepods *Ergasilus, Achtheres,* and *Salminicola*. Fish Disease Leaflet No. 4. U.S. Bureau of Sport Fisheries and Wildlife, Washington, D.C. 4 p.

Bowen, J. T., and R. E. Putz. 1966. Parasites of freshwater fish. IV. Miscellaneous. 3. Parasitic copepod *Argulus* Fish Disease Leaflet 3. U.S. Bureau of Sport Fisheries and Wildlife, Washington, D.C. 4 p.

Bower, S. M. 1988. Protozoan parasites of Pacific oysters *(Crassostrea gigas)* in British Columbia. J. Shellfish Res., 7: 150–151.

Bowser, P. R., and A. D. Munson. 1986. Seasonal variation in channel catfish virus antibody titers in adult channel catfish. Prog. Fish-Cult., 48: 198–199.

Bowser, P. R., R. B. Landy, G. A. Wooster, and J. G. Babish. 1988. Efficacy of elevated dietary fluoride for the control of *Renibacterium salmoninarum* infection in rainbow trout *Salmo gairdneri*. J. World Aquacult. Soc., 19: 1–7.

Bowser, P. R., J. H. Schachte, Jr., G. A. Wooster, and J. G. Babish. 1990. Experimental treatment of *Aeromonas salmonicida* infections with enrofloxacin and oxolinic acid: field trials. J. Aquat. Anim. Health, 2: 198–203.

Boyle, J., and J. Blackwell. 1991. Use of polymerase chain reaction to detect latent channel catfish virus. Am. J. Vet. Res., 52: 1965–1968.

Brackett, J. 1991. Potential disease interactions of wild and farmed fish. Bull. Aquacult. Assoc. Can., 91(3): 79–80.

Brady, Y. J., and S. Vinitnantharat. 1990. Viability of bacterial pathogens in frozen fish. J. Aquat. Anim. Health, 2: 149–150.

Bricknell, I. R., A. E. Ellis, and A. L. S. Munro. 1991. The current status of a Scottish furunculosis vaccine. ICES Council Meeting Papers. International Council for the Exploration of the Sea, Copenhagen. 8 p.

Brown, L. L., L. J. Albright, and T. P. T. Evelyn. 1990. Control of vertical transmission of *Renibacterium salmoninarum* by injection of antibiotics into maturing female coho salmon *Oncorhynchus kisutch*. Dis. Aquat. Org., 9: 127–131.

Bullock, G. L., and J. J. A. McLaughlin. 1970. Advances in knowledge concerning bacteria pathogenic to fishes (1954–1968). pp. 231–242, In: S. F. Snieszko (Ed.). A symposium on diseases of fishes and shellfishes. American Fisheries Society Special Publication No. 5. American Fisheries Society, Washington, D.C.

Burreson, E. M., J. A. Meyers, R. Mann, and B. J. Barber. 1989. Susceptibility of MSX-resistant strains of the eastern oyster and of the Japanese oyster to *Perkinsus marinus*. J. Shellfish Res., 8: 462.

Burrows, R. E. 1949. Prophylactic treatment for control of fungus *(Saprolegnia parasitica)* on salmon eggs. Prog. Fish-Cult., 11: 97–103.

Carballo, M., and M. J. Muñoz. 1991. Effect of sublethal concentrations of four chemicals on susceptibility of juvenile rainbow trout *(Oncorhynchus mykiss)* to saprolegniosis. Appl. Environ. Microbiol., 57: 1813–1816.

Carmichael, G. J., J. R. Tomasso, B. A. Simco, and K. B. Davis. 1984. Characterization and alleviation of stress associated with hauling largemouth bass. Trans. Am. Fish. Soc., 113: 778–785.

Cawthorn, R. J., S. Backman, J. O'Halloran, H. Mitchell, D. Groman, and D. Speare. 1991. *Perkinsus* sp. (Apicomplexa) in farmed Atlantic salmon *(Salmo salar)*. Bull. Aquacult. Assoc. Can. 91(3): 61–63.

Chandler, A. C., and C. P. Read. 1961. Introduction to parasitology. John Wiley & Sons, New York. 822 p.

Chintala, M. M., and W. S. Fisher. 1989. Comparison of oyster defense mechanisms for MSX-resistant and susceptible stocks held in Chesapeake Bay. J. Shellfish Res., 8: 467–468.

Chun, S. K. 1983. Fish diseases and their control in high density culture of eel. Bull. Korean Fish. Soc., 16: 103–110.

Cisar, J. O., and J. L. Fryer. 1969. An epizootic of vibriosis in chinook salmon. Bull. Wildl. Dis. Assoc., 5: 73–76.

Clark, T. G., H. W. Dickerson, and R. C. Findly. 1988. Immune response of channel catfish to ciliary antigens of *Ichthyophthirius multifiliis*. Dev. Comp. Immunol., 12: 581–594.

Couch, J. A. 1993. Observations on the state of marine disease studies. pp. 511–530, In: J. A. Couch and J. W. Fournie (Eds.). Pathobiology of marine and estuarine organisms. CRC Press, Boca Raton, Florida.

Cox, W. R., and D. J. Rainnie. 1991. Oxytetracycline for the treatment of vibriosis in Atlantic salmon. Bull. Aquacult. Assoc. Can., 91(3): 50–52.

Crosby, M. P., and Roberts. 1990. Seasonal infection intensity cycle of the parasite *Perkinsus marinus* (and an absence of *Haplosporidium* spp.) in oysters from a South Carolina salt marsh. Dis. Aquat. Org., 9: 149–155.

Davis, H. S. 1947. Studies of the protozoan parasites of fresh-water fishes. Fish. Bull., 41: 1–29.

Davis, H. S. 1953. Culture and diseases of game fishes. University of California Press, Berkeley. 332 p.

Dempster, R. P., P. Morales, and F.X. Glennon. 1988. Use of sodium chlorite to combat anchorworm infestations of fish. Prog. Fish-Cult., 50: 51–55.

Dunn, E. J., A. Polk, D. J. Scarrett, G. Olivier, S. Lall, and M. F. A. Goosen. 1990. Vaccines in aquaculture: the search for an efficient delivery system. Aquacult. Eng., 9: 23–32.

Durborow, R. M., P. W. Taylor, M. D. Crosby, and T. D. Santucci. 1991. Fish mortality in the Mississippi catfish farming industry in 1988: causes and treatments. J. Wildl. Dis., 27: 144–147.

Egusa, S. 1969. *Vibrio anguillarum,* a bacterium pathogenic to salt water and freshwater fishes. Fish Pathol., 4: 31–44.

Elliott, D. G., R. J. Pascho, and G. L. Bullock. 1991. Developments in the control of bacterial kidney disease of salmonid fishes. Dis. Aquat. Org., 6: 201–215.

Ellis, A. E., I. R. Bricknell, and A. L. S. Munro. 1992. Current status of a Scottish furunculosis vaccine. Bull. Aquacult. Assoc. Can., 92(1): 11–15.

Elston, R. 1992. Potential therapeutic use of amoxycillin for furunculosis treatment. Bull. Aquacult. Assoc. Can., 92(1): 68.

Elston, R. A., E. L. Elliot, and R. R. Colwell. 1982. Conchiolin infections and surface coating *Vibrio:* shell fragility, growth depression and mortalities in cultured oysters and clams, *Crassostrea virginica, Ostrea edulis,* and *Mercenaria mercenaria.* J. Fish Dis., 5: 265–284.

Evelyn, T. P. T., J. E. Ketcheson, and L. Prosperi-Porta. 1986. Use of erythromycin as a means of preventing vertical transmission of *Renibacterium salmoninarum.* Dis. Aquat. Org., 2: 7–11.

Eversole, A. G., and D. W. Field. 1982. Prophylactic treatment of elvers for *Trichodina* infestation. Prog. Fish-Cult., 44: 142–143.

Ewart, J. W., R. Cole, and J. Tinsman. 1988. Growth and survival of hatchery produced MSX resistant oyster stocks in Delaware Bay. J. Shellfish Res., 7: 543.

Ewing, M. S., S. A. Ewing, and K. M. Kocan. 1988. *Ichthyophthirius* (Ciliophora): population studies suggest reproduction in host epithelium. J. Protozool., 35: 549–552.

Fevolden, S. E., T. Refstie, and K. H. Roeed. 1991. Selection for high and low cortisol stress response in Atlantic salmon *(Salmo salar)* and rainbow trout *(Oncorhynchus mykiss).* Aquaculture, 95: 53–65.

Figueras, A. J., S. A. Kanaley, S. E. Ford, and E. M. Burreson. 1988. Development of enzyme-linked immunosorbent assays for detection of molluscan parasites. J. Shellfish Res., 7: 118–119.

Fijan, N. 1968. Progress report on acute mortality of channel catfish fingerlings caused by a virus. Bull. Off. Int. Epiz., 69: 1167–1168.

Fijan, N. N., T. L. Wellborn, Jr., and J. P. Naftel. 1970. An acute viral disease of channel catfish. Technical Paper 43. U.S. Bureau of Sport Fisheries and Wildlife, Washington, D. C. 11 p.

Fish, F. F. 1940. Formalin for external protozoan parasites. Prog. Fish-Cult. (Old Ser.), 48: 1–10.

Ford, S. E. 1985. Effects of salinity on survival of the MSX parasite *Haplosporidium nelsoni* (Haskin, Stauber, and Mackin) in oysters. J. Shellfish Res., 5: 85–90.

Ford, S. E., and A. J. Figueras. 1988. Effects of MSX *(Haplosporidium nelsoni)* parasitism on reproduction of the oyster, *Crassostrea virginica.* J. Shellfish Res., 7: 119.

Ford, S. E., and H. H. Haskin. 1988. Comparison of *in vitro* tolerance of the oyster parasite *Haplosporidium nelsoni* (MSX) and hemocytes from the host, *Crassostrea virginica.* Comp. Biochem. Physiol., 90A: 183–187.

Fryer, J. L., and J. S. Rohovec. 1993. Bacterial diseases of fish. pp. 53–83, In: J. A. Couch and J. W. Fournie (Eds.). Pathobiology of marine and estuarine organisms. CRC Press, Boca Raton, Florida.

Fryer, J. L., and J. E. Sanders. 1981. Bacterial kidney disease of salmonid fish. Ann. Rev. Microbiol., 35: 273–298.

Fryer, J. L., J. S. Rohovec, and R. L. Garrison. 1978. Immunization of salmonids for control of vibriosis. Mar. Fish. Rev., 40(3): 20–23.

Giles, J. S., H. Hariharan, and S. B. Heaney. 1991. Evaluation of six antimicrobial agents against fish pathogenic isolates of *Vibrio ordaili* and *Aeromonas salmonicida.* Bull. Aquacult. Assoc. Can., 91(3): 53–55.

Goven, B. A., D. L. Dawe, and J. B. Gratzek. 1981. Protection of channel catfish *(Ictalurus punctatus)* against *Ichthyophthirius multifiliis* (Fouquet) by immunization with varying doses of *Tetrahymena pyriformis* (Lwoff) cilia. Aquaculture, 23: 269–273.

Gratzek, J. B., J. P. Gilbert, A. L. Lohr, E. B. Shotts, Jr., and J. Brown. 1983. Ultraviolet light control of *Ichthyophthirius multifiliis* Fouquet in a closed fish culture recirculation system. J. Fish Dis., 6: 145–153.

Groman, D. B., and G. W. Klontz. 1983. Chemotherapy and prophylaxis of bacterial kidney disease with erythromycin. J. World Maricult. Soc., 14: 226–235.

Hardie, L. J., T. C. Fletcher, and C. J. Secombes. 1990. The effect of vitamin E on the immune response of the Atlantic salmon *(Salmo salar* L.). Aquaculture, 87: 1–13.

Hardie, L. J., T. C. Fletcher, and C. J. Secombes. 1991. The effect of dietary vitamin C on the immune response of the Atlantic salmon *(Salmo salar* L.). Aquaculture, 95: 201–214.

Haskin, H. H., and S. E. Ford. 1987. Breeding for disease resistance in molluscs. pp. 431–441, In: K. Tiews (Ed.). Selection, hybridization and genetic engineering in aquaculture, Vol. 2. Schriften der Bundesforschungsansalt für Fischerei, Hamburg.

Haskin, H. H., and S. E. Ford. 1989. Low salinity control of *Haplosporidium nelsoni* (MSX). J. Shellfish Res., 8: 468–469.

Hayashi, K., S. Kobayashi, T. Kamata, and H. Ozaki. 1964. Studies on vibrio disease of rainbow trout *(Salmo gairdneri irideus).* II. Prophylactic vaccination against vibriodisease. J. Fac. Fish. Prefect. Univ. Mei, 6:181–192.

Heartwell, C. M., III. 1975. Immune response and antibody characterization of the channel catfish *(Ictalurus punctatus)* to naturally pathogenic bacterium and virus. Technical Paper No. 85. U.S. Bureau of Sport Fisheries and Wildlife, Washington, D.C. 34 p.

Hedrick, R. P., and T. McDowell. 1987. Passive transfer of sera with antivirus neutralizing activity from adult channel catfish protects juveniles from channel catfish virus disease. Trans. Am. Fish. Soc., 116: 277–281.

Hedrick, R. P., J. M. Groff, and T. McDowell. 1987. Response of channel catfish to waterborne exposures of channel catfish virus. Prog. Fish-Cult., 49: 181–187.

Hjeltnes, B., K. Andersen, and H.-M. Ellingsen. 1989. Vaccination against *Vibrio salmonicida.* The effect of different routes of administration and of revaccination. Aquaculture, 83: 1–6.

Hogans, W. E. 1989. Mortality of cultured Atlantic salmon, *Salmo salar* L. parr caused by an infection of *Ergasilus labracis* (Copepods: Poecilostomatoida) in the lower Saint John River, New Brunswick, Canada. J. Fish Dis., 12: 529–531.

Holm, K. O., and T. Jorgensen. 1987. A successful vaccination of Atlantic salmon, *Salmo salar* L., against 'Hitra disease' or coldwater vibriosis. J. Fish Dis., 10: 85–90.

Huang, H.-J., D. V. Aldrich, and K. Strawn. 1986. Laboratory and pond evidence for shrimp *(Penaeus stylirostris)* predation by damselfly nymphs *(Ischnura ramburi).* J. World Maricult. Soc., 16: 347–353.

Itami, T., and R. Kusuda. 1980a. Studies on spray vaccination against vibriosis, in cultured ayu. 1. Effect of bentonite and pH on vaccination efficacy. Bull. Jpn. Soc. Sci. Fish., 46: 533–536.

Itami, T., and R. Kusuda. 1980b. Studies on spray vaccination against vibriosis, in cultured ayu. 2. Duration of vaccination efficacy and effect of different vaccine preparations. Bull. Jpn. Soc. Sci. Fish., 46: 699–703.

Itami, T., Y. Takahashi, and Y. Nakamura. 1989. Efficacy of vaccination against vibriosis in cultured kuruma prawns *Penaeus japonicus.* J. Aquat. Anim. Health, 1: 238–242.

Johnson, A. K. 1961. Ichthyophthiriasis in a recirculating closed-water hatchery. Prog. Fish-Cult., 20: 129–132.

Johnson, S. K. 1976. Laboratory evaluation of several chemicals as preventatives of ich disease. pp. 91–96, In: Proceedings of the 1976 Fish Farming Conference and Annual Convention of the Catfish Farmers of Texas, College Station. Texas A&M University, College Station.

Katavic, I., J. Jug-Dujakovic, and B. Glamuzina. 1989. Cannibalism as a factor affecting the survival of intensively cultured sea bass *(Dicentrarchus labrax)* fingerlings. Aquaculture, 77: 135–143.

Keith, I., W. D. Paterson, D. Airdrie, and L. Boston. 1988. Commercial development of a vaccine to prevent gaffkemia in lobsters: lab and field results. Bull. Aquacult. Assoc. Can., 88(2): 90.

Kern, F. G. 1988. Recent changes in the range of ''MSX'' *Haplosporidium nelsoni.* J. Shellfish Res., 7: 543–544.

Klesius, P., K. Johnson, R. Durborow, and S. Vinitnantharat. 1991. Development and evaluation of an enzyme-linked immunosorbent assay for catfish serum antibody to *Edwardsiella ictaluri.* J. Aquat. Anim. Health, 3: 94–99.

Kudo, R. R. 1954. Protozoology. Fourth edition. Thomas, Springfield, Illinois. 966 p.

LaPatra, S. E., J. L. Fryer, W. H. Wingfield, and R. P. Hedrick. 1989a. Infectious hematopoietic necrosis virus (IHNV) in coho salmon. J. Aquat. Anim. Health, 1: 277–280.

LaPatra, S. E., K. A. Roberti, J. S. Rohovec, and J. L. Fryer. 1989b. Fluorescent antibody test for the rapid diagnosis of infectious hematopoietic necrosis. J. Aquat. Anim. Health, 1: 29–36.

LaPatra, S. E., W. J. Groberg, J. S. Rohovec, and J. L. Fryer. 1990. Size-related susceptibility of salmonids to two strains of infectious hematopoietic necrosis virus. Trans. Am Fish. Soc., 119: 25–30.

Leteux, F., and F. P. Meyer. 1972. Mixtures of malachite green and formalin for controlling *Ichthyophthirius* and other protozoan parasites of fish. Prog. Fish-Cult., 34: 21–26.

Li, Y., and R. T. Lovell. 1985. Elevated levels of dietary ascorbic acid increase immune responses in channel catfish. J. Nutr., 115: 123–131.

Lightner, D. V., R. M. Redman, and T. A. Bell. 1983a. Infectious hypodermal and hemato-poietic necrosis, a newly recognized virus disease of penaeid shrimp. J. Invert. Path., 42: 62–70.

Lightner, D. V., R. M. Redman, T. A. Bell, and J. A. Brock. 1983b. Detection of IHHN virus in *Penaeus stylirostris* and *P. vannamei* imported into Hawaii. J. World Maricult. Soc., 14: 212–225.

Lightner, D. V., R. M. Redman, R. R. Williams, L. L. Mohney, J. P. M. Clerx, T. A. Bell, and J. A. Brock. 1986. Recent advances in penaeid virus disease investigations. J. World Maricult. Soc., 16: 267–274.

Lightner, D. V., R. M. Redman, T. A. Bell, and R. B. Thurman. 1988. Geographic disper-son of the viruses IHHN, MBV, and HPV as a consequence of transfers and introductions of penaeid shrimp to new regions for aquaculture purposes. J. Shellfish Res., 7: 554–555.

Lillehaug, A. 1989a. A cost-effectiveness study of three different methods of vaccination against vibriosis in salmonids. Aquaculture, 83: 227–236.

Lillehaug, A. 1989b. Oral immunization of rainbow trout, *Salmo gairdneri* Richardson, against vibriosis with vaccines protected against digestive degradation. J. Fish Dis., 12: 579–584.

Lillehaug, A. 1990. A field trial of vaccination against cold-water vibriosis in Atlantic salmon (*Salmo salar* L.). Aquaculture, 84: 1–12.

Lillehaug, A. 1991. Vaccination of Atlantic salmon (*Salmo salar* L.) against cold-water vibriosis—duration of protection and effect on growth rate. Aquaculture, 92: 99–107.

Liu, P. R., J. A. Plumb, M. Guerin, and R. T. Lovell. 1989. Effect of megalevels of dietary vitamin C on the immune response of channel catfish *Ictalurus punctatus* in ponds. Dis. Aquat. Org., 7: 191–194.

Lu, Y., P. C. Loh, and J. A. Brock. 1989. Isolation purification and characterization of infectious hypodermal and hematopoetic necrosis virus (IHHNV) from penaeid shrimp. J. Virol. Methods, 26: 339–344.

Mace, P. M. 1983. Bird predation on juvenile salmonids in the Big Qualicum Estuary, Vancouver Island. Canada Technical Report of Aquatic Sciences No. 1176. Department of Fisheries and Oceans, Ottawa. 89 p.

MacFarlane, R. D., G. L. Bullock, and J. J. A. McLaughlin. 1986. Effects of five metals on susceptibility of striped bass to *Flexibacter columnaris*. Trans. Am. Fish. Soc., 115: 227–231.

Matthiessen, G. C., S. Y. Feng, and L. Leibovitz. 1990. Patterns of MSX (*Haplosporidium nelsoni*) infection and subsequent mortality in resistant and susceptible strains of the east-ern oyster, *Crassostrea virginica* (Gmelin, 1791), in New England. J. Shellfish Res., 9: 359–365.

McCarthy, D. H., T. R. Croy, and D. F. Amend. 1984. Immunization of rainbow trout, *Salmo gairdneri* Richardson, against bacterial kidney disease: preliminary efficacy evalu-ation. J. Fish Dis., 7: 65–71.

McCraren, J. P., M. L. Landolt, G. L. Hoffman, and F. P. Meyer. 1975. Variation in response of channel catfish to *Henneguya* sp. infections (Protozoa: Myxosporidea). J. Wildl. Dis., 11: 2–7.

Menezes, J., A. Ramos, and T. G. Pereira. 1987. Lymphocystis disease: an outbreak in *Sparus aurata* from Ria Formosa, south coast of Portugal. Aquaculture, 67: 222–225.

Menezes, J., M. A. Ramos, T. G. Pereira, and A. Moreira da Silva. 1990. Rainbow trout culture failure in a small lake as a result of massive parasitosis related to careless fish introductions. Aquaculture, 89: 123–126.

Meyer, F. P. 1966a. Parasites of freshwater fishes. II. Protozoa. 3. *Ichthyophthhirius multifiliis*. Fish Disease Leaflet No. 2. U.S. Bureau of Sport Fisheries and Wildlife, Washington, D.C. 4 p.

Meyer, F. P. 1966b. Parasites of freshwater fishes. IV. Miscellaneous. 6. Parasites of catfishes. Fish Disease Leaflet No. 5. U.S. Bureau of Sport Fisheries and Wildlife, Washington, D.C. 7 p.

Meyer, F. P., and J. D. Collar. 1964. Description and treatment of a *Pseudomonas infection* in white catfish. Appl. Microbiol., 12: 201–203.

Meyer, F. P., and G. L. Hoffman. 1976. Parasites and diseases of warmwater fishes. Resource Publication No. 127. U.S. Fish and Wildlife Service, Washington, D.C. 20 p.

Meyer, F. P., and R. A. Schnick. 1989. A review of chemicals used for the control of fish diseases. Rev. Aquat. Sci., 1: 693–710.

Meyers, J. A., and E. M. Burreson. 1989. The role of oyster scavengers in the spread of the oyster disease *Perkinsus marinus*. J. Shellfish Res., 8: 469–470.

Moffitt, C. M. 1992. Survival of juvenile chinook salmon challenged with *Renibacterium salmoninarum* and administered oral doses of erythromycin thiocyanate for different durations. J. Aquat. Anim. Health, 4: 119–125.

Moffitt, C. M., and T. C. Bjornn. 1989. Protection of chinook salmon smolts with oral doses of erythromycin against acute challenges of *Renibacterium salmoninarum*. J. Aquat. Anim. Health, 3: 227–232.

Moore, A. A., M. E. Aimers, and M. A. Cardella. 1990. Attempts to control *Flexibacter columnaris* epizootics in pond-reared channel catfish by vaccination. J. Aquat. Anim. Health, 2: 109–111.

Mott, D. F. 1978. Control of wading bird predation at fish-rearing facilities. pp. 131–132, In: A. Sprunt IV, J. C. Ogden, and S. Winckler (Eds.). Wading birds. Research Report No. 7 of the National Audubon Society. National Audobon Society, New York.

Nguenga, D. 1988. A note on infestation of *Oreochromis niloticus* with *Trichodina* sp. and *Dactylogyrus* sp. pp. 117–119, In: R. S. V. Pullin, T. Bhukaswan, K. Tonguthai, and J. L. Maclean (Eds.). Second International Symposium on Tilapia in Aquaculture, Bangkok, Thailand, March 16–20, 1987. ICLARM Conference Proceedings No. 15. International Center for Living Aquatic Resources Management, Manila.

Nikl, L., L. J. Albright, and T. P. T. Evelyn. 1991. Influence of seven immunostimulants on the immune response of coho salmon to *Aeromonas salmonicida*. Dis. Aquat. Org., 12: 7–12.

Nikl, L., L. J. Albright, and T. P. T. Evelyn. 1992. Immunostimulants hold promise in furunculosis prevention. Bull. Aquacult. Assoc. Can., 92(1): 49–52.

Nusbaum, K. E., and J. M. Grizzle. 1987a. Adherence of channel catfish virus to sperm and leukocytes. Aquaculture, 65: 1–5.

Nusbaum, K. E., and J. M. Grizzle. 1987b. Uptake of channel catfish virus from water by channel catfish and bluegills. Am. J. Vet. Res., 48: 375–377.

Paperna, I., I. Sabnai, and A. Colorni. 1982. An outbreak of lymphocystis in *Sparus aurata* L. in the Gulf of Aqaba, Red Sea. J. Fish Dis., 5: 433–437.

Perkins, F. O. 1993. Infectious diseases of molluscs. pp. 255–287, In: J. A. Couch and

J. W. Fournie (Eds.). Pathobiology of marine and estuarine organisms. CRC Press, Boca Raton, Florida.

Plumb, J. A. 1971a. Channel catfish virus disease in the southern United States. Proc. Southeast. Assoc. Game and Fish Comm., 25: 489–493.

Plumb, J. A. 1971b. Channel catfish virus research at Auburn University. Auburn University Progress Report Series 95, Agriculture Experiment Station, Auburn, Alabama. 4 p.

Plumb, J. A. 1971c. Tissue distribution of channel catfish virus. J. Wildl. Dis., 7: 213–216.

Plumb, J. A. 1972. Channel catfish virus disease. Fish Disease Leaflet 18 (revised). U.S. Bureau of Sport Fisheries and Wildlife, Washington, D.C. 4 p.

Plumb, J. A. 1973. Effects of temperature on mortality of fingerling channel catfish *(Ictalurus punctatus)* experimentally infected with channel catfish virus. J. Fish. Res. Board Can., 30: 568–570.

Plumb, J. A. 1988. Vaccination against *Edwardsiella ictaluri*. Fish Vaccination, 1988: 152–161.

Plumb, J. A. 1993. Viral diseases of marine fish. pp. 25–52, In: J. A. Couch and J. W. Fournie (Eds.). Pathobiology of marine and estuarine organisms. CRC Press, Boca Raton, Florida.

Plumb, J. A., and E. E. Quinlan. 1986. Survival of *Edwardsiella ictaluri* in pond water and bottom mud. Prog. Fish-Cult., 48: 212–214.

Plumb, J. A., and D. J. Sanchez. 1983. Susceptibility of five species of fish to *Edwardsiella ictaluri*. J. Fish Dis., 6: 261–266.

Plumb, J. A., and T. E. Schwedler. 1982. Enteric septicemia of catfish (ESC): a new bacterial problem surfaces. Aquacult. Mag., 8(4): 26–27.

Plumb, J. A., G. Maestrone, and E. Quinlan. 1987. Use of a potentiated sulfonamide to control *Edwardsiella ictaluri* infection in channel catfish *(Ictalurus punctatus)*. Aquaculture, 62: 187–194.

Post, G., and K. R. Vesely. 1983. Administration of drugs of hyperosmotic or vacuum infiltration or surfactant immersion ineffective for control on intradermally encysted *Ichthyophthirius multifiliis*. Prog. Fish-Cult., 45: 164–166.

Pottinger, T. G., T. A. Moran, and P. A. Cranwell. 1992. The biliary accumulation of corticosteroids in rainbow trout, *Oncorhynchus mykiss,* during acute and chronic stress. Fish Physiol. Biochem., 10: 55–66.

Ragone, L. M., and E. M. Burreson. 1989. The effect of low salinity exposure on *Perkinsus marinus* infections in the eastern oyster, *Crassostrea virginica*. J. Shellfish Res., 8: 470.

Reitan, O., N. A. Hvidsten, and L. P. Hansen. 1987. Bird predation on hatchery reared Atlantic salmon smolts, *Salmo salar* L., released in the River Eira, Norway. Fauna Norv., Ser. A., 8: 35–58.

Roberts, R. J., and C. Sommerville. 1982. Diseases of tilapias. pp. 247–262, In: R. S. V. Pullin and R. H. Lowe-McConnell (Eds.). The biology and culture of tilapias. International Center for Living Aquatic Resources Management, Manila.

Robertson, L., P. Thomas, C. R. Arnold, and J. M. Trant. 1987. Plasma cortisol and secondary stress responses of red drum to handling, transport, rearing density, and a disease outbreak. Prog. Fish-Cult., 49: 1–12.

Robertson, L., P. Thomas, and C. R. Arnold. 1988. Plasma cortisol and secondary stress responses of cultured red drum *(Sciaenops occelatus)* to several transportation procedures. Aquaculture, 68: 115–130.

Rodgers, C. J. 1990. Immersion vaccination for control of fish furunculosis. Dis. Aquat. Org., 8: 69–72.

Rogers, W. A., and J. L. Gaines, Jr. 1975. Lesions of protozoan diseases in fish. pp. 117–141, In: W. E. Ribelin and G. Migaki (Eds.). The pathology of fishes. University of Wisconsin Press, Madison.

Ross, A., J. E. Martin, and V. Bressler. 1968. *Vibrio anguillarum* from an epizootic in rainbow trout *(Salmo gairdneri)* in the U.S.A. Bull. Off. Int. Epiz., 69: 1139–1148.

Rushton-Mellor, S. K. 1992. Discovery of the fish louse, *Argulus japonicus* Thiele (Crustacea: Branchiura), in Britain. Aquacult. Fish. Manage., 23: 269–271.

Sakai, M., S. Atsuta, and M. Kobayashi. 1989. Attempted vaccination of rainbow trout *Oncorhynchus mykiss* against bacterial kidney disease. Bull. Jpn. Soc. Sci. Fish., 55: 2105–2109.

Samuelsen, O. B. 1989. Degradation of oxytetracycline in seawater at two different temperatures and light intensities, and the persistence of oxytetracycline in the sediment from a fish farm. Aquaculture, 83: 7–16.

Schramm, H. L., Jr., M. W. Collopy, and E. A. Okrah. 1987. Potential problems of bird predation for fish culture in Florida. Prog. Fish-Cult., 49: 44–49.

Scott, P. W., and B. Fogle. 1983. Treatment of ornamental koi carp *(Cyprinus carpio)* infected with anchor worms *(Lernaea cyprinacea)*. Vet. Rec., 113: 421.

Seenappa, D., M. C. Nandeesha, P. W. Basarkar, and C. N. Srinivasan. 1985. Effect of pH on survival and hatching of anchor worm *L. bhadraensis*. Environ. Ecol., 3: 107–109.

Selosse, P. M., and S. J. Rowland. 1990. Use of common salt to treat ichthyophthiriasis in Australian warmwater fishes. Prog. Fish-Cult., 52: 124–127.

Shariff, M., and R. J. Roberts. 1989. The experimental histopathology of *Lernaea polymorpha* Yu, 1938 infection in naive *Aristichthys nobilis* (Richardson) and a comparison with the lesion in naturally infected clinically resistant fish. J. Fish Dis., 12: 405–414.

Shariff, M., Z. Kabata, and C. Sommerville. 1986. Host susceptibility to *Lernaea cyprinacea* L. and its treatment in a large aquarium system. J. Fish Dis., 9: 393–401.

Sheldon, W. M., Jr., and V. S. Blazer. 1991. Influence of dietary lipid and temperature on bactericidal activity of channel catfish microphages. J. Aquat. Anim. Health, 3: 87–93.

Shimura, S. 1983. Seasonal occurrence, sex ratio and site preference of *Argulus coregoni* Thorell (Crustacea: Branchiura) parasitic on cultured freshwater salmonids in Japan. Parasitology, 86: 537–552.

Shotts, E. B., V. S. Blazer, and W. D. Waltman. 1986. Pathogenesis of experimental *Edwardsiella ictaluri* infections in channel catfish *(Ictalurus punctatus)*. Can. J. Fish. Aquat. Sci., 43: 36–42.

Sindermann, C. J. 1970. Principal diseases of marine fish and shellfish. Academic Press, New York. 369 p.

Sindermann, C. J., and D. V. Lightner. 1988. Disease diagnosis and control in North American marine aquaculture. Elsevier, New York. 431 p.

Singhal, R. N., S. Jeet, and R. W. Davies. 1986. Chemotherapy of six ectoparasitic diseases of cultured fish. Aquaculture, 54: 165–171.

Snieszko, S. F., and G. L. Bullock. 1968. Freshwater fish diseases caused by bacteria belonging to the genera *Aeromonas* and *Pseudomonas*. Fish Disease Leaflet No. 11. U.S. Bureau of Sport Fisheries and Wildlife, Washington, D.C. 7 p.

Snieszko, S. F., and A. J. Ross. 1969. Columnaris disease of fishes. U.S. Bureau of Sport Fisheries and Wildlife, Fish Disease Leaflet 16. 4 p.

Sparks, A. K. 1993. Invertebrate diseases—an overview. pp. 245–253, In: J. A. Couch and J. W. Fournie (Eds.). Pathobiology of marine and estuarine organisms. CRC Press, Boca Raton, Florida.

Stewart, J. E. 1993. Infectious diseases of marine crustaceans. pp. 53–83, In: J. A. Couch and J. W. Fournie (Eds.). Pathobiology of marine and estuarine organisms. CRC Press, Boca Raton, Florida.

Stickney, R. R. 1988. Aquaculture on trial. World Aquacult., 19: 16–18.

Stickney, R. R., and D. B. White. 1974. Lymphocystis in tank-cultured flounder. Aquaculture, 4: 307–308.

Strout, R. G. 1962. A method for concentrating hemoflagellates. J. Parasitol., 48: 100.

Subasinghe, R. P., and C. Sommerville. 1986. Acquired immunity of *Oreochromis mossambicus* to the ciliate ectoparasite *Ichthyophthirius multifiliis* (Fouquet). pp. 279–283, In: J. L. Maclean, L. B. Dizon, and L. V. Hosillos (Eds.). Proceedings of the First Asian Fisheries Forum, Manila, Philippines, May 26–31. Asian Fisheries Society, Manila.

Tanaka, M., M. Yoshimizu, M. Kusakari, and T. Kimura. 1984. Lymphocystis disease in kurosoi *Sebastes schlegeli* and hirame *Paralichthys olivaceus* in Hokkaido, Japan. Bull. Jpn. Soc. Sci. Fish., 50: 37–42.

Thune, R. L., J. P. Hawke, and R. J. Siebeling. 1991. Vibriosis in the red swamp crawfish. J. Aquat. Anim. Health, 3: 188–191.

Tomasso, J. R., G. J. Carmichael. 1988. Handling and transport-induced stress in red drum fingerlings *(Sciaenops ocellatus)*. Contrib. Mar. Sci., 30: 133–138.

Tsoumas, A., D. J. Alderman, and C. J. Rodgers. 1989. *Aeromonas salmonicida:* development of resistance to 4–quinolone antimicrobials. J. Fish Dis., 12: 493–507.

Turgeon, Y., and Y. Elazhary. 1992. Furunculosis control with anti-*Aeromonas salmonicida* hyperimmune serum (SHAAS). Bull. Aquacult. Assoc. Can., 92(1): 64–67.

van As, J. G., L. Basson, and J. Theron. 1984. An experimental evaluation of the use of formalin to control trichodiniasis and other ectoparasitic protozoans on fry of *Cyprinus carpio* L. and *Oreochromis mossambicus* (Peters). S. Afr. J. Wildl. Res., 14: 42–48.

Ventura, M. T., and J. M. Grizzle. 1987. Evaluation of portals of entry of *Aeromonas hydrophila* in channel catfish. Aquaculture, 65: 205–214.

Ventura, M. T., and I. Paperna. 1985. Histopathology of *Ichthyophthirius multifiliis* infections in fishes. J. Fish Biol., 27: 185–203.

Villee, C. A., W. F. Walker, Jr., and F. E. Smith. 1963. General zoology. Saunders, Philadelphia. 848 p.

Walker, R. 1962. Fine structure of lymphocystis virus of fish. Virology, 18: 503–505.

Walker, R., and K. E. Wolf. 1962. Virus array in lymphocystis cells of sunfish. Am. Zool., 2: 566.

Wellborn, T. L., N. N. Fijan, and J. P. Naftel. 1969. Channel catfish virus disease. Fish Disease Leaflet No. 18. U.S. Bureau of Sport Fisheries and Wildlife, Washington, D.C. 3 p.

Wilson, E. A., E. N. Powell, M.A Craig, T. L. Wade, and J. M. Brooks. 1990. The distribution of *Perkinsus marinus* in Gulf coast oysters: its relationship with temperature, reproduction, and pollutant body burden. Int. Rev. Gesamten Hydrobiol., 75: 533–550.

Winton, J. R., C. K. Arakawa, C. N. Lannan, and J. L. Fryer. 1988. Neutralizing monoclonal antibodies recognize antigenic variants among isolates of infectious hematopoietic necrosis virus. Dis. Aquat. Org., 4: 199–204.

Wise, J. A., S. F. Harrell, R. L. Busch, and J. A. Boyle. 1988. Vertical transmission of channel catfish virus. Am. J. Vet. Res., 49: 1506–1509.

Wolf, K. 1988. Fish viruses and fish viral diseases. Cornell University Press, Ithaca. 476 p.

Wolf, K., and R. W. Darlington. 1971. Channel catfish virus: a new herpesvirus of ictalurid fishes. J. Virol., 8: 525–533.

Wolf, K., and M. E. Markiw. 1982. Ichthyophthiriasis: immersion immunization of rainbow trout *(Salmo gairdneri)* using *Tetrahymena thermophila* as a protective immunogen. Can. J. Fish. Aquat. Sci., 39: 1722–1725.

Wong, G., S. L. Kaattari, and J. M. Christensen. 1992. Effectiveness of an oral enteric coated *Vibrio* vaccine for use in salmonid fish. Immunol. Invest., 21: 353–364.

9 Harvesting, Hauling, and Processing

HARVESTING INTENSIVE CULTURE SYSTEMS

One of the primary advantages of intensive culture is the ease with which harvesting can be accomplished in systems of virtually all types. In closed recirculating water systems as well as in open systems utilizing tanks or raceways, harvesting is often merely a matter of draining the culture chambers and collecting the animals in dip nets. In large raceways and circular tanks, the water may be partially drained and the animals herded into a relatively small volume with the use of movable screens. As the number of animals is reduced through dip netting, the volume confining those remaining can be further decreased by movement of the screens and additional lowering of the water level is necessary. If continuous harvesting is practiced,[175] grader screens can be used in raceways that will allow submarketable individuals to escape while crowding marketable individuals into a small space from which they can be dip netted, often without requiring raceway draining (Figure 103).

In most cases animals harvested from intensive culture systems can be loaded directly into hauling tanks for transport to the processing plant. All such systems, including those located in buildings, should be designed to provide easy access to all culture chambers by hauling vehicles; or some suitable technique, such as the employment of fish pumps, should be used to move the fish from the tanks to the hauling truck. A fish pump can remove fish from culture chambers after they have been crowded. Fish pumps can move fish relatively long distances, can sort the fish in some cases, and when properly designed and operated, will not damage the animals (Figure 104).

Harvesting aquaculture animals from cages can be a relatively simple matter. Cages can be towed to shallow water or to a dock where the fish can be removed with dip nets or fish pumps (Figure 105). If dip nets are used, the fish are typically placed in baskets and carried by hand or lifted by means of a gantry fitted with a block and tackle to the hauling truck or to a frame attached to the dock. A scale hung between the block and tackle and the basket will provide a means of weighing each basket of fish as it is being loaded.

---

[175] Continuous harvesting involves collecting harvestable individuals from each culture chamber on a routine basis and restocking juveniles several times a year. This allows year-round harvesting and keeps culture systems constantly in operation. The technique will only work in conjunction with species that are compatible at various sizes (i.e., not cannibalistic).

Figure 103. Catfish being crowded with a grader that allows submarketable fish to escape while harvestable fish are retained so they can be removed with dip nets.

Lifting cages from the water is not generally feasible since most of them are not sufficiently strong to be removed from the water while filled with harvestable fish. If a cage ruptures during the process, the fish will be lost. An exception is small cages (typically no larger than 1 m^3) of the type used for research (Figure 106). Those cages are not usually heavily stocked and are commonly constructed of

Figure 104. A fish pump being used to harvest rainbow from a raceway. Submarketable fish are automatically returned to the raceway while marketable fish are shunted down a pipeline and directed to the processing plant seen in the background.

Figure 105. A catfish cage culture operation in a lake. For harvesting, the cages are towed to the dock and tilted using the block and tackle. The fish are dipped out and carried in baskets to the hauling truck. As fish are removed, the cage is tilted increasingly and lifted to crowd the fish.

sturdy materials that will accommodate removal from the water while the fish are present.

Net-pen harvesting is more involved than that associated with cages, because net pens are much larger in both surface area and depth. Typically, harvesting involves pulling up the netting to reduce the volume of the net pen (Figure 107), dip netting out some of the fish, pulling up more net to concentrate the remaining fish, dipping more fish, and so forth until all the fish have been harvested. Submarketable individuals are usually placed in a different net-pen for additional growout. If the facility is attached to the land, it may be possible to drive vehicles relatively close to pens that are being harvested. For offshore facilities, the fish are transported to the shore on boats.

HARVESTING EXTENSIVE CULTURE SYSTEMS

The harvesting of ponds is facilitated when construction has resulted in regular shapes with properly sloped banks, proper depth, easy access to at least one side by vehicles, large drain lines that allow rapid and complete emptying, and the incorporation of a harvest basin near the drain. Many ponds lack one or more of those features. Each is important, but the regularity of pond shape is of least concern. In some instances it is more convenient to lay out a series of ponds that conform to natural variations in terrain than to greatly alter the site to accommodate square or rectangular ponds. It is important to have smooth and clean pond bottoms. If culture

Figure 106. Experimental cages (approximately 1 m high and 0.5 m in diameter) can often be lifted from the water while full of fish. Most commercial cages would rupture under similar circumstances.

ponds are not fitted with drains, the water can be pumped out of them at the time of harvest.

It is common practice in the catfish industry to use the continuous harvesting approach (which is, in reality, intermittent harvesting). Ponds can be kept in production for a period of 3 or more years by harvesting marketable fish every few weeks and restocking with replacement fingerlings a few times a year. Many other species of fish are maintained in ponds through a single growing season and are then harvested *en masse*. No matter which technique is used, there will come a time when every culture pond will have to be drained. As time passes, ponds that are used for continuous harvesting accumulate organic matter, which leads to impaired water quality and reduced productivity. For example, phosphorus levels increase from year to year in undrained ponds (Hollerman and Boyd 1985). In addition, stunted fish will continue to eat but may never reach market size. As their numbers increase (a result of intermittent restocking), the impact on culture economics can become significant. In most cases, even ponds used for continuous harvesting are

Figure 107. When a netpen is harvested, the netting is pulled up and draped over the railing.

drained and completely harvested every few years. The bottoms are allowed to dry, are disked to promote oxidation of organic matter, and are then placed back in production for another cycle.

Harvesting should be planned in advance, and precautions should be taken to avoid stress to the extent possible, particularly when the harvested animals are to be live-hauled to market. Feeding should be discontinued at least 24 hr before harvesting, automatic feeders and other obstacles to seining should be removed, and water quality should be examined to ensure that optimum conditions are maintained. Of particular importance is temperature. Harvesting should be undertaken during the coolest part of the day to help avoid stress. Thus, harvest operations should be conducted as early in the morning as possible during warm weather.

While seining of full ponds may be undertaken when subsamples of fish are collected or in conjunction with the continuous harvest technique (Figure 108), harvesting is usually conducted in association with partial draining prior to initial seining, even when harvest basins are present. Typically, the water level is reduced, perhaps by half, and a seine is then passed through the pond to capture a portion of the fish present. The volume of the pond is then further reduced and, depending on the size of the pond, subsequent seine hauls may be made before the final harvest is undertaken in the harvest basin. In ponds with no harvest basins, seine hauls are made until the pond volume is reduced to 10% or less of original volume; then the remainder of the water is removed, and the remaining fish are harvested by hand.

Small ponds can be seined by hand (Figure 108), while tractors or trucks are required to pull seines in large ponds (Figure 109). When the fish have been concentrated in a harvest basin, they may be dip-netted into baskets or pumped (Figure 110), eventually ending up in live-hauling trucks.

Harvest seines should be approximately 1.5 times as long as the pond is wide to

Figure 108. Seining a full pond can be difficult, though the method can be used for sampling purposes and when the continuous harvesting method is used.

Figure 109. Removing channel catfish from a commercial culture pond. (Photograph courtesy of Marty Brunson)

Figure 110. A large portable fish pump being used to harvest tilapia.

ensure that they will bow out during harvesting operations (Figure 108). For additional capacity, seines may be equipped with a bag located in the middle (Figure 111). Seine bags provide additional capacity while reducing escapement. The depth of the seine should be at least twice the depth of the water in the pond being harvested. The mesh size should be small enough to retain harvestable animals but not smaller, since reduction in mesh size increases resistance of the net in the water making it more difficult to pull. In addition, the cost of netting increases with decreasing mesh size.

The upper rope on a seine (float or head line) typically consists of a rope to which floats made from cork, plastic, Styrofoam, or some other buoyant material are strung at intervals. The bottom rope (lead line), is designed to keep the seine in contact with the sediments. Lead lines may be ropes to which lead weights are attached at intervals or ropes that have a lead core. In ponds with muddy bottoms, traditional lead lines are often not efficient since they tend to burrow into the sediments and dig up mud, which weighs down the seine and causes it to roll up. In soft-bottomed ponds, seines with mud lines tend to be more effective. A mud line is composed of a number of relatively small diameter ropes tied loosely together. The ropes are made from a material that readily absorbs water (e.g., cotton). Mud lines tend to maintain contact with pond bottoms without digging in or lifting off the sediments; thus escapement under seines so equipped is reduced. Such lines tend to wear rapidly if used in ponds with firm sediments.

Harvesting is the most labor-intensive activity associated with an aquaculture

Figure 111. Seines having a bag (indicated by the floats behind the main body of the seine) may be more efficient than those without that feature.

operation. Several persons are required on the typical seine crew, even if trucks or tractors are used to do the bulk of the work. The aquaculturist may be required to employ additional help during harvesting, and the added expense should be taken into consideration during business planning.

SPECIALIZED HARVESTING TECHNIQUES

Harvesting sessile marine animals such as oysters and mussels requires techniques different from those used for motile species. Oysters reared on the bottom can be harvested manually by picking intertidal oysters up at low tide or by tonging or dredging for subtidal animals. A mechanical oyster harvester developed in South Carolina was described by Collier and McLaughlin (1983, 1984). Oysters grown in trays and on long lines suspended from rafts are manually harvested. Mussels are grown on the bottom, on poles, and on longlines. Harvesting technique is based on how the mussels are grown, but generally mirrors the methods used for oysters. Clams can be dredged or grown in trays and manually harvested. Scuba divers may also be used for harvesting benthic animals such as abalone.

Crawfish are harvested by trapping, though mechanical harvesting devices are being developed (Romaire 1988). During the harvest season, which extends for 60 to 180 days in Louisiana (Romaire 1988), traps, such as the pillow trap shown in Figure 112, are set out at intervals over the pond bottom. Various factors influence the efficiency with which crawfish are trapped. The catch can be affected by water quality, the amount of forage present, crawfish density, climate, trap design, the

Figure 112. A crawfish pillow trap being emptied.

bait being used and, of course, trap density (Romaire 1989). Romaire and Pfister (1983) compared catch rates with trap densities of 25, 50, 75, and 100 traps/ha and found that 3.3 times more crawfish were captured at the highest density as compared with the lowest. The catch rates at 50 and 75 traps/ha were 2.8 and 1.6 times the rate at 25 traps/ha. A variety of trap designs have been used. Pfister and Romaire (1983) compared the effectiveness of 10 different trap designs in retaining captured crawfish.

Traps are baited with dead fish or commercially manufactured bait and are usually checked once or twice a day. One impetus for developing manufactured baits relates to the increasing cost of the fresh fish that have traditionally been used, such as gizzard shad (Cange et al. 1985, Burns and Avault 1986). Rach and Bills (1987) evaluated the effectiveness of three types of commercial crawfish baits as compared with dead fish. They indicated that commercial baits were easier to handle, did not have noxious odors, and did not require refrigeration—all advantages over dead fish. Meriwether (1988) found that commercial crawfish bait containing about 20% protein resulted in the highest crawfish yields.

In small ponds the culturist may wade through the pond to check traps, but in most instances boats are used. The boat may be operated by one or two people. If two people are involved, one drives the boat while the other handles the traps. As each trap is approached, a newly baited trap is placed in the water and the trap that has been in the pond is removed, emptied, and rebaited. It will be placed in position at the location of the next trap in the line. With the technique described, it is possible to keep the boat in continuous motion during harvesting.

Following harvest, crawfish are often held in flowing water for sufficiently long to allow their digestive tracts to be purged. The process also leads to improvement in overall appearance and increases the market value of the animals. Lawson and Drapcho (1989) described the process and reported that purging in spray systems works as well as when flowthrough systems are employed. Purging in spray systems requires no more than 40 hr (Lawson et al. 1990). At a spray rate of 3.6 l/min/m^2 and a stocking density of 24.4 kg/m^2 of crawfish, mortality is less than 5% during the purging period.

Crawfish producers began producing soft-shell crawfish in 1985 (Culley and Doubinis-Gray 1989). The technique involves trapping immature animals, placing them in culture trays at high density, and feeding them. Individuals that are about to molt are identified and removed to molting trays to prevent cannibalism. Newly molted crawfish are packaged and frozen. Soft-shell crawfish bring a premium price and can be eaten whole, whereas processing for tail meat results in only about 15% dress-out. Huner (1988) estimated that 1,500 tons of soft-shell crawfish could be absorbed by existing markets.

Some ponds have drains that empty into a receiving ditch immediately adjacent to the pond levee. In such ponds, an alternative approach to collecting fish or shrimp in a harvest basin during pond draining is to place a bag of appropriate mesh size over the effluent end of the drain pipe. As water is released from the pond, many of the aquaculture animals are be swept into the bag. The technique works fairly well with shrimp and other species that are not particularly strong swimmers in strong currents.

ANESTHESIA

Anesthetics are not normally used in conjunction with harvest at market size, but they are frequently used when fish are being handled. Handling can occur while determining weights for examining growth and calculating new feeding rates, when treating fish for diseases, while grading and moving fish into new culture chambers, during surgery, and in some cases during spawning. Anesthesia prior to hauling is one means of reducing stress responses in largemouth bass (Carmichael et al. 1984).

Quinaldine and tricaine methanesulfonate (MS-222) have been widely used as fish anesthetics in the past. Hunn and Allen (1974) indicated that channel catfish can be anesthetized with 30 mg/l of quinaldine or 150 mg/l of MS-222. Schoettger and Steucke (1970a) reported that 10 to 20 mg/l of quinaldine or 100 to 150 mg/l of MS-222 are effective for anesthetizing northern pike, muskellunge, and walleye. Rainbow trout can be anesthetized with 5 mg/l of quinaldine or 20 to 30 mg/l of MS-222 (Schoettger and Steucke 1970b). Quinaldine at 2 mg/l or MS-222 at 21 mg/l have been effective for anesthetizing striped bass for spawning (Bonn et al. 1976).

Quinaldine has come into disfavor because of potential toxicity to people. MS-

222 is effective but quite expensive and can cause significant reduction in pH (Allen and Harman 1970). Thus, over the past few years, efforts have been made to find additional anesthetics for use with aquaculture species. Suitable anesthetics should be economical, and the animals should recover quickly and with low mortality if recommended dosages are used. Also, there should be no long-term residues in the tissues of anesthetized fish.

Bardach et al. (1972) reported that Chinese carps are often transported in 6.7 to 7.7 μg/l of sodium barbital or 1 to 4 g/l of urethane when water temperatures are within the range of 25.5 to 32°C. Sodium bicarbonate has also been used effectively as an anesthetic for brook trout and common carp (Brooke et al. 1978). Those authors felt that the result was due to the release of carbon dioxide, which we now know is an effective anesthetic.

Carbon dioxide anesthesia can be produced by adding carbon dioxide gas or carbonic acid. Iwama et al. (1991) found that adding $NaHCO_3$ with carbon dioxide gas reduced stress on juvenile steelhead trout. Carbon dioxide anesthesia in combination with low temperature was found to be effective in conjunction with transporting adult carp (Yokoyama et al. 1989). Effective concentrations of carbon dioxide for inducing and maintaining anesthesia in carp were discussed by Yoshikawa et al. (1988). Partial pressures of 200 to 250 mm Hg were found effective for inducing anesthesia, and the fish could be maintained under anesthesia at partial pressures of from 100 to 125 mmHg.

Benzocaine was evaluated as an anesthetic for common carp, tilapia, and various salmonids (Ferreira et al. 1984, Gilderhus 1989, 1990, Gilderhus et al. 1991). The effectiveness of the anesthetic varied with not only the concentrations used (25 to 100 mg/l), but also with temperature, which is related to the different metabolic rates of the fishes involved at various temperatures (Ferreira et al. 1984). For example, striped bass can be effectively anesthetized with 55 mg/l of benzocaine at 22°C as compared with 80 mg/l at 11°C (Gilderhus et al. 1991). Recovery from the anesthetic took longer as temperature decreased. Gilderhus (1990) evaluated benzocaine as an anesthetic for chinook salmon and Atlantic salmon and found that the chemical worked within 3.5 min at concentrations of 25 to 30 mg/l. The fish could tolerate 30 mg/l of the anesthetic for 20 min, but mortalities occurred after 25 min. Fifteen minute exposures were lethal to chinook salmon at a benzocaine concentration of 35 mg/l and to Atlantic salmon at 40 mg/l.

Allen (1988) examined benzocaine residues in largemouth bass and rainbow trout tissues to determine how quickly the level declined following anesthesia. Control levels were reached within a few hours. Fish meal manufactured from Pacific salmon that had been anesthetized with benzocaine or MS-222 contained residues of 45.1 μg/g and 47.7 μg/g, respectively.

Limsuwan et al. (1983) and Plumb et al. (1983) discussed the use of etomidate (ethyl—1-methylbenzyl-imidazole-5-carboxylate) as an anesthetic. Both papers examined the anesthetic in conjunction with channel catfish, golden shiners, and bluegills; and Plumb et al. (1983) also examined striped bass. Catfish were able to tolerate exposure of 0.4 to 3.6 mg/l for 80 min (Plumb et al. 1983). Anesthesia was induced at concentrations of 0.8 to 1.4 mg/l. Amend et al. (1982) found that four species of ornamental fish (zebra danio, black tetra, angelfish, and southern platyfish) could be anesthetized with 2.0 to 4.0 mg/l etomidate. Falls et al. (1988) evaluated etomidate as an anesthetic for red drum and determined that a dose of 1.8 mg/l was effective. Exposure to 0.8 mg/l required unacceptably long times for onset and recovery from anesthesia, while 8 mg/l was excessive.

Carp have been effectively anesthetized with 2-phenoxyethanol at 400 to 600 mg/l (Yamamitsu and Itazawa 1988). Irrigation with anesthetic-free fresh water led to rapid recovery. Suzuki and Sekizawa (1979) examined the rate at which another anesthetic, 2-amino-4-phenylthiazole, dissipated from the bodies of rainbow trout.

LIVE-HAULING

Some species are placed immediately on ice after harvest (e.g., shrimp) or can be hauled in refrigerated trucks or even at ambient temperature. Lobsters and crabs are hauled alive on ice in many instances. Molluscs can be hauled without refrigeration for at least short distances, and crawfish are often hauled alive in bags or other types of containers. In each of the examples above, with the exception of shrimp, the animals reach the processing plant alive. Most cultured fish should also reach processing plants alive, but they cannot be hauled out of water. Whether fingerlings are being transferred from one pond to another or one farm to another, and regardless of the purpose for which fingerlings or larger fish are being transported, they should be moved in water and should be handled and hauled under conditions that keep stress to a minimum.

While the average person in the United States and Canada may not even realize that fish are sharing the highways with them in live-hauling trucks, the fact is that at least 61 species of fish are transported on the highways of those two countries. According to a survey conducted by Carmichael and Tomasso (1988) fully 50% of the fish transported in the United States and Canada are from five species: channel catfish, rainbow trout, largemouth bass, walleye, and bluegill.

Many farmers own their own small hauling tanks that can be placed on the bed of a pickup truck (Figure 113) for use on the farm and for moving fish short distances. Some fish farmers possess their own large live-hauling trucks (Figure 114), but most depend on firms that specialize in custom live hauling. The live-hauling truck should be available at the time the fish are concentrated for removal from the culture chamber. In ponds, seined fish may be placed in live cars (pens constructed of netting or wire and placed in the pond), from which they are transferred to the hauling tanks as previously described.

Live-hauling tanks should be insulated if changes in water temperature are undesirable. Hauling tank water can become too hot for salmonids during summer in relatively short periods of time, so most hauling tanks used for coldwater fish are insulated. For long-distance hauling, insulated tanks should always be used, no matter what species is being hauled. The tanks can be constructed of fiberglass, wood, or metal, with insulated tanks having Styrofoam or some other type of appropriate material located between two layers of construction material. Insulated tanks

Figure 113. A fiberglass live-hauling tank on the bed of a pickup truck. The tank shown is not insulated. Note the two agitators on top (one lying on its side, the other in operating position).

Figure 114. Ten-wheel and even 18-wheel specialized live-hauling trucks are used by government agencies, some individual fish farmers, and custom live-hauling firms for transporting fish.

are more expensive and much heavier than uninsulated tanks. Refrigeration units, operated by generators, can be used in conjunction with large live-hauling trucks to maintain temperature.[176] Ice can be added to maintain water temperature in uninsulated tanks, but care should be taken not to add too much ice, or the water temperature may be reduced beyond what is desired. Frequent checks should be made to determine that the proper temperature is being maintained.

A live-hauling tank is not merely a water-tight box in which aquatic animals are placed. It must provide the proper environmental conditions to maintain the fish in good condition until they reach their final destination, whether that is another pond, a bait store, or a processing plant. Because of the high concentration of animals that are often carried in live-hauling tanks, some type of aeration must be provided. Mechanical agitators connected to the 12-volt electrical system of the truck are popular. Aeration can also be provided through airstones supplied from tanks of compressed air, compressed oxygen, or liquid oxygen. Gas cylinders and liquid oxygen tanks must be secured to the vehicle in accordance with state law. Compressed air and oxygen cylinders can be carried on pickup trucks, but liquid oxygen systems are usually reserved for larger vehicles. The driver should frequently check to determine that the aeration system is functioning properly.

As mentioned in the previous section, anesthesia is sometimes used in conjunction with live hauling, but fish being transported to a processing plant should not be anesthetized with chemicals that can leave residual levels in the flesh of the animals. Carbon dioxide may an acceptable anesthetic for fish going to processors since it is a natural byproduct of respiration. In addition to considering the use of anesthesia, the culturist should withhold feed at least 24 hr in advance of hauling. In conjunction with hauling largemouth bass, Carmichael et al. (1984) recommended not feeding the fish for 72 hr before transporting them; treating them for diseases; hauling them at a cool temperature, and adding salt,[177] antibiotic, and a mild anesthetic to hauling tank water to reduce stress. Hauling of largemouth bass would normally be in conjunction with stocking programs, though the rearing and sale of largemouth bass for human consumption is legal in California.[178]

Live hauling trucks are utilized for transporting fingerlings and marketable fish, while fish fry and invertebrate larvae are most simply shipped in polyethylene bags partially filled with water and topped off with oxygen, then sealed. When placed in an insulated box, properly packaged fish and invertebrates will remain alive and well for at least 24 hr. With modern transportation methods and schedules, it is now possible to ship fish virtually anywhere in the world by this method.

---

[176] Heaters can also be used in instances where the temperature would be reduced significantly during live hauling.

[177] Sodium chloride at 2 g/l (2 ppt) has often been used, though salt concentrations approaching the level in the blood may be more appropriate. Some live haulers prefer to use 8 g/l.

[178] George Ray, personal communication.

DEPURATION

Aquatic species, including aquaculture organisms, are sometimes contaminated with pathogens, trace metals, and organic chemicals when harvested. While it is desirable to harvest and market only healthful animals, it is possible in some cases to render contaminated organisms safe through a process, discussed in Chapter 6, known as depuration. The process involves placing contaminated animals in disinfected, recirculating water for a period sufficiently long to allow them to expel the microorganism or chemical contaminant that made them unwholesome. Water used for depuration can be sterilized with ultraviolet light, ozone, or chlorine (Rodrick 1988). Rodrick and Schneider (1989) evaluated sterilization of recirculated water used to depurate shellfish through the use of ultraviolet light and ozone and found that ultraviolet radiation was sometimes ineffective.

Depuration has also been used to eliminate foreign matter from animals. For example, sand can be substantially removed from oysters if they are depurated for 24 hr (Chellappan 1991).

FEE FISHING

Fee-fishing operations may be independent of other aquacultural activities, or they may be integrated into the production of food animals for wholesale. Fee fishing provides sportfishermen with an opportunity to catch high-quality fish and ensures a reasonable degree of success. This concept has been widely employed by salmonid and catfish producers but has not yet been developed for marine fish or aquacultured invertebrates, though fee-fishing operations for those types of animals could become popular in the future.

In some fee-fishing operations one or more ponds on an aquaculture facility are opened to the public. Other facilities are strictly fee-fishing lakes, the owners of which purchase the fish that they stock or produce them at a facility located elsewhere or at least out of sight from the fee fishermen. Fee-fishing lakes are often more natural in appearance than production ponds. They may be irregular in shape, and can be landscaped to provide an esthetically pleasing experience for the anglers. Picnic tables and other amenities may also be provided. Anglers are expected to keep all the fish caught and pay on the basis of the weight of fish they remove from the lake. In some cases an entry fee may also be required.

Typically, the operator of a fee-fishing operation will stock the ponds with catchable fish. A few very large fish may also be stocked, and there have even been cash awards or prizes offered to the angler who catches a particular tagged fish. Most of the fish are small but of appropriate size for consumption. As fish are caught and removed, the operator will restock with more catchable animals. Since most fee-fishing operations make their money on the basis of weight of fish taken, it makes sense to keep ponds well stocked and to limit feed so that the fish are hungry and will readily attack a baited hook. A good fishing experience is not only good for profits; it also helps ensure that customers will return for additional enjoyment.

Fee-fishing operations tend to work well in the vicinity of large cities and in other areas where public fishing opportunities may be limited, overcrowded by anglers, or not well maintained. People living in urban environments may find it difficult to mount major fishing expeditions, but they can often be tempted to fish with some regularity if a well maintained fee-fishing operation is located within easy commuting distance.

Well-operated fee-fishing facilities offer an array of services to anglers. Bait is commonly sold, and the facility may rent tackle. Sales of tackle can be lucrative as well. Many fee-fishing operators will clean the catch and ice it down for an additional fee. Cleaning stations where anglers can clean their catch before leaving the facility may also be made available. Public restrooms (often in the form of portable toilets) are a must, and refreshments should be made available through vending machines or at a small sales stand.

PROCESSING AND MARKETING

The presence of commercial processing facilities in various parts of the United States is a function of the volume of aquaculture produce reared in the immediate vicinity and the proximity of commercial fisheries producing the same or a similar commodity. In general, producers of such marine species as salmon and other finfish, shrimp, clams, mussels, abalone, and oysters should have little or no difficulty with processing as long as the aquaculture is conducted in regions where the same type of species is being produced commercially. The same is true for crawfish and a few other freshwater species. Processing plants established for the commercial catch can also accommodate aquacultured products. For seasonal crops such as salmon, a processing plant that might otherwise operate only a few months a year could be operated year-round if cultured fish are processed during periods when the wild catch is not available.

The two most widely cultured finfish species in the United States, channel catfish and rainbow trout, are not commercially fished to any extent, so processing plants for those species have been established in the regions where the fish are being most widely produced. Catfish processing plants are concentrated in Mississippi, and the large trout producers in Idaho also operate processing plants. For catfish and trout produced well away from the production centers, small processing plants may be operated by individual farmers. Some of those farmers may accept fish from other farms in the region. In many cases production levels are not high enough to support a large processing plant that is independent from a production facility.

While some processing plants handle a variety of species, most limit their activities to one or a small number of species. A crawfish processing plant would not, for example, also process catfish or oysters. Specialized equipment is required for each species being processed; also, personnel require different training depending on the species being processed.

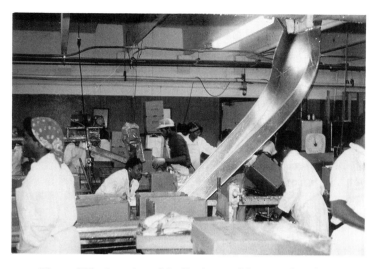

Figure 115. A portion of the line in a catfish processing plant.

Some species are processed largely by hand, while automated equipment is used for the processing of others. Automated processing equipment has been designed for many species of aquaculture importance, but economics may require the use of hand labor. Economic considerations include the need to provide jobs, reduce labor costs, and factor in the costs the equipment that will be used. Catfish processing plants vary in production methods from hand labor alone to using machines for heading, gutting, and skinning. Further processing (e.g., production of steaks or fillets) is often done by hand even in automated plants (Figure 115).

Value-added processing and various levels of automation can also be seen in trout processing plants (Figure 116). A number of different products are offered by the trout industry, including trout stuffed with crab and frozen.

As typified by the catfish industry, fish often arrive at the processing plant alive (Figure 117). Once each truckload of fish has been checked for off-flavors[179] the fish are offloaded into raceways supplied with flowing high-quality water. The time the fish spend in the raceway depends on how many fish are awaiting processing but is usually no more than a few hours.

When the time comes for processing, the fish are crowded into a basket that is suspended from an overhead gantry (Figure 118). The basket of fish is allowed to drain for a few seconds, and the fish are then stunned with electric current before being dropped into a hopper that leads them to the processing line. The fish are processed within minutes of being stunned.

The various forms of processed fish that are produced in a plant are typically chilled in water maintained just above freezing. Once chilled the processed fish may

---

[179] See Chapter 6 for a discussion of off-flavors and how the catfish industry deals with the problem.

Figure 116. A view in a trout processing plant.

be packed in ice for shipment to restaurants or retail outlets, or they may be frozen prior to shipment.

A large amount of crawfish is sold live to restaurants, where they are boiled and served whole. A significant amount also finds its way to processors, where the animals are also delivered alive. They may be processed immediately or maintained at about 5°C until they are processed (Moody 1988). Processing involves washing,

Figure 117. A live-hauling truck being unloaded at a catfish processing plant in Mississippi.

Figure 118. After being crowded to one end of a holding raceway, catfish are removed in a basket attached to a gantry and stunned with electric shock.

blanching, peeling, and deveining. The tails are the edible portion, comprising only about 15% of the weight of an individual crawfish. Processing is followed by marketing either chilled or frozen. Blanching helps destroy proteolytic enzymes thereby extending shelf life of the product. Properly blanched and frozen crawfish tails can be stored for 6 to 8 months (Moody 1988). There are over 100 crawfish processing plants in Louisiana (Moody 1989).

Oyster processing continues to require a great deal of manual labor. Each oyster is typically opened and the meat removed from the shell by hand. Sometimes a machine is used to create an opening at the hinged end of the oysters to provide access for the worker to insert an oyster knife. Workers are paid on the basis of the number of oysters that they can shuck during a shift.

A number of other examples of how processing plants operate could be provided, but the major point is that the animals should be alive (or very fresh) when they are processed, they need to be properly handled to ensure that proper product quality is maintained, and they need to be processed into forms that the consumers find attractive.

For the individual farmer who does not have access to a processing plant, marketing can be a significant problem. A small processing area can be established to prepare products for distribution to local restaurants and grocery stores, but costs for establishing and maintaining them may be very high. As mandatory seafood inspection is implemented in the United States, it may become impractical for individuals to maintain and operate their own processing facilities.

Whether the aquaculturist is actively involved in processing or not, it is incumbent upon aquatic animal producers to deliver the best possible quality product to the marketplace. The culturist who is involved with the production of fish or shellfish for human consumption should remember that the job is not complete until someone wraps a lip around the product. Culturists of bait minnows, koi, goldfish, and tropical fishes do not have to face taste tests, but they too have an obligation to market only animals that are in top condition if they want to maintain their list of customers.

LITERATURE CITED

Allen, J. L. 1988. Residues of benzocaine in rainbow trout, largemouth bass, and fish meal. Prog. Fish-Cult., 50: 59–60.

Allen, J. L., and P. D. Harman. 1970. Control of pH in MS–222 anesthetic solutions. Prog. Fish-Cult., 32: 100.

Amend, D. F., B. A. Goven, and D. G. Elliot. 1982. Etomidate: effective dosages for a new fish anesthetic. Trans. Am. Fish. Soc., 111: 337–341.

Bardach, J. E., J. H. Ryther, and W. O. McLarney. 1972. Aquaculture. Wiley-Interscience, New York. 868 p.

Bonn, E. W., W. M. Bailey, J. D. Bayless, K. E. Erickson, and R. E. Stevens (Eds.). 1976. Guidelines for striped bass culture. Southern Division, American Fisheries Society, Washington, D.C. 103 p.

Brooke, H. E., B. Hollender, and G. Lutterbie. 1978. Sodium bicarbonate, an inexpensive fish anesthetic for field use. Prog. Fish-Cult., 40: 11–13.

Burns, C., and J. W. Avault, Jr. 1986. Artificial baits for trapping crawfish (*Procambarus* spp.): formulation and assessment. J. World Maricult. Soc., 16: 368–374.

Cange, S. W., C. Burns, J. W. Avault, Jr., and R. P. Romaire. 1985. Testing artificial baits for trapping crawfish. La. Agri. 29(2): 3 ff.

Carmichael, G. J., and J. R. Tomasso. 1988. Survey of fish transportation equipment and techniques. Prog. Fish-Cult., 50: 155–159.

Carmichael, G. J., J. R. Tomasso, B. A. Simco, and K. B. Davis. 1984. Characterization and alleviation of stress associated with hauling largemouth bass. Trans. Am. Fish. Soc., 113: 778–785.

Chellappan, N. J. 1991. Processing of oyster meat for freezing. Fish. Technol. Soc. Fish Technol. Kochi, 28: 122–124.

Collier, J. A., and D. M. McLaughlin. 1983. A mechanical oyster harvester for South Carolina estuaries. J. World Maricult. Soc., 14: 297–301.

Collier, J. A., and D. M. McLaughlin. 1984. A mechanical oyster harvester for South Carolina estuaries. J. Shellfish Res., 4: 85.

Culley, D. D., and L. Doubinis-Gray. 1989. Soft-shell crawfish production technology. J. Shellfish Res., 8: 287–291.

Falls, W. W., G. K. Vermeer, and C. W. Dennis. 1988. Evaluation of etomidate as an anesthetic for red drum, *Sciaenops ocellatus*. Contrib. Mar. Sci., 30: 37–42.

Ferreira, J. T., H. J. Schoonbee, and G. L. Smit. 1984. The anesthetic potency of benzocaine-hydrochloride in three freshwater fish species. S. Afr. J. Zool., 19: 46–50.

Gilderhus, P. A. 1989. Efficacy of benzocaine as an anesthetic for salmonid fishes. N. Am. J. Fish. Manage., 9: 150–153.

Gilderhus, P. A. 1990. Benzocaine as a fish anesthetic: efficacy and safety for spawning-phase salmon. Prog. Fish-Cult., 52: 189–191.

Gilderhus, P. A., C. A. Lemm, and L. C. Woods III. 1991. Benzocaine as an anesthetic for striped bass. Prog. Fish-Cult., 53: 105–107.

Hollerman, W. D., and C. E. Boyd. 1985. Effects of annual draining on water quality and production of channel catfish in ponds. Aquaculture, 46: 45–54.

Huner, J. V. 1988. Soft shell crawfish industry. pp. 28–42, In: L. H. Evans, and D. O'Sullivan (Eds.). Proceedings, First Australian Shellfish Aquaculture Conference, Perth, Australia, Oct. 23. Curtin University of Technology, Perth.

Hunn, J. B., and J. L. Allen. 1974. Urinary excretion of quinaldine by channel catfish. Prog. Fish-Cult., 36: 157–159.

Iwama, G. K., T. Y. Yesaki, and D. Ahlborn. 1991. The refinement of the administrations of carbon dioxide gas as a fish anesthetic: the effects of varying the water hardness and ionic content in carbon dioxide anesthesia. ICES Council Meeting Papers. International Council for the Exploration of the Sea, Copenhagen, Denmark. 29 p.

Lawson, T. B., and C. M. Drapcho. 1989. A comparison of three crawfish purging treatments. Aquacult. Eng., 8: 339–347.

Lawson, T. B., H. Lalla, and R. P. Romaire. 1990. Purging crawfish in a water spray system. J. Shellfish Res., 9: 383–387.

Limsuwan, C., J. M. Grizzle, and J. A. Plumb. 1983. Etomidate as an anesthetic for fish: its toxicity and efficacy. Trans. Am. Fish. Soc., 112: 544–550.

Meriwether, F. H. 1988. A preliminary comparison of manufactured and natural crawfish baits in crawfish/rice ponds. J. World Aquacult. Soc., 19: 166.

Moody, M. W. 1988. Crawfish processing. J. Shellfish Res., 7: 212.

Moody, M. W. 1989. Processing of freshwater crawfish: a review. J. Shellfish Res., 8: 293–301.

Pfister, V. A., and R. P. Romaire. 1983. Catch efficiency and retentive ability of commercial crawfish traps. Aquacult. Eng., 2: 101–118.

Plumb, J. A., T. E. Schwedler, and C. Limsuwan. 1983. Experimental anesthesia of three species of freshwater fish with etomidate. Prog. Fish-Cult., 45: 30–33.

Rach, J. J., and T. D. Bills. 1987. Comparison of three baits for trapping crayfish. N. Am. J. Fish. Manage., 7: 601–603.

Rodrick, G. E. 1988. Bacterial and viral elimination in commercial plants. J. Shellfish Res., 7: 132.

Rodrick, G. E., and K. Schneider. 1989. Depuration of vibrios from Florida shellfish. J. Shellfish Res., 8: 450–451.

Romaire, R. P. 1988. Overview of harvest technology used in commercial crawfish culture. J. Shellfish Res., 7: 210–211.

Romaire, R. P. 1989. Overview of harvest technology used in commercial crawfish aquaculture. J. Shellfish Res., 8: 281–286.

Romaire, R. P., and V. A. Pfister. 1983. Effects of trap density and diel harvesting frequency on catch of crawfish. N. Am. J. Fish. Manage., 3: 419–424.

Schoettger, R. A., and E. W. Steucke, Jr. 1970a. Quinaldine and MS–222 as spawning aids for northern pike, muskellunge and walleyes. Prog. Fish-Cult., 32: 199–201.

Schoettger, R. A., and E. W. Steucke, Jr. 1970b. Synergistic mixtures of MS–222 and quinaldine as anesthetics for rainbow trout and northern pike. Prog. Fish-Cult., 32: 202–205.

Suzuki, A., and Y. Sekizawa. 1979. Residue analyses on 2–amino–4–phenylthiazole, a piscine anesthetic, in fishes. 4. GC/MS analysis on rainbow trout. Bull. Jpn. Soc. Sci. Fish., 45: 167–171.

Yamamitsu, S., and Y. Itazawa. 1988. Effects of an anesthetic 2–phenoxyethanol on the heart rate, ECG and respiration in carp. Bull. Jpn. Soc. Sci. Fish., 54: 1737–1746.

Yokoyama, Y., H. Yoshikawa, S. UIeno, and H. Mitsuda. 1989. Application of CO_2-anesthesia combined with low temperature for long-term anesthesia in carp. Bull. Jpn. Soc. Sci. Fish., 55: 1203–1209.

Yoshikawa, H., Y. Ishida, S. Ueno, and H. Mitsuda. 1988. The use of sedating action of CO_2 for long-term anesthesia in carp. Bull. Jpn. Soc. Sci. Fish., 54: 545–551.

APPENDIX A
Conversion Factors for Units of Weight and Measurement in the English and Metric Systems (Bold Face Units Are Those Most Often Utilized by Aquaculturists.

Unit of Measurement	Conversion Factor[a]
Acre	43,450 square feet
	0.04 hectare
Acre-foot	43,560 cubic feet
	1,233.5 cubic meters
Angstrom (Å)	10^{-10} meter
Centimeter	0.39 inch
	0.01 meter
	10 millimeters
Cubic centimeter (cm^2)	0.06 cubic inch
	1 milliliter
Cubic foot (ft^3)	7.48 gallons (U.S. liquid)
	0.03 cubic meter
	28.32 liters
Cubic inch (in^3)	16.39 cubic centimeters
Cubic meter (m^3)	35.3 cubic feet
	264.2 gallons (U.S. liquid)
	1.3 cubic yards
	1,000 liters
Foot (ft)	0.3 meter
	12 inches
Gallon, U.S. liquid (gal)	231 cubic inches
	0.13 cubic foot
	3.8 liters

Unit of Measurement	Conversion Factor[a]
Gram (g)	0.04 ounce (avoirdupois)
Hectare (ha)	2.47 acres
	10,000 square meters
Inch (in)	2.54 centimeters
Kilogram (kg)	2.2 pounds (avoirdupois)
	1,000 grams
Kilometer (km)	0.6 mile
	1,000 meters
Liter (l)	0.26 gallon (U.S. liquid)
	1.06 quarts (U.S. liquid)
Meter (m)	39.37 inches
	1,000 millimeters
	100 centimeters
Micron (μ)	10^{-6} meter
Mile, statute (mi)	5,280 feet
	1.6 kilometer
Mile, nautical	6,080 feet (U.S. Navy)
Millimeter (mm)	0.04 inch
	0.1 centimeter
Ounce, U.S. fluid (oz)	1.8 cubic inches
	29.57 cubic centimeters
	29.57 milliliters
Ounce, apothecary (oz)	31.1 grams
Ounce, avoirdupois (oz)	28.47 cubic centimeters
	28.47 milliliters
Pint, U.S. liquid (pt)	0.47 liter
	473.2 cubic centimeters
Pound, avoirdupois (lb or p)	453.6 grams
Pound (apoth.) (lb or p)	373.2 grams
Quart, U.S. dry (qt)	1.1 liters
Quart, U.S. liquid (qt)	0.9 liters
Square centimeter (cm^2)	0.16 square inch
	100 square millimeters
Square foot (ft^2)	0.09 square meter
Square inch (in^2)	645.2 square millimeters
Square meter (m^2)	10.8 square feet
Square yard (yd^2)	0.8 square meter

Unit of Measurement	Conversion Factor[a]
Ton, metric (t)	1,000 kilograms
	2,205 pounds
Ton, short (t)	2,000 pounds
	907.2 kilograms
Ton, long (t)	2,240 pounds
	1,016 kilograms
Yard (yd)	0.91 meter
	3 feet

[a] All values have been rounded to the nearest 0.1 or 0.01 unit. This is generally adequate for aquaculture conversions; however, more precise conversion factors may be required under certain circumstances.

APPENDIX B
Equivalent Temperatures Between 0 and 40°C on the Celsius and Fahrenheit Scales[a]

Celsius	Fahrenheit	Celsius	Fahrenheit
0	32.0	21	69.8
1	33.8	22	71.6
2	35.6	23	73.4
3	37.4	24	75.2
4	39.2	25	77.0
5	41.0	26	78.8
6	42.8	27	80.6
7	44.6	28	82.5
8	46.4	29	84.2
9	48.2	30	86.0
10	50.0	31	87.8
11	51.8	32	89.6
12	53.6	33	91.4
13	55.4	34	93.2
14	57.2	35	95.0
15	59.0	36	96.8
16	60.8	37	98.6
17	62.6	38	100.2
18	64.4	39	102.2
19	66.2	40	104.0
20	68.0		

[a]Formulas for conversions between Celsius and Fahrenheit are as follows: $C = \frac{5}{9}(F - 32)$; $F = \frac{9}{5}C + 32$.

INDEX